Physiology
of Exercise

Physiology of Exercise

for Physical Education and Athletics
Third Edition

Herbert A. deVries
University of Southern California

Wm. C. Brown Company Publishers
Dubuque, Iowa

Consulting Editors
PHYSICAL EDUCATION
Aileene Lockhart
Texas Woman's University

HEALTH
Robert Kaplan
Ohio State University

PARKS AND RECREATION
David Gray
California State University, Long Beach

Contents

Preface

Over the years of working with undergraduate and graduate students in the physiology of exercise laboratory, it has become apparent that theory and practice are not always related in the student's mind. Too often, the scientific method remains an *ivory tower* concept. Unfortunately, some coaches base their practices on methods of the highly successful athlete, whose success may be totally unrelated to the "fads" used in his training. Because of such practices and conditions, I have made an effort to bring theory and practice into a closer and more meaningful relationship—to add the *how to* approach while at the same time developing respect for scientific investigations that provide the *why* for the *how to*.

Physiology of Exercise for Physical Education and Athletics is concerned with human responses and adaptations to muscular activity. The text provides a basis for the study of physical fitness and athletic training. The text is written primarily for the upper-division *undergraduate student* who has a background in basic anatomy and general physiology. Physics, chemistry, or mathematics beyond that of the high school level are not necessary for comprehension. In addition, those who aspire to be *athletic coaches* will find within these pages the scientific basis for their profession. Since emphasis is placed upon the "holes" in our patchwork quilt of knowledge and since a substantially updated and expanded bibliography is provided at the end of each chapter, the *graduate student* who wishes to "chip away" at the frontiers of knowledge in this discipline will be aided. Those who use exercise as a therapeutic modality will also find guiding principles in this text.

In the first two editions of this textbook a tripartite organization was developed because of my appreciation for the fact that students enter this field with greatly diverse backgrounds in the basic sciences. Thus, Part 1 selectively reviewed the most pertinent areas of basic physiology. For classes where the background in general physiology was strong, it was then possible for the instructor to direct the emphasis toward applications that were presented in Parts 2 and 3. This approach has been maintained because of its basic pedagogic soundness and its obvious advantages where homogeneity of lower-division preparation is not common.

In this, the third edition, a clearer division and better definition of the second and third parts is possible. In the years since the second edition, it has become clear that physical fitness can make substantial contributions to our lifelong good health. The body of evidence that was previously only suggestive is rapidly becoming compelling. Therefore, the text has been reorganized to recognize that the study of exercise physiology has two major interests for the physical educator. The first concern is directed to the enhancement of *health and physical fitness* for the general population and the second is directed toward optimizing *performance* in the various types and levels of competitive athletics. Thus, the third edition is organized quite naturally into three parts. The first deals with the basic physiology that provides the groundwork for application to our field. Although this is organized on a systemic basis, the potential applications are pointed out and emphasized. Recognizing the basic differences in interest between those who identify primarily as teachers of health and physical education and those whose future lies largely in the coaching of athletics, the second section deals with the physiology of exercise directed toward health and fitness, while the third section is dedicated to the scientific improvement of athletic performance.

The first section includes a new chapter to provide orientation to the philosophy and methods used in exercise physiology. The second section includes three new chapters dealing with physical fitness testing, exercise prescription, and the use of exercise physiology in the prevention and rehabilitation of cardiovascular disease.

As in the first two editions, emphasis is on areas that are of practical importance, many of which have been neglected in the past. Whole chapters are devoted to muscle soreness, therapeutic and prophylactic effects of exercise, and the female in sports. Neurophysiology is also presented in sufficient detail so that its principles may be applied to the everyday procedures of physical education.

The research horizons in exercise physiology have been greatly expanded in recent years by the advent of the electron microscope and new techniques in microbiology and histochemistry. Use of these tools in the hands of a new generation of highly trained exercise physiologists has created an increasing body of knowledge in cellular and molecular biology applied to human performance. Since much of the emerging material resulting from these new directions of investigation is very pertinent to and of direct practical value in physical education and athletics, no exercise physiology text could be complete without this new material. Thus, chapters two, three, and four present the results of such research including the important knowledge relating muscle-fiber type to performance in a fashion that the undergraduate physical education student can comprehend.

There has also been a gradual increase in the use of electromyography by physical educators in the effort to better understand muscle physiology. Consequently, the well-prepared physical educator needs sufficient exposure to these procedures to be able to read the professional literature critically. Chapter four provides this background simply but in sufficient depth to allow comprehension.

Just as each chapter ends with a summary, so, in the second and third editions a final chapter has been added to enable the coach and athlete to synthesize the first twenty seven chapters. *The Unified Athlete: Monitoring Training Progress* is designed to encourage the young coach to take his/her exercise physiology background to the athletic field instead of leaving it behind in the ivory tower of college days.

I benefitted greatly from the constructive critiques with respect to the first and second editions and I value most highly the comments from you—my colleagues—who use this text on the firing line—the classroom. I solicit your continuing help to make this text more valuable to you and your students.

Grateful appreciation is extended to my associates at the University of Southern California, particularly to Dr. J. Tillman Hall, Chairman of the Physical Education Dept., and to Dr. J.P. Meehan, Chairman of the Physiology Dept., and to Dr. Aileene Lockhart for her guidance and invaluable advice.

Part 1

Basic Physiology Underlying the Study of Physiology of Exercise

1

The Why
of Physiology
# of Exercise

The undergraduate student in physical education often regards physiology of exercise as one of the more difficult and rigorous courses in the curriculum. The question may well be asked, "Is all this scientific preparation really necessary so that I can teach physical education and coach track, swimming, football, etc.?" This is a fair question and deserves an answer in depth. In this chapter we shall provide answers to that question, along with an overview of intensely interesting, provocative, and highly practical material on the physiological basis of all human movement, whether performed for work, for play, for physical conditioning, or for training for athletic competition.

Physiology of exercise is a subdivision of general physiology that concerns itself largely with the improvement of human functional capacities. Interest in improved functional capacity may be directed toward the enhancement of health and physical fitness for the general population or toward optimizing performance in the various types and levels of competitive athletics.

Why Physical Fitness?

As little as twenty to thirty years ago, discussions about physical fitness frequently evolved into arguments about definition. The discussion often ended with this question: Physical fitness for what? The implication of that question, which not all of us accepted even then, was that we need only be fit enough to meet the necessary physical challenges of our workaday world, with maybe a little bit left over for good measure! Obviously, then, since most of us in this overly mechanized and industrialized civilization met very few physical challenges, low levels of functional capacity (fitness) were acceptable. Application of the concept of "pursuit of excellence" to optimizing physical fitness awaited the knowledge boom of the past three decades. We now have enough scientific evidence to answer the question, Physical fitness for what? with the simple statement, Optimal levels of physical fitness are conducive to life-long good health. The second section of the text presents much of the important scientific evidence that supports this statement.

Improving Human Athletic Performance

With respect to the development of the best possible performance in our athletes, we must recognize that coaching is both art and science. The art lies in the application of sound psychological and sociological principles in the development of motivation and in the ingenuity displayed in designing workouts to gain desired ends without inducing boredom or unhappiness in athletes. But no matter how good an artist the coach may be, all is for naught if he or she does not have a sound grounding in exercise physiology, which has taught the coach the nature of the body's responses to training stimuli, both immediate (acute response) and long-term (chronic adaptation).

If a coach has overworked the athletes for a period of time, no amount of art will prevent staleness from setting in. Conversely, ignorance of the physiological bases of good training practice, such as progressive resistance training for power or interval training for endurance, will prevent realization of the athletes' full potential—and will probably result in a poor win-loss record!

Even more important than the win-loss record, however, is the maintenance of good health in athletes. Well-trained professional coaches must know the physiological effects of environmental factors such as heat and cold on their charges. Every year several athletes die on the football field from heatstroke. Most of these deaths could be prevented if all coaches were thoroughly trained in the basics of environmental physiology. Also important for the health of the athlete is the maintenance of good nutritional practice. Too many coaches, even today, rely on old wives' tales instead of scientific evidence with respect to the diet of their athletes.

The use of drugs to improve performance is a hazardous practice at best, and it is doubly deplorable because the hazard is rarely accompanied by any significant improvement in performance. Athletes have died in ignorance because they used amphetamines while bicycle racing on a hot day.

Professionalism in Physical Education and Athletics

During World War II we trained lay members of the army (enlisted men who had been well-known players in baseball, football, or other sports) to be noncommissioned physical training leaders in six to eight weeks. These soldiers functioned quite well in leading calisthenics programs and athletic competitions. Were these people equivalent to physical education professionals? Most definitely not! While they knew the what of the program, they had not had the professional training in the basic exercise sciences, such as exercise physiology, kinesiology, and motor learning, so they did not in most cases understand the why. Thus they were dependent on commissioned officers who were professionally trained physical educators for the development of the overall training program and guidance in its implementation.

The difference between a professional and a lay person is that the member of a profession has "professed" a commitment to a learned discipline with a well-defined body of knowledge. This profession implies, in turn, the application of the scientific method to the professional body of knowledge, usually within a well-structured college or university curriculum. Thus the professional physical educator learns the basic principles, which are grounded in the scientific method. All practice then (to the extent that scientific data are available) is based on scientifically derived principles. Untrained lay persons can only practice what has been handed down to them, since they do not understand the underlying principles that should govern their practice. For example, lay coaches can only do what they have seen their coaches (or other athletes) do, whether right or wrong.

While the lay person can only function at the cookbook level, that is, follow the instructions of a professionally trained individual, the person trained in the basic sciences, such as exercise physiology, proceeds from first principles. We, as members of a learned profession, must always seek out the mechanisms underlying our practice. In so doing, we derive at least four practical advantages: (1) we achieve better prediction of results; (2) we can better control the conditioning and training process, thus protecting the health of our charges; (3) we grow more efficient in terms of results gained per unit of time spent; and (4) we may even satisfy our intellectual curiosity with respect to cause-and-effect relationships in our field (this, of course, is research).

"Get Some Exercise"

Most of us have heard a physician or a well-meaning friend advise: "Jack, what you need is to get some exercise." Implicit in such a prescription is the thought that it makes no difference whether one lifts weights for an hour, swims a mile, or jogs three miles. It is long past time for our profession to grab the reins of leadership in this domain. Let's educate the public and the health-related professions, too, to understand that the admonition to "get some exercise" is analogous to a physician's writing a prescription that simply says, "Administer some drugs." Just as physicians choose from many drugs when they prescribe aspirin for a headache, so there are many exercise training and conditioning modalities, each of which can be modified or administered in terms of intensity, frequency, and duration. Many of us at work in our exercise physiology research laboratories are developing dose-response relationships for the various types of exercise. Physical education students must learn what is presently available in our pharmacopeia of exercise. We must learn the scientific answers to how much is enough, how much is too much, and how much is best for any given individual. What is presently known in this area is presented in chapter 13.

The future holds great promise for expansion of the physical education profession from one that at present limits its audience largely to schoolchildren to one that will cater to the needs of people of all ages and both sexes. Who in a sales position would voluntarily limit one's clientele to 17% of the total potential population (persons six to eighteen instead of birth to seventy)? Interestingly, the lay public's acceptance of the need for adult fitness seems to have advanced far more rapidly than our profession's leadership in training the personnel for such programs. Obviously, such personnel must be well grounded in exercise physiology to prevent hazardous situations from arising among middle-aged and older people who may have unrecognized disease problems, as well as flabby muscles and poor cardiovascular function as a result of sedentary living.

Sex and Age Differences in Response to Training

Only in recent years has the female of the species been invited to participate to any great extent in physical conditioning programs or competitive athletics. Women's athletics that were worthy of the name did not exist prior to World War I, and women did not begin Olympic competition until 1928. Because of women's very recent entry into the sports world, professionals are only now beginning to accumulate data in their laboratories about the similarities and differences in response between males and females to the stimuli of exercise and heavy training regimens. It has been a long, slow process, but we now know that females respond in a fashion similar to men to a conditioning program and derive the same health benefits. With respect to high level competition, women come much closer to the men's records than men would have supposed only a few years ago. Yet, there are some dramatic differences in female responses. For example, women and their leaders had long avoided the use of weight training for fear of developing bulging and unsightly muscles. However, the truth of the matter is that the female's response to weight training is very different from the male's, probably because of endocrine differences. Intense weight training at a level that brings about large and well-defined muscles in the male only serves to enhance the strength and power of the female, with only slight increases in muscle bulk, which in most cases produce an even more attractive female figure! The old wives' tale that prohibited heavy weight training for women because it would lead to masculineness was probably the result of fallible human observation that some highly successful female athletes were unusually muscular. What probably happened was that those women who were extremely muscular by genetic endowment (mesomorphs) were more likely to succeed and consequently were more likely to pursue athletic careers.

Again, with respect to age, as recently as one decade ago it was assumed on insufficient scientific evidence that the older individual was virtually untrainable! How old was considered "older"—would you believe forty? Evidence from our own laboratory dispells this myth. We have shown that seventy- and eighty-year-olds, if healthy, have the same relative capacity for training as the young. That is to say, the *percentage* improvement in performance with training was every bit as good in the elderly as in the young. Our findings have now been corroborated by several other investigators and will be discussed in detail in chapter 16.

Scientific Method

Physical educators or coaches who claim membership in the profession are morally bound to base their practice within that profession on the best (most reliable and authoritative) information available. Obviously hearsay evidence is not in the same class with evidence derived from application of the scientific method in controlled experiments. The credibility of various sources of evidence can be ordered from poorest to best as follows:

1. *Hunch or guess.*
2. *Hypothesis*—a tentative supposition (based on hunches or guesses) provisionally adopted to explain certain facts and to guide in the investigation of others. An hypothesis is set up to be tested, accepted, or rejected on the basis of further observation or experiment.
3. *Theory*—based on some scientific evidence, but insufficiently verified to be accepted as fact. Theories provide the basis for developing *principles*.
4. *Principle*—a settled rule of action based on theories that are well supported by research findings. Principles are the guidelines for decision making in our professional activities.

True professionals are set apart from lay practitioners in their ability and inclination to challenge the source of information. For example, the lay coach hears that a certain athlete, who has broken the world record for the 1,500-meter run, was a confirmed vegetarian. Not having been trained in the application of the scientific method and also unable or unwilling to read the available research literature, the coach assumes there was a cause-and-effect relationship between the outstanding performance and the athlete's vegetarian habits. The coach then attempts to make vegetarians out of all athletes. The professionally trained coach sees this explanation of superior performance for what it is—no better than a hunch or guess, at best. The coach goes to the professional literature, consulting reliable sources in physiology, biochemistry, and nutrition. Finding no support for a relationship between superior running ability and a vegetable diet, the coach rejects the hunch that the record performance was causally related to the athlete's vegetarianism.

In the long run, coaching practice that is based on scientifically derived principles rather than on unproven hunches or personal opinions will be considerably more successful. For this reason, this text will present the pertinent sources of research information in the hope that students will learn to be discerning consumers of information about physical education and coaching practices.

Overview of Text

This text is organized into three parts. The first part deals with the basic physiology that provides the groundwork for application in our field. Although the information is organized by systems of the body, the potential applications are pointed out and emphasized.

Recognizing the basic differences in interest between those who identify primarily as teachers of health and physical education and those who will be involved largely in the coaching of athletics, the second part deals with the physiology of exercise in relation to health and fitness. The third part is dedicated to the scientific improvement of athletic performance.

Basic Physiology of Exercise. Of all the tissue comprising the human body, by far the greatest proportion is muscle—some 40% of body weight. Furthermore, all movement must be implemented through the skeletal muscles. Therefore we begin our study at this level. Since a knowledge of structure (anatomy) is necessary to an understanding of function (physiology), the second chapter deals with gross, microscopic, and submicroscopic structure. Recent evidence provided by the electron microscope even allows us to explore the fascinating relationship between structure and function.

The third chapter tells the story of how energy from the food we eat is made available to individual muscle cells. Recent research has made it abundantly clear how important this information is to the best possible preparation of athletes for competition.

Chapter 4 covers the physiology of gross muscle contraction, dealing with such practical problems as muscle fatigue, muscle length and tension, and speed and force of contraction in relation to power output, a concern in all physical activity. Even the design of derailleur bicycles is based on the principles presented here.

Nervous control of the muscles, which is responsible for the beautifully coordinated movements possible in champion athletes, is discussed in chapter 5. The groundwork is laid here for understanding such practical problems as how strength is limited, how muscle sense is developed, and why some forms of stretching are superior to others.

The next two chapters (6 and 7) describe how the heart and circulation act as the transport system for bringing in necessary oxygen and nutrients and removing the products of metabolism. Here the groundwork is laid for understanding how training affects the heart and blood vessels and how different types of exercise affect the work of the heart.

The remaining two chapters in the first part deal with breathing function and the transport of gasses by the blood, including such practical matters as the best breathing patterns, second wind, the use of hyperventilation in athletics, the effects of smoking on wind, and aerobic capacity.

Physiology Applied to Health and Fitness. In the second section of the book we discuss in chapter 11 the large, new body of knowledge that spells out the potential health benefits of physical fitness.

The following chapter describes in some detail the methods used in exercise physiology laboratories around the world in the measurement of aerobic capacity and in stress testing.

Cardiovascular disease has increased at such an alarming rate that chapter 14 is devoted exclusively to the potential contributions of physical conditioning in the prevention and rehabilitation of its victims.

The chapter on metabolism and weight control provides the scientific bases for successful weight maintenance and reduction, and discusses the old wives' tales and misinformation that have made obesity such a problem in the United States. Nutrition is discussed, and simple means for self-evaluation of diet are provided.

Chapter 16 on age and exercise presents available evidence on the physiological effects of aging, together with evidence that suggests that much of what has been accepted as the effect of aging is really the result of our sedentary life style, which leads to atrophy of muscles and other tissues and dysfunction in general.

This section closes with definitions of the scientific principles that should govern physical conditioning programs and the testing methods available to evaluate physical fitness.

Physiology of Training and Conditioning Athletes. In this section, all available scientific evidence that can be applied to the problems of coaching is culled selectively and presented in the form of a structured approach to systematic improvement of athletic performance. To do this sport by sport would be to revert to a cookbook philosophy; moreover, generalization would be difficult. Therefore the more scientific approach is adopted in which the individual elements of human performance are discussed. Thus the section is organized into chapters dealing with strength, endurance, muscular efficiency, flexibility, muscle soreness, warming up, environmental factors, nutrition, special aids, female athletes, and training concepts and practices.

So, having briefly glimpsed what lies ahead and understanding the need for exposure to the physiology of exercise, the student is now ready to press on to the study of one of the most fascinating and intriguing subdivisions of physiology—human physiological responses, both immediate and long-term, to the demands of exercise.

2

Structure of Muscle Tissue

All human activity, whether in work or sport, depends ultimately on the contraction of muscle tissue for its driving forces. There are three types of muscle tissue in the human body:

1. Smooth, nonstriated muscle, which is found in the walls of the hollow viscera and blood vessels.
2. Striated, skeletal muscle, which provides the force for movement of the bony, leverage system.
3. Cardiac muscle, which is found only in the heart.

Smooth muscle receives its innervation from the autonomic nervous system and ordinarily contracts independently of voluntary control. The fibers of smooth muscle are usually long, spindle-shaped bodies, but their external shape may change somewhat to conform to the surrounding elements. Each fiber usually has only one nucleus.

Skeletal muscle, which is innervated by the voluntary or somatic nervous system, consists of long, cylindrical muscle fibers. Each fiber is a large cell with as many as several hundred nuclei, and it is structurally independent of its neighboring fiber or cell. Skeletal, or striated, muscle, as the name implies, is most easily distinguished by its cross-striations of alternating light and dark bands.

Cardiac muscle in all vertebrates is a network of striated muscle fibers. It differs structurally from the other two types of muscle tissue mainly in the interweaving of its fibers to form a network, called a *syncytium,* which differentiates it from skeletal muscle, which is also striated. It further differs from smooth muscle in that it has cross-striations, which smooth muscle does not have. Cardiac muscle contracts rhythmically and automatically, without outside stimulation. Whereas skeletal muscle is made up of discrete fibers that can contract individually (but with other members of its motor unit), cardiac muscle is composed of a network of fibers that responds to innervation with a wavelike contraction that passes through the entire muscle.

Gross Structure of Skeletal Muscle

If we dissected a limb such as the upper arm and removed the skin, subcutaneous adipose tissue, and the superficial fascia, we would lay bare the biceps brachii muscle and note that it is covered in its entirety by a deep layer of fascia that binds the muscle into one functional unit. This outermost sheath of connective tissue is called the *epimysium,* and it merges at the ends of the muscle with the connective tissue material of the tendon. Thus the force of muscular contraction is transmitted through the connective tissues, binding the muscle to the tendon, then through the tendon to the bony structures, to bring about movement.

In cross section, it may be seen that the interior of the muscle is subdivided by septa into bundles of muscle fibers (fig. 2.1). Each bundle contains upwards of a dozen, possibly as many as 150, fibers. Each bundle is called a *fasciculus,* and it has a more or less complete connective tissue sheath that is called the *perimysium.* The structures discussed so far are visible to the naked eye.

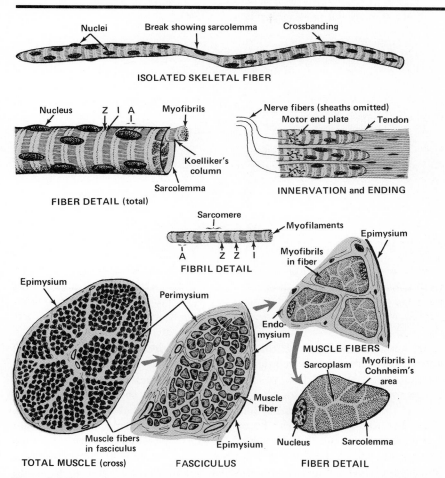

Figure 2.1 Cross section through a typical skeletal muscle showing (below) the breakdown from the gross muscle to the single muscle fiber and (above) relationships of the microscopic structures to the muscle fiber. (From Arey, L.B. *Human Histology,* 1968. Courtesy of W.B. Saunders Company, Philadelphia.)

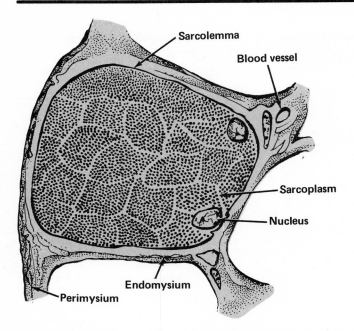

Sarcolemma

Blood vessel

Sarcoplasm

Nucleus

Endomysium

Perimysium

Figure 2.2 Cross section of human skeletal muscle fiber drawn with camera lucida. Each dot represents a myofibril (x 1500). (From Copenhaver, W.M.; Bunge, R.P.; and Bunge, M.B. *Bailey's Textbook of Histology,* 1971. Courtesy of Williams & Wilkins Company, Baltimore.)

Microscopic Structure of Skeletal Muscle

More detailed study requires the aid of a microscope, so that the structure of an individual fiber and its relationships to other fibers, to form fasciculi, may be seen (fig. 2.2). Each fiber is surrounded by a connective sheath called the *endomysium.* The need for these connective tissue sheaths—endomysium around the single fiber, perimysium around the fasciculus, and epimysium about the whole muscle—can be understood better when it is realized that one fiber may not run through the whole length of a muscle, or even through a fasciculus (fig. 2.3). Therefore it becomes necessary to transmit the force of contraction from fiber to fiber to fasciculus, and from fasciculus to fasciculus (since these sometimes do not run through the length of a large muscle either) to the tendons, which act upon the bones. This function is provided by the connective tissues described above.

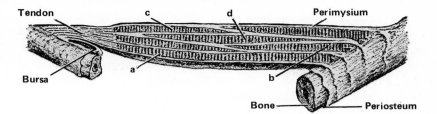

Figure 2.3 Diagram of attachment of muscle to skeleton and relation of fibers to each other within a fasciculus: (a) fiber extends the length of fasciculus; (b) fiber begins at periosteum but ends in muscle; (c) fiber begins at tendon but ends in muscle; (d) both ends of fiber within the muscle (redrawn from Braus). (From Copenhaver, W.M.; Bunge, R.P.; and Bunge, M.B. *Bailey's Textbook of Histology,* 1971. Courtesy of Williams & Wilkins Company, Baltimore.)

The dimensions of individual fibers may vary, according to most investigators, from 10 to 100 microns (1,000 microns = 1 mm) in diameter and from 1 mm to the length of the whole muscle. Thus the thickness of a large fiber is roughly comparable to that of a fine human hair, although the smaller fibers cannot be seen by the unaided eye.

Each fiber constitutes one muscle cell. Each muscle has fibers of characteristic size, and the thickness of each fiber is related to the forces involved in the function of the muscle. Thus the fibers of the extrinsic ocular muscles are small in diameter, whereas those of the quadriceps femoris are large.

Structure of the Muscle Cell or Fiber

The cell membrane of the muscle cell is called the *sarcolemma.* This membrane is extremely thin, and seems almost structureless, even under the electron microscope. Inside the sarcolemma are the many nuclei, mainly situated peripherally, close to the sarcolemma. Corresponding to the cytoplasm of other cells is the *sarcoplasm,* which is the more fluid part of the cell. Running longitudinally within the sarcoplasm are slender columnlike structures called *myofibrils,* which have alternating segments of light and dark color. The presence of the myofibrils impart to the fiber as a whole the appearance of lengthwise striations. The cross-striations, however, are far more obvious because the dark segments of the many myofibrils are arranged in lateral alignment. All light segments are likewise aligned with each other.

Structure of Muscle Tissue

Muscle Fiber Types

For many years, anatomists and histologists have classified muscles as red or white according to whether red or white fibers predominated in the makeup of the gross muscle structure. In this classification, the red fibers were considered to be better suited to long-term, slow contractions, as required of postural, antigravity muscles, while the white fibers were considered to be differentiated for speed of contraction and thus were to be found predominantly in the flexor muscles.

In recent years, with the advent of modern histochemical techniques, examination of chemical constituents at the cellular level became possible and provided the means to correlate functional activity of individual fibers with their morphology. Thus identifying skeletal muscle fiber types has become more sophisticated and at the same time has provided laboratory-derived information that helps us understand why one individual may be better suited to endurance-type athletics while another excels in sprint-type athletics.

Various workers have named from two to as many as eight different fiber types, based on the new laboratory techniques. Part of the difficulty of naming the types has arisen because some investigators have used human muscle while others have used animal muscle. We now know that there are interesting differences between human and animal muscles with respect to fiber types. In addition, nomenclature was based on at least four different approaches: (1) the anatomical appearance; red vs. white, etc.; (2) muscle function; fast-slow or fatiguable vs. fatigue-resistant; (3) biochemical properties, such as high or low aerobic capacity; and (4) histochemical properties, such as the enzyme profile of the fiber.

Agreement seems to be at hand on a classification system based on three fiber types (table 2.1) that can best serve our purposes with respect to human muscle (7, 15). The older, fast twitch (white) vs. slow twitch (red) system has become inadequate because there are two subtypes of fast twitch fibers, with both physiological and histochemical differences. Most important from the standpoint of exercise physiology, they respond differently to training. For these reasons we shall use the nomenclature of Peter (15) and co-workers as shown in table 2.1.

First, we must take up some basic considerations with respect to muscle makeup. It has been shown (2) that surgical crossing of the motor nerve supplying a fast muscle such as flexor hallucis longus with that of a slow muscle such as soleus resulted in a reversal of the muscles' contractile properties. Therefore we suspected that all muscle fibers within one motor unit must be of the same fiber type. That has been verified (3).

Table 2.1
Classifications and Characteristics of Human Skeletal Muscle Fibers

A. Nomenclature			
1. Dubowitz and Brooke (7)	Type I	Type IIa	Type IIb
2. Peter et al. (15)	Slow, oxidative (SO)	Fast, oxidative, glycolytic (FOG)	Fast, glycolytic (FG)
3. Older systems	Red Slow twitch (ST)	White Fast twitch (FT)	
B. Characteristics			
1. Speed of contraction	Slow	Fast	Fast
2. Strength of contraction	Low	High	High
3. Fatiguability	Fatigue-resistant	Fatiguable	Most fatiguable
4. Aerobic capacity	High	Medium	Low
5. Anaerobic capacity	Low	Medium	High
6. Size	Small	Largest	Large
7. Capillary density	High	High	Low

While the percentages of the three fiber types present in the various muscles of one individual may be quite different, the fiber type (with respect to fast or slow twitch, at least) for any one muscle is established early in life and apparently does not change thereafter (5, 6). However, training can bring about considerable improvement in both aerobic capacity and glycogen content of the muscle, as will be discussed in the next chapter. Also there appears to be no sex difference with respect to fiber type percentages (5).

The significance of the fiber type composition for athletics becomes readily apparent from a consideration of table 2.1. The individual endowed with a high percentage of slow twitch, oxidative (SO) fibers, which are slow but highly fatigue-resistant, would be a good candidate for distance running or other endurance events. On the other hand, a person whose genetic makeup produced high percentages of fast twitch (FT) fibers would be predisposed toward success in power and sprint events.

These expectations have been supported by recent laboratory work in which samples of muscle tissue in different individuals under varying conditions were taken with biopsy needles. Early work in this area by Gollnick and associates (9) showed interesting relationships between successful performance in endurance athletic events and the percentage of SO fibers. More recent work by Costill and colleagues (6) supported these findings. They found that the gastrocnemius muscles of fourteen championship-caliber distance runners were characterized by high percentages of SO fibers (79%), while the same muscles in trained middle distance runners had only 62%, only slightly more than untrained men at 58%. Interestingly, they also found that the SO fibers of the elite runners were 22% larger than their FT fibers, which is contrary to what is usually found (table 2.1). This is probably a selective training effect resulting from the heavy use of the type SO fibers.

In the studies of human muscle using only two categories of fibers, slow twitch (ST) and fast twitch (FT), there is agreement that these two categories do not change their relative proportions as the result of training; only their size and oxidative capacities improve. However, recent work (16) using the newer classification of three muscle types (table 2.1) suggests that changes within the FT fibers are the important responses to training. Thus research suggests that humans can adapt to different muscular activities by way of a shift from fast twitch, glycolytic to fast twitch, oxidative, glycolytic fibers in response to distance running, and from fast twitch, oxidative, glycolytic to fast twitch, glycolytic muscles in response to weight training. This is an attractive hypothesis in that it would bring human adaptive responses in line with those observed in lower mammals by several different groups of investigators.

Only very recently have we found evidence to show that intact human skeletal muscle does behave in a fashion related to the behavior of animal skeletal muscle with respect to fiber types. Thorstensson and co-workers (17, 19) in Sweden have shown significant correlations of .48 and .50 between the percentage of FT fibers an athlete has in the knee extensor muscles and the maximal torque produced (a measure of strength). Similar correlations were observed between percentage of FT fibers and the speed of muscle contraction. Furthermore, the Thorstensson studies (18) showed an even stronger correlation (.86) between percentage of FT fibers and the rate at which fatigue sets in.

With respect to fiber type, our genetic endowment determines to a large extent whether we should pursue endurance-type sports or sports that demand sprint or power in their performance. In either case, ultimate success depends on many other factors such as training, conditioning, and dedication.

Structure of the Myofibril and the Contractile Mechanism

The advent of the electron microscope and its wide usage in recent years has provided greater insight into both the structure and function of the myofibril. Though the story is not complete in all details, the sliding filament model of muscle contraction is now widely accepted as best explaining all the experimental data (14).

The *sarcomere* is the functional unit of the myofibril, and it extends from Z line to Z line, as shown in figures 2.4 and 2.5. Each sarcomere is composed of two types of interdigitating parallel filaments that run the length of the

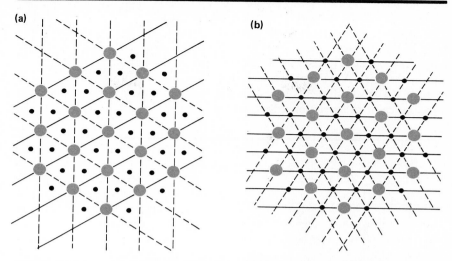

Figure 2.4 **(A)** End-on view of double hexagonal lattice of thick and thin filaments, characteristic of overlap region in A bands of vertebrate striated muscle, showing the three sets of lattice planes in the crystallographic directions. The actin filaments at the trigonal points of the lattice will tend to fill in the space between the dense planes of filaments at the hexagonal lattice points, thereby decreasing the intensity of the X-ray reflections given by the thick filaments on their own. **(B)** Similar view, showing lattice planes in *crystallograhpic* directions. In this case, both the actin and myosin filaments lie in the same lattice planes and so their contributions to the intensity of the corresponding X-ray reflection are additive. Hence, as the amount of material at the trigonal points is increased, the intensity of the reflections decreases and that of the reflections increases. (From Huxley, H.E. *Proc. R. Soc. London,* 178:131, 1971.)

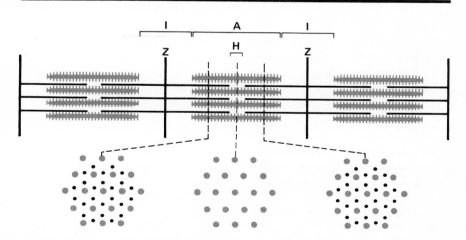

Figure 2.5 Diagrammatic representation of the structure of striated muscle, showing overlapping arrays of actin- and myosin-containing filaments, the latter with projecting cross-bridges on them. For convenience of representation, the structure is drawn with considerable longitudinal foreshortening; with filament diameters and side spacings as shown, the filament lengths should be about five times the lengths shown. (From Huxley, H.E. *Proc. R. Soc. London,* 178:131, 1971.)

myofibril. One type is about twice as thick as the other, and its length is equal to the length of the A band (the dark band seen as part of the striation effect). The second and thinner type of filament is longer, and extends inward from both Z membranes, almost to the center of the sarcomere. The amount by which the two ends of the thin filament fail to meet constitutes a lighter band, within the dark A band which is called the H zone. The area between the ends of the thick filaments is less dense, and therefore gives the light band appearance of the striation effect, which is known as the I band. Thus the

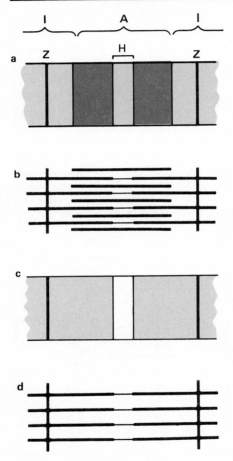

Figure 2.6 Diagram illustrating the changes taking place when the muscle protein myosin is extracted chemically. Differences in optical density before and after extraction are shown in **A** and **C**, with a logical schematic explanation in **B** and **D**. (From Huxley, H.E., and Hanson, J., in *The Structure and Function of Muscle*, G.H. Bourne, ed., 1960. Courtesy of the Academic Press, New York.)

light and dark striped effect of striated muscle rests on a rational basis of bands of greater and lesser optical density, as can be seen in figure 2.5. The cross-sectional views show the relationship of each thick filament to a hexagon of six thin filaments, each hexagon, however, being shared by three thick filaments.

Chemical extraction of the muscle protein *myosin* results in the disappearance of the dark A band (fig. 2.6), and extraction of the muscle protein *actin,* similarly affects the I band. These facts are very strong evidence that the thick filaments consist of myosin and the thin filaments are composed of actin and, also *tropomyosin.*

A sliding movement of the actin and myosin filaments during contraction of the myofibril has been well demonstrated, and it seems that the A bands remain the same length while the I bands change only in shortening below 90% of the myofibrils' resting length (fig. 2.7). The exact nature of the changes in the H zone are not as yet clearly understood, although the disappearance of the H zone during contraction is well established. The contractile process depends upon the presence of adenosine triphosphate (ATP) and its splitting by dephosphorylation into adenosine diphosphate (ADP) and phosphate. This splitting of an ATP bond furnishes large amounts of energy.

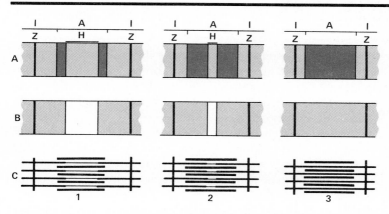

Figure 2.7 Diagram illustrating the structural changes associated with contraction *(3)* and extension *(1)* from resting length *(2)*. **A** shows the band patterns of intact fibrils. **B** shows the band patterns after extraction of myosin. **C** shows the positions of the filaments. (From Huxley, H.E., and Hanson, J., in *The Structure and Function of Muscle,* G.H. Bourne, ed., 1960. Courtesy of the Academic Press, New York.)

Earlier in vitro experiments had shown that three necessary elements of muscle function—relaxation, contraction, and rigor mortis—could be explained as follows:

1. In the presence of ATP (unsplit), actomyosin breaks down into a noncontractile state of dissociated actin and myosin, thus causing relaxation.
2. When ATP splits to form ADP + P, actomyosin threads reform, and the reformed threads contract in the presence of more ATP.
3. If the reformed actomyosin thread is removed from the presence of ATP, it resists extension.

These in vitro observations seem to explain the necessary facts observed in vivo:

1. Relaxation, in the presence of unsplit ATP.
2. Contraction, in the presence of unsplit ATP, when some ATP is undergoing dephosphorylation.
3. Rigor mortis, caused by total dissipation of ATP after its splitting has already caused the precipitation of inextensible actomyosin threads.

Now we must account for the mechanics involved in bringing about the sliding of the filaments and the biochemical events that initiate and provide the energy for this interdigitation of the actin and myosin filaments. The most acceptable explanation for the process of contraction at the cellular level is as follows:

1. An electrical impulse conducted by the motor nerve activates the motor end plate of the muscle fiber, which in turn brings about the release of a substance that depolarizes the resting muscle membrane. This depolarization is what is recorded and measured by electromyographic methods as muscle action potentials.
2. The action potential in turn sets off two independent electrical currents, one of which is a weak longitudinal current; the other is transverse and moves inward into the fiber along a system of tubules (see fig. 2.8).
3. The inwardly invading current releases internal tightly bound calcium (11).
4. In the fiber's resting state, inactivity is maintained because a complex of two other proteins, *troponin* and *tropomyosin,* when in combination with actin, prevents the normal course of interaction between actin and myosin filaments. When the calcium ions are released because of the electrical excitation, they bind strongly to the troponin-tropomyosin complex and thus suppress the inhibitory action upon the interaction of the actin and myosin, which are then free to combine.

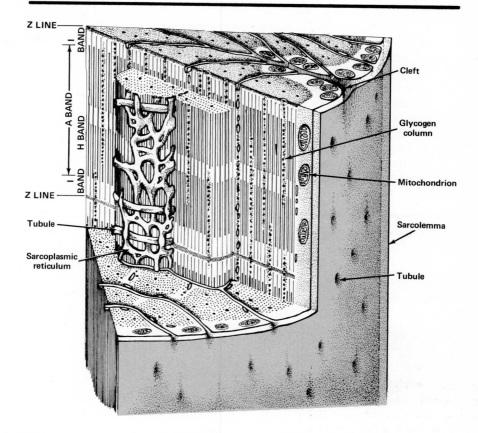

Figure 2.8 Structure of muscle fiber consists of a number of fibrils, which in turn are made up of orderly arrays of thick and thin filaments of protein. A system of transverse tubules opens to the exterior of the fiber. The sarcoplasmic reticulum is a system of tubules that does not open to the exterior. The two systems, which are evidently involved in the flow of calcium ions, meet at a number of junctions called dyads or triads. Mitochondria convert food to energy. The sarcolemma is a membrane surrounding the fiber. (From Hoyle, G. *How Is Muscle Turned On and Off?* Copyright © 1970 by Scientific American, Inc. All rights reserved.)

5. The combination of actin-myosin acts as an enzyme (*catalyst*), which is called actomyosin adenosine-triphosphatase or *actomyosin ATPase*. Actomyosin ATPase catalyzes the breakdown of ATP to ADP + P, which in turn furnishes the energy for contraction (8).

6. Having the contractile structure and its source of energy defined, we now need to explain the mode of action of the cross-bridges in making the filaments slide to shorten the sarcomere and thus the myofibril—the fiber and the whole muscle. The most plausible explanation is offered by H. E. Huxley (14), who was involved in the original formulation of the sliding filament theory (10), along with the independent work of A. F. Huxley (12). Figure 2.9 illustrates the process referred to as the swinging cross-bridge model.

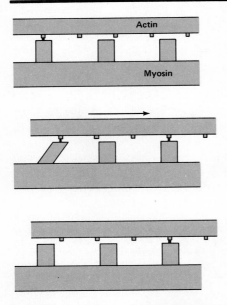

Figure 2.9 Diagram showing, very schematically, possible mode of action of cross-bridges. A cross-bridge attaches to a specific site on the actin filament, then undergoes some configurational change, which causes the point of attachment to move closer to the centre of the A band, pulling the actin filament along in the required manner. At the end of its working stroke, the bridge detaches and returns to its starting configuration, in preparation for another cycle. During each cycle, probably one molecule of ATP is dephosphorylated. Asynchronous attachment of other bridges maintains steady force. (From Huxley, H.E. *Proc. R. Soc. London* 178:131, 1971.)

Blood Supply and Lymphatics

In keeping with the high level of metabolic activity of muscle tissue, muscle is extremely well supplied with capillaries; each muscle fiber is supplied with several capillaries. To furnish this rich vascularization, larger branches of arteries penetrate the muscle by following the paths of the septa between fasciculi (in the perimysium). The arteries furnish arterioles to the fasciculi, and the arterioles give off capillaries at sharp angles to individual fibers. The veins follow the arteries, typically, and even the smallest veins have valves. Lymphatic capillaries are found only at the fascicular level of organization.

Nerve Supply

Because each muscle fiber represents a single cell and is a discrete functioning unit, it must be innervated individually. This is not to say, however, that one nerve cell may not innervate more than one muscle cell, by sending twigs from the same nerve fiber to several or many muscle fibers. The cell bodies of the neurons (nerve cells) lie in the ventral horns of the spinal cord. The axons of these cells form the nerve fibers of the efferent fibers of the peripheral nerves, which innervate the muscles. Each nerve, as seen grossly in dissection, represents the association of many axons or nerve fibers, just as a gross muscle represents many muscle fibers. The nerve fibers, as do the arterioles, travel in the perimysium and branch several times, thus permitting one neuron to innervate more than one muscle fiber.

The ratio of nerve fibers to muscle fibers varies with the degree of precision that is required of the muscle. In one of the muscles that move the eye, a ratio of one nerve fiber to one muscle fiber has been found. However, one nerve cell may innervate as many as 150 or more muscle fibers.

The neuron and its axon (or nerve fiber) with its twigs, plus the muscle fibers supplied by all the twigs, form the basic neuromuscular unit: the *motor unit.* After the repeated branchings, referred to above, the nerve fibers (or their branches) lose their myelin sheaths and enter individual muscle fibers. The neurolemma sheath of the nerve fiber apparently becomes continuous with the sarcolemma of the muscle fiber, and the nerve fiber branches into several clublike *terminal endings.* These terminal endings are embedded in sarcoplasm, just under the sarcolemma, and are called the *motor end plate* or the *myoneural junction.*

The muscle fibers making up one motor unit do not lie contiguously; they are usually scattered throughout a considerable volume of the gross muscle structure (fig. 2.10). This fact has important implications for our discussion of electromyography in a later chapter.

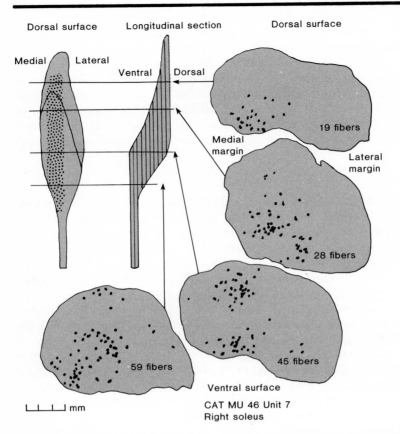

Dorsal surface   Longitudinal section   Dorsal surface

Medial   Lateral

Ventral   Dorsal

19 fibers

Medial
margin

Lateral
margin

28 fibers

59 fibers   45 fibers

Ventral surface

└─┴─┴─┘ mm

CAT MU 46 Unit 7
Right soleus

Figure 2.10 Distribution of muscle fibers in a motor unit. The diagrams on the upper left represent the whole soleus muscle as seen from the dorsal surface and in longitudinal section. The extent of the ventral aponeurosis of origin and of the dorsal aponeurosis of insertion are indicated by heavier lines on the longitudinal section, and the arrangement of primary fiber fasciculi is denoted by the parallel lines. Crosshatched area represents the approximate territory of the muscle unit projected onto the muscle outline. Cross sections taken through various levels along the longitudinal axis of the muscle are shown in outline tracings. Muscle fibers belonging to the studied unit have been plotted as dots. The approximate levels of sectioning are indicated on the whole muscle diagrams. Fiber counts are given on each section. (From Burke, R.E., et al. *J. Physiol.* 238:503–14, 1974.)

SUMMARY

1. Three types of muscle tissue are present in the human body.
   a. Smooth, nonstriated: usually found in viscera and blood vessels
   b. Skeletal, striated: found in the somatic muscles
   c. Cardiac, striated syncytium: found only in the heart
2. Gross structure of skeletal muscle is at three levels.
   a. Whole muscle, surrounded by epimysium
   b. Muscle bundle or fasciculus, surrounded by perimysium, which constitutes the septum seen grossly in cross section
   c. Muscle fiber, surrounded by endomysium
3. In the microscopic structure of skeletal muscle, fiber diameter is roughly related to the load of work the muscle ordinarily performs.
4. Each skeletal muscle fiber represents one multinucleated cell in which the sarcolemma is the cell membrane and the sarcoplasm roughly corresponds to the cytoplasm of other cells.
5. Newer systems of classification for human muscle fiber types include three basically different skeletal muscle fibers: (a) slow twitch, oxidative (SO), (b) fast twitch, oxidative, glycolytic (FOG), and (c) fast twitch, glycolytic (FG). Genetic endowment determines the proportion of slow twitch to fast twitch fibers (SO to FOG + FG). But available evidence suggests that endurance training may cause a shift from FG to FOG and power training may cause a shift from FOG to FG.
6. The supply of capillaries to muscle tissue is very abundant to support the large metabolic demands of exercise.
7. When a neuron in the ventral horn of the spinal cord is stimulated, its axon, all of the axon's branches, and all of the muscle fibers supplied by the branches function together simultaneously as a motor unit.
8. A motor unit may consist of one neuron with its axon (nerve fiber) innervating one muscle fiber, or it may consist of one neuron with its axon branching and innervating as many as 150 or more muscle fibers.
9. Although the muscle fibers of one motor unit function together, they do not constitute a unified structure. In fact they are distributed throughout a considerable volume of muscle tissue in many cases.

REFERENCES

1. Bailey, Kenneth. Muscle protein. *Br. Med. Bull.* 12:183–87, 1956.
2. Buller, A.J.; Eccles, J.; and Eccles, R. Interactions between motor neurons and muscles in respect of the characteristic speeds of their responses. *J. Physiol.* 150:417–39, 1960.
3. Burke, R.E.; Levine, D.N.; Tsairis, P.; and Zajac, F.E. Physiological types and histochemical profiles in motor units of the cat gastrocnemius. *J. Physiol.* 234:723–48, 1973.

4. Burke, R.E.; Levine, D.N.; Saloman, M.; et al. Motor units in cat soleus muscle: physiological, histochemical and morphological characteristics. *J. Physiol.* 238:503–14, 1974.

5. Costill, D.L.; Daniels, J.; Evans, W.; Fink, W.; Krahenbuhl, G.; and Saltin, B. Skeletal muscle enzymes and fiber composition in male and female track athletes. *J. Appl. Physiol.* 40:149–54, 1976.

6. Costill, D.L.; Fink, W.J.; and Pollock, M.L. Muscle fiber composition and enzyme activities of elite distance runners. *Med. Sci. Sports* 8:96–100, 1976.

7. Dubowitz, V., and Brooke, M.H. *Muscle biopsy: a modern approach.* Philadelphia: W.B. Saunders Company, 1973.

8. Ebashi, S., and Endo, M. Calcium ion and muscular contraction. *Prog. Biophys. Mol. Biol.* 18:125–83, 1968.

9. Gollnick, P.D.; Armstrong, R.B.; Saubert, C.W.; Piehl, K.; and Saltin, B. Enzyme activity and fiber composition in skeletal muscle of untrained and trained men. *J. Appl. Physiol.* 33:312–19, 1972.

10. Hanson, J., and Huxley, H.E. The structural basis of contraction in striated muscle. *Symp. Soc. Exp. Biol. Med.* 9:228–64, 1955.

11. Hoyle, G. How is muscle turned on and off? *Sci. Am.* 222:84–93, 1970.

12. Huxley, A.F. Muscle structure and theories of contraction. *Prog. Biophys. Chem.* 7:257–318, 1957.

13. Huxley, H.E. The ultrastructure of striated muscle. *Br. Med. Bull.* 12:171–73, 1956.

14. ———. The structural basis of muscular contraction. *Pro. R. Soc. Med.* 178:131–49, 1971.

15. Peter, J.; Barnard, R.; Edgerton, V.; Gillespie, C.; and Stempel, K. Metabolic profiles of three fiber types of skeletal muscle in guinea pigs and rabbits. *Biochemistry* 11:2627–33, 1972.

16. Prince, F.P.; Hikida, R.S.; and Hagerman, F.C. Human muscle fiber types in power lifters, distance runners and untrained subjects. *Pfleugers Arch.* 363:19–26, 1976.

17. Thorstensson, A.; Grimby, G.; and Karlsson, J. Force-velocity relations and fiber composition in human knee extensor muscles. *J. Appl. Physiol.* 40:12–16, 1976.

18. Thorstensson, A., and Karlsson, J. Fatiguability and fibre composition of human skeletal muscle. *Acta Physiol. Scand.* 98:318–22, 1976.

19. Thorstensson, A.; Larsson, L.; Tesch, P.; and Karlsson, J. Muscle strength and fiber composition in athletes and sedentary men. *Med. Sci. Sports* 9:26–30, 1977.

3

Energetics
of Muscular Contraction
and Adaptations to Training
at the Cellular Level

Energetics of Muscular Contraction

As described in the second chapter, the breakdown of adenosine triphosphate (ATP) to adenosine diphosphate (ADP), $ATP \rightarrow ADP + P +$ approximately 8,000 calories of energy (or 8 kcal), furnishes the immediate source of energy for the contractile mechanism. This chapter discusses the processes by which ingested food energy is converted and utilized to regenerate the energy of the high energy bonds of the ATP, which ultimately make the muscle cell (fiber) contract.

In the past, considerations in depth of the biochemistry of muscular contraction did not seem justified because these theoretical concepts could not be applied as down-to-earth physical education and athletic training principles. However, there are now several compelling reasons for the physical educator and coach to become familiar with at least the rudiments of muscle contraction at the cellular level.

1. There is recent evidence that some very important effects of athletic training occur at the cellular level of organization in terms of modification of intracellular structure and the enzyme systems that are so important to energy supply.
2. There is now excellent laboratory data bearing on the need for diet modification to maximize the athlete's cellular energy supply.
3. Researchers in physical education have become sophisticated in the use of the electron microscope and biochemical procedures, and understanding their research reports will require more background on the part of our professional readership.

In short, the material in this chapter is now essential to a scientifically based program of physical education and athletic training.

The metabolic processes that supply the energy needs of muscle contraction ordinarily take place in the presence of adequate O_2 to oxidize the carbohydrate sources of energy completely to CO_2 and H_2O. This constitutes *aerobic* muscle activity, which in general is exercise that is low enough in intensity to be carried on for at least five minutes or longer. Energy for exercise that is so intense that exhaustion ensues within one to two minutes or less must be supplied largely by *anaerobic* processes (without O_2), because O_2 cannot be transported via the lungs and cardiovascular system rapidly enough to supply such a demand.

In general, four processes occurring within the muscle cell have to do with the chemistry of muscle contraction, and three of them are common to aerobic and anaerobic contraction. It will be noted that the first three reactions are reversible. That is to say, the reactions are such that, as some molecules of ATP are being broken down to provide energy for muscle fiber contraction, other molecules of ADP and P are being regenerated (at a cost of energy provided by the next reaction down, $CP \rightarrow C + P$). A balance obviously must be struck between the rate of breakdown and the rate of regeneration, or the muscle effort would run out of gas. Thus each reaction in the chain depends on energy supply from reactions below to remain in balance while supplying energy to the reaction above. Or we may say that each succeeding reaction supplies energy for the reversal of the preceding reaction, as follows:

$$\boxed{\text{Energy for muscle contraction}}$$

Anaerobic metabolism (must also precede aerobic metabolism)

$$\begin{cases} \text{ATP} \rightarrow \text{ADP} + \text{P (inorganic phosphate)} \\ \text{ATP} \leftarrow \text{ADP} + \text{P} \end{cases} \quad (1)$$

$$\boxed{\text{Energy for resynthesis of ATP}}$$

$$\begin{cases} \text{Phosphocreatine} \rightarrow \text{Creatine} + \text{P} \\ \text{Phosphocreatine} \leftarrow \text{Creatine} + \text{P} \end{cases} \quad (2)$$

$$\boxed{\text{Energy for resynthesis of phosphocreatine}}$$

Aerobic metabolism

$$\begin{cases} \text{Glycogen} \rightarrow \text{Glucose} \rightarrow \text{Pyruvic Acid} \rightarrow \text{Lactic Acid} \quad (3) \\ \text{Lactic Acid} \rightarrow \text{Pyruvic Acid} \rightarrow \text{(Krebs cycle)} \\ + \text{O}_2 \rightarrow \text{CO}_2 + \text{H}_2\text{O} \quad (4) \end{cases}$$

This diagram is a great oversimplification in that it does not show the intermediate reactions and enzyme systems that are necessary. Figure 3.1 provides some details.

Aerobic and anaerobic metabolism share the common paths of *glycogenolysis* and *glycolysis*. Glycogenolysis is defined as the breakdown of the large glycogen molecule into many glucose molecules, and glycolysis is defined as the splitting of the glucose molecule into two pyruvic acid molecules. In the absence (or relative shortage) of O_2, the reactions can only proceed through equation 3 in the diagram with the production of lactic acid as the end product, plus the freeing of small amounts of energy. The fourth reaction, with its oxidative (therefore aerobic) pathway, provides greater amounts of energy.

Regeneration of ATP Energy from Carbohydrate Food. We may think of the overall conversion of carbohydrate in our food to energy in terms of the following simple equation, which summarizes the diagram found on page 32.

$$\text{C}_6\text{H}_{12}\text{O}_6 + 6\text{O}_2 \rightarrow 6\text{CO}_2 + 6\text{H}_2\text{O} + \text{Energy}$$

Glucose   Oxygen   Carbon   Water
dioxide

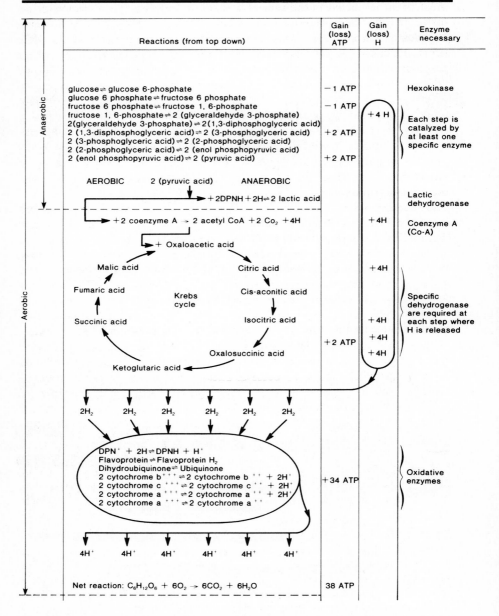

Figure 3.1 The breakdown of glucose (basic building block of all carbohydrates) to furnish the energy for regeneration of ATP.

While this is indeed straightforward, it could only happen in such direct fashion by raising the temperature very high and actually *burning* the carbohydrate in a very hot flame, a process that is not consistent with living tissue. To bring about the conversion of food energy to ATP energy for muscle contraction at body temperature, many intermediate steps catalyzed by enzyme systems are necessary. The above equation is quite correct in that it summarizes the whole process, but it tells us nothing about how this process is actually accomplished. It is the enzyme systems that promote the stepwise chemical breakdown of the foodstuffs at body temperature to provide the energy for muscle contraction. Enzymes are proteins that have the ability to promote specific chemical reactions without themselves being degraded or changed in the process. Thus the same enzyme protein can function over and over again in the same metabolic process.

The final products of carbohydrate digestion in the alimentary tract are the *monosaccharides* (6 carbon atom sugars) *glucose, fructose,* and *galactose.* These are the results of the breakdown of such larger carbohydrate molecules as starch and the usual sugars included in the diet (*disaccharides*). The three monosaccharides are interconvertible, and as they pass through the liver, they are converted almost entirely to glucose for transport in the blood to the muscle cells and other tissues of the body.

As the glucose molecule enters the muscle cell through the sarcolemma, a process greatly aided by the presence of insulin, it is immediately phosphorylated. That is to say, the 6 carbon atom monosaccharide picks up a phosphate radical on its number 6 carbon atom and becomes glucose-6-phosphate. Note in figure 3.1 that this process of phosphorylation requires the presence of the enzyme *hexokinase,* as well as the breakdown of one ATP to form ADP + P. Since our discussion has to do with formation of ATP, this first step would seem to be in the wrong direction, but we will wait for the whole story.

Once phosphorylated, the glucose molecule is trapped in the muscle cell because it requires the enzyme *phosphatase* to dephosphorylate the glucose and the muscle cell has no phosphatase, although the liver cells and some other tissues do. This is immediately important from the practical standpoint in that we now know that the energy stored away in one muscle is not available to another, which may be in the process of exhausting its energy due to locally heavy work. We will return to this concept later in the consideration of muscular endurance.

The glucose-6-phosphate, once in the cell, can be either utilized directly for energy, as shown in figure 3.1 (p. 32), or stored, depending on cell activity levels. To be stored it must be converted to glycogen, which is a large chain or network of glucose molecules called a *polymer.* The polymerization process involves several steps and requires several enzymes. The glycogen is deposited as granules in glycogen columns as shown in figure 2.8 (p. 23). Storage in the form of large, polymerized molecules is necessary so that the intracellular osmotic pressure is not unduly raised, which it would be if carbohydrate were stored as glucose units. The amount of glycogen stored in the muscle cell determines its endurance for exercise under certain conditions, as we will discuss later in this chapter. The breakdown products of protein and fat digestion and lactic acid can also be converted to glucose and thence to glycogen for storage. We shall now direct our attention to the more detailed pathways by which carbohydrate can be broken down to produce the energy necessary for rebuilding the ATP, which is what is needed by the myofilaments to bring about muscle contraction. Basically, we are concerned with aerobic and anaerobic sources of energy. However, we must resist the temptation to think of muscular activity as *either* aerobic or anaerobic. Even light exercise brings both mechanisms into play, as will be described later.

Figure 3.1 (p. 32) illustrates the whole process of carbohydrate metabolism schematically. While it is not suggested that the student commit this to memory, some measure of understanding is necessary. Glucose becomes available as an energy substrate either by the glycogenolysis of glycogen stored in the muscle fiber or by the transportation of blood glucose into the muscle fiber. The glucose molecule first undergoes glycolytic breakdown into two pyruvic acid molecules, as shown in figure 3.1. The enzyme hexokinase is necessary to add a phosphate (PO_4 or P) radical to make glucose-6-phosphate. This process costs the breakdown of one ATP to ADP + P. Each of the further steps of glycolysis is also catalyzed by at least one enzyme specific to that step, and one more ATP is broken down to add the second P to form fructose-1, 6-phosphate (the 1, 6 means that there are two phosphates in the molecule, one at the number 1 carbon atom and one at the number 6 carbon atom). Note, however, that this cost of two ATPs broken down is recouped with a net total gain of two ATP molecules during the process of glycolysis, which results in formation of two molecules of pyruvic acid. In the absence of O_2 (or insufficient O_2), the end products of glycolysis—pyruvic acid and H-atoms—combine to form lactic acid. This last step is important because, if the end products of a reaction accumulate, the reaction is ended before all of the energy substrate can be used. Since the lactic acid diffuses freely into the tissue fluids and blood, it is removed, and thus anaerobic metabolism can proceed until the energy substrate is used up.

In the presence of sufficient O_2, the system does not back up at the pyruvic acid step, because pyruvic acid can now continue on its aerobic course by entering a metabolic system called the *Krebs cycle* (also called *citric acid cycle* or *tricarboxylic acid cycle*). The course of aerobic metabolism is much more advantageous, because up to this point the anaerobic route has netted us only two ATP molecules generated from one molecule of glucose. As will be seen, complete aerobic breakdown will provide at least thirty-eight molecules of ATP per molecule of glucose.

The next stage in glucose breakdown requires the conversion of the two pyruvic acid molecules into two molecules of acetyl coenzyme A (acetyl CoA). Acetyl CoA combines with the oxaloacetic acid to become citric acid, and the Krebs cycle is underway. The net result of the Krebs cycle is to degrade the acetyl portion of the acetyl CoA to CO_2 and H atoms. The steps of the Krebs cycle depend upon specific enzymes called *dehydrogenases* to break off the H atoms, which are subsequently oxidized, liberating large amounts of energy for formation of ATP molecules. Note that although only two ATP molecules are regenerated in the Krebs cycle itself, twenty H atoms are released.

The H atoms that are released must be converted to H ions, written H+, before they can react with O_2 to form water and large amounts of energy. This is accomplished by the oxidative enzymes of the respiratory chain. (These enzymes and those of the Krebs cycle reside in the mitochondria of the cell.) This is schematized at the bottom of figure 3.1, which shows the final breakdown of the glucose molecule. The net chemical reaction for the whole aerobic process depicted in figure 3.1 is as follows:

$$C_6H_{12}O_6 + 6O_2 + 38ADP + 38P \rightarrow 6CO_2 + 6H_2O + 38ATP$$

Since each high energy phosphate bond represents about 8 kcal and each gram molecular weight of glucose (180 gm) has an ultimate energy value of about 4 kcal/per gram, the efficiency of this process of storage of energy as ATP is about $\dfrac{38 \times 8}{180 \times 4} = \dfrac{304}{720} = 42\%$.

Regeneration of ATP Energy from Fat and Protein. As will be seen in later discussion, exercise of intensities up to about 70% of maximum proceeds largely by way of energy gained from fat metabolism. Thus the regeneration of ATP energy from fat is quite important in exercise physiology. Energy can also be provided for work by protein metabolism, but this becomes important in exercise physiology only in a state of negative energy balance (starvation or the semistarvation of rigorous weight reduction). Figure 3.2 shows the derivation of energy from the metabolism of fat, carbohydrate, and protein.

Use of Fat for Energy. Fat is basically a combination of three fatty acids, each of which is attached to one of the three carbon atoms of a glycerol molecule. Chemically this combination (fat) is called a *triglyceride.*

The first step in the utilization of fat for energy is the hydrolysis into the original components, glycerol plus three fatty acids. The glycerol is easily changed by enzymic action into glyceraldehyde, which can be used for energy via the phosphogluconate pathway that occurs in the liver or in the fat cells, but not in muscle. The fatty acids undergo a process called *beta oxidation,* which involves a stepwise breakdown of the long-chain fatty acid molecule into acetyl CoA molecules, which then enter the Krebs cycle and proceed just as the acetyl CoA from carbohydrate.

Use of Protein for Energy. Proteins are very large complex molecules and consist of from 20 to 100,000 amino acids, which are the basic building blocks of all proteins. The digestive enzymes of the stomach start the breakdown of protein, and it is carried further in the small intestine to the polypeptide level (combination of several amino acids) and completed to the amino acid level in passage through the wall of the small intestine into the blood. The amino acids are then deaminated in the liver (removal of the ammonia radical), becoming keto acids. These resulting keto acids are then converted into substances that can enter the Krebs cycle and thence proceed to complete degradation and energy formation as has been described for carbohydrate.

It is important to recognize that just as all three foodstuffs can be utilized for energy, they are also convertible into fat for storage when the food ingested exceeds the need. In the case of protein, some amino acids must first go through gluconeogenesis to become carbohydrate before deposition on the hips and other body parts as fat pads. ⟍ *protein to glucose*

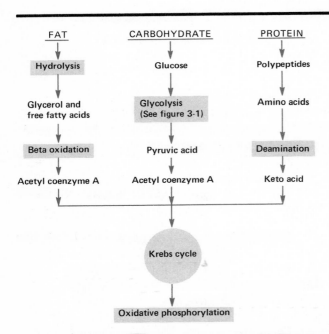

Figure 3.2 Derivation of energy from the metabolism of the three basic foodstuffs by way of the final common path: the Krebs cycle and oxidative phosphorylation.

Basic Physiology Underlying the Study of Physiology of Exercise

Adaptations to Training and Conditioning at the Cellular Level

Only in recent years have we begun to realize how very specific are the cellular adaptations to training or conditioning. During the course of evolution, different muscles have developed different abilities. For example, muscle composed largely of fast twitch (FT) fibers can work hard and quickly but cannot maintain forces efficiently, whereas muscles with a large complement of slow twitch (ST) fibers work somewhat slower but have great ability to maintain a force for a long period.

Furthermore, we have learned that even the nature of the required energy substrate varies with the intensity-duration characteristics of the exercise. Thus, when exercise is continued for many hours, the limits are set by the supply of adequate food in the form of fat and carbohydrate. For somewhat more intense work that is to be maintained for one to three hours, the limit is set by the level of glycogen stored within the muscle at the start of exercise. The glycogen may be used either relatively slowly via complete oxidation (aerobic metabolism) to form CO_2, water, and many molecules of ATP, or it may be used very quickly and anaerobically to make much less ATP per molecule of glucose. In the latter case, lactic acid accumulates, which can so increase the acidity of the tissues that intermediate metabolic processes are severely limited by the increased hydrogen ion concentration. For shorter periods—a few minutes up to thirty minutes—availability of O_2 may be the limiting factor. For very heavy workbouts that can be maintained only a matter of seconds, the rate of creatine phosphate breakdown is probably limiting. But the ultimate maximum speed of a muscle is set by the rate at which actomyosine ATPase can break down ATP.

For these reasons, we shall discuss cellular adaptations to conditioning in three different areas: (1) endurance exercise, (2) sprint or speed-type exercise (involving anaerobic metabolism), and (3) strength and power training.

Endurance Training (Aerobic Metabolism). Cardiovascular-respiratory system adaptations to training have occupied the interest of exercise physiologists for many years, but we now recognize that endurance exercise also induces major adaptations in the skeletal muscle, which helps explain the improvement shown with training in capacity for prolonged work.

Myoglobin. Animal studies have shown that endurance exercise increases the concentration of myoglobin in the skeletal muscle. In rats, this increase amounts to 80% (27). Myoglobin probably facilitates O_2 utilization in muscle by enhancing O_2 transport through the cytoplasm (sarcoplasm) to the mitochondria (13).

Mitochondrial Enzymes. A number of investigators working with both rodents and humans have found that the enzymes of the respiratory chain (fig. 3.1) increase some twofold in skeletal muscle in response to endurance exercise (13). The enzymes of the Krebs cycle also increase significantly, but by differing amounts ranging from 35% to 100%. Thus the protein composition of the mitochondria changes with training, since the enzymes (protein) do not all increase proportionately. These changes are related to the daily work load and the total training time (3).

There are also major increases in the levels of enzymes involved in the activation, transport, and beta oxidation of long-chain fatty acids, which thus improve the organism's ability to utilize fat as an energy source. As might be expected, these enzymic changes approximately double the capacity of trained muscles to oxidize a variety of energy substrates (13).

Mitochondrial Size and Number. Probably the most important training effect of endurance exercise is the increase in mitochondrial protein, which has been reported by several different groups of investigators (2,7,22). Both the size and the number of mitochondria in the muscle cell are increased as the result of training, if the intensity of the training stimulus is sufficiently severe. It will be recalled from the preceding discussion of muscle metabolism that the very important enzymes of both the Krebs cycle and the respiratory chain reside in the mitochondria of the muscle cell. Holloszy (12) has shown that training rats on a treadmill produces a twofold increase in their capacity to oxidize pyruvate. Oxidation of pyruvate represents the entire aerobic portion of the muscle cell's metabolic process; this tells us that the cell is now able to work at double the load, since the glycolytic pathway is probably not a limiting factor.

Differences in Response by Fiber Type. In animals, endurance exercise has been shown to result in increases in mitochondrial enzyme levels in all three fiber types. Although the capacity of FT muscle (white) for oxidative metabolism increases, the muscle is not converted to ST muscle (red), as had been suggested by earlier workers. In human muscles, recent evidence suggests that endurance training brings about a shift from FG (fast twitch, glycolytic) to FOG (fast twitch, oxidative, glycolytic) fiber types (28). This increase in oxidative capacity would, of course, serve to improve aerobic performance (endurance).

Glucose-Alanine-Glucose Cycle. In figure 3.1. only two routes are shown for the removal of the pyruvate with its consequent energy production and formation of ATP. However, recent evidence (5) has shown the importance of a third route. Pyruvate can be converted in muscle tissue to the amino acid alanine in the presence of the enzyme alanine transaminase. The alanine is transported by the blood to the liver where its carbon skeleton is reconverted to glucose, which becomes available as energy at a later time in the form of blood sugar (glucose). Since endurance exercise increases alanine transaminase in trained muscles 50%-80% (26), it is possible that this adaptation could result in conversion of a greater proportion of pyruvate to alanine and less to lactate, thus creating a more favorable environment for muscle contraction.

Glycolytic Enzymes. Endurance exercise results in only small and probably physiologically insignificant changes in the glycolytic enzymes, with the exception of hexokinase, which tends to vary directly with respiratory capacity. It is thought that the glycolytic process (labeled anaerobic in fig. 3.1) is not limiting in endurance exercise, although the supply of glycogen for energy substrate may be. The limits appear to be set at the level of the Krebs cycle, and the respiratory chain and the training adaptations with respect to enzyme response are consistent with this concept.

Energy Substrate Availability. The supply of muscle glycogen available in the muscle fiber for endurance exercise can be greatly enhanced by a combination of appropriate exercise and diet. This will be discussed in greater detail later in this chapter.

Physiological Implications of Cellular Adaptations. As Holloszy (13) has pointed out, "It now seems clear that as skeletal muscle adapts to endurance exercise it becomes more like heart muscle in its mitochondrial and cytoplasmic enzyme patterns." Thus, in addition to greater glycogen stores, there is a slower depletion of glycogen, lower lactate levels, and a greater oxidation of free fatty acids during endurance exercise. It is also reasonable to believe that the increases in muscle mitochondria and myoglobin account for approximately 50% of the increase in maximal O_2 consumption that is typical as a result of endurance training. The other 50% is likely accounted for by better O_2 transport by the cardiovascular system (13).

Sprint Training (Anaerobic Metabolism). The effects of sprint training upon anaerobic metabolism are less well defined than the well-documented effects of endurance training on aerobic metabolism. In short-term, heavy exercise, the only significant energy pools available are (1) the breakdown of the phosphagens, ATP and CP, and (2) glycolysis, the breakdown of glucose to pyruvate and lactic acid as described in figure 3.1.

Limiting Factors. The energy available from the breakdown of all possible ATP and CP has been shown to be approximately equivalent to 1.2 liters of O_2 utilization (20). Thus, in an event such as the 100-meter dash, which requires supramaximal O_2 consumption rates on the order of 20 or more liters per minute, the phosphagen breakdown can support maximal effort at that speed for only about four seconds. The energy available from anaerobic breakdown of glycogen must furnish the balance.

Two separate groups of investigators (14, 20) have shown that in exercise heavy enough to be maintained for only one to two minutes (anaerobic), the rate of phosphagen regeneration is likely to be the limiting factor. Under these conditions the CP level is completely depleted, and after that time the work can no longer be continued at the same high rate. However, under such conditions the blood lactate also builds up to very high levels, and it remains an open question whether the limits for sprint performance are set by lack of energy substrate (phosphagen depletion) or by accumulation of metabolic end products such as lactate.

Effects of Training. In a recent study on rats, Hickson, Heusner, and Van Huss (11) compared the effects of sprint and endurance training. Interestingly they found that similar enzyme adaptations occurred over time with the two very different training regimens. They found no increase in the glycolytic enzymes in the rat after sprint training, and their work suggests that the preexisting levels of glycolytic capacity in the rat muscles studied were adequate to supply the energy needs even during the stress of very heavy exercise. Unexpectedly they found an increase in mitochondrial enzymes (those involved in aerobic metabolism) in response to sprinting as well as to endurance training.

On the other hand, Gollnick and associates (8) trained six young men for five months, four days a week, one hour a day, at 75%–95% of maximal work load. Any training that can be sustained for one hour would be expected to challenge largely the aerobic pathways, but they found a doubling in PFK activity as well as in succinic dehydrogenase. Since PFK is an enzyme that is thought to reflect glycolytic capacity best, these data suggest a considerable increase in anaerobic capacity in humans even though the exercise challenge was largely aerobic. Interestingly they also found that oxidative (aerobic) capacity was increased in both FT and ST muscle fibers, but glycolytic capacity (anaerobic) was improved only in the FT fibers.

Russian investigators (33) have found increases in creatine phosphate, and this may be the most important training adaptation.

Strength and Power Training. It is in strength and power training that the gross effects of cellular changes are best exhibited. The hypertrophic changes in response to very heavy work have been observed since time immemorial and scientifically elucidated over a period of at least 100 years.

Hypertrophy. Most of our information with respect to cellular changes in hypertrophy has come from animal studies, largely those done on rats. The most common animal model for the study of hypertrophy has been the surgical severing of the rat gastrocnemius, which produces an overload on the remaining ankle extensors, the soleus and the plantaris, where the processes of hypertrophy can be clearly observed (6). In this type of experiment very rapid responses are common. The rat soleus typically increases in weight by 40% within six days. This rapid growth is due largely to an enlargement of muscle fibers (hypertrophy), although occasionally muscle fiber splitting (hyperplasia) is also seen. In the rat as well as in man, the hypertrophy is accompanied by increased muscle strength.

Muscle hypertrophy has been demonstrated in rats under these conditions even when the pituitary gland has been removed and in diabetic animals, thus implying that growth hormone, insulin, and even testosterone are not essential to muscle hypertrophy. This finding does not carry over to human muscle hypertrophy, however, as will be discussed in later chapters.

Another interesting finding in the rat model is that hypertrophy is still induced in the overworked muscle when the animal is fasted and other relatively unused muscles are in a process of wasting (atrophy) (6).

The increase of muscle weight reflects an increase in protein, which is largely sarcoplasmic protein in contradistinction to the enzyme and mitochondrial protein that is seen to increase in endurance conditioning. By eight hours after surgery, protein synthesis in rats is enhanced, and indeed within one hour of surgery the active transport of certain amino acids is increased. The magnitude of this latter effect is related to the amount of contractile activity in the muscle.

Most interesting from a therapeutic viewpoint is the fact that simple, passive stretching of rat muscle was found to retard protein degradation and to stimulate amino acid transport. These factors could lead to less atrophy and better physical therapy in individuals who are immobilized in a cast or by occupation (astronauts). This finding suggests that increased tension, whether active or passive, is the necessary stimulus affecting muscle protein balance. Such a hypothesis appears to be consistent with all observed facts to this time.

Biochemical adaptations. Very interesting experiments are underway at McMaster University in Canada where the effects of weight training are being elucidated. In one experiment (24), nine college age men were studied by needle biopsy under control conditions and before and after five months of weight training and five weeks of immobilization. Training resulted in an 11% increase in arm circumference and a 28% increase in elbow extension strength. These changes were accompanied by significant increases in the phosphagens ATP and CP by 18% and 22% respectively. Immobilization significantly reduced CP concentration by 25%. These findings bring theory and practice closer together.

Mitochondrial Density. MacDougall and co-workers (25) at McMaster University recently reported on six weight lifters before and after six months of weight training. Training resulted in a significant 26% reduction in mitochondrial volume density of the trained elbow extensors, a 25% reduction in the ratio of mitochondrial volume to myofibrillar volume (the contractile elements), and a 12% increase in volume density of the sarcoplasm. Fiber area was increased by 39% for FT and 31% for ST fibers. Thus heavy resistance training leads to dilution of the mitochondrial volume density by virtue of an increase in myofibrillar size and cytoplasmic volume.

Fiber Type and Power. Komi and colleagues (23) studied anaerobic performance capacity in eighty-nine athletes. They found that the main determinants of successful performance in explosive events like jumping were related to muscle fiber composition. The correlation between vertical velocity (measured by timing a short run up stairs) and the percentage of FT fibers was .37.

Summary of Cellular Training Adaptation. While all the answers are not yet in, the presently available evidence suggests that there are three basic modes of adaptation to training at the cellular level: (1) improvement of aerobic capacity, (2) enhancement of anaerobic capacity, and (3) hypertrophy.

Endurance training at prolonged exercise of moderate intensity improves aerobic capacity mainly through increased levels of myoglobin, mitochondrial enzymes (both in size and number), glycogen stores, and generally greater oxidative capacity in both FT and ST fibers.

Sprint training at high speeds in intensive workouts for short periods, as commonly used in interval training, increases the enzyme PFK, which is reflective of glycolytic capacity, and most important, creatine phosphate which is likely to limit anaerobic work.

Strength and power training involving near maximal voluntary contraction (MVC) strength increases the cross-sectional area of muscle fibers, a result also observed grossly as hypertrophy.

Energy Substrate and Training

Until quite recently our only evidence regarding the energy substrate utilized by working muscles was indirect—by calculation of respiratory quotients (RQ) as described in the later chapter on exercise metabolism. In 1962 Bergstrom introduced a biopsy needle which he and Hultman and their co-workers subsequently used to good advantage in making direct observations of the level of various energy substrates remaining in an active muscle after various types, intensities, and durations of muscle work. From such studies, rates of utilization and various practical implications for athletes and coaches have become available.

Figure 3.3 shows the rate of glycogen depletion in working muscle cells when they are fatigued by bicycle exercise at various work loads (31). When the individual is working at 70%–80% of maximal aerobic capacity (VO_2 max) then exhuastion occurs when the muscle fiber's glycogen supply is depleted. This suggests that at such work loads, the muscle can utilize only glycogen stored in the muscle cells for its energy substrate, although at lighter loads glucose and free fatty acids transported by the blood form the source of energy. At maximal work loads, the load cannot be sustained long enough to bring about glycogen depletion. At lighter loads of 60% or less of aerobic capacity, the limiting factor is probably the availability of blood-borne glucose and free fatty acids (FFA). Under the conditions of lighter loading, the liver store of carbohydrate may become an important factor (30). It must be understood that only the duration of the effort is set by the energy stores available; the rate of work that can be maintained depends upon O_2 transport and possibly on rate-limiting enzymic processes.

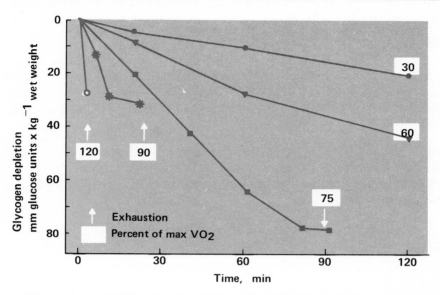

Figure 3.3 Glycogen depletion in the quadriceps muscle during bicycle exercise of different intensities. (From Saltin, B., and Karlsson, J., in *Muscle Metabolism during Exercise,* B. Pernow, and B. Saltin, eds., 1971. Courtesy of Plenum Press, New York.)

Although glycogen depletion and muscular exhaustion are closely related in bicycle work this is not the case in distance running. Costill and co-workers (4) have shown that complete glycogen depletion is not seen after intense distance running even though the maximal O_2 requirement may be 80%–90%. The most likely explanation for this interesting difference between running and cycling is probably a difference in muscle fiber recruitment, since the Costill group also found marked depletion of ST fibers with very little depletion of FT fibers. Thus, although total muscle glycogen during running was depleted by only 56%, there was a 93% decline in periodic acid-Schiff (PAS) dark-stained ST fibers (measure of glycogen storage). Apparently the FT fibers cannot be fully recruited in running as they are in cycling with respect to the vastus lateralis muscle, which the researchers sampled.

The situation is quite different with respect to isometric exercise. It has been calculated that the actual utilization of glycogen under maximum isometric contraction is only about one-tenth that available to the muscle, and consequently glycogen stores cannot be a limiting factor in isometric contraction (16). Indeed, fatigue occurring isometrically at any load above 20% of maximum cannot be explained on the basis of glycogen depletion, and loads below that are unimportant in human athletic performance (1). A more definitive answer regarding the question of isometric muscle fatigue must await further research.

Training Effects on Cellular Energy Substrate Level. Using the muscle biopsy technique, Karlsson and associates (21) have shown increased levels of ATP concentration in muscle as a result from seven months of military training, which included distance running two to three times per week. ATP concentration in muscle, of course, would be very important during periods of anaerobic work when work-load intensity is too heavy to permit O_2 transport to keep up with tissue demand.

With respect to carbohydrate energy sources, the percentage of energy that is supplied by glycogen is apparently affected to some extent by its relative availability to the muscle cell as a result of diet. Pruett (29) has shown that the percentage contribution of carbohydrate varies with diet and work load as follows:

Work load	Standard diet	High fat	High carbohydrate
50% VO_2 max	40%	35%	50%
70% VO_2 max	53%	50%	60%

Hultman, working with Bergstrom and others, has reported some findings that are of importance to all concerned with endurance type athletics (15, 16, 17, 18). Most important, he has shown that endurance under heavy aerobic work loads is determined largely by the level of glycogen storage in the muscle cells. Thus, for such work, the rate or intensity is set by O_2 transport capacity (to be discussed in following chapters), but the duration or total work that can be accomplished is set by the level of glycogen stored within the cell.

Of even greater practical interest are the data found in figure 3.4. Here one can see the very important *overshoot phenomenon;* that is, when a muscle is worked hard enough to bring about glycogen depletion, then it develops the ability to store greater than normal amounts of glycogen. Second, it can be seen that recharging of the stored glycogen depends greatly upon the type of diet. On a carbohydrate-rich diet, glycogen resynthesis was complete in twenty-four hours, whereas on a carbohydrate-free diet of the same caloric content, resynthesis was complete only after eight to ten days. It was also shown that the glycogen content of the liver was dependent upon diet. If CHO is not supplied in the diet, liver glycogen can decrease rapidly to values that will not sustain work for more than about one hour (by glycogenolysis after muscle glycogen has been depleted).

Unfortunately, the overshoot phenomenon works only for the muscle that has been worked, and the glycogen storage level varies considerably from muscle to muscle.

It has also been shown (32) that in rats FOG fibers have a glycogen repletion rate three times that of FG fibers and twice that of SO fibers. If this

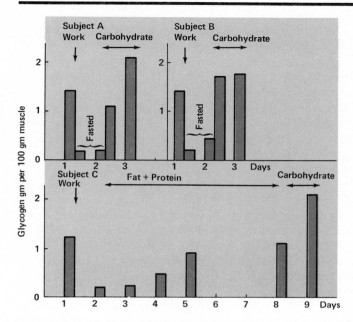

Figure 3.4 Muscle glycogen before and after work. Two subjects (upper graph) fasted for one day and got carbohydrate diet on day 2; the third subject (lower graph) got fat and protein diet during eight days, and thereafter one day of carbohydrate. (From Hultman, E. *American Heart Association Monograph*, no. 15, 1967, p. 106. Courtesy of the American Heart Association, New York.)

is true for human muscle also, then recovery for endurance exercise should be considerably more rapid than that for high intensity exercise such as sprints. This seems consistent with coaching experience.

A short fast of up to six hours was found to have no effect on glycogen level, so that a long wait for an athletic competition after a meal will have no adverse effects, unless stomach growls annoy teammates.

The exact mechanism of the overshoot phenomenon is not yet fully understood. Jeffress, Peter, and Lamb (19) have reported that an increase in glycogen synthetase, an enzyme necessary to the storage of glycogen, results from training, but this alone cannot fully explain the overshoot.

Training also has a *glycogen-sparing* effect in that the better trained individual can supply more blood-borne energy substrate, primarily free fatty acids that supply the bulk of the energy for resting and low level, long duration exercise (10).

For the reader who would like more details on the subject of cellular adaptations to exercise, excellent reviews are available (9, 13, 33).

SUMMARY

1. ATP energy, which is the immediate source of energy at the acto-myosin filament level, is replenished by the breakdown of phosphocreatine. The energy for regeneration of phosphocreatine is in turn supplied by *glycolysis,* the breakdown of glucose to pyruvic acid. Under *anaerobic* conditions the pyruvic acid is converted to lactic acid, whose accumulation eventually brings about cessation of exercise. Under *aerobic* conditions the first three reactions are in turn refueled by energy gained from the complete oxidation of the pyruvic acid to carbon dioxide and water.

2. Both fat and protein can also be used for energy, but only fat is important in exercise metabolism.

3. The chemical breakdown of foodstuffs to supply energy for the regeneration of ATP from ADP and P is accomplished by a complex process in which many protein enzymes are necessary as catalysts. The overall reaction describing the entire chain of events can be written as follows:
$$C_6H_{12}O_6 + 6O_2 + 38ADP + 38P \rightarrow 6CO_2 + 6H_2O + 38ATP$$
The synthesis of 38 ATP molecules represents the accumulation of about 304 kcal of energy from the oxidation of some 720 kcal of glucose (1 gm molecular weight).

4. Endurance-type training using prolonged exercise at moderate intensities improves aerobic capacity mainly through increased levels of myoglobin, mitochondrial enzymes (both size and number), glycogen stores, and in general greater oxidative capacity for both FT and ST fibers.

5. Sprint-type training, accomplished through high speed, intensive workouts for short periods, as commonly used in interval training, results in increases in the enzyme PFK, which reflects glycolytic capacity, and most important, creatine phosphate, which is likely to be the limiting factor for anaerobic work.

6. Strength and power training involving near maximal voluntary contraction (MVC) strength results in increases in the cross-sectional area of muscle fibers, which are also observed grossly as hypertrophy.

7. At heavy work loads of 70%–80% of capacity, glycogen stored in the muscle cell provides the energy for contraction, and endurance time depends on the level of glycogen storage in the muscle at the beginning of exercise. This is not true, however, for distance running or isometric exercise.

8. At maximal work loads the work cannot be sustained long enough to deplete the stored glycogen, and the end point is probably determined by the level of lactic acid accumulation, or CP depletion.

9. When muscular work is such that glycogen depletion is brought about, the muscle responds by large increases of glycogen storage. This response

is complete within twenty-four hours on a carbohydrate-rich diet, but may take eight to ten days on a carbohydrate-free diet of equal caloric content.

10. Training also has a *glycogen-sparing* effect in that better trained individuals are able to utilize more free fatty acid as energy substrate.

REFERENCES

1. Ahlborg, B.; Bergstrom, J; and Hultman, E. Muscle metabolism during isometric exercise performed at constant force. *J. Appl. Physiol.* 33:224–28, 1972.
2. Barnard, R.J.; Edgerton, V.R.; and Peter, J.B. Effect of exercise on skeletal muscle: 1. Biochemical and histological properties. *J. Appl. Physiol.* 28:762–66, 1970.
3. Benzi, G., et al. Mitochondrial enzymatic adaptation of skeletal muscle to endurance training. *J. Appl. Physiol.* 38:565–69, 1975.
4. Costill, D.L.; Gollnick, P.D.; Jansson, E.D.; Saltin, B.; and Stein, E.M. Glycogen depletion pattern in human muscle fibers during distance running. *Acta Physiol. Scand.* 89:374–83, 1973.
5. Felig, P., and Wahren, J. Amino acid metabolism in exercising man. *J. Clin. Invest.* 50:2703–14, 1971.
6. Goldberg, A.L.; Etlinger, J.D., Goldspink, D.F.; and Jablecki, C. Mechanism of work-induced hypertrophy of skeletal muscle. *Med. Sci. Sports* 7:185–98, 1975.
7. Gollnick, P.D.; Ianuzzo, C.D.; and King, D.W. Ultrastructural and enzyme changes in muscles with exercise. In *Muscle Metabolism during Exercise,* eds. B. Pernow and B. Saltin, pp. 69–71. New York: Plenum Press, 1971.
8. Gollnick, P.D.; Armstrong, R.B.; Saltin, B.; Saubert, C.W.; Sembrowich, W.L.; and Shepherd, R.E. Effect of training on enzyme activity and fiber composition of human skeletal muscle. *J. Appl. Physiol.* 34:107–11, 1973.
9. Gollnick, P.D. Biochemical adaptations to exercise: anaerobic metabolism. In *Exercise and Sport Sciences Reviews,* ed. J. H. Wilmore. New York: Academic Press, 1973.
10. Havel, R.J. Influence of intensity and duration of exercise on supply and use of fuels. In *Muscle Metabolism during Exercise,* eds. B. Pernow and B. Saltin, pp. 315–25. New York: Plenum Press, 1971.
11. Hickson, R.C.; Heusner, W.W.; and Van Huss, W.D. Skeletal muscle enzyme alterations after sprint and endurance training. *J. Appl. Physiol.* 40:868–72, 1976.
12. Holloszy, J.O. Effects of exercise on mitochondrial oxygen uptake and respiratory enzyme activity in skeletal muscle. *J. Biol. Chem.* 242:2278–82, 1967.
13. ———. Adaptation of skeletal muscle to endurance exercise. *Med. Sci. Sports* 7:155–64, 1975.
14. Hultman, E.; Bergstrom, J.; and McLennan-Anderson, N. Breakdown and resynthesis of phosphorylcreatine and adenosine triphosphate in connection with muscular work in man. *Scand. J. Clin. Lab. Invest.* 19:56–66, 1967.
15. Hultman, E. Physiological role of muscle glycogen in man with special reference to exercise. In *Am. Heart Assoc. Monograph,* no. 15, pp. 99–112, 1967.
16. ———. Muscle glycogen stores and prolonged exercise. In *Frontiers of Fitness,* ed. R. J. Shephard, pp. 37–60. Springfield: Charles C. Thomas, 1971.

17. Hultman, E., and Nilsson, L.H. Liver glycogen in man; Effect of different diets and muscular exercise. In *Muscle Metabolism during Exercise,* eds. B. Pernow and B. Saltin, pp. 143–51. New York: Plenum Press, 1971.

18. Hultman, E.; Bergstrom, J.; and Roch-Norlund, A.E. Glycogen storage in human skeletal muscle. In *Muscle Metabolism during Exercise,* eds. B. Pernow and B. Saltin, pp. 273–88. New York: Plenum Press, 1971.

19. Jeffress, R.N.; Peter, J.B.; and Lamb, D.R. Effects of exercise on glycogen synthetase in red and white skeletal muscle. *Life Sci.* 7:957–60, 1968.

20. Karlsson, J.; Diamant, B.; and Saltin, B. Muscle metabolites during submaximal and maximal exercise in man. *Scand. J. Clin. Lab. Invest.* 26:385–94, 1971.

21. Karlsson, J.; Nordesjo, L.O.; Jorfeldt, L.; and Saltin, B. Muscle lactate, ATP, and CP levels during exercise after physical training in man. *J. Appl. Physiol.* 33:199–203, 1972.

22. Kiessling, K.H.; Piehl, K.; and Lundquist, C.G. Effect of physical training on ultrastructural features in human skeletal muscle. In *Muscle Metabolism during Exercise,* eds. B. Pernow and B. Saltin, pp. 97–101. New York: Plenum Press, 1971.

23. Komi, P.V.; Rusko, H.; Vos, J.; and Vihko, V. Anaerobic performance capacity in athletes. *Acta Physiol. Scand.* 100:107–14, 1977.

24. MacDougall, J.D.; Ward, G.R.; Sale, D.G.; and Sutton, J.R. Biochemical adaptation of human skeletal muscle to heavy resistance training and immobilization. *J. Appl. Physiol.* 43:700–703, 1977.

25. MacDougall, J.D.; Sale, D.G.; Moroz, J.R.; and Howald, H. Mitochondrial volume density in human skeletal muscle following heavy resistance training. Abstracted in *Med. Sci. Sports* 10:56, 1978.

26. Mole, P.A.; Baldwin, K.M.; Terjung, R.L.; and Holloszy, J.O. Enzymatic pathways of pyruvate metabolism in skeletal muscle: adaptations to exercise. *Am. J. Physiol.* 224:50–54, 1973.

27. Pattengale, P.K., and Holloszy, J.O. Augmentation of skeletal muscle myoglobin by a program of treadmill running. *Amer. J. Physiol.* 213:783–85, 1967.

28. Prince, F.P.; Hikida, R.S.; and Hagerman, F.C. Human muscle fibre types in power lifters, distance runners, and untrained subjects. *Pfluegers Arch.* 363:19–26, 1976.

29. Pruett, E.D.R. Glucose and insulin during prolonged work stress in men living on different diets. *J. Appl. Physiol.* 28:199–208, 1970.

30. Rowell, L.B. The liver as an energy source in man during exercise. In *Muscle Metabolism during Exercise,* eds. B. Pernow and B. Saltin, pp. 127–41. New York: Plenum Press, 1971.

31. Saltin, B., and Karlsson, J. Muscle glycogen utilization during work of different intensities. In *Muscle Metabolism during Exercise,* eds. B. Pernow and B. Saltin, pp. 289–99. New York: Plenum Press, 1971.

32. Terjung, R.L.; Baldwin, K.M.; Winder, W.W.; and Holloszy, J.O. Glycogen repletion in different types of muscle and in liver after exhausting exercise, *Am. J. Physiol.* 226:1387–91, 1974.

33. Yakovlev, N.N. Biochemistry of sport in the Soviet Union: beginning, development and present status. *Med. Sci. Sports* 7:237–47, 1975.

4

The Physiology
of Muscle Contraction

Physiology of Gross Muscle Contraction

Muscle tissue is specifically differentiated for the purpose of contraction; thus its most important physiological property is *contractility*. However, it possesses other properties common to protoplasm in general: *irritability* and *conductivity*. Irritability indicates that muscle tissue responds to adequate stimuli with its typical response, contraction, and conductivity means that an adequate stimulus will be propagated throughout any one muscle fiber in skeletal muscle, and from fiber to fiber in smooth and cardiac muscles for reasons described in chapter 2.

Excised muscle may be stimulated electrically, chemically, and mechanically. The intact muscle is normally stimulated by its motor nerve only, but it can also be stimulated electrically through the skin, as is frequently done by physicians and physical therapists, and mechanically, as when a bruise elicits a contracture (charley horse).

The Muscle Twitch and Its Myogram.   It has been customary in elementary physiology textbooks to describe the simple muscle twitch as the basis for understanding the process of muscle contraction. This concept is likely to be helpful only if the student realizes from the outset that this is not typical of muscle contraction either in the intact body, or even in an excised muscle that is normally innervated.

In the typical laboratory experiment, the gastrocnemius muscle of a frog is excised and hung from a ring stand so that its contraction, in terms of the movement of its free end, is recorded on the rotating drum of a kymograph. The muscle is then stimulated electrically by a single shock to the entire sciatic nerve trunk, or to the muscle itself, and it responds under these conditions by a single twitch. The record of the events occurring during this contraction is called a *myogram* (fig. 4.1).

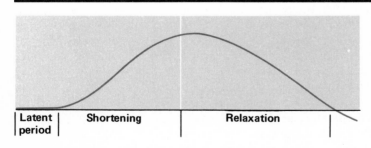

Figure 4.1 The muscle twitch and its myogram.

At this point the student must understand the two basic differences between the laboratory preparation and normal stimulation and contraction processes as they occur in the intact animal; otherwise difficulty will arise in conceptualizing other aspects of neuromuscular function.

1. In the laboratory preparation, nearly all of the nerve fibers for the whole muscle are innervated simultaneously because all the fibers of the sciatic nerve are shocked at the same time, but the innervation of normal human muscle is asynchronous, motor unit by motor unit, or we might say each nerve fiber is stimulated individually at varying points in time.

2. In the laboratory preparation, the twitch is the response to a single stimulation (shock), and this probably never occurs in the intact organism, for we know that innervation of human muscle is accomplished by volleys of nerve impulses, ranging from five or six per second to as many as eighty or ninety.

Having recognized the foregoing artifacts, there is still much we can learn from the myogram of the muscle twitch. After the stimulus is applied, approximately 0.01 second elapses, before contraction of the frog muscle begins. This interval is called the *latent period,* and it has been found to be much shorter— 0.001 second—if the muscle is completely unloaded (recording by optics). The shortening of the muscle is called the *contraction phase,* which typically takes approximately 0.04 second in the frog gastrocnemius. The lengthening of the muscle back to its resting length occupies about 0.05 second, and this is called the *relaxation phase.* These values, of course, vary from species to species and from muscle to muscle within the same species, as was discussed in chapter 2 in regard to fast and slow twitch muscle fibers.

Summation of Contractions and Tetanus. With the same muscle preparation discussed above, and by application of a second stimulus to the nerve trunk within the period of the single twitch, tension can be increased considerably (fig. 4.2). The best results are usually gained when the stimulus for the second contraction occurs at the high point of the first. The second contraction is very similar to the first, but it starts with the elevated level of tension (or shortening) supplied by the first contraction. The explanation for this phenomenon, *summation of contractions,* seems to be that the short duration of the single twitch does not allow sufficient time for the structural rearrangements within fibers (discussed in chapter 2) to go to completion (29).

If the same preparation is stimulated repeatedly with a series of shocks too closely spaced to allow a complete relaxation phase, a prolongation of the contraction occurs, which is called *tetanus* (fig. 4.3). If a relaxation phase persists, as is shown in the curves *A, B, C, D* of the illustration, it is termed a *partial* or *incomplete tetanus.* If the relaxation phases are completely eliminated, as in the curve *E,* it is called a *complete tetanus.* In a complete tetanus, the developed tension may be three to four times that of the single twitch.

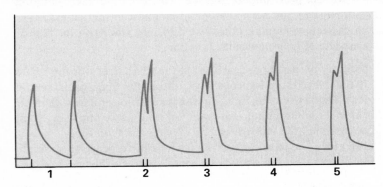

Figure 4.2 Summation of twitches. (From Zoethout. *Introduction to Human Physiology,* 1948. Courtesy of The C.V. Mosby Company, St. Louis.)

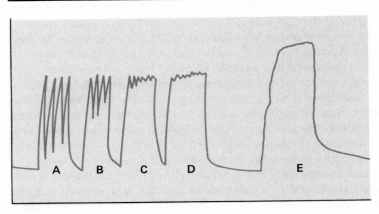

Figure 4.3 Muscle curves showing the genesis of tetanus: *A, B, C, D,* incomplete tetani; *E,* complete tetanus. Faradic shocks of the same intensity were used throughout. (From Zoethout. *Introduction to Human Physiology,* 1948. Courtesy of The C.V. Mosby Company, St. Louis.)

Temperature Effects upon Muscle Contraction. The effects of temperature upon intact human muscle have been known for a long time (27, 28). The contraction time for the human gastrocnemius muscle increases, with cooling, by as much as 21%–82%, depending upon the length of time the muscle is cooled and the temperature drop. The relaxation time under comparable conditions of cooling increases from 51% to 150%. Heating the muscle, on the other hand, causes small but significant improvements in the speed of contraction (12%) and in the speed of relaxation (22%). It should be pointed out that the important factor here is the deep-muscle temperature.

Basic Physiology Underlying the Study of Physiology of Exercise

More recent work has corroborated the early work in this area and has provided further elucidation that has practical significance. Petijan and Eagan (21) have shown that exercise-induced heating is more effective than is the same amount of passive heating in improving the rate of muscular relaxation after stimulation. Furthermore, they also provided evidence that training induces an increased rate of muscle relaxation after exercise.

It may also be noted that if the gastrocnemius muscle is cooled, the relaxation phase is slowed down two to three times as much as the contraction phase. It has been postulated that this difference may explain poor performance or muscle injury after improper or insufficient warm-up. The rationale for this hypothesis is that a slowly relaxing antagonist may be driven into its relaxation phase by a relatively faster contracting agonist. Thus there would be an opposition of forces of the paired muscles around any given joint, which might result in sore muscles or impaired performance—certainly a reasonable hypothesis.

The author would like to present another equally reasonable hypothesis in this vein. A representative time for one stride in sprinting is approximately 0.4 second, and Tuttle's work (27) indicated a slowing of the total twitch (gastrocnemius) in cooling down to 0.54 second after 5 minutes and to 0.88 second after 20 minutes of cooling. This indicates the possibility that contraction of the agonist, or prime mover, may occur while the muscle is still in a somewhat contracted condition from the previous stimulation; this results in a summation of contractions without a relaxation phase for the muscle concerned. That this condition may very likely result in contracture or sore muscles will be discussed in detail in a later chapter.

The words of A.V. Hill summarize the effects of temperature. "The speed of everything that can be measured in muscle is diminished two or three times by a fall in temperature of 10° C (15)." Hill also points out that mammalian muscle can be quickened approximately 20% by elevating body temperature 2° C, and he suggests that a good runner might do 100 yards in 8 seconds under these conditions.

The All-Or-None Law. If a muscle fiber (or motor unit) is stimulated by a single impulse at or above threshold value, it responds by a contraction, or twitch, that is maximal for any given set of conditions of nutrition, temperature, etc. In other words, stimulation by impulses much larger than threshold value will result in no increase in either the shortening or the force of contraction. The muscle fiber contracts maximally, or it does not contract at all, and this fact is referred to as the *all-or-none law* of muscle contraction. Obviously this law applies to the motor unit, since all its component fibers are innervated by the same nerve fiber and impulse (it does not, of course, apply to whole muscles).

Gradation of Response. Because of the all-or-none law for the contraction of motor units and fibers, other explanations must be sought for the fact that whole muscles are capable of exquisitely fine gradations in speed and in force of contraction. The ability of a large muscle to function in providing the force necessary to lift a maximal weight in weight training and also to thread a needle requires explanation. There are two explanations of how the nervous system brings this about.

First, and probably most important, the gradation of muscular effort is brought about by the innervation of varying numbers of the motor units within the whole muscle. A skeletal muscle may consist of several hundred (or well over a thousand in a large muscle) motor units. The size of the impending task is evaluated through sensory channels, and an appropriate number of motor units is stimulated to respond to the task. Occasionally an error is made in evaluating the severity of the job, as in lifting a mock barbell whose real weight is far less than its size and apparent material indicate. In this event an embarrassingly large number of motor units is brought into play. A similar situation occurs when a weight slips out of grasp in the act of throwing and the arm is "thrown out" because of the imbalance between the resistance and the muscular effort brought to bear.

The second factor known to operate in grading muscular response is the variation of the frequency of stimulation. Thus innervation in any motor unit is usually in volleys of stimuli or nervous impulses, ranging from several to eighty or ninety per second. In reference to the principle of tetanic contraction, it will be recalled that frequent stimulation, which results in complete tetanus, may increase the tension by as much as four times that of a single twitch. This, then, is the second factor operative in bringing about graded responses. It may seem that this second factor violates the all-or-none law, but the law applies only to the response of a motor unit to a single stimulus.

These two factors appear to act simultaneously and cooperatively to bring about the very fine gradation in response seen in even the large skeletal muscles.

Muscle Fatigue. It has long been known that when a muscle is caused to contract repeatedly and with very short rest intervals (one to two seconds), a decrement in response can be seen both in the intact muscle and in the excised muscle. Figure 4.4 illustrates this fatigue phenomenon in the excised muscle. The first few contractions demonstrate the treppe effect (in a rested muscle); then, after a period of normal contractions, the response of the muscle grows less—both in the contraction and the relaxation phase—until finally no visible reaction to further stimulation is obtained.

This should denote several factors of practical importance to the physical educator and the coach. First, the effect upon the relaxation phase is earlier, and larger in magnitude, than the effect upon the contraction phase. This lack of relaxation is referred to as *contracture,* which plays an important role in the discussion of muscle soreness. Second, there is good evidence (19) that at

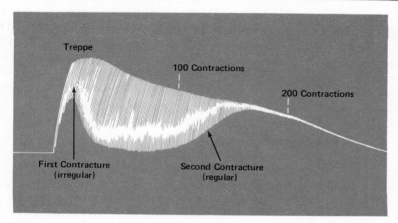

Figure 4.4 Effect of repeated stimulation. The muscle was stimulated electrically at intervals of one second. Note the effect of fatigue on the ability of the muscle to "relax." (From Howell. *A Textbook of Physiology*, 11th ed., 1930. Courtesy of W.B. Saunders Company, Philadelphia.)

the point of complete fatigue, where no further visible response results from the stimulation, the muscle action potentials are undiminished. This indicates that the nervous transmission of the impulse through the myoneural junction can be absolved of the blame for fatigue. Furthermore, nerve fibers have been found to be practically indefatigable, and thus the site of fatigue would seem to be either in the muscle contractile mechanism or in the coupling of the action potential to the contraction process. Although older work seemed to indict the myoneural junction as the site of fatigue, at least in excised muscle, the finding of undiminished action potentials by Merton (when the muscle is no longer capable of response) is strong evidence to the contrary.

Our discussion refers only to *local muscular fatigue* in one muscle or in one functional muscle group. General fatigue of the entire organism must be considered from a different vantage point, much greater in scope. This will be discussed at greater length in succeeding chapters, which will consider fatigue in relation to work load, physical fitness, accumulation of metabolites, and depletion of energy stores.

Types of Contraction. *Muscular contraction* is an unfortunate term in that it implies shortening. In actuality, innervation of a muscle often results in an expenditure of energy to produce force that is used either in maintaining a held position or in opposing lengthening as a muscle resists a superior force or slows down the effect of gravity. The nomenclature for these various types of muscular contraction is not standard, and therefore, at least three pairs of terms are necessary for the study of exercise physiology.

The Physiology of Muscle Contraction

Isotonic versus Isometric Contraction. This is the most widely used terminology for differentiating between the shortening contraction, *isotonic* (one level of tension throughout the contraction), and the contraction of holding a position, *isometric* (one length throughout the contraction).

Isokinetic contraction is muscle contraction in which the muscle shortens at a constant velocity determined by instrumentation—utilizing servomechanism controls.

Phasic versus Static (or Tonic) Contraction. *Phasic* and *static* serve the same purpose and are virtually synonymous with isotonic and isometric, respectively.

Concentric versus Eccentric Contraction. *Concentric* and *eccentric* differentiate between shortening and lengthening types of contraction. *Concentric contraction* is synonymous with isotonic contraction. *Eccentric contraction,* however, refers to a situation in which a muscle is innervated and responds with an expenditure of energy to produce force that is less than the opposing outside force; although the muscle tries to shorten, it is actually lengthened during its contraction phase. Examples of this are use of the biceps brachii in letting the body down slowly from a chin-up and use of the inward rotators of the humerus in arm wrestling. The loser's inward rotators, although attempting to shorten, have been forced to lengthen by the superior force of the opponent.

It is customary to measure the work output of muscular contraction by the method of the physicist: $W = F \times D$; W is the work done, and F is the force acting through a distance, D. Thus the computation of the work load for isotonic or concentric contraction in lifting a 100-pound weight 2 feet is $W = 100 \times 2$, or $W = 200$ foot-pounds of work done. In isometric or static contraction, however, since no movement is involved and $D = 0$, no work is done, and all of the energy of muscular contraction goes into the development of heat. In eccentric contraction, where the 100-pound weight is slowly lowered 2 feet—through the forced stretching of muscles resisting the force of gravity—it is suggested that the same formula be used, and that this work load of 200 foot-pounds be termed negative work.

Mechanical Factors in Muscular Activity. For any given strength and nutritive condition of a muscle, there are at least three very practical considerations for physical educators and coaches regarding the external force that can be produced from that muscle:

1. The angle of pull of the muscle
2. The length of the muscle at any given time
3. The velocity of muscle shortening

Angle of Pull. The angle between the lengthwise axis of the muscle and the lengthwise axis of the bone it is causing to move is referred to as angle *a*. In the diagram (fig. 4.5), force triangles are drawn to show the relationship of the internal muscular force exerted (side *B*), which is held the same throughout figure 4.5, to the net force available to do the work (side *A*) and to the wasted force (side *C*). It is easily seen that there is an optimal value of the angle of pull that is closely approximated by the condition in 4.5, part 2. When the muscle pulls at right angles to the bone that it is moving, sides *A* and *B* will coincide, and all of the muscle's force becomes available to do useful work. At angles of pull greater or less than the optimal value, the wasted force, side *C*, becomes larger, and the externally available force, side *A*, becomes smaller for any given value of muscular effort (side *B*).

For example, the most difficult points in a chin-up seem to be at the very bottom and top of the exercise, which are represented by parts 1 and 3 of figure 4.5, respectively. Subjects who are able to start chinning themselves can probably get past the midpoint where the angle of pull is favorable, but will have difficulty stretching their necks over the bar at the top.

Length of Muscle. At any given time during contraction, the length of the muscle determines how much internal force or tension it can generate. It has been demonstrated that, in an isolated muscle fiber, the tension developed in a tetanus is maximum at the maximum resting length, and decreases with greater and with lesser lengths (24). The same general conditions have been observed in intact skeletal muscle (1).

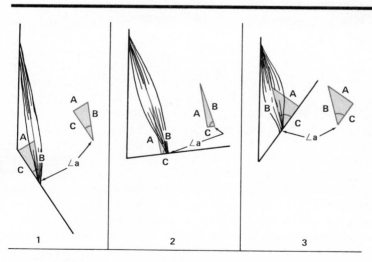

Figure 4.5 Effect of the angle of pull of a muscle upon the external force *A* provided for equal amounts of internal muscular force *B*. Side *C* in each case represents wasted internal force.

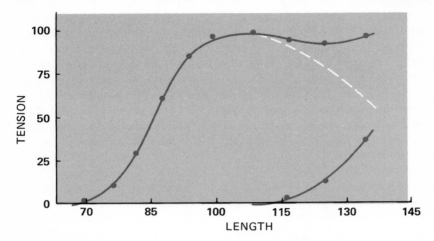

Figure 4.6 Relation between length and tension. One hundred is rest length and maximum tension. Monotonic increasing curve at lower right represents effect of passive stretch on tension. Upper solid curve is obtained from system as a whole. Broken line represents behavior of contractile elements when passive stretch curve is subtracted from upper curve. (From Zierler, K.L., Mechanism of muscle contraction and its energetics, in *Medical Physiology*, 13th ed., V.B. Mountcastle, ed., in press. Courtesy of The C.V. Mosby Company, St. Louis.)

The relationship between muscle length and tension is one of the few in physiology in which cellular structure and cellular function can be shown to be related. Figure 4.7 shows the results of ingenious experiments to define the length-tension relationship within a single sarcomere by Gordon, Huxley, and Julian (14). We can see clearly that the electronmicrographs of sarcomere length nicely support the sliding filament theory discussed in chapter 2. Thus at position 1, where few cross-bridges can interact, the developed tension is near zero. As the cross-bridges attain more binding sites, tension rises to near maximum at point 2. At points 3 and 4 the situation is still near optimal, and tension declines only slowly. Points 5 and 6, showing rapid decline in tension, are thought to result from two factors: (1) the extensive overlap of the actin and myosin filaments interferes with the formation of cross-bridges, and (2) the rigidity of the thick myosin filaments probably absorbs some of the force that is generated. Note the close agreement between the length-tension relationship in gross muscle (fig. 4.6) with the length-tension relationship in the single sarcomere as developed in figure 4.7. The small differences can probably be accounted for by the connective tissue present in the gross muscle.

Basic Physiology Underlying the Study of Physiology of Exercise

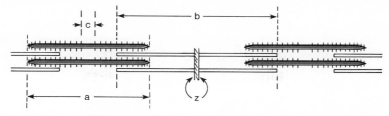

A Schematic diagram of filaments, indicating nomenclature for the relevant dimensions.

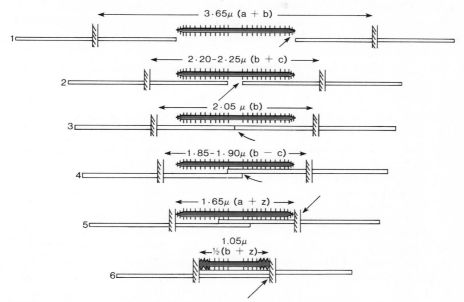

B Critical stages in the increase of overlap between thick and thin filaments as a sarcomere shortens.

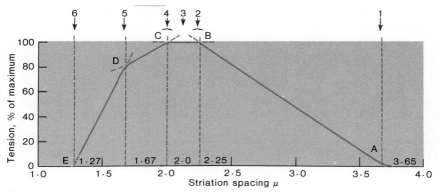

C Tension levels in sarcomere as striation spacing narrows.

Figure 4.7 Relationship of tension developed to the position of actin and myosin filaments within a sarcomere. (From Gordon, A.M.; Huxley, A.F.; and Julian, F.J. *J. Physiol.* 184:170–92, 1966.)

Thus the principle for physical education is, "Put the muscle on stretch to obtain the greatest force of which the muscle is capable." However, the stretch factor that provides the greatest internal forces may be working at odds with the angle of pull (described above), which determines how much of the internal muscle force will be externally available for useful work. Thus the intelligent analyst of athletic activity must consider the interaction of both factors to achieve the best results.

Velocity of Muscle Shortening. The speed with which a muscle shortens affects the external force available (fig. 4.8). It has been found that as the speed of shortening increases, the force decreases in exponential fashion (11). This would indicate that force falls off disproportionately as the speed of contraction increases, and this in turn would dictate an optimum speed for a needed amount of force.

A.V. Hill (15) believes the optimum speeds for muscular efficiency and power are approximately the same: 20%–30% of the maximum speed at which the muscle can shorten under zero load. Further research is needed to determine how these figures could be applied to power events, such as the shot put.

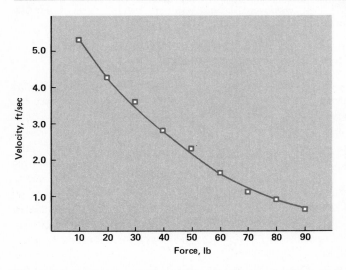

Figure 4.8 Force-velocity relationship in healthy young men. (From Damon, E.L. *An Experimental Investigation of the Relationship of Age to Various Parameters of Muscle Strength.* Doctoral dissertation, Physical Education, USC, 1971.)

Basic Physiology Underlying the Study of Physiology of Exercise

Improvement of Force and Work Delivery by Dynamic Stretching of Muscle. Cavagna, Dusman, and Margaria (4) have provided us with some new concepts in muscle physiology that have direct application to athletic performance in certain events. Having found that the efficiency of running was considerably better (40%) than that usually calculated for simple ergometer work (25%), they looked for the source of this increased efficiency in the possible storage of energy in elastic tissues during negative work, which could be released to aid the subsequent positive contraction. Thus in running, as the extended leg strikes the ground, the quadriceps is forced into lengthening to accept the weight of the body and in so doing accumulates elastic energy, which is immediately used to augment the following positive work of striding. They found that the maximal positive work (W') done by a muscle shortening immediately after being stretched is greater than the work (W) done when shortening from a state of isometric contraction, and further found that the improvement represented by the ratio W'/W increases with the speed of stretching and shortening and also with the length of the muscle. Not only is this true for the work produced, but the same principle holds true for the force production.

Thus for such events as the discus or the shot put the ultimate results will be better the longer and more rapid the preparatory windup. Such a windup also requires optimal development of flexibility, of course.

Electromyography in Analysis of Muscle Function

Basic Concepts of Electrical Phenomena and Electromyography. It has been known, at least since the middle of the nineteenth century, that the contraction of muscle tissue is accompanied by an electrical change that can be recorded and measured. The electrical change is called a *muscle action potential* (MAP), and the recording of muscle action potentials (or their currents) is called *electromyography* (EMG). The MAP arises at the muscle cell membrane (or sarcolemma) and passes lengthwise through the fiber in wavelike form, as the fiber is stimulated to contract. The science of recording and analyzing MAPs probably received its greatest impetus in the related science of *electrocardiography,* in which the events of the cardiac cycle are recorded on electrocardiograms (ECGs) and examined for abnormality. Physicians also use electromyography clinically in the diagnosis of various types of muscular diseases such as spasticity and paralysis.

Most EMG instrumentation and procedures have been developed for the use of physicians in diagnosing abnormal neuromuscular function and may be thought of as *qualitative* rather than *quantitative* in nature since the greatest concern is with the recording and analysis of the *wave form* of the MAP from single discrete motor units.

Quantitative electromyography is the study of the amount of electrical activity that is present in a given muscle under varying conditions. Obviously it is not electrical activity per se that is of interest here; rather it is the fact that EMG recordings accurately reflect muscle activation at a level of sensitivity at which palpation and other methods fail to produce evidence.

Investigators in exercise physiology and kinesiology (sometimes physical therapy and physical medicine) are mainly concerned with occurrences in the whole muscle rather than in isolated motor units. For this reason, surface electrodes on the skin over the belly or the motor point of the muscle are used. Thus the firing of many motor units is observed simultaneously, and a better statistical sampling is had than by the use of needle electrodes. This results, however, in a wave form that is a summation of the randomly organized activity of the many motor units observed, and it tells us nothing about any one motor unit. Also, the MAPs—in passing through the muscle tissue, fascia, subcutaneous fat, and skin—are severely attenuated (decreased in magnitude). Therefore, if we wish to know what is going on in a resting or relatively inactive muscle, we need recordings of extremely high sensitivity (the author has recorded meaningful differences in activity of resting muscles in which the difference between two conditions is less than 1 μv). The difference between recordings of individual motor units with needle electrodes and recordings of summated potentials from many motor units with surface electrodes is shown in figure 4.9.

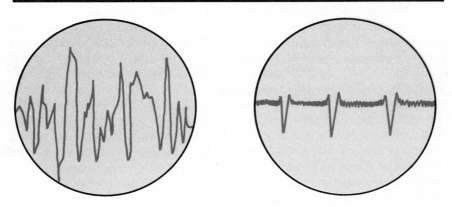

Figure 4.9 Differences between recordings of single motor-unit potentials with needle electrodes (right) and summated potentials from many motor units with surface electrodes (left).

It is important to realize that when many motor units fire randomly, some will, by chance, fire simultaneously, and the size of the wave form recorded at that point will be larger. Furthermore, as a muscle is required to produce more tension, recruitment of ever greater numbers of motor units occurs; by the laws of chance, more will fire simultaneously, and the wave form will grow larger in amplitude as more tension is produced.

The information to be gained from the pattern of motor unit spikes shown in the right-hand side of figure 4.9 is immediately obvious. We can count the spikes and measure the amplitude, but we have absolutely no quantitative knowledge of what the whole muscle is doing. We know only what is going on in one motor unit. As it turns out, the physiological significance of the electrical activity in the total gross muscle rests on the mean amplitude of the summated MAPs. The problem here is immediately recognized if we pose the question: What is the mean amplitude of the summated MAPs shown in the left side of figure 4.9? What do we measure to get a true mean value for a function that varies constantly and randomly over time as does the *interference pattern* (so called from electronics parlance) of figure 4.9? To do this requires integration (a procedure of the calculus). This concept of integration is simply depicted in figure 4.10. First, the integration process can be applied only to the electrically positive or negative halves of the spikes, because otherwise the end result would be zero, with the positive being balanced out by the negative

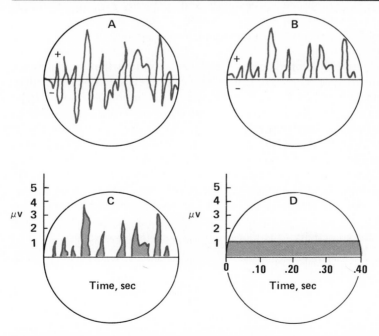

Figure 4.10 Integration of muscle action potentials to get true mean value of their amplitude.

aspects of the wave form. Thus, as the first step shown in figure 4.10 B, we have eliminated the negative swinging part of the spikes. Next, we must think of the spikes as having area, as in figure 4.10 C. The dimensions of this area are, vertically, microvolts of electrical activity (μv) and, horizontally, time in fractions of a second. The next step is to convert the area under the random spikes into a neat geometric figure such as a rectangle whose area equals height times length (or in this case μv times seconds). This conversion to the rectangle is accomplished by *planimetry* (tracing the curve with an engineering instrument that provides the area under any curve). Now we have the area of figure 4.10 D in an approximate rectangle, whose dimensions are a height of 1.0 μv and a time of 0.40 seconds, yielding an area of 1.0 $\times$ 0.40, or 0.40 μv second. This is the reading we would obtain from the planimeter. To get the mean μv level then we need only divide as follows.

$$\frac{\mu v \times sec}{sec} \text{ or } \frac{1.0 \ \mu v \times 0.40 \ sec}{0.40 \ sec} = 1.0 \ \mu v$$

Now we can say that over the 0.40-second observation period the mean amplitude of the MAPs was 1.0 μv. Fortunately electronic integrators are available that accomplish the whole procedure without even the necessity for planimetry.

Only through integration can accurate data regarding the physiological import of the electrical activity in the muscle be evaluated. Even such a tedious process as measuring spike amplitudes is of little value for precise data because the amplitude is constantly and randomly varying, and to get a true mean one must know over what period of time the spike has acted. A fat spike is more important than a thin spike of equal amplitude in determining the true mean value.

Equipment was developed for the author's laboratory that simplifies the integration procedure. This equipment will record integrated EMG potentials with a sensitivity of 0.3 μv. Using this equipment, the author and his colleagues accumulated evidence that the electrical activity from muscle tissue—even at the lowest levels—is indeed entirely similar to that at the higher levels. Figure 4.11 shows the high relationship between muscle tension in the elbow flexors and electrical output at the very lowest tension levels when all possible care has been taken to reduce electrode resistance to an essential minimum. We have also shown a highly significant correlation of 0.58 between resting oxygen consumption and the level of EMG in one representative muscle group (the elbow flexors). These two lines of evidence confirm that even at the very lowest levels of electrical activity, it is motor unit activity that is measured (10).

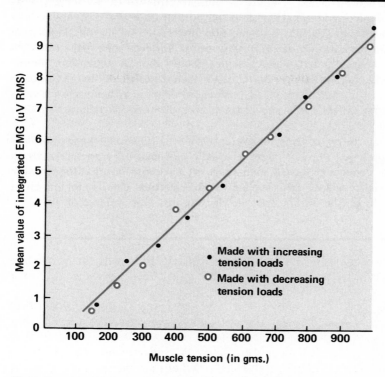

Figure 4.11 Integrated EMG output from elbow flexors as a function of muscle tension.

Applications of EMG to Physiological Problems.

Estimation of Tension Developed within a Muscle. Many situations arise in the physiology of exercise laboratory when it is desired to know the tension developed within a given muscle. It has been shown by Lippold (18) and co-workers that the EMG voltage is proportional to the force of contraction in isometric contraction and that in movements of constant velocity the electrical activity is proportional to the tension developed. In movements of constant tension the electrical activity is proportional to the velocity (constant velocity) (3). This work has been extended in our laboratory to include accelerated movements (5), and it was found that even here the electrical activity is proportional to the *effort impulse value*, a measurement of effort suggested by Starr (26). Thus we may conclude that in any kind of physical activity, a reasonably good estimate of the tension or effort developed within the muscle can be obtained through appropriate EMG instrumentation and techniques (integration is necessary, of course). This would be difficult if not impossible to attain by any other means.

Estimation of Strength.   Where the musculature is relatively normal, measurement of maximal strength by cable tension, strain gauge, or dynamometer provides one dimension that is unquestionably related to the functional state of the tissue. As a *physiological measurement,* however, maximal strength is notoriously contaminated by psychological factors such as motivation. Muscle tonus is another dimension for evaluating muscle function. Although this term is widely used by members of all professions dealing with muscle tissue, a definition that satisfactorily encompasses all experimental evidence available does not as yet exist.

A very interesting concept for evaluation of the functional state of muscle has been proposed (8, 12) in terms of the *efficiency of electrical activity* (EEA). The concept is illustrated in figure 4.12. It is obvious that the stronger individual (flatter slope) needs less activation (electrical activity) for any given muscle loading. The author has shown that this EMG slope, or EEA, as

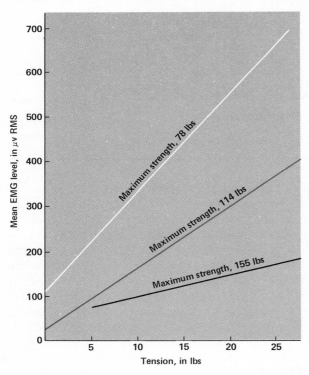

Figure 4.12 Electrical activity in a muscle as a function of the force of contraction. Note the differences on the rate of increase in activity between subjects of varying levels of strength.

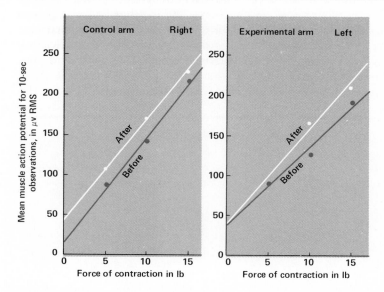

Figure 4.13 Changes in the experimental and control elbow flexors as the result of placing the left elbow flexors in a cast for four weeks. *Left:* right biceps (control arm). *Right:* left biceps (experimental arm). (From deVries, H.A. *Am. J. Phys. Med.* 47:10, 1968.)

plotted in figure 4.12 is well related (in young subjects) to measured strength ($r = -.75$ to $-.90$). The EEA is also sensitive to hypertrophy and atrophy of muscle through use and disuse as shown in figures 4.12 and 4.13.

EMG Evaluation of Hypertrophy versus Neural Factors in the Time Course of Strength Gain. Moritani and deVries (20) recently developed a method for separating the effects of (1) strength "learning" through improvement of neural factors and (2) true muscular hypertrophy. In figure 4.14 hypertrophy is reflected in a flatter slope of the EMG voltage-force relationship because fewer available motor units need be recruited at any given load. On the other hand, learning is reflected not in change of slope, but in the innervation of more motor units, with an accompanying increase in the maximum EMG voltage.

Using this method, the researchers showed that young men training with weights achieve the largest part of the strength gain through learning to better innervate the muscle for the first three weeks, after which time the gains are largely attributable to hypertrophy (fig. 4.15).

EMG Estimation of Endurance—Fatigue Parameters. When a muscle contracts isometrically against constant force, the electrical activity in that muscle increases with time as shown in figure 4.16. This phenomenon is thought to be the result of the fatigue process impairment of muscle fiber function so that recruitment must take place to compensate for the constantly decreasing force available per fiber. If the fatigue process is brought about quickly with loads of 30%–50% maximal voluntary contraction (MVC), the

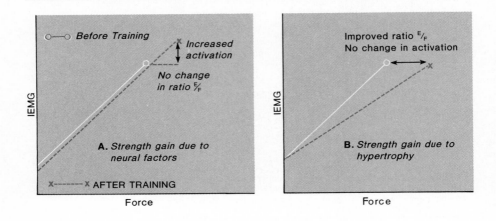

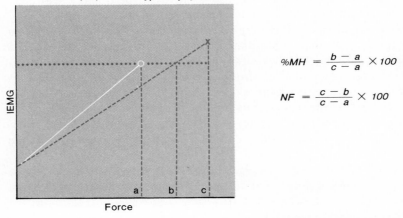

$$\%MH = \frac{b - a}{c - a} \times 100$$

$$NF = \frac{c - b}{c - a} \times 100$$

Figure 4.14 Schema for evaluation of percentage contributions of neural factors (NF) and muscular hypertrophy (MH) to the gain of strength through progressive resistance exercise. *E/F* ratio is EMG voltage (IEMG) to force relationship. (From Moritani, T., and deVries, H.A. *Am. J. Phys. Med.*, 58:115–130, 1979.)

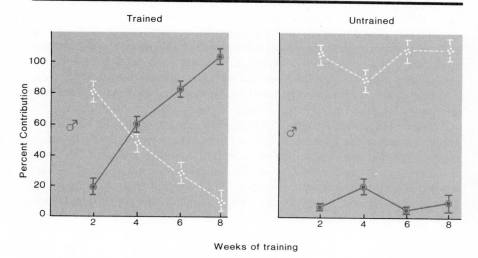

Trained

Untrained

Figure 4.15 The time course of strength gain showing the percentage contributions of neural factors (•‡•————————•‡•) and hypertrophy (◉————————◉) in the trained and untrained arms of two male subjects. (From Moritani, T., and deVries, H.A. *Am. J. Phys. Med.*, 58:115–130, 1979.)

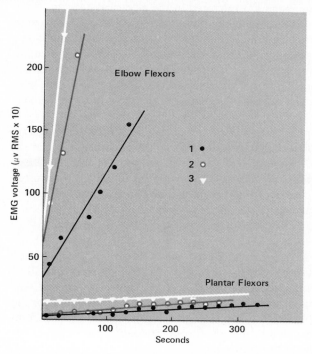

Figure 4.16 EMG fatigue curves in three test runs on the same subject with increasing fatigue of elbow flexors and plantar flexors. (From deVries, H.A. *Am. J. Phys. Med.* 47:175, 1968.)

plot of EMG voltage as a function of time usually approximates linearity (later data in the laboratory show the curve to be an exponential with a considerable period of approximate linearity). The slope of this curve can be used under appropriate conditions for estimating the rate at which muscle fatigues (9). Test-retest reliability of such curves was found to be $r = .93$. The validity of this approach was evaluated by correlating the slope coefficient of the fatigue curve against measured maximal endurance time, and the correlations were found to be $r = .79$ and .82 using isometric tensions of 40% and 50% of MVC, respectively. The important point to remember here is that in plotting either the EEA curve or the fatigue curve only a few data points are required that can be obtained without the subject going beyond 25%–50% of maximum capacity in either case. Obviously, this eliminates the necessity for the usually invalid assumption that the investigator has obtained a true maximum effort from a subject under testing conditions since motivation and other psychological factors are no longer pertinent.

Note in figure 4.16 the dramatic difference in the rate at which flexor muscles and the extensor muscles involved in posture show fatigue. This undoubtedly reflects the fiber type make-up of these muscles. The flexors generally are composed of more FG fibers, which are easily fatigued (see chaps. 2 and 3).

Observation of Other Physiological Phenomena by EMG. Many interesting physiological phenomena are mirrored in the electrical activity of the skeletal muscles. One of the more easily observed is that of the increase in electrical activity in irrelevant muscles when any one skeletal muscle is strained. Thus, if we load the right elbow flexor group to 50%–75% or more of MVC, we can detect measurable activity of at least five μv in the previously inactive left elbow flexors. The author has shown that even straining the eyes to read fine print raises the electrical activity in the elbow flexors by about 100% (6).

Another interesting phenomenon that the author routinely demonstrates for his students is the effect of hyperventilation upon the activity of irrelevant forearm muscles. The *blowing off* of CO_2 as discussed in later chapters raises pH and increases irritability of muscle tissue, which is displayed by twofold to fivefold increases of electrical activity (best shown in the most peripheral flexor muscles). This, of course, is the reason for the commonly observed tendency of the fingers to curl (flex) toward a fist during moderate to severe hyperventilation.

Another use which has been made of this type of EMG is the observation of the onset of the *shivering response*. It must be noted, however, that all of the interesting phenomena alluded to in this section require an order of sensitivity beyond that which is ordinarily available in apparatus designed for the physician because the physician doesn't need it. Furthermore, without integration procedures, even a change in wave form that doubles the energy is not always easily discernible on the oscilloscope.

Muscle Tonus. The classic experiments of Sherrington (25) on stretch reflexes resulted in the theory that muscle tonus was the result of a partial tetanus of muscle tissue in which there is a *constant* state of activity by a small portion of the muscle fibers. This constant state of activity was thought to be the result of *myotatic* (stretch) reflexes that originated in the muscle spindles and reflexly innervated the alpha motor neurons (chap. 5). Forbes (13) speculated that this muscle tonus was brought about not by constant contraction of the same fibers, but by a *rotation of duty* among many different motor units. Thus skeletal muscles were thought never to be in a state of complete rest.

By these concepts, postural tonus was supposed to result from constant feedback from a postural muscle that was stretched by slight swaying, which deformed the muscle spindles and brought about afferent impulses to the spinal cord that reflexly innervated the motor units of the stretched muscle to contract. The resulting contraction checked the sway, and balance was restored.

However, with the advent of electromyographic equipment of ultrahigh sensitivity pioneered by Jacobson (16) with the help of Bell Telephone Laboratories, it was found that skeletal muscles in relaxed subjects (supine or sitting) could be electrically silent for minutes at a time. Since it was well known that any muscular contraction, no matter how small, is accompanied by MAPs, this required modification of the traditional concept of muscle tonus. The findings of electrical silence in relaxed, resting muscles in man have been corroborated by many other laboratories (although not always with the same rigorous level of sensitivity). Thus Jacobson (16, p. 229), as early as 1938, suggested redefining muscle tonus in skeletal muscles as ". . . a state of slight contraction, more or less constant, or irregular, often present, but *sometimes absent* in health."

In the light of recent research, the most acceptable definition of general muscle tonus is that of Basmajian (2, p. 41):

. . . the general tone of a muscle is determined both by the passive elasticity or turgor of muscular (and fibrous) tissues and by the active (though not continuous) contraction of muscle in response to the reaction of the nervous system to stimuli. Thus, at complete rest, a muscle has not lost its tone even though there is no neuromuscular activity in it.

Resting Muscle Tonus. Over a period of many years, since electromyographic recording of MAPs has become a common laboratory procedure, many investigators have reported electrical silence in relaxed, resting muscles. However, it is an obvious truism that to demonstrate the absence of a physiological phenomenon, it is necessary to use equipment that has the capacity to render observable any and all phenomena under consideration. Obviously, the use of a magnifying glass to observe bacteria might result in a false conclusion that no such organisms exist. It must always be considered a more hazardous conclusion to show the absence of an event than its presence.

Seen in this light, most evidence has not been conclusive because of deficiencies in equipment. Investigators have at times been satisfied that electrical silence existed when their sensitivity did not allow detection of anything under fifty microvolts. Furthermore, the use of needle electrodes to find electrical silence is akin to sampling the opinion of only a few persons to predict the outcome of a national election. Obviously, the smaller the sampling, the greater the potential error in the conclusion.

Nevertheless, electrical silence has been demonstrated by a few investigators such as Jacobson, whose work is unimpeachable. The conclusion that muscles can indeed be entirely inactive rests upon their work. The author has found that, without special training, ten of twenty-nine physical education students could achieve complete relaxation (electrical silence) in the right quadriceps femoris muscle for a period of two minutes. Under similar conditions, only four of the twenty-nine could achieve electrical silence in the right elbow flexors. Jacobson has demonstrated that the ability to relax voluntarily can be improved by training.

Nothing that has been said should be construed as denying the activity of the neuromuscular system as a participant in general muscle tonus. Electrical silence is achieved only by relaxed subjects under the best laboratory conditions. How often, how long at a time, or in how many muscles this may occur in everyday activity is quite another question. The best we can say at present is that it is possible for some persons to relax selected muscles completely on some occasions and under certain conditions. It is felt that the definition for general muscle tonus stated by Basmajian, and referred to earlier, is the best summation of the present state of our knowledge.

Postural Muscle Tonus. There is little agreement upon the mechanisms of postural tonus, and, more specifically, upon the presence or absence of neuromuscular activity in the extensor muscles involved in maintaining the erect posture in man.

In a comprehensive review of the work done in this area, Ralston (22) and Ralston and Libet (23) concluded that during easy standing there is no electrical activity, even in such postural muscles as the erector spinae and lower leg muscles unless the subject sways sufficiently.

It seems rather difficult to accept the hypothesis of "hanging on the ligaments" as it were, in anything that approaches normal erect posture. Therefore, the author reinvestigated this question with highly sensitive equipment (7). Highly significant differences were found in muscular activity in the lower leg between resting and easy standing (standing in a very relaxed fashion). All subjects showed a marked increase in activity in two lower leg muscles when they changed from the resting (seated position) to the standing position.

Furthermore, no convincing evidence of electrical silence was found, even for brief periods, in either muscle in any subject. This agrees completely with the work of Jacobson (17), the only other investigator whose equipment achieved the same sensitivity. It would seem that those who found electrical silence in these muscles did so because of deficiencies in equipment.

As a consequence of these experiments (7), it would seem that the traditional concepts of postural tonus are still tenable and that the more recent research that purports to find electrical silence in postural muscles rests on insufficient evidence from equipment of insufficient sensitivity.

Effects of Exercise upon Resting Muscle Tonus. It has long been thought that one of the benefits of exercise is the improvement of muscle tonus. It has also been observed that the lack of exercise from having a cast on a fractured leg or from forced bed rest results in a degree of flaccidity. Although these statements are probably true, they seem to be unsupported by experimental evidence at this time.

Physiology of Muscle Tonus. The most logical explanation of muscle tonus, based upon the research cited above; would seem to rest upon a *passive component* and an *active component*.

Passive Component. The passive component is composed of the elasticity of muscle and connective tissues, plus the tissue turgor, which may be defined as the pressure with which body fluids tend to distend their surrounding tissues. This factor is present regardless of the state of innervation.

Active Component. The active component rests on the gamma loop, to be discussed in chapter 5 and is hypothetical in large part. The most logical explanation that encompasses all experimental results is as follows.

1. Muscle tonus rests on a reflex basis.
2. The afferent limb of the reflex arises in the receptor of the muscle spindle and enters the spinal cord, where it synapses directly with the alpha ventral horn cells.
3. A constant facilitation effect is brought about by impulses from subcortical nuclei, reticular substance, and vestibular and cerebellar pathways.
4. Firing of the alpha ventral horn cells innervates a small proportion of the motor units in all muscles concerned with posture, probably (although not demonstrably) in rotating fashion, as postulated by Forbes (13).
5. In muscles other than the postural muscles, and if the subject is sufficiently relaxed, complete inactivity may exist at least for short periods of time.

SUMMARY

1. The *myogram* illustrates the contraction of a whole muscle (muscle twitch) when it is artificially innervated by a single stimulus. Its response consists of a *latent period* of 0.001–0.01 second, a *contraction phase* of approximately 0.04 second, and a *relaxation phase* of approximately 0.05 second.

2. If, during one *muscle twitch,* a second stimulus occurs, a *summation of contractions* will occur that results in greater tension than either twitch separately.

3. A *tetanus* or *tetanic contraction* results from repeated stimulation of muscle when the intervals between stimuli are too short to allow complete relaxation. If *no* relaxation occurs between stimuli, it is a complete tetanus, and the tension developed may be three to four times that of a single twitch.

4. Low muscle temperatures result in slow contraction and relaxation phases, and this effect is relatively much greater in the relaxation phase. High temperatures have a converse effect, and these two facts have important implications for warm-up and the prevention of muscular distress.

5. The *all-or-none law* states that a muscle fiber (or motor unit) contracts either maximally, or not at all, in response to a single stimulus.

6. *Gradation of response* in muscle tissue is brought about by two methods: (a) innervation of varying numbers of motor units, and (b) variation of the frequency of stimulation.

7. The *site of local muscular fatigue* seems to be peripheral and probably involves either the contractile mechanism itself or the coupling of the muscle action potential (MAP) to the contractile mechanism—or both.

8. The *external muscular force* available for useful work is the result of three component factors: (a) the angle of pull of the muscle, (b) the length of the muscle, and (c) the velocity of shortening.

9. Whenever a muscle is innervated, one or more MAPs are produced, which can be recorded and measured by *electromyography* (EMG).

10. By recording MAPs, muscle activation may be measured quantitatively. The author refers to this use of EMG as *quantitative EMG,* in contradistinction to the methods and instrumentation commonly applied by the physician who is usually more concerned with the observation of the *wave form* of MAPs from single motor units.

11. When surface electrodes are used in EMG, the *interference pattern* observed on the oscilloscope is the result of the summation of several or many different motor units firing randomly in time.

12. To make meaningful use of EMG data from the interference pattern requires integration procedures to calculate true mean amplitudes. Integration may be accomplished by *planimetry* or by electronic integrators.

13. That the integrals so achieved (or the true mean amplitudes calculated from them) have physiological significance has been well verified in various laboratories.

14. The methods of quantitative EMG have been applied to many aspects of exercise physiology:

    a. It has been shown that the tension developed within a muscle can be estimated (i) under isometric tension, (ii) under constant velocity movement, and (iii) under accelerated movements.

    b. Strength and functional quality of muscle tissue can be evaluated in young subjects by the slope of the EMG voltage regression upon force of contraction.

    c. The endurance (or fatigue rates) of muscles can be objectively evaluated by the slope of the EMG voltage regression upon time when the subject holds an isometric contraction of 40%–50% maximal voluntary contraction.

    d. Highly sensitive EMG instrumentation also renders visible the reactions of muscle tissue to such homeostatic displacements as occur during hyperventilation.

15. Resting muscle tonus is the result of *active* and *passive components*. The active component is due to neuromuscular activity and is not always present, whereas the passive component is due to tissue pressure and the elastic quality of muscle and connective tissue and is continuous during the life of the organism.

16. Postural muscle tonus has been shown to have both components in all muscles that are involved in maintaining the erect posture. Investigations that have failed to show the active component probably suffered from deficiencies in equipment.

REFERENCES

1. Banus, M.G., and Zetlin, A.M. The relation of isometric tension to length in skeletal muscle. *J. Cell. Comp. Physiol.* 12:403–20, 1938.
2. Basmajian, J.V. *Muscles Alive.* Baltimore: Williams & Wilkins Co., 1962.
3. Bigland, B., and Lippold, O.C.J. The relation between force, velocity, and integrated electrical activity in human muscles. *J. Physiol.* 123:214–24, 1954.
4. Cavagna, G.A.; Dusman, B.; and Margaria, R. Positive work done by a previously stretched muscle. *J. Appl. Physiol.* 24:21–32, 1968.
5. Damon, E.L. An experimental investigation on the equating of isometric, concentric, and eccentric muscular efforts. In preparation.
6. deVries, H.A. Neuromuscular tension and its relief. *J. Assoc. Phys. Ment. Rehabil.* 16:86–88, 1962.
7. ———. Muscle tonus in postural muscles. *Am. J. Phys. Med.* 44:275–91, 1965.

8. ———. Efficiency of electrical activity as a physiological measure of the functional state of muscle tissue. *Am. J. Phys. Med.* 47:10–22, 1968a.

9. ———. Method for evaluation of muscle fatigue and endurance from electromyographic fatigue curves. *Am. J. Phys. Med.* 47:125–35, 1968b.

10. deVries, H.A.; Burke, R.K.; Hopper, R.T.; and Sloan, J.H. Relationship of resting EMG level to total body metabolism with reference to the origin of tissue noise. *Am. J. Phys. Med.* 55:139–47, 1976.

11. Fenn, W.O., and Marsh, B.S. Muscular force at different speeds of shortening. *J. Physiol.* 85:277–97, 1935.

12. Fischer, A., and Merhautova, J. Electromyographic manifestations of individual stages of adapted sports technique. In *Health and Fitness in the Modern World,* chap. 13. Chicago: The Athletic Institute, 1961.

13. Forbes, A. Spinal reflexes. *Physiol. Rev.* 2:401, 1922.

14. Gordon, A.M.; Huxley, A.F.; and Julian, F.J. The variation in isometric tension with sarcomere length in vertebrate muscle fibres. *J. Physiol.* 184:170–92, 1966.

15. Hill, A.V. The design of muscles. *Br. Med. Bull.* 12:165–66, 1956.

16. Jacobson, E. *Progressive Relaxation.* Chicago: University of Chicago Press, 1938.

17. ———. Innervation and tonus of striated muscle in man. *J. Nerv. Ment. Dis.* 97:197–203, 1943.

18. Lippold, O.C.J. The relation between integrated action potentials in a human muscle and its isometric tension. *J. Physiol.* 117:492–99, 1952.

19. Merton, P.A. Problems of muscular fatigue. *Br. Med. Bull.* 12:219–21, 1956.

20. Moritani, T., and deVries, H.A. Neural factors vs. hypertrophy in the time course of muscle strength gain. *Am. J. Phys. Med.* 58:115–130, 1979.

21. Petajan, J.H., and Eagan, C.J. Effect of temperature and physical fitness on the triceps surae reflex. *J. Appl. Physiol.* 25:16–20, 1968.

22. Ralston, H.J. Recent advances in neuromuscular physiology. *Am. J. Phys. Med.* 36:94–120, 1957.

23. Ralston, H.J., and Libet, B. The question of tonus in skeletal muscle. *Am. J. Phys. Med.* 32:85–92, 1953.

24. Ramsey, R.W., and Street, S. The isometric length-tension diagram of isolated skeletal muscle fibers of the frog. *J. Cell. Comp. Physiol.* 15:11–33, 1940.

25. Sherrington, C.S. Integrative action of the nervous system. New Haven: Yale University Press, 1923.

26. Starr, I. Units for the expression of both static and dynamic work in similar terms and their application to weight lifting experiments. *J. Appl. Physiol.* 4:21–29, 1951.

27. Tuttle, W.W. The effects of decreased temperature on activity of intact muscles. *J. Lab. Clin. Med.* 26:1913–15, 1941.

28. ———. The physiologic effects of heat and cold on muscle. *Athletic J.* 24:45, 1943.

29. Wilkie, D.R. The mechanical properties of muscle. *Br. Med. Bull.* 12:177–82, 1956.

5

The Nervous System
and Coordination
of Muscular Activity

The nervous system may be divided in two principal ways. First, we may think of it as divided structurally into *central* and *peripheral* components. The central component consists of the brain and spinal cord, and the peripheral component consists of all the ganglia (groups of nerve cells not in the spinal cord) and nerve fibers (axons). Second, it is divided functionally into the *somatic* and *autonomic* systems, and both systems have central as well as peripheral components. The autonomic system controls the internal environment, and innervates the smooth muscles of the gastrointestinal tract, the blood vessels, and so on, as well as the endocrine glands.

The autonomic system has two divisions: (1) the *sympathetic division,* whose central outflow is from the thoracic and lumbar regions of the spinal cord, and (2) the *parasympathetic division,* which originates in the cranial nerves and in the sacral region of the spinal cord. In a very general way, these two divisions of the autonomic system are antagonistic and balance each other. The sympathetic has to do with "fight or flight" adjustments of the organism to a dangerous situation, to the stress of sport, and other tension-producing activities. The parasympathetic, on the other hand, is related in a general way to vegetative functions, such as digestion.

The *somatic system* consists of the central and the peripheral components of the nervous system that have to do with peripheral reception of nervous impulses, conduction to the spinal cord or brain, organization of motor patterns, and conduction back to the skeletal muscles to bring about the desired movements.

The Neuron and the Motor Unit

The *neuron,* a single nerve cell, forms the basic structural unit of the nervous system; it is specialized for its function by having a high degree of irritability and conductivity. There are billions of neurons in the nervous system, and usually several neurons are interconnected by *synapses* to form pathways for conduction of nervous impulses. The neurons that conduct sensory impulses from the periphery to the central nervous system are called *sensory* or *afferent* neurons; the neurons that conduct impulses from the central nervous system to the muscles and other effectors are called *motor* or *efferent* neurons. Although neurons are microscopic in width, one cell may extend in length from the cerebral cortex almost to the caudal end of the spinal column, or an equal distance from the spinal cord to a muscle in the foot, a distance of approximately three feet.

The typical motor neuron (fig. 5.1) has two types of processes from its cell body: (1) *dendrites,* which receive impulses and conduct them to the cell body, and (2) an *axon,* which conducts the impulses away from the cell and accounts for the great length of some neurons. The cell body of the motor neuron, which innervates skeletal muscle, lies in the gray matter in the ventral horn of the spinal cord, and its axon joins many axons from other motor neurons (and

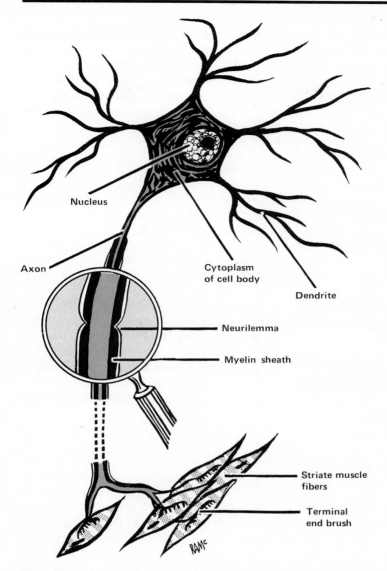

Figure 5.1 A diagrammatic drawing of a neuron. At the top is the cell body and its numerous branchings, the dendrites. They make up the soma of the neuron. The axon, of which there is only one, extends downward. The point at which the axon leaves the soma is the axon hillock. Axons, and sometimes dendrites, may be covered with a myelin sheath, and, outside the nervous system, with a neurilemma. (From Morgan, C.T., and Stellar, E. *Physiological Psychology,* 1950. Used by permission of McGraw-Hill Book Company.)

many sensory axons) to form a spinal nerve. This spinal nerve is thick enough to be seen and handled grossly in dissection.

The motor nerve, after entering the muscle through the epimysium, branches and rebranches until one axon enters a fasciculus. The axon then branches into many twigs, each of which innervates one muscle fiber. Thus the cell body in the ventral horn, plus the axon and all its twigs, together with the many muscle fibers innervated by the twigs form one *motor unit.*

The Reflex Arc and Involuntary Movement

A great deal of muscular activity is accomplished by reflex control. A *reflex* is most simply defined as an involuntary motor response to a given stimulus. An illustration is the automatic, unthinking response to touching a hot surface. In its simplest form, a reflex consists of a discharge from a sensory nerve ending, called a *receptor* or *sensory end organ,* whose impulses are propagated over the sensory nerve fiber to a *synapse,* or junction, in the spinal cord with a motor neuron. When the motor neuron is stimulated to discharge, impulses are propagated over its axon to the *effector* (muscle or gland), bringing about the reflex response. The sensory pathway is *afferent* and the motor pathway is *efferent*.

This simple reflex arc (fig. 5.2) occurs in the *myotatic* or stretch reflex, and because there is only one synapse in the cord, it is called a *monosynaptic reflex*. Other reflexes, such as the *flexion reflex* (used in removing the hand from a hot surface), involve at least three neurons and two synapses. These are called *disynaptic reflexes* for three or more neuron arcs, or *multisynaptic* for four or more neuron arcs (fig. 5.3). The short neuron that serves to connect the sensory and motor neurons in this case is called an *internuncial* or *association* neuron.

The preceding discussion is an oversimplification of what actually occurs. For example, so that the flexion reflex might proceed with dispatch, the antagonistic extensor muscle group must be prevented from acting. This inhibition of the antagonists is accomplished by kinesthetic impulses from the muscle spindles of the flexor muscle and these synapse, not only with the flexor motor neurons directly, but also indirectly with the motor neurons of extensor muscles of the same joint. This process is called *reciprocal inhibition*.

Applying this same flexion reflex to the foot (in stepping on a sharp object), we find that not only does the reflex result in withdrawal of the foot with its concomitant reciprocal inhibition of the extensor muscle of the same leg, but it may also result in the extension of the contralateral (opposite) leg to support the body during the flexion reflex. This action of the contralateral limb is called the *crossed extensor reflex*; it is the result of *facilitation* that is also brought about by action of the afferent neuron from the muscle spindles, but upon the motor neuron of the extensor muscles of the opposite limb.

This is still not a complete account of even these simple reflexes, and the interested reader is referred to a text in neurology for further discussion. Even

Basic Physiology Underlying the Study of Physiology of Exercise

the simplest reflex involves organization and integration of responses into meaningful movement patterns.

Knowledge of the reflex mechanism of the stretch reflex will greatly facilitate the student's understanding of the scientific principles involved in recent research in flexibility and in the muscle soreness phenomenon, which are discussed in part 3.

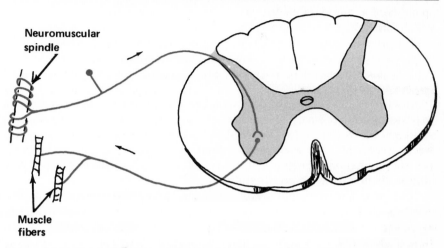

Figure 5.2 Diagram of a two-neuron reflex—from a spindle in a muscle back to other fibers of the same muscle. Arrows indicate direction of conduction. (From Gardner, E. *Fundamentals of Neurology,* 1958. Courtesy of W.B. Saunders Company, Philadelphia.)

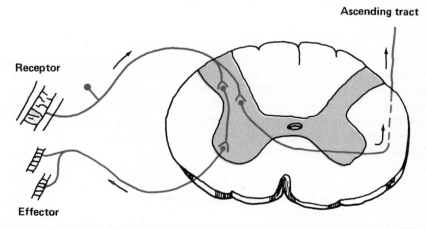

Figure 5.3 Diagram illustrating how impulses from a cutaneous receptor reach an effector (skeletal muscle) by a three-neuron arc at the level of entrance. (From Gardner, E. *Fundamentals of Neurology,* 1958. Courtesy of W.B. Saunders Company, Philadelphia.)

The Nervous System and Coordination of Muscular Activity

Intersegmental and Suprasegmental Reflexes

Reflexes are not necessarily limited to causing action at the same spinal cord level as the incoming afferent stimulus. For example, in the scratch reflex of the dog, the stimulus (a fleabite, for example) may occur at a level where the afferent part of the reflex arc enters the cord in the upper thoracic area, and the response is brought about by the hind leg muscles, whose efferents leave the cord at the sacral level. The association neuron travels down the dog's spinal cord a considerable distance before synapsing with the motor neuron, which forms the efferent pathway to bring about the scratching response. This is illustrated in figure 5.4, which also shows the possibilities for contralateral responses as well as upward or downward conduction pathways in the spinal cord. Reflexes that involve more than one spinal segment (or level) are called *intersegmental reflexes.*

So far we have considered only spinal reflexes. When attention is directed to postural mechanisms, it is seen that, although these do not require conscious volitional control, they are abolished in the absence of the centers and pathways of the brain. For example, in a faint, or in anesthesia, postural reflexes fail. The spinal or segmental portion of the reflexes involved in posture may still be elicited with the proper stimulus.

Thus it is apparent that higher brain centers are necessary for postural reflexes, and for this reason we refer to this group of reflexes as *suprasegmental reflexes.* The stretch or myotatic reflex forms the basic component of the postural reflexes, but it depends upon facilitation from higher centers in order that incoming sensory impulses may achieve threshold value, and thus bring about total postural responses. Furthermore, higher brain centers seem to be necessary to coordinate and control these mechanisms.

Proprioception and Kinesthesis

For the optimal coordination of motor patterns to take place in the brain and spinal cord, a constant supply of sensory information must be available to feed back the results of movement as it progresses (26). This feedback of sensory information about movement and body position is termed *proprioception.* The receptors for proprioception are of two types: vestibular and kinesthetic.

The *vestibular receptors* are found in the nonauditory labyrinths of the inner ear. Each of these two labyrinths, one on each side in the temporal bone of the skull, consists of a small chamber, the *vestibule,* which communicates with three small canals known as *semicircular canals* (fig. 5.5). Within the semicircular canals is a fluid called *endolymph.*

The inertia of the endolymph, which results in its remaining stationary at the first part of a movement of the body and also its continued movement when the body has returned to a resting position, causes disturbance of a sensory receptor organ, the *crista.* This disturbance is transmitted to the brain

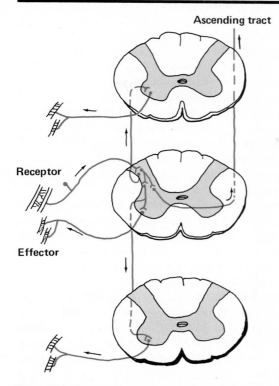

Ascending tract

Receptor

Effector

Figure 5.4 Diagram of the connections by which impulses from a receptor reach motor neurons at different cord levels. (From Gardner, E. *Fundamentals of Neurology*, 1958. Courtesy of W.B. Saunders Company, Philadelphia.)

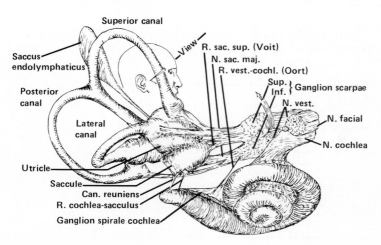

Superior canal

Saccus endolymphaticus

Posterior canal

Lateral canal

Utricle

Saccule

Can. reuniens

R. cochlea-sacculus

Ganglion spirale cochlea

View

R. sac. sup. (Voit)

N. sac. maj.

R. vest.-cochl. (Oort)

Sup. } Ganglion scarpae
Inf. }

N. vest.

N. facial

N. cochlea

Figure 5.5 Drawing showing structural relations of innervation of the human labyrinth (by Max Brodel). (From *The Anatomical Record* 59:403–18, 1934.)

by way of the vestibular branch of the eighth cranial nerve. Thus the sensory information from the cristae of the semicircular canals provides the data regarding movement, more specifically, rotational acceleration or deceleration of movement as in twisting or tumbling. Movement in itself is not recognized; for example, moving at almost the speed of sound in an airliner produces no sensation, unless a change of direction or velocity occurs.

Along with the semicircular canals, the *utricle* and *saccule* of the vestibule complete the vestibular system. Apparently, the saccule has little if any function in equilibrium or position sense; it seems to be involved in sensory perception of vibration only. The utricle, however, is the sense organ that provides the data necessary for positional sense. The *otolith organ* in the utricle responds to linear acceleration and to tilting, and thus seems to be the source of data that inform us of our posture in space.

Although no sensation of movement is engendered in the body of a passenger in an airliner (in smooth air), the *orientation* of the body in space, as in standing upright or lying down, is clearly recognized, even with the eyes closed. This recognition of spatial orientation is the result of interpretation of sensory information received from the otolith organ in the utricle.

The kinesthetic sense and its receptors are of even greater interest. At least five types of receptors serve the muscle sense, or *kinesthesis*: (1) the muscle spindle,* (2) the Golgi tendon organ, (3) the pacinian corpuscle, (4) Ruffini receptors, and (5) free nerve endings. It has long been known that these receptors provide the individual with muscle sense. The individual can tell what his limbs or body segments are doing at any given time without having to look. For instance, the normal individual has no difficulty making accurately controlled movements, such as bringing his finger from arm's length to touch the end of his nose, even when blindfolded. Furthermore, he can usually make reasonably accurate guesses as to the weight of an object by lifting it (see fig. 5.6).

*Spindle afferents, although extremely important in their sensory input to reflex behavior, do not contribute to subjective sensations.

Figure 5.6 Schema of the innervation of mammalian skeletal muscle based on a study of cat hindlimb muscles. Those nerve fibers shown on the right of the diagram are exclusively concerned with muscle innervation; those on the left also take part in the innervation of other tissues. Roman numerals refer to the groups of myelinated (I, II, III) and unmyelinated (IV) sensory fibers; Greek letters refer to motor fibers. The spindle pole is cut short to about half its length, the extracapsular portion being omitted. blood vessel (b.v.); capsule (c.); epimysium (epi.); extrafusal muscle fibers (ex M.F.); nuclear-bag muscle fiber (n.b.m.f.); nuclear-chain muscle fiber (n.c.m.f.); nodal sprout (n.s.); motor end plate; primary ending; p_1, p_2, two types of intrafusal end plates; perimysium; paciniform corpuscle; secondary ending; trail ending; vasomotor fibers. (From Barker, D. In *Handbook of Sensory Physiology*, 1973, vol. III, pt. 2. Courtesy of Springer-Verlag, Berlin, Heidelberg, New York.)

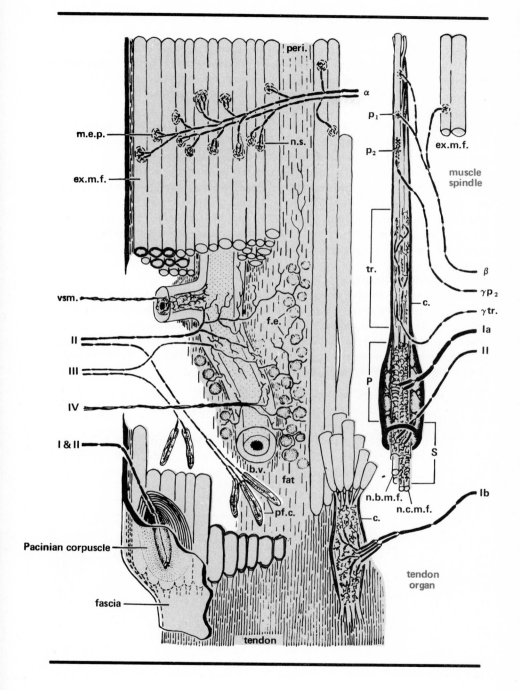

This muscle sense or kinesthesis, as it is properly called, is important in providing sensory information for skilled movement. However, recent research indicates that this system, particularly the muscle spindle and its attendant nerve supply, may be even more important in functioning as *autogenetic governors* of motor-nerve activity (7). The muscle spindles have their own motor-nerve system, which is important enough to constitute about one-third of the total efferent fibers that enter skeletal muscle (16).

Therefore let us consider the muscle spindle first, and in some detail (fig. 5.7). The spindle is large enough to be visible to the naked eye, and it is fusiform, as the name implies. That muscle spindles are widely distributed throughout muscle tissue is clearly illustrated by the large number of efferent fibers that serve them. However, the distribution varies from muscle to muscle. In general, the muscles used for complex movements (such as finger muscles) are abundantly supplied with up to thirty spindles per gram of muscle, while muscles involved in only very gross movements such as latissimus dorsi may have only one to two per gram. In some of the cranial muscles no spindles were found (2).

Each spindle consists of a connective tissue sheath 4 to 10 mm long that contains from five to nine *intrafusal muscle fibers* (IF) in mammalian tissue. These IF fibers are quite different from the typical fibers (extrafusal or EF) of muscle described in chapter 2, which are involved in bringing about gross muscle contraction. Spindle structure is shown in figures 5.6 and 5.7. It is important to note that the spindles are always oriented parallel to the EF fibers. There are two types of IF fibers:

1. Usually there are two *nuclear bag fibers,* which are thicker and longer and in which as many as a dozen nuclei are crowded into the middle section.

2. There are three to seven *nuclear chain fibers,* which are so named because their nuclei, although also located in the midsection, are fewer in number and are arranged in single file, forming a thinner, shorter fiber.

Figure 5.7 Schematic diagrams illustrating the structure and innervation of the mammalian muscle spindle as found in cat hindlimb muscles. **A** shows the general proportions of the receptor; the ends of the poles are curled round so as to fit into the figure. **B** shows the morphology of the equatorial region and about half of one pole, and the same figure is used in **C** with the addition of the sensory and motor innervation. ax.sh., axial sheath; c., capsule; cap., capillary; ex.mf., extrafusal muscle fibers; i.m.f., intrafusal muscle fibers; i.m.t., intramuscular nerve trunk; myo.reg., myotube region; n.b., nuclear bag; n.c., nuclear chain; n.b.m.f., nuclear-bag muscle fiber; n.c.m.f., nuclear-chain muscle fiber; P, primary ending; p_1, p_2, two types of intrafusal motor end plates; peri.sp., periaxial space; S, secondary ending; tr., trail ending. Sensory nerve fibers indicated by Roman numerals, motor fibers by Greek letters. (From Barker, D. In *Handbook of Sensory Physiology,* 1973, vol. III, pt. 2. Courtesy of Springer-Verlag, Berlin, Heidelberg, New York.)

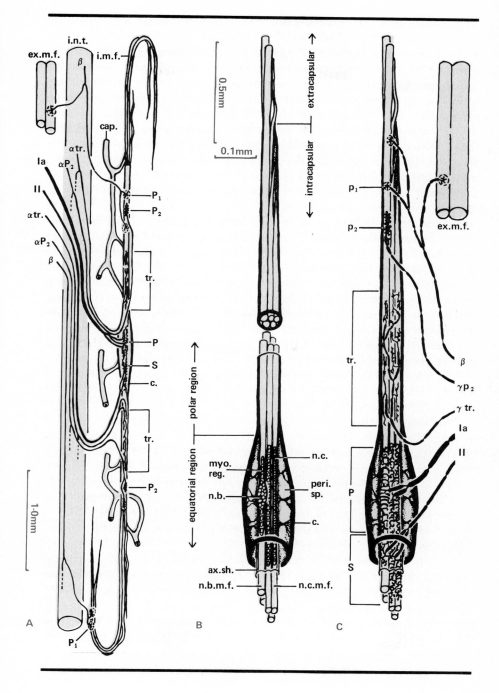

At each end of the IF fiber are the motor poles, which are the contractile elements composed of striated myofibrils.

The sensory end organs within the spindles are of two types, the *annulo-spiral ending* (also called primary ending) and the *flower spray ending* (also called secondary ending). The large annulospiral type Ia afferent nerve axon (only one per spindle) sends individual twigs to wrap around the middle of each nuclear bag fiber and other less well developed spiral terminals to most of the nuclear chain fibers. The flower spray endings are so named because they arborize in spraylike formation from a smaller afferent nerve fiber (type II) and wrap themselves around nuclear chain fibers almost exclusively. There may be one or two flower spray endings per spindle, and they attach not at the nuclear area but at one or both sides of it in the transition zone between the motor pole and the nuclear area.

Both of these endings are deformed by the stretching of the intrafusal fibers. Because these intrafusal fibers lie lengthwise, parallel with the skeletal (extrafusal) fibers, an externally applied stretch results in stretching the intrafusal as well as the extrafusal fibers. The consequent deformation of the annulospiral ending evokes an afferent discharge in its large, type Ia sensory nerve (fast-conducting); simultaneously, the flower spray ending discharges into its type II sensory nerve (slow-conducting). This afferent discharge from the spindle results in a motor response: contraction of the muscle that was stretched. This response, called a *stretch* or *myotatic reflex,* is typified by the tendon jerk elicited by the physician when he strikes the patellar tendon.

The annulospiral ending responds to both phasic and static stretching, whereas the flower spray ending responds to static stretch but is relatively insensitive to phasic stretch (4).

The Golgi tendon organ (fig. 5.6), a somewhat simpler end organ, is found in the musculotendinous junction, and throughout the perimysial connective tissues. It should be noted that this ending is in series with the skeletal muscle fibers, and therefore is deformed by tension in the tendon whether by passive stretching or by active shortening of the muscle. It therefore discharges under both conditions, whereas the spindle discharges only when stretched. The spindle ceases to fire when contraction begins because it is in parallel with the extrafusal muscle fibers, and is thus unloaded as soon as the extrafusal fibers shorten in contraction (see fig. 5.8).

A further difference in function between the spindle and the tendon organ is that the spindle facilitates—indeed, may cause—contraction, whereas the tendon organ seems to be a protective device, inhibitory not only to its muscle of origin but to the entire functional muscle group as well (21). The stretch reflex that originates in the muscle spindles is the basis for unconscious muscular adjustments of posture where a slight stretching of the extensor muscles at the knee, for instance, is immediately corrected by reflex shortening to prevent collapse. The tendon reflex that prevents overstressing of the tissues is called the *inverse myotatic reflex*. Granit (7) points out that "the muscle machine is working under self-regulation from autogenetic governors, first aiding it to contract, then damping the discharge from its motoneurons."

The pacinian corpuscle is a large encapsulated end organ, made up of several layers of fibrous tissue in which nerve endings ramify. These receptors, although widely distributed throughout the body, do not lie within the muscle tissue proper. They are found concentrated in the region of the joints and in the sheaths of tendons and muscles, and consequently are pressed upon when muscles contract. They are excited by the deformation of deep pressure, and may be more important than the spindles and tendon organs in detecting passive movement or position of a body segment in space.

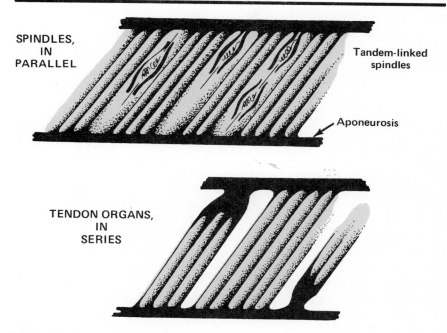

Figure 5.8 The principle of "in series" and "in parallel" arrangement of muscle receptors with reference to extrafusal (nonspindle) muscle fibers. Spindles lie between muscle fibers, either with both poles ending in interfascicular connective tissue, with one pole attached to a tendon, or in some short muscles with both poles inserting into muscle tendons or aponeuroses. In muscles with long extrafusal fasciculi, several spindles may be linked in tandem by one or more long intrafusal fibers that pass from one capsule to another. This histological distinction of series and parallel arrangements has a physiological counterpart in response characteristics of the afferent during contractions. (From Eldred, E. *Am. J. Phys. Med.* 46:83, 1967. Courtesy of the Williams & Wilkins Company, Baltimore.)

Ruffini receptors are scattered throughout the collagenous fibers of joint capsules and are differentially activated by joint movement. Consequently, they are probably most important in sensing joint position and motion. As can be seen in figure 5.9, each individual receptor ending monitors a well-defined and restricted range of the total movement. The sensing of the complete movement is thus the result of integration by the nervous system of the bits and pieces of information provided by the many Ruffini receptors.

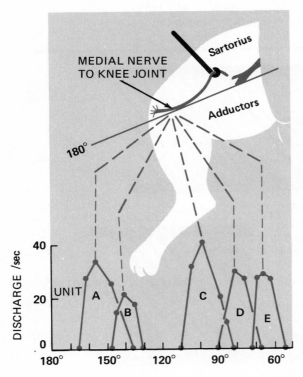

Figure 5.9 Range fractionation in signalling of joint position. Five joint receptor units were isolated from the medial nerve to the knee joint of the cat. Each unit discharged only over a restricted range of knee positions and had a maximum at one fairly sharply defined point. The discharge was constant at a stable position of the joint, but became elevated during movements within the response range for that unit. (From Eldred, E., adapted from S. Skoglund. *Am. J. Phys. Med.* 46:69, 1967. Courtesy of the Williams & Wilkins Company, Baltimore.)

Basic Physiology Underlying the Study of Physiology of Exercise

Unmyelinated, free nerve endings are thought to be widely distributed through muscles, tendons, joints, fascia, and ligaments, but the only definite histological information seems to indicate their distribution to these tissues is not direct but indirect—by serving the blood vessels that supply these tissues. Their function as kinesthetic receptors is also largely unknown.

In summary, proprioception involves sensing five aspects of muscle status:

1. *Active contraction,* by the tendon organ discharge
2. *Passive stretch,* by spindle discharge only since tendon organ threshold is high
3. *Tension,* by tendon organ discharge
4. *Position,* by discharge of Ruffini receptors
5. *Pressure sensation* from pacinian corpuscles

The sense of limb position rests largely on the last two elements of proprioception.

The Alpha and Gamma Systems for Muscular Control

Recent evidence has shown that approximately a third of all the efferent nerve fibers that enter a muscle have no connection with skeletal, extrafusal muscle fibers (16); rather, these nerve fibers have been demonstrated to innervate the intrafusal fibers of the muscle spindles. The importance of this has been suggested, and it is of practical significance to understanding muscular control. The smaller motor-nerve fibers that make up this one-third of the total efferent nerve fibers are called *gamma efferent fibers,* in contrast to *alpha efferent fibers,* which are larger and are the motor-nerve fibers for the extrafusal or skeletal muscle fibers.

In figure 5.10 the monosynoptic response to stretch can be traced as follows:

1. Stretching the main muscle fibers of the quadriceps results in stretching the muscle spindle and its intrafusal fibers.
2. The annulospiral nerve ending in the intrafusal fibers is distorted and causes propagation of nervous impulses to the cord via the large Ia spindle afferent.
3. The large spindle afferent synapses directly with the large alpha efferent, which brings about muscular contraction (stretch reflex) of the quadriceps.

It will also be noted in figure 5.10 that the flower spray receptor activates its smaller, slower conducting type II afferent, which goes through an internuncial neuron to end in an inhibitory synapse with the alpha motor neuron pool. Activation of the flower spray ending also brings about a facilitating effect on the alpha motor neuron going to the antagonistic flexor muscle shown as the hamstring group in the diagram. Typically, the flower spray ending, if

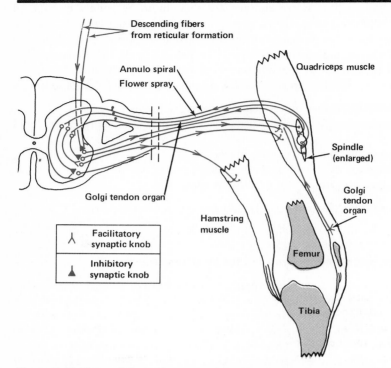

Figure 5.10 The myotatic and inverse myotatic reflexes as autogenetic governors of movement at the knee joint. Note that supraspinal influence, both facilitatory and inhibitory, is brought to bear on the gamma efferent neuron, thus setting the bias of the spindle.

it is in an extensor muscle, facilitates contraction of the antagonistic flexors and inhibits its own extensor muscle. When this ending lies in a flexor, it helps to bring about the stretch reflex or shortening of the flexor. Threshold for stimulation of the flower spray ending is much higher so the action is usually controlled by the annulospiral ending until stretch is of considerable magnitude.

As the extrafusal fibers contract, the intrafusal fibers in the spindle go slack (see fig. 5.11); thus the deformation of the annulospiral ending is relieved, and consequently the innervation of the large spindle afferent ceases, and the stretch reflex is over. In voluntary movement, however, the situation is somewhat more complex. As the extrafusal fibers contract, the intrafusal fibers are innervated by their motor nerves, the gamma efferents, which causes a shortening of the intrafusal fibers and a consequent resetting of the spindle length, as it were, so that they are again (and constantly) sensitive to any stretching

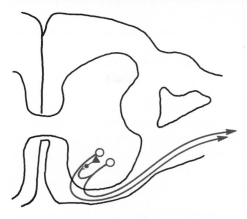

Figure 5.11 The Renshaw internuncial neuron or Renshaw cell is the short interneuron with the inhibitory synaptic knob, which is activated by a recurrent collateral from an alpha motor nerve.

of spindles. Thus a system is set up in which the innervation of the intrafusal fibers sensitizes the spindle, and the spindle firing causes more contraction of the extrafusal fibers. This is an alternative method for initiating movement and provides the feedback information as to *length* of the muscle.

One other system (fig. 5.11) should also be noted. When the alpha efferent neuron fires, a recurrent axon collateral is also innervated, which synapses with an internuncial neuron in the cord called a *Renshaw cell*. This cell has the property of synapsing with and inhibiting other motoneurons. Its system constitutes what is termed a *feedback loop*. In other words, the fact that the alpha neuron is initiating contraction in the main muscle is *fed back* and is used to inhibit other muscle fibers from making the contraction too strong. This is negative feedback, in that action tends to limit itself from getting out of control. The process by which a physiological (or mechanical) mechanism controls itself by feeding back information that reflexly governs the action is called a *servomechanism*.

Another factor enters into the reflex control of muscular movement. It will be recalled from our discussion of kinesthetic receptors that the Golgi tendon endings are also sensitive to stretch. The response, however, is inhibitory in nature, and evidence is available that these endings inhibit not only the muscle from which the afferent impulse arises, but the entire functional muscle group (21). The tendon endings have a higher threshold to stretch than the spindles, and consequently do not fire until considerably more force is expended. However, they are very sensitive to tension brought about by active contraction of the muscle fibers with which they are in series. Even .1 gr of force results in impulse propagation from the Golgi tendon organ (GTO) (15). Consequently, the GTO is thought to be important in providing feedback information with respect to tension produced by the muscle. It must also be considered important as a peripheral source of inhibition to protect the muscle against too great an overload.

As has been pointed out by investigators in this area such as Granit (7), Holmgren and Merton (14), and Mountcastle (23), muscles seem to be controlled by a system of autogenetic governors. Muscular activity seemingly may be initiated by either the alpha or gamma efferents. Indeed, it is now thought that coactivation is the rule (25). A constant feedback of information by the muscle spindles and tendon organs provides the central nervous system with the data needed to control the movement as it progresses. The spindles are, in the words of Granit, "the private measuring instruments of the muscle's servomechanisms. They do not record length so much as differences in length between the extrafusal muscle fibers and intrafusal fibers of the spindles. The tendon organs apply the brakes to muscular contraction to prevent the development of too great a tensile force, which could result in injury to muscle and/or tendon."

The work of Granit and Kaada (8) indicates that the gamma efferent system is tonically activated from central regions, such as the diencephalic reticular system, the motocortex, and the anterior lobe of the cerebellum. This tonic activity of the gamma efferents results in a degree of tonus in the muscle spindles, which in turn tonically innervates the large spindle afferents for postural reflexes and results in postural tonus of the extensor muscle groups throughout the body.

Furthermore, the stretch reflex has some interesting qualities for muscle stretching in developing flexibility or warming-up athletes. There is evidence that the stretch reflex has two components, one of which is the result of phasic (jerky) stretching and the other the result of static (maintained) stretching (16, 19, 23). The phasic response seems to be faster and stronger and is typified by synchronous discharges of the spindles, whereas the static response is slower and weaker and is typified by asynchronous discharge of the spindles.

From the practical standpoint, then, we may say that the *amount* and *rate* of response of a stretch reflex are proportional to the *amount* and *rate* of stretching. In other words, the use of bouncing or jerking movements in stretching will cause the muscle to contract with a vigor proportional to that of the bouncing and jerking. It is rather obvious that this is not desirable, either in warming-up cold muscles for athletic participation or in work that is designed to improve flexibility.

There is also a difference between the stretch reflexes of the flexor and extensor muscles. The phasic component of the reflex is well developed in both, but the static stretch component is well defined only in the extensors, where it is needed for the sustained muscular activity of maintaining posture.

It would seem intelligent for the physical educator and coach to apply known principles. If muscle tissues are to be loosened up, lengthened, or relaxed, the use of a sustained pull on the muscle would tend to eliminate the phasic component of the stretch reflex, and the pull should be of sufficient force to reach the threshold of the tendon organs or flower spray endings. This will then initiate the inverse myotatic reflex, which will inhibit the muscle under stretch and thus further aid in stretching the muscle.

These reflex patterns can be applied in other areas. It seems reasonable to believe, for example, that the upper limits of muscle strength and power must be set by the level of inhibition brought about centrally by the Renshaw cell and peripherally by the GTO and the flower spray endings. Learning to disinhibit may be an important part of strength training.

In the light of the preceding discussion it is easy to see how muscle cramp can be brought about by making a maximally vigorous contraction of a muscle in a shortened position. This can be demonstrated by maximal contraction of a fully flexed biceps. What happens is that in the shortened position, no tension can be placed on the tendon organ (see the length-tension diagram in chapter 4). So we have maximum innervation with minimal inhibition, and the result in a large percentage of trials is cramping. Fortunately, this newfound knowledge can be applied to relieve the cramp by simply forcing the muscle into its longest position, thereby creating tension in the tendon and the GTO, with resulting inhibition that relieves the cramp.

Higher Nerve Centers and Muscular Control

So far the discussion of muscular control has centered around the involuntary reflex systems; now we will consider the voluntary control of muscular activity by the brain. It will be convenient to consider this in three parts: (1) the *pyramidal system,* (2) the *extrapyramidal system,* and (3) the *proprioceptive-cerebellar system.* Although much is known about the function of the brain in controlling muscular activity, a much greater portion awaits further research. Furthermore, a complete discussion of what is known of the motor functions of the brain is beyond the scope of this text and unnecessary for its purposes. Therefore the discussion is somewhat brief and is confined to those aspects that are of interest to the physical educator.

The Pyramidal System. The cerebral cortex has been mapped out in fair detail, over a period of time, by two methods: (1) by relating clinically observed motor defects with the lesions seen in surgery and autopsies, and (2) by electrical stimulation of the cortex of experimental animals and observation of the resulting motor effects. This research has resulted in cytoarchitectural maps of the human cortex. The most commonly used map is that of Brodmann, and our areas of interest are shown in figure 5.12.

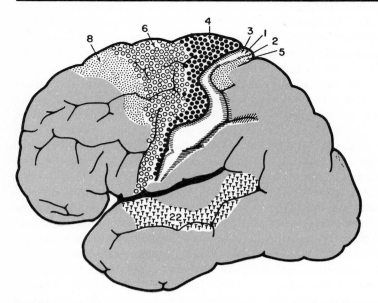

Figure 5.12 Diagram of the areas of the human cerebral cortex involved in the extrapyramidal system. In the frontal lobe are areas 4, 6, and 8. In the postcentral region are areas 1, 2, 3, and 5. Down in the temporal lobe, area 22 is concerned in the extrapyramidal pathways. (From Morgan, C.T., and Stellar, E. *Physiological Psychology,* 1950. Used by permission of McGraw-Hill Book Company.)

The pyramidal system originates in large nerve cells, shaped like pyramids, that lie mainly in area 4 of the cortex. It had once been thought that the giant Betz cells were the entire source of the pyramidal tracts, but recent research has indicated they are only about 2% of the total motoneurons of the pyramidal system. The axons from the motoneuron cell bodies in area 4 form large descending motor pathways, called *pyramidal tracts,* that go directly (in most cases) to synapses with the motor neurons in the ventral horn of the spinal cord, with which we have already dealt. The neurons whose cell bodies are in the brain are commonly called *upper motor neurons,* and those in the spinal cord are called *lower motor neurons.* Seventy to eighty-five percent of the nerve fibers of the pyramidal tract cross from one side to the other, some at the level of the medulla, others at the level of the lower motoneuron.

Area 4 of the cortex is also referred to as the *motor cortex,* or because of its location, as the *precentral gyrus.* The mapping of the motor cortex clearly demonstrates a neat, orderly arrangement of stimulable areas. It is of interest that the area of cortex devoted to a given body part is not proportional to the amount of tissue served but rather to the complexity of the movement potential of the body part. It has been shown that the hands and the muscles of vocalization involve a disproportionate share of motoneurons.

It should be pointed out that the motor cortex is oriented by *movement*, not by *muscle*. That is, stimulation of the motor cortex results not in a twitch of one muscle but in a smooth, synergistic movement of a group of muscles.

The Extrapyramidal System. The extrapyramidal system has its origin mainly in area 6, according to Brodmann's nomenclature. This area is rostral to area 4, the motor area, and is sometimes called the *premotor cortex*. Evidence, however, indicates that some of the fibers descending in the extrapyramidal tracts also originate with cell bodies in other areas, notably areas 8 and 5 and also in areas 3, 1, and 2, which are usually considered the sensory areas for somesthetic impulses.

The descending tracts that are composed of the axons from the neurons of the premotor cortex are considerably more complex. These fibers do not go directly to synapses with the lower motoneurons, but by way of relay stations, which are called *motor nuclei*. The most important nuclei are *corpus striatum, substantia nigra,* and *red nucleus.* Some fibers also go by way of the pons to the cerebellum.

For the physical educator, the most important differences between the pyramidal and extrapyramidal systems are the functional differences. Whereas electrical stimulation of area 4 produces *specific movements,* stimulation of area 6 produces only large, general *movement patterns.* Consequently, it is believed that learning a new skill in which conscious attention must be devoted to the movements (as in learning a new dance step, where every movement is contemplated) involves area 4. As skill progresses (dancers no longer concentrate on their feet but rather on very general patterns of movement, so that they can interpret the music better), the origin of the movement is thought to shift to area 6. However, area 4 still participates as a relay station, with fibers connecting area 6 to area 4.

The Proprioceptive-Cerebellar System. We have already dealt with some of the sensory functions of this system, namely, the kinesthesis and vestibular function in proprioception. In general, the pathway of vestibular proprioception leads to the cerebellum, either directly or by way of the vestibular nucleus in the medulla. Of the fibers that conduct kinesthetic information, some go to the thalamus and cortex to provide sensory knowledge of movement at the conscious level. Others go to the cerebellum. The confluence of sensory data on position, balance, and movement upon the cerebellum is indicative of the importance of this organ to movement.

The cerebellum rivals the cerebral cortex in complexity but, unfortunately, it is not so well understood. Removal of the cerebellum of experimental animals results in a loss of function that has three distinct components: (1) impairment of volitional movements, (2) disturbances of posture, and (3) impaired balance control. Consequently, its functions may be deduced.

Without entering into anatomical detail we may say that the cerebellum receives constant sensory information from receptors in muscles, joints, tendons, and skin, and also from visual, auditory, and vestibular end organs.

Despite all this sensory reception, however, no conscious sensation is aroused in the cerebellum. This function is served by the sensory cortex of the cerebral hemispheres. The cerebellum is intimately connected with the motor centers, from the motor cortex all the way to the spinal cord, and may be considered to modify muscular activity from the beginning to the end of a movement pattern.

Posture, Balance, and Voluntary Movement

It is now appropriate to consider how all of the foregoing discussion relates to a better understanding of human muscular activity, which is the purpose of this text.

Posture.   Human upright posture is mainly brought about through suprasegmental reflexes. The major part of these suprasegmental reflexes is the basic stretch, or myotatic reflex, which has been described; let us therefore direct our attention to the postural mechanism that occurs about the knee joint. If the joint starts to collapse (flex) immediately as the muscle spindles of the quadriceps are stretched, an impulse is generated that is propagated over large afferents to the cord where synapse is made with a motoneuron, which innervates fibers in the quadriceps to contract and thus reextend the joint into its proper position. This reflex, however, occurs only in the presence of facilitation by the vestibular nucleus in the medulla, whose fibers in the cord tend to maintain a state of excitation of the motoneurons innervating the muscles, so that a relatively small afferent impulse can reach threshold value. Furthermore, postural reflexes also depend upon effects of the extrapyramidal system and the autogenetic governor system to bring about smoothly coordinated contraction of the proper strength.

Balance.   This aspect of muscular activity is best discussed by reference to the *righting reflex,* which illustrates the underlying principles rather well. If one drops a cat from a small height in supine position (upside down), observations can be made of the sequential events that invariably lead to its righting itself and landing on all four feet. The first reaction is a turning of the head toward the floor in the attempt to normalize the sensations coming from the otolith organs, which inform the cat it is not oriented in space as it would wish to be. This turning of the head innervates muscle spindles, tendon receptors, and other nerve endings in the neck muscles that initiate kinesthetic impulses and reflexly bring about the execution of a half twist, usually long before the cat hits the ground.

As the cat turns right side up, its visual receptors bring the necessary sensory data to the cerebellum for organization of muscle activity in the extensors to bring about a gradual acceptance of the force involved in landing. As the cat lands, the force that flexes the legs invokes the stretch reflex in the extensors, and if the cat was an unwilling subject, the muscles of propulsion will already be underway. These principles apply equally well in diving, trampolining, and all other activities in which balance is a factor.

Voluntary Movement. Let us now take a very simple voluntary movement and analyze the neural activity involved in bringing it about. Assume the right arm is at the side-horizontal position and the desired movement is to bring the right index fingertip to the end of the nose. Since this is not a usual movement, it will probably originate in the arm section of area 4 of the motor cortex and proceed by way of the pyramidal tracts to synapse with the lower motoneuron in the cord and out to the appropriate muscles by way of the brachial plexus. At the same time, kinesthetic impulses traverse the afferent pathways to the cerebellum and bring about the proper control and coordination so that the shoulder muscles are activated to support the arm as it is moved through horizontal flexion adduction.

Furthermore, the same kinesthetic impulses act reflexly at the segmental level to relax the antagonists through reciprocal inhibition, and the gamma efferent system is busy all the while innervating the muscle spindles so that constant measurements of the movement's progress can be fed back. As the movement accelerates, more and more motor units are innervated and the rate of impulse transmission to each motor unit increases, with each motor unit participating at the proper time in the sequential development of the movement (1). Finally, the movement has to be decelerated in reverse fashion.

This may seem rather complex, but this discussion is really a superficial treatment of a relatively simple movement. Highly skilled movements made every day in the gymnasium may be infinitely more complex and may defy detailed analysis.

Practical Considerations

Much of the discussion in this chapter has been technical; however, an understanding of the foregoing material can be of value in very practical situations in physical education and coaching. Some of these situations will now be considered.

Effects of Hypnosis and Emotional Excitement on Performance. It is well known that emotional excitement can greatly increase muscular strength and endurance. Every coach and athlete is aware of the improvement of performance in the game situation, compared with that under practice conditions. Upon occasion, however, a coach may also see examples of a decrease in performance due to overexcitability of a young, inexperienced athlete.

Explanations of these phenomena rest upon a knowledge of the neural factors involved (as well as humoral factors, to be discussed later). Gellhorn (6) explains improved performance during emotional excitement in terms of an excitation of the hypothalamus that accompanies excitement. He cites studies that have shown that hypothalamic stimulation (which by itself does not elicit muscular movement) increased the intensity and complexity of a cortically induced movement. Furthermore, many muscles that were not activated by cortical stimulation alone became active, and even activity in other extremities was noted. Gellhorn attributes these effects to summation processes that take place at two sites:

1. In the spinal cord, as the result of the interaction of the subthreshold, extrapyramidal discharges from the hypothalamus and of the efferent discharges from the motor cortex.
2. In the motor cortex itself, due to the intensification of pyramidal discharges as the result of hypothalamic-cortical discharges (5).

This discussion serves to explain both the commonly seen improvement and the less commonly seen decrease in athletic performance, sometimes referred to as *tying up,* due to emotional excitement. The rationale for improved performance is obvious from the above; an explanation for tying up, though less obvious, is also at hand. Nervous overflow to other muscles, and even to other extremities, could well be expected to result in innervation of synergistic muscles, and in some cases, antagonistic muscles may be innervated in a portion of the desired athletic movement. When the latter occurs, the level of performance declines.

Thus, to summarize the effect of emotional excitement upon muscular performance, we may say that under normal conditions only some of the available motor units are activated, but that this number can be augmented by activity of the hypothalamus due to emotional excitement. If the activity of the hypothalamus and humoral factors, such as an increased level of adrenaline in the muscle tissue, results in an activation of other muscles, some of which are antagonists, tying up and a decline in performance level may result.

Although great feats of strength and endurance have been claimed as the result of hypnotic trance states, demonstrations by competent investigators under controlled laboratory conditions have been less spectacular. Johnson and Kramer (17) were unable to demonstrate significant changes in the strength or power of ten athletes under hypnotic conditions, although muscular endurance was enormously improved for one professional athlete.

Roush (24), in a study noted for its large number of subjects (twenty) and rigid criteria of the state of trance, found significant improvements in grip strength, arm strength, and muscular endurance, but Johnson's (17) summary of the available research points out that the use of hypnosis to improve muscular strength or endurance is unreliable, even though great improvements have been found in a few individuals. It is also Johnson's opinion that the beneficial effects, when found, are the result of the neurophysiological mechanisms described in regard to emotional excitement. In any event, hypnosis is certainly not a practice to be indulged in by coaches and physical educators, unless it is supervised or directed by medical or other qualified personnel.

The Cross-Education Effect. Before the turn of the century, psychologists demonstrated that training one limb resulted in significant improvements not only in the exercised limb but in the symmetrical, unexercised limb as well. This phenomenon, the *cross-education effect,* was found to apply to both the learning of skills and the improvement of strength.

Cross education has been thoroughly investigated for its usefulness in physical education and rehabilitation by Hellebrandt and her co-workers (9, 10, 11) and by Walters (27). Their work demonstrated the cross-education effect in relation to the gross motor activities of physical education. As a result of training one limb, they found that significant improvements in strength, endurance, and skill occurred not only in the trained limb but also in the contralateral (opposite), untrained limb. To produce a well-defined effect, it was necessary to go into an overload training condition; however, it seems that the nondominant arm could be trained as well by cross education with an overload condition as by direct practice with an underload. Furthermore, the nondominant arm sometimes gained as much in skill by cross education as it did in direct practice, but this was not true for the dominant arm. The cross-education effect on strength has been shown to result entirely from a greater ability to innervate previously inactive muscle fibers, with no evidence of muscle hypertrophy (22). See figure 4.15.

The rationale for this cross-education effect has not yet been clearly elucidated; however, from what is known of the motor pathways from the cortex, logical deductions can be made. As was pointed out in the description of the pyramidal system, 70%–85% of the descending nerve fibers cross from one side to the other in their descent to synapse in the cord. Some of the remainder seem to descend ipsilaterally, and it is known that ipsilateral effects are obtainable by stimulation of the posterior premotor area of the cortex. Thus it seems likely that the cross-education effect is brought about by an overflow of nervous energy from neurons in the motor cortex, which innervate the crossed pyramidal fibers, to a smaller number of neurons that supply the uncrossed fibers. Sufficient overflow apparently occurs only under the conditions of strong volition that are involved in overloading. It should be pointed out that the untrained limb was frequently observed to produce isometric contractions during the training of its symmetrical partner. Technically, the untrained limb was not unexercised.

The implications of this work for the maintenance of muscle tone and prevention of atrophy in immobilized muscles is obvious. From the foregoing rationale, it will also be obvious that intact motor innervation is necessary for achievement of the cross-education effect.

The Dynamogenic Effect of Cocontractions. Hellebrandt and Houtz (11), after observing isometric contractions in the unexercised limb during the cross-education experiments, postulated that the concurrent contraction of homologous (corresponding) parts (such as right and left arms) might function in augmenting the work output of a tiring or a weakened muscle. Their experiments, as well as the earlier experiments of Karpovich (18), demonstrated that this is indeed the case. When a limb was fatigued ergographically by rhythmic contractions, bringing the other unfatigued limb into synchronous movement resulted in a marked improvement of the work output of the fatigued limb. This phenomenon has much practical significance for corrective physical education and may prove of value in athletics as well.

Reaction Time and Movement Time. Of great interest and importance to all concerned with physical education and athletics is the speed with which an individual can react in a game situation. For example, this factor partly determines how successful a basketball player can be on defense. When the offensive player makes a move, the difference between a slow and a fast reaction by the defensive player (possibly 0.10 second) can result in the offensive player's getting a lead of several additional feet simply because the defensive player must react to the offensive move. In swimming races a difference in reaction to the pistol shot can result in several feet gained or lost. On such slim margins hang the fruits of victory in closely fought contests.

Reaction time is defined, for purposes of physical education, as the interval between presentation of the stimulus and the first sign of response. The measurement is easily made by electric circuitry, which applies current to an electric timer (chronoscope) at the presentation of the stimulus and cuts the current when the subject's hand is removed from a push-button switch. The stimulus may be visual, audible, or tactile.

Movement time is defined as the interval between the start and the finish of a given movement. The movement may be terminated if it ends in striking an object, or it may be nonterminated if the chronoscope is stopped by interruption of a light beam or similar device that allows follow-through.

An interesting question here is that of the specificity versus the generality of these measures. In other words, if the reaction time is swift for removing a hand from a switch in response to a stimulus, can the motion be expected to be equally swift when performed by a leg on the same or the opposite side? Well-controlled studies by Henry and Rogers (13), Clark and Glines (3), and Lotter (20) seem to indicate a relatively high degree of specificity by limb and movement. Thus an individual may be quick in reacting with an arm but slow when reacting with the legs.

In regard to sex, adequate data for the college age group indicates that reaction and movement times are slower in women by approximately 14% and 30%, respectively (12, 13).

Motor Set versus Sensory Set. *Set* has been defined as the direction of the subject's attention preliminary to an anticipated movement. Thus *motor set* is obtained by directing the subject's attention to the movement response as in thinking of the starting movement in track. *Sensory set* is the direction of attention to the stimulus, as in concentrating upon the gunshot for the start.

It has been the accepted procedure in athletics to use the motor set on the assumption that it results in faster reaction; however, as a consequence of the *memory drum theory,* Henry and Rogers (13) hypothesized that attempts to institute conscious control of movement will interfere with the programming, thus increasing reaction time and resulting in a slower start. Their experiment on college age men and women indicated that both sexes react approximately 2.6% slower and move 2.1% slower when using the motor set, compared with the sensory set. It should be pointed out, however, that they found 20% of the subjects had a natural preference for the motor set and that these subjects performed better with the motor set.

Although this phase of performance research cannot yet be considered closed, the evidence weighs in favor of the sensory set, contrary to widespread opinion.

SUMMARY

1. The nervous system may be divided in two ways: (a) structurally, into *central* and *peripheral* components, and (b) functionally, into *somatic* and *autonomic* systems.

2. The basic structural unit of the nervous system is the nerve cell, called the *neuron.* The cell body has two types of processes: *dendrites,* which bring impulses to the cell body and the *axon,* which conducts nerve impulses away.

3. The *reflex arc,* in its simplest form, is a discharge from a sensory receptor that is transmitted by an *afferent* pathway to a synapse in the spinal cord with a motor neuron, which conducts the *efferent* impulse to an *effector* organ (muscle or gland).

4. Reflexes may occur entirely at one level of the spinal cord (*segmental reflexes*), or they may traverse upward or downward in the cord (*intersegmental reflexes*), or they may be influenced by higher brain centers (*suprasegmental reflexes*).

5. Coordination of movement depends upon feedback information of what the muscles are doing. This feedback is called *proprioception* and consists of two types of information: (a) *kinesthetic,* from receptors in the muscles, tendons, and joints, and (b) vestibular, from receptors of the non-auditory labyrinths of the inner ear.

6. Muscular control is mediated directly through the large *alpha motor nerve fibers* and indirectly through the smaller *gamma nerve fibers*. The alpha motor nerve fibers and their skeletal muscles, together with the gamma motor nerve fibers and their intrafusal fibers, form an interlocking servomechanism in which each reacts to the other's actions, thus bringing about very fine control of movement.

7. The *pyramidal system* is comprised of area 4 of the motor cortex, and the nerve fibers emanating therefrom, which descend through the pyramidal tracts to synapse directly or indirectly with the lower motoneuron in the cord. This system is involved mainly in conscious, specific, volitional movement.

8. The *extrapyramidal system* originates mainly in area 6 of the *premotor cortex*. Its descending fibers go to the cord indirectly, by way of various relay stations called *nuclei*. It is involved in general, diffuse motor patterns.

9. The *proprioceptive-cerebellar system*, although it arouses no conscious sensory sensation, is the clearinghouse for sensory data necessary for the coordination of movement patterns, which is mainly accomplished here.

10. *Upright posture* is maintained largely through the operation of *stretch reflexes*, also called *myotatic reflexes*. These reflexes make the minute adjustments in extensor tone that is necessary when extensor muscles start to slacken.

11. *Emotional excitement* or *hypnotic suggestion* can lead to an augmentation of the number of motor units participating in a given motor activity and thus improve performance. Both, however, can lead to a decrease in performance levels under certain conditions, for example, by *tying up* an overly emotional or inexperienced athlete due to overexcitement.

12. The *cross-educational effect* is well established in respect to strength, endurance, and skill. It requires intact motor innervation and overload conditions of the trained limb for its best manifestation. The *dynamogenic effect of muscular cocontractions* in augmenting the work output of fatigued muscles has also been established, and the rationale is probably similar for both phenomena.

13. *Reaction time* and *movement time* appear to be unrelated, and neither is a general quality; that is, quick eye-hand reactions are not necessarily evidence that other reactions in the same individual will be similarly fast.

14. Tests of a hypothesized *memory drum theory* of neuromotor reaction have produced evidence that tends to overthrow older opinions of motor versus sensory set. Evidence indicates that, for most people, a sensory set results in quicker reactions.

REFERENCES

1. Bigland, B., and Lippold, O.C.J. Motor unit activity in the voluntary contraction of human muscle. *J. Physiol.* 125:322–35, 1954.
2. Bourne, G.A. *The Structure and Function of Muscle,* vol. 1. New York: Academic Press, 1960.
3. Clarke, H.H., and Glines, D. Relationships of reaction, movement and completion times to motor strength, anthropometric and maturity measures of thirteen-year-old boys. *Res. Q.* 33:194–201, 1962.
4. Eldred, E. Peripheral receptors: their excitation and relation to reflex patterns. *Am. J. Phys. Med.* 46:69–87, 1967.
5. Eldred, E.; Granit, R.; Holmgren, B.; and Merton, P.A. Proprioceptive control of muscular contraction and the cerebellum. *J. Physiol.* 123:46–47, 1954.
6. Gellhorn, E. The physiology of the supraspinal mechanisms. In *Science and Medicine of Exercise and Sports,* ed. W.R. Johnson, p. 737. New York: Harper & Row, 1960.
7. Granit, R. Reflex self-regulation of muscle contraction and autogenetic inhibition. *J. Neurophysiol.* 13:351–72, 1950.
8. Granit, R., and Kaada, B.R. The influence of stimulation of central nervous structures in muscle spindles in the cat. *Acta Physiol. Scand.* 27:130–60, 1952.
9. Hellebrandt, F.A. Cross education: ipsilateral and contralateral effects of unimanual training. *J. Appl. Physiol.* 4:136–44, 1951.
10. Hellebrandt, F.A.; Parrish, A.M.; and Houtz, S.J. Cross education: the influence of unilateral exercise on the contralateral limb. *Arch. Phys. Med.* 28:76–84, 1947.
11. Hellebrandt, F.A.; Houtz, S.J.; and Kirkorian, A.M. Influence of bimanual exercise on unilateral work capacity. *J. Appl. Physiol.* 2:446–52, 1950.
12. Henry, F.M. Influence of motor and sensory sets on reaction latency and speed of discrete movements. *Res. Q.* 31:459–68, 1960.
13. Henry, F.M., and Rogers, D.E. Increased response latency for complicated movements and "memory drum" theory of neuromotor reaction. *Res. Q.* 31:448–58, 1960.
14. Holmgren, B., and Merton, P.A. Local feedback control of motoneurons. *J. Physiol.* 123:47–48, 1954.
15. Houk, J., and Henneman, E. Responses of Golgi tendon organs to active contractions of soleus muscle of the cat. *J. Neurophysiol.* 30:466–81, 1967.
16. Hunt, C.C. The effect of stretch receptors from muscle on the discharge of motoneurons. *J. Physiol.* 117:359–79, 1952.
17. Johnson, W.R., and Kramer, G.F. Effects of stereotyped nonhypnotic, hypnotic, and post-hypnotic suggestions upon strength, power and endurance. *Res. Q.* 32:522–29, 1961.
18. Karpovich, P.V. Physiological and psychological dynamogenic factors in exercise. *Arbeitsphysiologic* 9:626, 1937.
19. Katz, B. Depolarization of sensory terminals and the initiation of impulses in the muscle spindle. *J. Physiol.* 111:261–82, 1950.
20. Lotter, W.S. Interrelationships among reaction times and speeds of movement in different limbs. *Res. Q.* 31:147–55, 1960.

21. McCouch, G.P.; Deering, I.D.; and Stewart, W.B. Inhibition of knee jerk from tendon spindles of crureus. *J. Neurophysiol.* 13:343–50, 1950.
22. Moritani, T., and deVries, H.A. Neural factors vs. hypertrophy in the time course of muscle strength gain. *Am. J. Phys. Med.* 58:115–30, 1979.
23. Mountcastle, V.B. Reflex activity of the spinal cord. In *Medical Physiology,* ed. Philip Bard. St. Louis: The C.V. Mosby Company, 1961.
24. Roush, E.S. Strength and endurance in the waking and hypnotic states. *J. Appl. Physiol.* 3:404–10, 1951.
25. Smith, J.L. Fusimotor loop properties and involvement during voluntary movement. *Exercise and Sports Sci. Rev.* 4:297–333, 1977.
26. Taub, E. Movement in nonhuman primates deprived of somatosensory feedback. *Exercise and Sports Sci. Rev.* 4:335–74, 1977.
27. Walters, C.E. The effect of overload on bilateral transfer of motor skill. *Phys. Ther. Rev.* 35:567–69, 1955.

6

The Heart
and Exercise

It is the function of the heart and the circulatory system to provide the flow of blood necessary to maintain homeostasis of the various tissues of the body. _Homeostasis_ can be defined as the sum total of regulatory functions that maintain a constant environment for the cells of the tissues. The _internal environment,_ i.e., the tissue fluid, must be held relatively constant in regard to nutrients and metabolites, oxygen and carbon dioxide, temperature and hormonal content. The blood must transport nutrients to the cells, wastes to the kidney, etc. The greatest concern in the physiology of exercise, however, is the transport of oxygen and carbon dioxide. During exercise, the supply of oxygen to the tissues is the most urgent tissue need; oxygen cannot be stored, in any real sense, and its supply or lack of it is usually the critical factor in any endurance exercise. The removal of carbon dioxide is intimately related to the blood's transport of oxygen, as will be seen in a later discussion.

In other words, in the study of exercise physiology we are primarily interested in the heart as the pump in the _cardiorespiratory system_ that maintains the proper pressure and flow of blood to active muscle tissues.

Review of the Cardiac Cycle

The salient anatomical features of the heart are reviewed in figure 6.1. Blood flows into the right atrium from the systemic circulation by way of the superior and inferior venae cavae. The thin-walled atrium acts as a combination storage basin and pump primer for blood flow through the tricuspid valve into the right ventricle, which provides most of the energy for blood flow through the

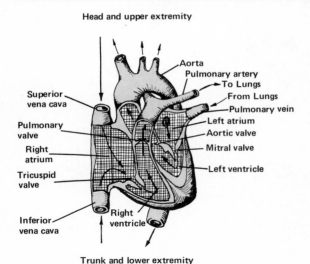

Head and upper extremity

Aorta
Pulmonary artery
To Lungs
Superior vena cava
From Lungs
Pulmonary vein
Left atrium
Pulmonary valve
Aortic valve
Right atrium
Mitral valve
Tricuspid valve
Left ventricle
Inferior vena cava
Right ventricle

Trunk and lower extremity

Figure 6.1 Details of the functional parts of the heart. (From Guyton, A.C. _Function of the Human Body,_ 1959. Courtesy of W.B. Saunders Company, Philadelphia.)

pulmonary valve and artery into the pulmonary circuit. Blood flows to and through the lungs and back to the heart through the pulmonary veins into the left atrium. Blood flow proceeds from the left atrium, through the mitral valve, and into the left ventricle. Contraction of the left ventricle provides the energy for blood flow through the aortic valve, through the aorta, and through the systemic arterial system to the capillary beds of the various tissues.

From the capillaries, the blood flow returns through veins of ever increasing size to the great veins: the superior and the inferior venae cavae. Thus the circulatory system is comprised of two loops, each with its own pump. The right heart and the pulmonary circuit form one loop, and the left heart and the systemic circuit form the other.

Origin and Transmission of the Heartbeat. The muscle tissue of the heart possesses _autorhythmicity_. It requires no innervation or stimuli from without to produce its regular contractions. Under normal conditions, the wave of excitation originates at the _sinoatrial_ (SA) _node_; however, all cardiac tissue has the property of autorhythmicity. If, under abnormal conditions, the rate of emission from the SA node should slow unduly, any area of the myocardium that has a faster inherent rate may assume the role of pacemaker. This is what occurs in abnormal heart rhythms.

Figure 6.2 illustrates the transmission of the wave of excitation from the SA node, by way of the syncytium of muscle fibers in the atrium to the atrioventricular (AV) node, and thence into the Purkinje system, which conducts the impulse throughout the ventricular myocardium.

Pressure Relationships of the Cardiac Cycle. Relating the events of the cardiac cycle to each other and placing them in time is best done by considering the _pressure relationships_ that are basic to valve action in the heart. It is

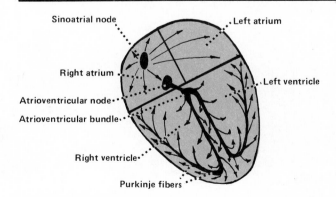

Figure 6.2 Transmission of the cardiac impulse from the sinoatrial (SA) node into the atria, then into the atrioventricular node, and finally through the Purkinje system to all parts of the ventricles. (From Guyton, A.C. *Function of the Human Body,* 1959. Courtesy of W.B. Saunders Company, Philadelphia.)

important to realize from the outset that all valve action is brought about by pressure differentials on the two sides of any given cardiac valve.

The pressure changes in figure 6.3 are recorded along with the events of the electrocardiogram, phonocardiogram (recording of heart sounds), and the heart volume curve. The vertical lines intersect these tracings to indicate events that occur simultaneously. The pressure curves depict the changes in the left heart, but the occurrences in the right heart are similar—though at a considerably lower pressure level.

Table 6.1 represents the events of an average cardiac cycle, under resting conditions, with a rate of seventy-four beats per minute. Under conditions of

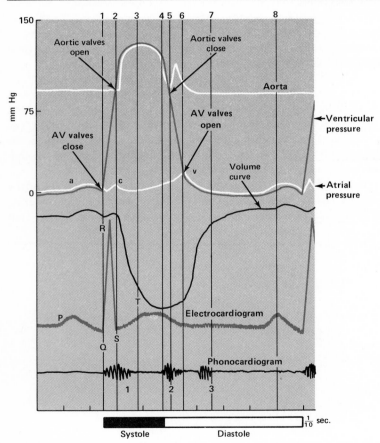

Figure 6.3 The cardiac cycle; superimposed curves of ventricular, atrial, and aortic pressures, together with a ventricular volume curve, an electrocardiogram, and a phonocardiogram. (From Best, C.H., and Taylor, N.B. *The Physiological Basis of Medical Practice,* 3d ed., 1943. Courtesy of Williams & Wilkins Company, Baltimore.)

Basic Physiology Underlying the Study of Physiology of Exercise

Table 6.1
Cardiac Cycle

Line	Nomenclature	Pressure Events	Valve Action	Average Duration
		SYSTOLE		
1		Ventricular pressure overcomes atrial pressure	AV valves close	
1-2	Period of isometric contraction (presphygmic)	Ventricular pressure increases	AV valves closed, aortic valve closed	0.05 sec
2		Ventricular pressure overcomes aortic pressure	Aortic valve opens	
2-3	Period of maximal ejection	Ventricular pressure continues to increase	Aortic valve open, AV valves closed	0.12
3-4	Period of reduced ejection	Ventricular pressure decreases	Aortic valve open, AV valves closed	0.14
			Systole Total	0.31 sec
		DIASTOLE		
4-5	Protodiastolic phase	Ventricular pressure drops		0.04 sec
5		Ventricular pressure falls below aortic pressure	Aortic valve closes	
5-6	Isometric relaxation phase (postsphygmic)	Ventricular pressure continues to drop rapidly	Aortic valve closed, AV valves closed	0.08
6		Ventricular pressure falls below atrial pressure	AV valves open	
6-7	Period of rapid filling	Ventricular pressure falls to zero	Aortic valve closed, AV valves open	0.09
7-8	Diastasis— period of slower filling	Atrial and ventricular pressure both low	Aortic valve closed, AV valves open	0.19
8-1	Atrial systole	Increase in both atrial and ventricular pressure due to atrial contraction	Aortic valve closed, AV valves open	0.10

Diastole Total 0.50 sec
Total Cardiac Cycle 0.81 sec
Heart Rate 74

exercise, which necessitate higher heart rates, the relative durations of the phases of systole and diastole are altered somewhat. The most notable change and the change most important to the physiology of exercise is that the period of *diastasis* is the first to be shortened, and it may be eliminated completely as the heart rate increases. Because diastasis is the period during which the entire myocardium is at rest, a decrease in resting time at higher heart rates results in losses of efficiency.

The Cardiac Output

The cardiac output is the volume of blood ejected by the heart per unit of time and is usually expressed in liters per minute. At rest, in the average-sized man, it is approximately 5 liters per minute, and it can be increased to 42 liters per minute in a well-trained athlete. The amount of this increase in cardiac output is one important limiting factor in athletic performance. The output of the heart is determined by two factors: the heart rate and the stroke volume (the amount of blood ejected with each beat). Consequently, the cardiac output equals the heart rate times the stroke volume.

Measurement of Cardiac Output. The most direct method of cardiac output measurement is based upon the *Fick principle*. If we know how much O_2 an individual is consuming per unit time and if we also know the concentration of O_2 in the arterial and mixed venous blood, then we can calculate the cardiac output as follows:

$$\text{Cardiac output in liters per minute} = \frac{O_2 \text{ consumption in milliliters per minute}}{\text{AV difference in } O_2 \text{ in milliliters per liter of blood}}$$

For example, in a resting subject we find an O_2 consumption of 250 ml/min. Arterial O_2 concentration is 20 volume percent and mixed venous concentration is 15 volume percent. First we convert the volume percent, which means the volume in milliliters carried by 100 ml of blood, to milliliters per liter (1,000 ml). Thus each liter of arterial blood carries 200 ml of O_2, while the mixed venous blood returns only 150 ml to the heart and the lungs. Therefore we calculate the cardiac output:

$$\text{Cardiac output in liters per minute} = \frac{O_2 \text{ consumption (ml)}}{\text{AV difference in } O_2 \text{ (ml/liters)}} = \frac{250}{50} = 5 \text{ liters.}$$

What we are really saying is that if we know the O_2 consumption has been at a rate of 250 ml/min and that the delivery of O_2 by the blood is such that it requires 1.0 liter blood to deliver 50 ml, then 250 ml would require the service of 5 liters of blood to deliver the 250 ml we know have been consumed. This is simple and direct, but unfortunately the calculation requires the value of O_2 in *mixed* venous blood, which can only be gotten by catheterization of the right heart to get a measure of truly mixed venous blood. This procedure involves inserting a tube into an arm vein and feeding it through the vein into the right ventricle or pulmonary artery. Although this is a standard hospital procedure for evaluation of cardiac patients, it is not often used in exercise physiology.

Noninvasive techniques have been developed which involve the rebreathing of CO_2 and the estimation of mixed venous CO_2 from the rate at which CO_2 approaches a plateau. (The Fick principle is applied just as it was above, but CO_2 values are substituted for O_2 values.) This method is attractive in that no trauma is involved for the subject and it is feasible in the well-equipped exercise physiology laboratory. It is reasonably accurate and reproducible under exercise conditions, but the author has found the reproducibility under resting conditions to be poor (3).

Control of Heart Rate. The rate of the heartbeat is determined by the frequency of impulse generation at the SA node. The activity of the SA node, in turn, is controlled by other factors, most important of which seems to be the effect of autonomic innervation upon the SA node. The autonomic nervous system supplies both parasympathetic and sympathetic fibers to the SA node. Parasympathetic fibers are supplied by the *vagus nerve,* and sympathetic fibers by the *accelerator nerve.* In both cases the innervation arises in the cardioregulatory centers of the medulla.

It has long been known that severing the vagus fibers to the SA node results in an immediate quickening of the heart rate, and thus we know that these fibers are inhibitory, and also that they exhibit *tone;* that is, they are chronically active. Further, the accelerator fibers also are constantly active, in the other direction; they have an accelerating effect called *accelerator tone.* Thus the heart rate is precisely adjusted by a balance of activity of the two divisions of the autonomic system.

For the adjustment of the heart rate to be meaningful and to serve the changing needs of the organism, it must reflect the demands of changing metabolic activity. To understand how this servomechanism works, we must know the origin of the afferent impulses that form the sensory side of the cardioregulatory reflexes, whose center lies in the medulla. These afferent nervous impulses come from a number of sources. The following impulses increase heart rate.

1. *Proprioceptive impulses* from the working muscles and joints
2. Impulses arising in the chemoreceptors of the *carotid body* and the *aortic body*
3. Impulses arising in the *cerebral cortex*
4. Impulses arising in the heart itself (see fig. 6.4)

The sources of depressor afferent stimuli are mainly from the *stretch receptors* of the *carotid sinus,* the *aortic arch*, and the heart itself (25). See fig. 6.5

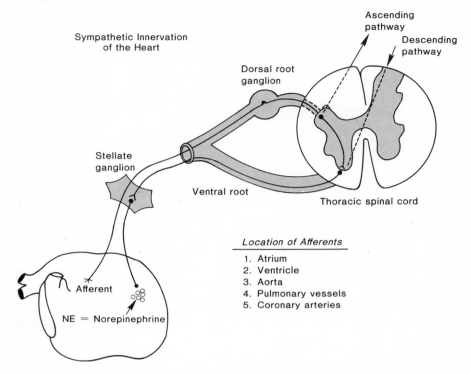

Figure 6.4 Schematic cross section of the spinal cord in the thoracic region showing the pathway taken for afferent and efferent sympathetic fibers. The known locations of sympathetic afferent nerve endings are listed in the figure. (From Stone, H.L. *Med. Sci. Sports* 9:253, 1977.)

The effects of the autonomic nervous system in controlling the heart rate in an exercise bout may be summarized in three phases.

1. The anticipatory rise in heart rate, frequently seen before exercise begins, is undoubtedly the result of reflexes that arise in the cerebral cortex, due to factors such as anxiety and excitement.
2. As exercise begins, stimuli from the working muscles influence the cardioregulatory centers (mainly, they inhibit the cardioinhibitory center), which results in a release from vagal inhibition, and consequently produces an increased heart rate (6, 10).
3. As exercise continues beyond thirty or forty seconds, increases in heart rate—if the exercise is demanding—are mainly brought about by increases of tone in the accelerator nerve, and this through stimulation of the chemoreceptors in the carotid and aortic bodies by decreasing pH and increasing concentrations of carbon dioxide.

On the other hand, the frequency of the impulses in the pressure receptors of the carotid sinus and the aortic arch rises with increasing blood pressure and exerts a mitigating influence to prevent too high a rise in blood pressure. The receptors also function in bringing about an increase in heart rate (and blood pressure) when their discharge slows because of decreasing blood pressure as occurs, for example, in a change of posture from lying to standing.

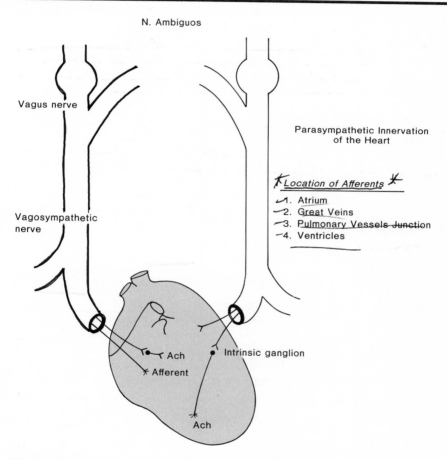

Figure 6.5 Schematic representation of the parasympathetic innervation of the heart. Vagal afferent and efferent fibers course in the vagosympathetic trunk in the cervical region. The cardiac branches leave the main nerve trunk near the heart. The locations of cardiac vagal afferent nerve endings are given in the figure. (From Stone, H.L. *Med. Sci. Sports* 9:253, 1977.)

In addition to the nervous control of the heart rate, at least two other factors are of considerable importance. An increase in adrenal activity also plays a large part in the increased heart rate of the third phase (described above). Both the nervous and the hormonal effects are superimposed, as it were, on the basic effects of body temperature on the heart rate. Increased body temperature results in increases in heart rate, and vice versa, in accordance with the reaction of all metabolic processes to temperature.

Control of Stroke Volume. Until relatively recently, the stroke volume of the heart had been considered a simple and direct function of the end diastolic volume of the ventricle, a conclusion based on the classic work of Frank and Starling. Starling's law of the heart stated that the mechanical energy set free on passage from the resting to the contracted state depends on the length of the muscle fibers at the end of diastole; and this is in agreement with the length-tension law of skeletal muscle discussed in chapter 3. However, X-ray motion pictures and X-ray kymograms have shown that the end diastolic volume of the heart is not larger in exercise than at rest, and may even be smaller (8). Furthermore, whether the results of the Frank and Starling experiments on thoracotomized (open chest) dogs can be extrapolated to normal intact dogs—let alone to intact man—is open to question (9). Because the classic work had been done on heart-lung preparations in dogs, these limitations were serious challenges to its validity, and further research has shed more light on the mechanisms that control the stroke volume of the heart (2, 11, 16, 17).

First of all, there is no reason to challenge Starling's law as a determinant of stroke volume in a heart-lung preparation. Second, there is every reason to believe that a physiological law that can be as elegantly displayed as Starling's law of the heart must have some use for the intact organism. However, the weight of accumulating evidence indicates that although end diastolic volume may be a determining factor under certain conditions (postural changes and gravitational changes), several other factors are probably more important in the adjustments of stroke volume to the demands of exercise.

It would seem that the stroke volume of the heart is controlled by a physiological interplay of at least four factors:

1. Effective filling pressure
2. Distensibility of the ventricle in diastole
3. Contractility
4. Systemic arterial blood pressure

The first and second items are, in a sense, expressions of Starling's law; the third item needs elaboration. The *contractility* of the heart means its ability to produce force per unit of time or, in other words, power. Randall (16) has summarized the evidence that demonstrates increased contractility as the result of sympathetic stimulation. Under sympathetic stimulation, systolic pressures rise much faster and reach higher levels because of the development of augmented myocardial fiber tension. This means that, in addition to the obvious advantage of a more forceful beat, there is greater time for diastolic filling because systole is completed more rapidly within each cardiac cycle, resulting in even more advantage to the succeeding cycle due to the complete filling. Roskamm (18) has shown clearly that increases in heart rate and contractility are the important factors for humans during heavy exercise.

The importance of the fourth factor, systemic arterial blood pressure, is rather obvious in that the magnitude of this pressure is the resistance against which the blood must be ejected into the aorta. As this resistance grows higher, the stroke volume must inevitably grow smaller for any given force of contraction. However, this factor achieves importance only at relatively heavy exercise loads.

To summarize, Starling's law of the heart, while entirely valid in the experimentally controlled heart, is not nearly so important in the physiological control of stroke volume at exercise as was once believed. Furthermore, the law should probably be amended, as suggested by Rushmer (19) ". . . the energy released during contraction is related to the initial length of the muscle fibers *under equal states of responsiveness*."

Starling's law is probably quite important when the venous return is altered due to such changes as those in posture, without concomitant changes in innervation or humoral content of the blood. During exercise, then, the most important changes in ventricular performance for providing the needed increase in cardiac output are (1) an accelerated heart rate and (2) an increased contractility due to increased sympathetic stimulation of the ventricular myocardium, which can result in a 100% increase in cardiac output (see fig. 6.6).

Importance of the Venous Return. With the accumulating evidence against a greater end diastolic volume during exercise, there may be a tendency to forget the need for increased venous return during exercise. It is an obvious truism that the heart cannot eject more blood than it receives. Therefore, although the increase in venous return can no longer be considered the single determining factor in increasing cardiac output, it is nevertheless an important limiting factor in cases where venous return may be curtailed in exercise, as in hot and humid environments. Furthermore, it is undoubtedly the major factor in increasing cardiac output when an erect man lies down; venous return is indeed temporarily increased due to the smaller effect of the force of gravity upon the circulatory system. The reduction of cardiac output in the change from the supine position to the standing position must be explained on a similar basis, and will be discussed at greater length in chapter 7.

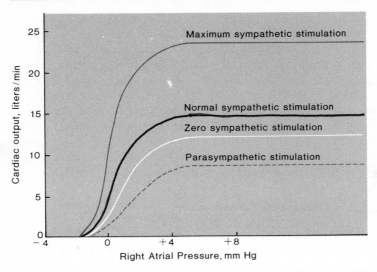

Figure 6.6 Effect on the cardiac output curve of different degrees of sympathetic and parasympathetic stimulation. (From Guyton, A.C. *Textbook of Medical Physiology,* 1976. Courtesy of W.B. Saunders Company, Philadelphia.)

Coronary Circulation and Efficiency of the Heart

Cardiac output is one of the limiting factors in determining the level at which work output can be maintained because skeletal muscle tissues depend on a constant supply of oxygen from the blood to maintain their metabolic needs. As a consequence, the supply of blood to the heart muscle might be supposed to be the determinant of the upper limits of cardiac output, and thus indirectly a determinant of the maximum exercise load to be sustained by the skeletal muscles.

This, as it turns out, seems to be very sound reasoning. Arterial blood usually contains approximately 19 ml of oxygen per 100 ml of blood. In mixed venous blood, this may be reduced to 12 to 14 ml per 100 ml of blood at rest, thus leaving some room for greater utilization in exercise. In the coronary veins under resting conditions, however, the oxygen content has already been reduced to 4 to 6 ml per 100 ml of blood. Thus the oxygen extraction from blood in the coronary vessels is relatively high even at rest, and consequently leaves little coronary venous oxygen reserve for the demands of exercise loads. For these reasons, the maximum sustained cardiac output is limited by two factors: the volume of the coronary blood flow and the efficiency of the heart muscle in performing its work.

Basic Physiology Underlying the Study of Physiology of Exercise

Coronary Circulation. Let us discuss first the control of the coronary blood flow and its adjustment to the demands of exercise. It is obvious that the blood flow through the heart is dependent upon two factors: (1) the difference in pressure (pressure gradient) between the entering arterial blood and the venous outflow, and (2) the resistance to the flow, which in turn depends on the state of vasoconstriction or vasodilatation of the coronary vessels. The first factor, the pressure gradient, is determined largely by aortic pressure, since the coronary arteries derive from the aortic sinus. The resistance to flow is probably the more important factor of the two. Decreased oxygen tension has been shown to have a very strong dilatory effect upon the coronary vessels.

During exercise several factors tend to increase myocardial O_2 requirements and therefore would operate to decrease coronary vascular resistance and thus increase coronary blood flow:

1. Heart rate (tachycardia) ← *means heartbeat*
2. Arterial pressure
3. Cardiac output
4. Left ventricular work
5. Myocardial contractility

Experiments on dogs have shown that about one-third of the increment in coronary flow and three-fourths of the decrease in coronary resistance during severe exercise can be accounted for by the tachycardia, unaffected by any other factor (28).

It seems, then, that increased coronary flow during exercise is brought about by at least two factors: (1) the increased arterial blood pressure provides a greater pressure gradient for driving blood through the myocardium, and (2) as the myocardial tissue increases its oxygen consumption, the lowering levels of oxygen result in dilatation of the coronary vessels.

Efficiency of the Heart. Before we enter this discussion, we must define some terms. *Efficiency,* in general, is the ratio of work production to energy input; thus, in this case:

$$\text{Efficiency} = \frac{\text{Work of the heart}}{\text{Energy input (in terms of } O_2 \text{ consumption)}}$$

The *work of the heart* can be measured, as can any work done by fluid pressure:

$$\text{Work} = \text{Pressure} \times \text{Volume moved}$$

It should be noted here that pressure and volume moved (or flow rate) are independent of each other; that is, an increase in pressure may be associated with an increase, a decrease, or no change at all in flow rate.

A simple law of physics, the law of LaPlace, can also help us understand the cardiac function:

Tension (in the cylinder wall) = Pressure × Radius of cylinder

This formula tells us that the tension in the muscle fibers of the myocardium must be proportionately greater as the pressure rises. In other words, the tension required of the heart to move a given volume of blood increases as the arterial blood pressure rises. The formula also tells us that to move the same quantity of blood under equal pressures, the large heart (of greater radius) must exert greater tension than the small heart. Because the energy requirements of the heart are determined largely by the tension demanded of it, this means that a small heart is more efficient than the large heart in working against pressure, other things being equal.

The work of Sarnoff and his associates (20, 21) illustrates the conformance of practice to theory in this regard. Ingeniously devised experiments allowed measurement of the oxygen utilization of the myocardium of dogs when aortic pressure was increased and cardiac output and heart rate were held constant. In these *pressure runs* it was found that a 175% work increase was accompanied by a 178% increase in oxygen consumption. On the other hand, in *flow runs* where the work output was increased 696% through increasing the cardiac output while holding aortic pressure and heart rate constant, an oxygen consumption increase of only 53% was noted. Even this 53% seemed to be accounted for by an inadvertent increase in aortic pressure. Thus a striking difference in cardiac efficiency is seen when the work output of the heart is increased by pressure-load increases compared with volume-load increases.

Another very important observation in this series of experiments was that if the amount of work done by the heart was held constant by holding the cardiac output and the mean aortic pressure constant while increasing the heart rate, the oxygen consumption of the myocardium increased. Consequently, we may say that *high heart rates are less efficient than low rates,* other things being equal.

The researchers concluded that the principal, if not the sole, determinant of myocardial oxygen utilization is the *tension-time index* (mean systolic aortic pressure × duration of systole).

Recent work has shown that a reasonably valid estimate of the myocardial O_2 consumption (equivalent to work load stress on the heart muscle) can be made even from the heart rate-blood pressure product using systolic pressure measured with a cuff (12, 13). This is important in any conditioning or testing program involving middle-aged or older adults, since the work of the heart is not always proportional to the work of the total body. Data from the author's laboratory will be presented in chapter 16 to illustrate this point.

Some practical conclusions can be drawn from the preceding discussion that have large implications for exercise physiology. For any given set of conditions, (1) the heart of a subject with abnormally high blood pressure must work much harder than that of the normal subject; (2) the slower the heart rate for any given work load, the more efficiently the cardiac work is performed.

Factors Affecting the Heart Rate

The heart rate at rest varies widely from individual to individual and also within the same individual from one observation to another under similar circumstances; therefore, it is almost meaningless to speak of a *normal* heart rate. We may, however, say that the *average* heart rate is 78 beats per minute without implying that a rate of 40 (observed in highly trained endurance athletes) or 100 is necessarily *abnormal*. Although heart rate during the stress of exercise or during the recovery period after exercise is a very valuable source of information for the exercise physiologist, the resting rate is affected by so many variables that it has very little meaning for the prediction of physical performance. Some of the factors that affect the resting rate will now be discussed.

Age. The heart rate at birth is approximately 130 beats per minute, and it slows down with each succeeding year until adolescence. The average rate in a resting adult male in standing position is approximately seventy-eight. The maximal attainable heart rate decreases with increasing age in the adult.

Sex. The resting heart rate in adult females averages five to ten beats faster than it does in adult males under any given set of conditions.

Size. In the animal world in general, it seems to be a biological rule that the heart rate varies inversely with the size of the species. For example, the canary has a rate of approximately 1,000 beats per minute, whereas that of an elephant is about 25 beats per minute. However, no consistent relationship between size and heart rate in adult humans has been demonstrated.

Posture. Posture has a very definite effect upon heart rate. Although the results of different investigators show variances, the typical response to the change from recumbent to standing position seems to be an increase of ten to twelve beats per minute. It was at one time thought that this change in heart rate due to posture was related to physical fitness, and consequently it has appeared as a test item in some tests of physical fitness; however, its value in this regard is open to serious doubt.

Ingestion of Food. The resting heart rate is higher while digestive processes are in progress than in the postabsorptive state. This is also true in exercise; a given exercise load elicits a greater heart rate after a meal, and this is one of many reasons that militate against heavy exercise immediately after a meal.

Emotion.   Emotional stress brings about a cardiovascular response that is quite similar to the response to exercise. An increase in heart rate is the most notable factor, and it occurs in all but the most experienced athletes as an anticipatory reaction. Dill (4) found a mean increase of nineteen beats per minute in the resting rate of teenage boys waiting to be tested in his laboratory. The effect of emotional excitement is most easily observed at rest, but it also occurs during exercise, where it tends to result in an excessive cardiovascular response (1). Under these conditions the response to a standard exercise load may be considerably greater, with the heart rate being elevated by the summation of the stimuli from exercise and from the emotional situation. The recovery period may also be unduly prolonged.

Body Temperature.   With increases in body temperature above normal, the heart rate increases. Conversely, with decreases in temperature, the rate slows, until a temperature of about 26°C is reached, at which temperature abnormal electrocardiograms are obtained that show danger of heart failure.

Environmental Factors.   Ambient temperature is one of the most important factors affecting the heart rate and the total cardiovascular response to exercise. In moderate exercise, an increase of from ten to forty beats per minute may occur, depending upon the magnitude of the temperature rise. At rest, small increases in heart rate are seen as temperature increases. However, humidity and air movement also are factors. For any given temperature and work load, the rise in heart rate will be greater if the humidity is high and the air is motionless.

Effects of Smoking.   It has been found that smoking even one cigarette significantly increases the resting heart rate, in either the sitting or the standing position (23).

The Heart Rate during and after Exercise

The ready availability of the pulse rate as a measure of what transpires internally has resulted in the accumulation of much interesting data that relate various exercise conditions and heart rate. We have pointed out that cardiac output is a determinant of how large an exercise load may be tolerated and that cardiac output is the result of two components: heart rate and stroke volume. Since heart rate is easily measured, it is indeed fortunate that research has shown heart rate to be the more important variable in response to the demands of exercise. The greater importance of heart rate is due to at least three reasons.

1. Stroke volume probably increases very little with an increase in metabolism until a level approximately eight times the resting level is reached.
2. Heart rate is proportional to the work load imposed.
3. Heart rate is proportional to the oxygen consumption during an exercise.

All three of these factors hold true only during the steady state, however, when the work is done aerobically.

It is obvious from preceding discussions that the slower the heart rate in response to a given exercise work load, the more efficient is the myocardium—again for at least three reasons.

1. The oxygen consumption of the heart increases with increasing heart rate, even though the work load is held constant (20).
2. As the heart rate increases, the filling time decreases.
3. Diastasis, the only resting period for the myocardium, is disproportionately shortened in faster rates, and may disappear entirely at high rates.

All things considered, the rate of the heartbeat furnishes data that quite accurately reflect the degree of stress created by an exercise work load; conversely, it provides insight into the adequacy of physiological responses to the exercise.

The Typical Heart Rate Response to Exercise.   As exercise begins, the pulse rate elevates very rapidly. If the exercise is light or moderate, a plateau (leveling off) is seen in thirty to sixty seconds, and this pulse rate is relatively constant until cessation of the exercise. This rate is proportional to the work load of the exercise in any individual. If the work load is heavy (ten or more times the resting metabolic rate), the rate increases until exhaustion supervenes (fig. 6.7). For the first two to three minutes after the end of the exercise, the heart rate decreases almost as rapidly as it increased. After this initial decrease, further decline in the heart rate occurs more slowly at a rate that is roughly related to the intensity and duration of the work.

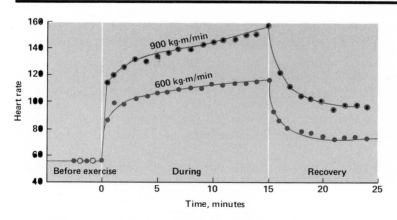

Figure 6.7 Heart rate changes in a moderately conditioned, middle-aged subject at work loads of 600 and 900 kg-m/min on a bicycle ergometer.

Response of the Heart Rate to Differences in Exercise

Static versus Dynamic Exercise. In exercises that involve a held position, or a straining of the musculature against a heavy load, as in weight lifting, only a very slight increase in heart rate is observed. At the other end of the continuum are exercises that involve rapid and vigorous alternating contractions, such as running and cycling, in which a large increase in heart rate occurs. This difference can be explained on two bases.

1. Venous return may be decreased in the straining exercise due to the increased intrathoracic pressure (chapter 7), and it may be increased in dynamic exercise due to the pumping action of the muscles.
2. The heart rate is proportional to the work load per unit of time, and slow, straining exercises seldom create a work load sufficient to bring about large responses in heart rate.

Intensity of the Exercise. Intensity is the number of foot-pounds (or kilogram-meters) of work per minute expended in an exercise. Since *Work = Force × Distance,* in a simple exercise such as bench-stepping, the work load and thus the intensity may be increased by increasing the height of the bench, by increasing the rate of stepping, or by increasing both. Both factors would, of course, increase the distance per unit of time, and the force could be increased by having the subject carry a weight on the back. Any and all of these factors increase the intensity of the exercise, which is the most important factor in determining the heart rate during exercise.

Duration of the Exercise. If a moderate work load is maintained over a considerable period of time, a secondary increase in heart rate can be observed after the plateau has been attained and held. This can best be explained in terms of fatigue of the skeletal musculature; larger numbers of motor units are recruited, which results in a greater metabolic demand for the same level of exercise work load and thus an increase in the heart rate. This secondary increase is usually progressive, and continues until exhaustion ends the exercise.

Rest Periods in Discontinuous Exercise. Our discussion has centered about work or exercise that is done continuously. However, because of the advent of interval training methods in track, swimming, and many other sports, a discussion of the importance of rest periods and of their interaction with work load in influencing the stressfulness of the exercise or workout is in order.

Although both resting and exercise heart rate decrease with training, the physiology involved is quite different. Experiments involving autonomic blockade (7, 27) suggest that the lowered resting rate is due to enhanced parasympathetic influence, but the decreased exercise rate is due to a decreased sympathetic drive.

For a light work load, a rest period of constant size can result in a return of the heart rate to the prework level after many work sessions; but if the intensity of the work session is increased beyond a certain point, recovery is no longer completed in each successive rest period, unless the length of the rest periods is also progressively increased. On the other hand, if the rest interval is held constant and the intensity is increased, a progressively ascending heart rate will be observed during the work or exercise, and exhaustion will result if the work is carried on too long. A coach will, of course, end a workout after sufficient fatigue has occurred in bringing about the desired training effect and before exhaustion ensues.

In industry, on the other hand, where a six- to eight-hour day is involved, the rest periods must be adequate to prevent a progressive heart rate increase, or the intensity of the work must be decreased. In general, then, the stress of a day's work, or of an interval of a training and conditioning program, is the result of an interaction of two components: (1) total work done and (2) total amount of rest periods. The final recovery of the heart rate will be the result of these two components.

Heart Rate as a Measure of Stress. It is frequently desirable to evaluate the degree of stress imposed upon an individual's cardiovascular system by work loads in athletics and in industry. This is particularly important where the total load is increased by unfavorable environmental conditions, such as high ambient temperatures (see chap. 24). Again, observation of the heart rate response to the work load provides the easiest and quickest evaluation.

Use of the cardiac cost concept provides reasonably valid information on the total stress an individual's cardiovascular system is subjected to under conditions in which the stress of exercise is complicated by the summation of the exercise stress and the stress imposed by the environment.

Cardiac cost = Total heartbeats during exercise minus resting rate for the same period of time

Recovery cost = Heartbeats between end of exercise and return to resting minus resting rate for the recovery period

Total cardiac cost = Cardiac cost during exercise plus cardiac cost during recovery

Effects of Athletic Training on the Heart

Heart Rate. As training progresses, the heart rate for any given work load decreases, as is illustrated in figure 6.8. We may also say that, other things being equal, the physically fit or athletically trained individual has a lower heart rate for any given exercise work load. Furthermore, at the maximum heart rate which is similar for the trained and the untrained states, the trained individual will be able to produce a greater work load.

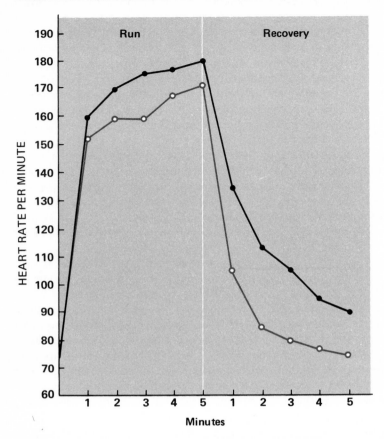

Figure 6.8 Effect of training on the heart rate of an oarsman doing a standard amount of exercise. Solid dots represent out-of-training values; open circles represent in-training values. (From Brouha, L.A. *Work and the Heart,* F.F. Rosenbaum and E.L. Belknap, eds., 1959, chap. 21. Courtesy of Harper & Row, Publishers, New York.)

Stroke Volume. A preponderance of evidence indicates that the increased maximum cardiac output in athletes is due largely to an increase in stroke volume. This is not to say that the immediate adjustment to exercise, which was discussed earlier, is the result of an increased stroke volume; rather, the stroke volume increase seems to be a long-term effect of training and is also manifest in the slower resting rate of the endurance-trained athlete. This greater stroke volume is the result of the greater contractility of the ventricle in the heart of the trained athlete (2, 18, 22, 26).

Basic Physiology Underlying the Study of Physiology of Exercise

Heart Size. The effect of exercise upon the heart in animals is well established. Many investigators have found increases in the weight of the heart in different species as the result of heavy endurance exercise. There is a temptation to extrapolate these data to humans, but there is little direct evidence, or agreement, on the effects of exercise upon heart size in humans.

Some investigators have found increases in heart size that they attribute to (1) physiological hypertrophy of the musculature, (2) greater diastolic size due to the greater filling at the slower heart rate, or (3) greater end systolic size at rest and a smaller size as exercise commences. Still other investigators have found that the heart size of athletes is within the normal range.

A change in heart size would be a normal physiological reaction, in all respects similar to the physiological hypertrophy of skeletal muscle, as a result of training. It is certainly not to be confused with the pathological increase of heart size that occurs as a compensatory mechanism for heart disease and that imposes greater loads upon the myocardium twenty-four hours a day for many years. There is no evidence that strenuous exercise can injure a normal heart.

Training Effects at the Cellular Level

In chapter **3** we have discussed the large and significant cellular adaptations that occur in skeletal muscle in response to endurance training. It would be reasonable to expect similar adaptations in the myocardium. While direct observations of cellular change in the skeletal muscles of humans have been made possible by the muscle biopsy technique, we must rely on extrapolation from animal experiments with respect to myocardial adaptations.

Surprisingly, ●scai and c●-workers have shown that in rats the heart does not appear to undergo any adaptive increase in cellular respiratory capacity in response to endurance exercise (14, 15). They found the levels of activity of various representative mitochondrial enzymes and the concentrations of cytochrome c and of mitochondrial protein to be unchanged in trained rats. More recent work on dogs has supported their work (24), so it appears unlikely that training adaptations in the human heart can be explained on this basis.

Since the heart muscle in the human, as in the animal preparations, shows large improvements in contractility with endurance training, we might expect some improvement in myofibrillar ATPase activity to accompany the training regimen, but three groups of investigators have shown that this is also not the case (5, 24, 26). At the present time the best available evidence suggests that the improved contractile function of the heart in response to endurance training lies in subtle changes in the mechanisms of calcium transport into and out of the sarcoplasmic reticulum (24, 26). See chapter 3 for the importance of calcium in initiating muscle contraction.

The Cardiac Reserve Capacity

Vigorous physical exercise provides the greatest challenge to the cardiovascular system. In the normal individual, only the increased metabolic demand of strenuous work loads can raise the output of the heart to its maximum values. Let us now consider the mechanisms available to the heart in its adjustment to supply the oxygen needs of exercised muscle tissue. It is convenient to think of these mechanisms, in total, as the *cardiac reserve*.

The first and probably the most important factor is the *reserve of heart rate*. In exercise, the rate can increase from its resting value of 70 to 80 beats per minute to a rate of 170 to 180 beats per minute. Although the heart rate can exceed these values, cardiac output will not be improved above approximately 180 because stroke volume will decrease as a result of the decreased diastolic filling times. Thus the cardiac output can be increased 2 to 2½ times by increases in heart rate.

The second factor is the *stroke volume reserve*. As was mentioned before, this factor probably comes into play only at a very high level of exertion. However, at very high levels of oxygen consumption, the heart has two means by which stroke volume may be increased:

1. By reducing the blood volume remaining at the end of systole, a more complete ejection as the result of greater contractility.
2. By increasing the diastolic filling as the result of a greater effective filling pressure and possibly a more distensible ventricle.
   ↳ CAN BE STRETCHED MORE

The third factor is the possibility for *increased oxygen utilization* by the active muscle tissues. At rest, the arterial oxygen level supplied to the muscles is approximately 19 ml per 100 ml of blood and the venous blood may have 13 to 14 ml per 100 ml of blood with a consequent utilization of 5 to 6 ml per 100 ml of blood supplied. However, when the muscle is exercising strenuously, 16 to 17 ml of oxygen per 100 ml of blood may be extracted, which results in a much larger arteriovenous oxygen difference—mainly at the expense of a lowered venous oxygen content in the blood flow from the active muscles.

Heart Murmurs

A heart murmur is the sound created by turbulent blood flow. The degree of turbulence necessary to create vibrations that can be heard is determined in the cardiovascular system by the velocity of blood flow, and by the amount of eddy currents caused by obstructions, restrictions, etc. Because the flow within the vascular system is normally streamlined (or laminar) and not turbulent, ordinarily no murmurs are heard; however, the velocity of flow in the roots of the aorta and in the pulmonary artery is sufficient to create murmurs during the rapid ejection phase of systole. These normal sounds, when they are heard or recorded, are termed *functional murmurs* and have no pathological significance.

On the other hand, cardiac valves are frequently damaged by disease, and the murmurs caused by regurgitation of blood through an insufficient valve, or by the increased velocity due to a damaged valve (which creates a restriction in the orifice), are valuable diagnostic signs to the physician. The physical educator should at least understand the significance of these murmurs.

The significance of murmurs that result from valvular insufficiency is that in pumping a given volume of blood to the tissues to meet metabolic demands the heart must, in a sense, pump the blood that regurgitated twice. This results in a greater-than-normal *volume load* on the heart, and it responds by increasing its stroke volume through an increase in size (and probably by hypertrophy). In doing this, however, a portion of the cardiac reserve is lost because greater tension is required in the myocardial wall to maintain a given blood pressure in the systemic arteries, according to the law of LaPlace. Because our discussion of the tension-time index indicated a rise in oxygen consumption with tension, it is apparent that the heart is now less efficient. This lessened efficiency may be so slight in some cases as to be a limiting factor only in very strenuous exercise, but in other cases it may severely limit the exercise tolerance.

In a murmur that is due to a valvular restriction, the result is an increased *pressure load,* and the heart responds by hypertrophy of the ventricular walls. Again, there is increased tension in the wall of the myocardium of even greater magnitude and a consequent lessening in efficiency. Obviously, participation of individuals with heart murmurs in strenuous exercise or athletics should be under the control of a physician. Present-day physicians and cardiologists recognize the advantages that accrue from exercise programs commensurate to the needs of individuals and their limitations. Both professions would be immeasurably helped in their efforts to attain a corrective physical education program if the physical educator and the physician were able to communicate, in quantitative terms, on exercise work loads. These terms might be kilogram meters per minute of work, liters of oxygen per minute, or the number of times the work load demand is greater than the resting metabolic demand. We will discuss this further in the second section of this text.

SUMMARY

1. The heart and the vascular system maintain *homeostasis* of the various tissues of the body. Most important to the physical educator is an understanding of oxygen and carbon dioxide transport to and from the skeletal muscles.

2. Although the nervous impulse that stimulates the heart to beat originates within the myocardium, the rate is controlled to a large extent by a balance of the effects of the parasympathetic and sympathetic divisions of the autonomic system, acting through the *vagus* and *accelerator nerves,* respectively.

3. All cardiac valve action is the result of differences in pressure on the two sides of each valve. The cardiac cycle can best be understood in light of the pressure changes that occur in the heart.

4. One of the most important factors that limits human physical performance is the ability to increase cardiac output. Cardiac output $=$ Heart rate $\times$ Stroke volume. Consequently, the cardiac output can be increased by an increase in rate, or stroke volume, or both.

5. The rise in heart rate with exercise seems to occur in three phases: (a) a preexercise, anticipatory rise due to cortical activity, (b) an early rise due largely to inhibition of vagal activity, and (c) a later rise that is probably the result of a combination of accelerator nerve activity and increased adrenal activity.

6. Starling's law undoubtedly operates under all conditions, but it is the principal determinant of stroke volume only in changes of posture and other manifestations of gravitational changes. Under conditions of exercise, stroke volume is more likely determined by an interplay of the following factors: (a) effective filling pressure, (b) the distensibility of the ventricle in diastole, (c) contractility, and (d) the systemic arterial blood pressure.

7. Coronary blood flow is the limiting factor in cardiac output responses, and thus it indirectly sets the upper limit of exercise tolerance. Probably the most important factors that affect the increased coronary flow necessary for exercise conditions are (a) increased arterial blood pressure and (b) lowering levels of oxygen in the myocardium to bring about vasodilatation of the coronary vessels.

8. Many extrinsic factors affect the resting heart rate, and in many cases, the rate during exercise. All of the following factors must be considered in observing the effects of exercise upon heart rate: age, sex, size, posture, ingestion of food, emotion, body temperature, environmental factors, and smoking.

9. The heart rate responds to light or moderate exercise loads with a rapid increase to a plateau, at which the rate is proportional, in any individual, to the work load and to the oxygen consumption. In very heavy exercise, the rate continues to increase without leveling off, until exhaustion ends the work bout. After exercise ends, the return to normal is very rapid for the first two to three minutes, then slows considerably to a rate of decrease that is roughly related to the intensity and duration of the exercise.

10. The heart rate responds differently to different types of exercise. In general, dynamic muscular activity brings about a much greater increase in heart rate than static, straining types of exercise. Increases in heart rate vary directly with the intensity and duration of the exercise, and inversely with the number of rest periods in a long, continued work bout.

11. *Cardiac cost* has been proposed as a method for evaluating the stress imposed upon the cardiovascular system by a combination of exercise and environmental stress. Total cardiac cost includes the increase in heart rate during the work bout and the increase during recovery.

12. Athletic training brings about a complex of changes in heart rate, stroke volume, and other factors, all of which together interact to bring about a more effective and efficient adjustment of the organism to the increased metabolic demands of exercise. There seems to be no evidence that indicates a normal heart is harmed by the stress of exercise.

13. Although important training effects on the cellular respiratory capacity have been shown for skeletal muscle (chap. 3), this is not the case for heart muscle. The improved contractility of heart muscle in endurance training appears to be the result of better calcium transport in the sarcoplasmic reticulum.

14. The *cardiac reserve* for conditions of strenuous exercise is the result of three factors: (a) increase in heart rate, (b) increase in stroke volume, (c) increased oxygen utilization of the active muscle tissues.

15. Heart murmurs are heart sounds that are brought about by increased turbulence in the blood flow through the heart or large vessels. This increased turbulence can be the result of increased velocity or of increased eddy currents in the blood flow. In some cases the velocity reaches the critical value without the existence of an abnormality, and the murmur is said to be *functional*. Heart murmurs are usually brought about by a valvular insufficiency that allows regurgitation of blood through a small opening at high velocity, or by restriction of a valvular orifice as a result of disease. Exercise or athletic participation under these conditions should be evaluated and controlled by the individual's physician.

REFERENCES

1. Antel, J., and Cumming, G.R. Effect of emotional stimulation on exercise heart rate. *Res. Q.* 40:6–10, 1969.

2. Asmussen, E., and Nielsen, M. Cardiac output during muscular work and its regulation. *Physiol. Rev.* 35:778–800, 1955.

3. deVries, H.A. Physiological effects of an exercise training regimen upon men aged 52–88. *J. Gerontol.* 25:325–36, 1970.

4. Dill, D.B. Regulation of the heart rate. In *Work and the Heart,* eds. F.F. Rosenbaum and E.L. Belknap. New York: Paul B. Hoeber, Inc., 1959.

5. Dowell, R.T.; Stone, H.L.; Sordahl, L.A.; and Asimakis, G.K. Contractile function and myofibrillar ATPase activity in the exercise-trained dog heart. *J. Appl. Physiol.* 43:977–82, 1977.

6. Fagraeus, L., and Linnarsson, D. Autonomic origin of heart rate fluctuations at the onset of muscular exercise. *J. Appl. Physiol.* 40:679–82, 1976.

7. Frick, M.H.; Elovainio, R.O.; and Somer, T. The mechanism of bradycardia evoked by physical training. *Cardiologia* 51:46–54, 1967.

8. Gauer, O.H. Volume changes of the left ventricle during blood pooling and exercise in the intact animal: their effects on left ventricular performance. *Physiol. Rev.* 35:143–55, 1955.

9. Gregg, D.E.; Sabiston, D.C.; and Thielen, E.O. Performance of the heart: changes in left ventricular end-diastolic pressure and stroke work during infusion and following exercise. *Physiol. Rev.* 35:130–36, 1955.

10. Hollander, A.P., and Bouman, L.N. Cardiac acceleration in man elicited by a muscle-heart reflex. *J. Appl. Physiol.* 38:272–78, 1975.

11. Katz, L.N. Analysis of the several factors regulating the performance of the heart. *Physiol. Rev.* 35:91–106, 1955.

12. Kemp, G.L.; Ellestad, M.H.; Beland, A.J.; and Allen, W.H. The maximal tread mill stress test for the evaluation of medical and surgical treatment of coronary insufficiency. *J. Thorac. Cardiovasc. Surg.* 57:708–13, 1969.

13. Kitamura, K.; Jorgensen, C.R.; Gobel, F.L.; Taylor, H.L.; and Wang, Y. Hemodynamic correlates of myocardial oxygen consumption during upright exercise. *J. Appl. Physiol.* 32:516–22, 1972.

14. Oscai, L.B.; Mole, P.A.; Brei, B.; and Holloszy, J.O. Cardiac growth and respiratory enzyme levels in male rats subjected to a running program. *Am. J. Physiol.* 220:1238–41, 1971.

15. Oscai, L.B.; Mole, P.A.; and Holloszy, J.O. Effects of exercise on cardiac weight and mitochondria in male and female rats. *Am. J. Physiol.* 220:1944–48, 1971.

16. Randall, W.C. Sympathetic control of the heart-peripheral mechanisms. In *Cardiovascular Functions*, ed. A.A. Luisada. New York: McGraw-Hill Book Co., 1962.

17. Richards, D. Discussion of Starling's law of the heart. *Physiol. Rev.* 35:156–60, 1955.

18. Roskamm, H. Myocardial contractility during exercise. In *Limiting Factors of Human Performance*, ed. J. Keul. Stuttgart: Georg Thieme, 1973.

19. Rushmer, R.F. Applicability of Starling's law of the heart to intact, unanesthetized animals. *Physiol. Rev.* 35:138–42, 1955.

20. Sarnoff, S.J.; Braunwald, E.; Welch, G.H.; Stainsby, W.N.; Case, R.B.; and Macruz, R. Oxygen consumption of the heart, with special reference to the tension-time index. In *Work and the Heart*, eds. F.F. Rosenbaum and E.L. Belknap. New York: Paul B. Hoeber, Inc., 1959.

21. Sarnoff, S.J., and Braunwald, E. Hemodynamic determinants of myocardial oxygen consumption. In *Cardiovascular Functions*, ed. A.A. Luisada. New York: McGraw-Hill Book Co., 1962.

22. Scheuer, J. The advantages and disadvantages of the isolated perfused working rat heart. *Med. Sci. Sports* 9:231–38, 1977.

23. Schilpp, R.W. A mathematical description of the heart rate curve of response to exercise, with some observations on the effects of smoking. *Res. Q.* 22:439–45, 1951.

24. Sordahl, L.A.; Asimakis, G.K.; Dowell, R.T.; and Stone, H.L. Functions of selected biochemical systems from the exercise-trained dog heart. *J. Appl. Physiol.* 42:426–31, 1977.

25. Stone, H.L. The unanesthetized instrumented animal preparation. *Med. Sci. Sports* 9:253–61, 1977.

26. Tibbits, G.; Koziol, B.J.; Roberts, N.K.; Baldwin, K.M.; and Barnard, R.J. Adaptation of the rat myocardium to endurance training. *J. Appl. Physiol.* 44:85–89, 1978.

27. Tipton, C.M.; Matthes, R.D.; Tcheng, T.K.; Dowell, R.T.; and Vailas, A.C. The use of the Langendorff preparation to study the bradycardia of training. *Med. Sci. Sports* 9:220–30, 1977.

28. Vatner, S.F.; Higgins, C.B.; Franklin, D.; and Braunwald, E. Role of tachycardia in mediating the coronary hemodynamic response to severe exercise. *J. Appl. Physiol.* 32:380–85, 1972.

7

The Circulatory System and Exercise

From the standpoint of exercise physiology, blood is primarily a tissue of respiration. Its importance lies in its ability to transport the respiratory gasses, oxygen, and carbon dioxide between the respiratory organs and the active tissues. Although blood subserves many other very important general physiological functions, this transport of oxygen and carbon dioxide may become a limiting factor in physical performance and consequently assumes major importance.

It is tempting to liken the blood flow through the vascular system to the flow of water through a plumbing system; this is a very poor analogy, however, because the pipes in a plumbing system serve only *passively* as conduits for the transport of fluids whereas the blood vessels are *active* participants in the adjustments made by the circulatory system for the demands of exercise. For example, an individual may increase oxygen intake and energy output to twenty times the basal rate, but the increase of energy output by the *active muscles* may be as much as fifty times the resting rate, and this difference, of course, is the result of a redistribution of blood (and thus oxygen) by the vascular system through a vasoconstriction in inactive tissues and a vasodilatation of the active muscle tissues.

Hemodynamics: Principles Governing Blood Flow

Pressure Gradient. Blood flows through the vessels of the circulatory system because of differences in pressure. It flows from a point of high pressure to a point of lower pressure, and the difference in pressure between the two points is called a *pressure gradient*. In the systemic circulatory system, the point of highest pressure is within the left ventricle of the heart during systole, and the pressure gradient between this point and the lowest pressure point in the right atrium is the driving force that brings about blood flow through the entire systemic circulation (although this is aided by the muscle pump, to be discussed below). An analogous situation exists in the pulmonary circulation.

Velocity of Blood Flow. The combination of systemic circulation and pulmonary circulation may be thought of as the two loops of a figure eight (fig. 7.1). The two loops are interconnected in series, and each has its own pressure gradient, provided by its side of the cardiac pump. It is apparent that this is in reality a closed, single circuit—although possessed of two pressure gradients in series—and it follows that the *same volume* of blood must pass *each* and *any point* in the system per unit time (if we assume that no retention of blood occurs at any point). This being so, the *velocity of blood flow* past any point obviously depends upon the total cross-sectional area of the vascular bed at that point; where the area is smaller, the velocity must be higher, and vice versa (fig. 7.2).

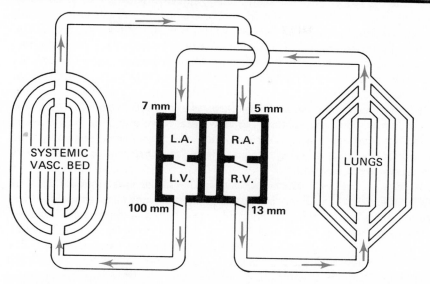

Figure 7.1 Interrelationships of systemic and pulmonary circulations and their pressure gradients.

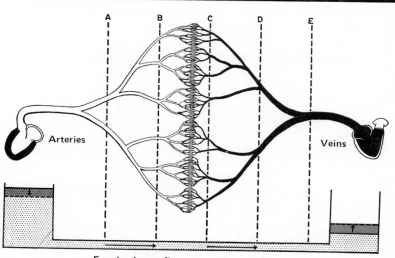

Equal volumes flow past each vertical line

Figure 7.2 Arborization of the systemic circulatory system is schematically represented with all vessels of the same caliber arranged vertically. This simplified illustration emphasizes the fact that the volume of fluid flowing past each vertical line in a unit of time must be equal to the quantity entering and leaving the system, just as in a single tube. (From Rushmer, R.F. *Cardiovascular Dynamics*, 2d ed., 1961. Courtesy of W.B. Saunders Company, Philadelphia.)

The aorta has a cross-sectional area of approximately 2.5 to 5.0 cm². As the large arteries bifurcate, the combined area of the branches considerably exceeds the area of the parent vessel. As this branching of the arterial system progresses, the total cross-sectional area of the circulatory system constantly increases, and consequently the velocity of flow decreases proportionately. In the capillary bed, where each vessel is only 10 μ (1/100 of a millimeter) in diameter, this multiplication has progressed to such an extent that the combined cross-sectional area of all capillaries is about 700 to 800 times the area of the aorta. This means that the flow rate or velocity has decreased in proportion, and the result is a very slow flow of 0.5 to 1.0 mm/second through the capillaries, thus allowing adequate time for the exchange of respiratory gasses, nutrients, and so on.

Resistance to Flow. Resistance may be thought of as the sum of the forces opposing blood flow and may be illustrated by breathing through various sizes and lengths of glass tubing. It is readily seen that exhaling becomes progressively more difficult as the length of the tube is *increased;* it also becomes more difficult as the diameter of the tube is *decreased.* Another factor, the *viscosity* of the fluid, is also important. Other things being equal, the more viscous the fluid, the greater the resistance to its flow. For example, the flow of molasses is much slower than the flow of water.

Poiseuille's Law. If we make the simplifying assumption that the walls of the blood vessels are rigid, we can state the relationships discussed above as follows:

$$\text{Volume of blood flow} = \frac{\text{Pressure gradient}}{\text{Resistance}}$$

This can also be stated:

$$\text{Resistance} = \frac{\text{Pressure gradient}}{\text{Volume of blood flow}}$$

$$\text{Pressure gradient} = \text{Volume of blood flow} \times \text{Resistance}$$

A theoretically derived mathematical law that expresses all these relationships is frequently used by physiologists. This is Poiseuille's law (fig. 7.3), an approximation of which, in simplified form, can be stated as follows:

$$\text{Volume of blood flow} = \frac{\text{Pressure} \times (\text{Vessel radius})^4}{\text{Vessel length} \times \text{Viscosity}} \quad \text{—thickness of blood}$$

Hydrostatic Pressure. In diving to the bottom of a swimming pool, an increase in pressure is felt upon the ears. This increased pressure is *hydrostatic pressure,* and it increases with depth. It is also present in any vertical tube that contains a liquid because of the increasing weight of the liquid with increasing height (fig. 7.3). This principle is also applied when we *weigh* the atmospheric air by balancing it against a column of mercury to obtain barometric pressure. It is important to remember the effect of hydrostatic pressure in physiology because changes in body position alter the blood pressure in various parts of the body. Thus, when blood pressure is taken by a physician, it is taken at the level of the heart. Finding the blood pressure at any other level of the body requires correction for the hydrostatic pressure effect.

For example, in an individual whose mean blood pressure is 90 mm mercury (Hg) at heart level, we should expect to find a decrease of pressure of 30 cm of blood at the ear level. Converting 30 cm of blood to millimeters of mercury (1 mm Hg = 13.6 mm blood), we would expect a mean blood pressure at the ear of approximately $90 - 22 = 68$ mm Hg. Conversely, at the ankle (in complete rest)—assuming a measurement of 125 cm from ankle to heart level in standing position (and neglecting such factors as frictional losses)—we should expect to find an increased pressure of some $125 \times 10/13.6$, or $90 + 92 = 182$ mm Hg blood pressure at the ankle. These figures correspond rather closely to the pressures found experimentally.

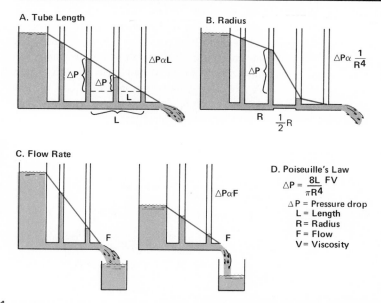

Figure 7.3 A, the drop in pressure (ΔP) during laminar flow of a homogeneous fluid through a rigid tube of constant caliber is directly proportional to the length of the tube. **B,** under the same conditions, the pressure drop is also inversely proportional to the reciprocal of the radius to the fourth power ($1/R^4$) and directly proportional to the volume flow (F) through the tube and to the viscosity (V) of the fluid. (From Rushmer, R.F. *Cardiovascular Dynamics,* 2d ed., 1961. Courtesy of W.B. Saunders Company, Philadelphia.)

The Microcirculation: Blood Flow through the Capillary Bed

The work of Zweifach (36, 37, 38) and Zweifach and Metz (39) has resulted in a better understanding of the capillary bed and allows much better rationalization of the important circulatory changes that occur in muscle tissue during exercise.

The microcirculation is thought of as a well-organized network in which the arterioles give rise to *metarterioles* (fig. 7.4), which possess a gradually dispersing, thin muscular coat. These metarterioles constitute major thoroughfares, or *preferential channels*, through the tissue, and they remain open even during resting conditions; eventually, they join a collecting venule. The *true capillaries* are endothelial tubes, with no smooth muscle, which arise from the metarterioles by way of a *precapillary sphincter* which closes down the true capillaries during resting conditions and relaxes during muscle activity to allow the increased circulation demanded by the higher metabolic rate.

There is a third alternative route for blood flow through the microcirculation, by way of *arteriovenous anastomoses* (AVA) that form a short and direct connection between (1) small arteries and small veins, (2) arterioles and venules, and (3) matarterioles and adjacent venules. These AVA have a well-developed muscular coat and are under sympathetic nervous control. They seem to be most common in the extremities, and they function in bringing about greater losses of heat when environmental temperature rises or when heat is produced within, as during exercise. This loss of heat is brought about because the opening of the AVA causes a greater amount of blood to flow through the venous collection system, which is closer to the surface than the arteries and arterioles and thus provides greater heat losses.

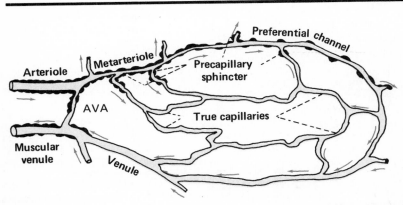

Figure 7.4 A schematic representation of the structural pattern of the capillary bed. The distribution of smooth muscle is indicated in the vessel wall. (From Zweifach, B. W. In *Proceedings of Third Conference on Factors Regulating Blood Pressure,* 1950. Courtesy of Josiah Macy, Jr., Foundation, New York.)

Basic Physiology Underlying the Study of Physiology of Exercise

We can think of the functions of the peripheral blood vessels as four-fold:

1. Maintenance of blood pressure by the arteries and arterioles.
2. Distribution of blood flow to the active tissues by the interaction of metarterioles and the precapillary sphincters in determining how much flow goes through the thoroughfare (preferential) channels and how much through the true capillaries.
3. Temperature control, largely by heat loss controlled through the AVA.
4. The return of blood to the heart, by way of the muscular venules to the small veins, etc.

Various tissues have smaller or larger ranges of metabolic activity. Muscle tissue has the greatest range, from the resting state to forty or fifty times the resting level of energy metabolism. Apparently for this reason, the ratio of true capillaries to thoroughfare channels also varies considerably from tissue to tissue. Zweifach has estimated the ratio in skeletal muscle as eight or ten to one, whereas in tissues of lesser metabolic activity this ratio may be as low as two or three to one.

Control of Blood Distribution

Control of the flow of blood to the various tissues of the systemic circulation is brought about by changes in the diameter of the small arteries and arterioles. The effectiveness of change in the bore of these vessels is seen by referring to Poiseuille's law. If the diameter of a small vessel is doubled, the amount of blood flow will not be doubled, but will be increased sixteen times because the flow varies as the fourth power of vessel radius. These changes in diameter are brought about by two mechanisms: *nervous regulation* and *chemical regulation.* Since the main function of the small arteries and arterioles is to control by their degree of constriction the resistance to flow, they are properly referred to as the *resistance vessels.* The veins are thin-walled but nevertheless also have smooth muscle in their walls, and thus can also actively change their diameters. However, changes in their wall tension do not have any great effect upon resistance to flow, but they are very important in altering the capacity of the postcapillary system. Thus they are important in determining the rate of return flow to the heart and may be referred to as *capacity vessels.*

Nervous Regulation. So far as we know, all of the nervous regulation of blood flow to the skeletal muscles is brought about by the sympathetic system. However, this system supplies two types of fibers: (1) *adrenergic,* which bring about vasoconstriction, and (2) *cholinergic,* which cause active vasodilatation. The normal state of the resistance blood vessels that supply skeletal muscle tissue is one of vasoconstrictor tone. Greater blood supply to an active muscle can be brought about by release of its vasoconstrictor tone or by active vasodilatation. Release of vasoconstrictor tone in the active tissues, with concomitant vasoconstriction in less active tissues—particularly the skin and viscera—seems to be the more important factor. Vasodilatation of resistance vessels appears to function only in emotional reactions or in the expectation of an exercise bout. Furthermore, active vasodilatation probably does not contribute to the support of an increased metabolic rate because it does not result in an increased flow through the true capillaries but only through the AVA or thoroughfare channels discussed above (5).

Muscular exercise causes a reflex increase in tension of the venous walls in both exercising and nonexercising limbs, which persists throughout the exercise and is proportional to the severity of the work (7). This closing down of the capacitance vessels along with the muscle pump and the abdominal-thoracic pump aids the venous return to the heart.

Chemical Regulation. Greater blood flow to active tissue is also brought about by the chemical results of metabolic activity. Lowered pH, increased CO_2 level, and other local changes in metabolites have such a potent effect upon the microcirculation that these effects can override the more centrally mediated vasoconstrictor stimuli whether blood-borne or neurogenic in origin.

The overall picture in exercise appears to be one in which nervous regulation closes down, to a large extent, the circulation to inactive tissues, and it also opens the precapillary sphincters of the microcirculation so that greater flow is provided through the true capillaries that serve the metabolic needs of the active muscle tissue. Simultaneously, or as soon as metabolic products of the activity accumulate, chemical regulation augments the blood flow by further local vasodilatation.

Blood Distribution in Rest and Exercise

As we described earlier, the volume of blood flow to a particular tissue depends upon the resistance it offers in relation to the pressure gradient of its blood flow. Thus the resistance to flow is frequently described in R units, where:

$$R = \frac{\text{Pressure gradient in millimeters mercury}}{\text{Flow rate in milliliters per second}}$$

This is a very convenient method. If we assume an average human at rest has a cardiac output of 5,400 ml/min, then with the usually accepted average resting mean arterial blood pressure of 90 mm Hg (diastolic $+$ ⅓ pulse pressure), the resistance for the total circulatory system—with essentially zero central venous pressure—becomes:

$$R = \frac{90 - 0}{5,400/60} = \frac{90}{90} = 1$$

This resistance of one R unit can be broken down into its component parts, according to the part of the circulatory system that interests us. Thus the resistance of the arterial system can be compared with that of the venous system under resting conditions, as follows:

Flow rate = 90 ml/sec
Central venous pressure = 0.0
Mean arterial pressure = 90 mm Hg
Mean pressure in middle of capillary = 25 mm Hg

Arterial resistance, then, equals: $\dfrac{90 - 25}{90} = 0.72\ R$

and venous resistance equals: $\dfrac{25 - 0.0}{90} = 0.28\ R$

Total = $1.00\ R$

Under a moderate exercise load, the same average human will increase cardiac output some three times, to 16,200 ml/min. Let us examine some typical figures and determine the resistance changes.

Flow rate = 16,200 ml/min = 270 ml/sec
Central venous pressure = 0.0
Mean arterial pressure = 120 mm Hg
Mean capillary pressure = 25 mm Hg

Arterial resistance equals: $\dfrac{120 - 25}{270} = 0.35\ R$

and venous resistance equals: $\dfrac{25 - 0}{270} = 0.09\ R$

Total = $0.44\ R$

These figures, which are quite realistic, illustrate what happens in the circulatory system during exercise. First, the total peripheral resistance is greatly reduced because the blood pressure gradients do not rise nearly as much as the flow rate. This, of course, is the result of vasodilatation in the active skeletal muscles, which far outweighs the accompanying vasoconstriction in less active tissues. Second, the decrease in resistance is even greater in the venous than in the arterial system. The decrease in the arterial system is the result of vasodilatation of the arterioles, while the decrease in the venous system results from the assistance to blood flow given by the muscle pump. Overall, it is this decreased total peripheral resistance that allows the heart to function at a greatly elevated output without strain.

It is also of interest to know where and to what extent compensatory vasoconstriction occurs to support the increased blood flow to the active muscles in exercise. It has been estimated that the resistance to flow in the muscles is reduced from 3.1 R units to 0.37 R during exercise, while the resistance in the portal arteries increases from 2.6 R to 12.0 R, and the portal venous and liver resistance increases from 1.0 R to 3.0 R (6).

The Circulatory System and Exercise

Blood Pressure

The importance of blood pressure as the driving force for the circulatory system has been emphasized. Now let us consider the practical problem of measurement, and how observed measurements may change under varying conditions.

Measurement of Blood Pressure. The common indirect method involves use of a pressure cuff whose pressure is read from a mercury or aneroid manometer. This device is called a _sphygmomanometer._ The sphygmomanometer's cuff is applied to the upper arm, as the subject sits comfortably, so that the cuff is approximately at heart level. The pressure required to occlude the brachial artery is noted by listening to the flow of blood below the cuff, and this is read as _systolic blood pressure. Diastolic blood pressure_ is recorded as the pressure at which the sounds resulting from occlusion become muffled, or disappear, as the cuff pressure is reduced. The systolic-diastolic difference is _pulse pressure._*

Blood pressure measurements made by the indirect method during exercise must be viewed with caution. Comparisons between indirect (sphygmomanometer) and direct (catheter) methods showed that systolic pressure was underestimated by mean values of 8 to 15 mm Hg by the indirect method and overestimated during recovery by 16 to 38 mm Hg (16). Other investigators (25) have found that the indirect method provides satisfactory data on systolic but not diastolic pressure. Some sources of variability in blood pressure follow.

Age. Figure 7.5 illustrates the increase of systolic and diastolic pressures with age. It also shows the more enlightened approach to the problem of _normality,_ in which ranges are given instead of single figures. It is of interest that not all cultures show this agewise increment in blood pressure. Henry and Cassel (15) have furnished impressive evidence for a psychosocial effect in which dissonance between the social milieu in later life and expectations based on early experiences brings about the often observed increase in blood pressure with age.

Sex. The blood pressure in women prior to menopause tends to be slightly lower—and after menopause somewhat higher—than it is in men of the same age.

Emotion. The problem of measuring blood pressure is greatly complicated by the fact that the slightest emotional involvement is reflected by significant rises in blood pressure. This fact is used to advantage in the _polygraph_ or lie detector because the act of lying creates emotional conflict that is sensed and recorded, along with other physiological variables that react to emotional changes.

Diurnal Variation. Blood pressure tends to rise from a low point during sleep to a high point (15 to 20 mm Hg higher) after the evening meal.

Ingestion of Food. After a large meal, there is normally a considerable rise in systolic, and sometimes a fall in diastolic, pressure.

Posture. In changing from supine to erect posture, the hydrostatic pressure increase requires greater arterial pressure, and the response by the cardiovascular system usually overshoots the mark so that systolic and diastolic pressures usually show an increase of 5 to 10 mm Hg. The pulse pressure usually shows a decrease, due to the relatively greater increase in diastolic pressure.

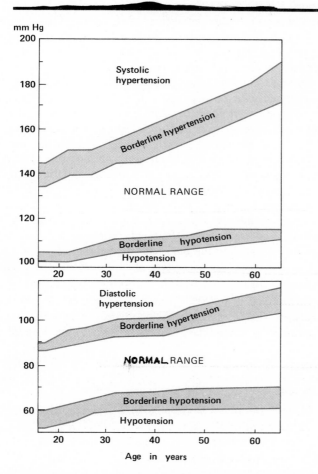

Figure 7.5 Range of systemic arterial blood pressure as a function of age. (From Rushmer, R.F. *Cardiovascular Dynamics,* 2d ed., 1961. Courtesy of W.B. Saunders Company, Philadelphia.)

Maintenance of Arterial Blood Pressure.   The formula,

$$\text{Pressure gradient} = \text{Volume of blood flow} \times \text{Resistance,}$$

shows clearly that blood pressure is maintained by the interaction of two factors: volume of blood flow and resistance. The volume of blood flow is in turn dependent upon (1) heart rate and (2) stroke volume. The resistance is largely determined by the vasoconstrictor tone of the arterioles.

Maintenance of Venous Return to the Heart.   The venous blood pressure at the ankle in the standing posture (under conditions of no muscular activity) is roughly equivalent to the hydrostatic pressure 92 mm Hg if we assume a vertical distance from ankle to right atrium of 125 cm. However, such a pressure, if maintained for any length of time, results in filtration of fluids from the vascular system into the tissue spaces, which results in considerable *edema* (swelling). This happens when a person is forced to stand motionless for a protracted period of time, but it does not occur if the muscles of the leg are active. If the venous pressure is measured in a leg whose muscles are active—as in walking, etc.—the pressure is greatly reduced (to 20 or 25 mm Hg) because of one-way valves in the veins that serve the musculature. Thus the contraction of muscle drives blood toward the heart, and during the relaxation phase these valves prevent a backward flow. This is the first of three major factors that maintain venous return: *the pumping action of contracting muscles*. The importance of this muscle pump, sometimes called a peripheral heart, is shown by the fact that it has been calculated to provide more than 30% of the energy required to circulate the blood during running (35).

A second factor is the *abdominothoracic pump*. The descent of the diaphragm in inspiration creates an increase in intra-abdominal pressure while simultaneously lowering intrathoracic pressure. This increased pressure gradient, from abdomen to right atrium, aids the venous return. During expiration, the backward flow is prevented by the valves in the veins of the muscles of the legs.

A third factor is the *shortening of the inferior vena cava* during the descent of the diaphragm, which results in its having smaller volume, which in turn aids the flow from abdomen to thorax. The lengthening inferior vena cava during expiration lowers pressure within its walls and allows a better pressure gradient for its filling, preparatory to the next descent of the diaphragm.

Arterial Blood Pressure during Exercise

The effects of exercise upon arterial blood pressure can best be described as the end result of the balance struck between the increased blood flow due to the increased cardiac output and the decreased peripheral resistance caused by the vasodilatation of the microcirculation. Consequently, the end result is considerably influenced by the *type* and *intensity* of the exercise and by the physical condition of the subject.

Type of Exercise. In rhythmic exercise that involves moderate to strenuous work loads, the typical response is an elevation of systolic pressure with little, if any, elevation in diastolic pressure; figure 7.6 shows a typical response for a young male. The mean arterial pressure is usually calculated at one-third of the way between diastolic and systolic pressures because of the shape of the arterial pressure wave form. Therefore, the mean pressure is much less affected by exercise.

In static or isometric exercise, where an expiratory effort is made against a closed glottis, the situation is quite different. The intrathoracic pressure is raised from 80 to 200 mm Hg or more, and this increased pressure is transmitted through the thin walls of the great veins; venous return to the right atrium is thus severely decreased, resulting in the following sequence of events.

1. There is a sharp increase in pressure, both systolic and diastolic, which reflects the increase in intrathoracic pressure.
2. After a period of several seconds, during which blood in the lungs furnishes the venous return, the decreased venous return brings about a decreased pulse pressure.
3. After release of the straining activity, there is an increase in both mean pressure and pulse pressure due to the improved venous return of blood, which had been blocked.

This sequence of events is called the *Valsalva effect.*

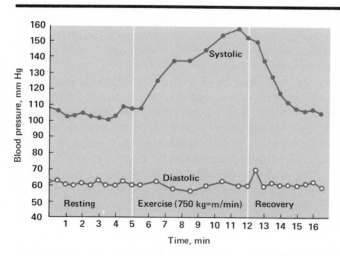

Figure 7.6 The typical time course of the arterial blood pressure response to rest-exercise-recovery in a healthy young man.

Although exercise physiologists have often cautioned against strength exercises with a large static component because of possibly harmful effects of the increased pressure upon the heart and large blood vessels, the work of Hamilton and others (14) shows that no increased difference in pressure across the walls of the heart and great vessels can exist; as the pressure increases within these walls, it increases proportionately outside them. In fact, the pressure difference decreases.

The vessels of the cerebral circulation are similarly protected in that the increased intrathoracic pressure is also transmitted to the cerebrospinal fluid. The only possible danger, then, is in peripheral vessels; but the peripheral vessels are small in diameter, and the laws of physics tell us they are consequently better able to withstand pressure.

However, it must be recalled from the earlier discussion that the work of the heart is determined to a great extent by the arterial pressure against which it is working. It will also be recalled that during exercise the redistribution of blood flow is accomplished by increased sympathetic adrenergic vasomotor tone in the inactive areas while this tone is overridden in the active muscles by local effects to increase the blood flow to the active muscles.

There is now considerable evidence that the systemic level of arterial blood pressure is set by the perfusion pressure required to get through that muscle tissue which is contracting most strongly (2, 3, 13, 21, 22, 23). That is to say, the level of systemic blood pressure is set not by the total amount of muscle working (level of total body work), but rather by the pressure required to perfuse that muscle which is working hardest even though this may be only one small muscle group (10). Furthermore, the type of exercise is very important. In isometric contraction, the muscle tissue pressure upon the arteries can be very high, thus requiring very high perfusion pressure, which in turn requires large increases in systemic pressure.

Lind and McNicol (22) performed an experiment in which their subjects walked on a treadmill at three miles per hour against a grade of 22%, which required an O_2 uptake of 2.8 liters/min. In spite of this very heavy work load sustained by the entire body musculature, when they then performed an isometric contraction, while walking, of 50% maximum on a handgrip dynamometer for one minute their systolic pressure rose by 45 mm Hg and diastolic by 40 mm Hg. This illustrates clearly the need to avoid isometric contractions and also high loadings of small muscles in any exercise situation where high cardiac work loads are undesirable as in cardiac rehabilitation exercise or in conditioning programs for older adults.

Blood Flow in Exercising Muscles

As soon as exercise begins, the metabolic demands of the active muscle tissue increase (by as much as fiftyfold in all-out activity). Many physiological mechanisms cooperate to supply the demand. In addition to the increased cardiac output and redistribution of blood from inactive to active tissues, changes in blood flow occur within the active tissues. In an interesting experiment in 1932—in which India ink was injected into dogs' muscles—Martin and others (24) showed that in the active gracilis muscle there were 2,010 open capillaries per square millimeter, compared with 1,050 in the same muscle when resting.

Concurrently with the increased flow rate through the tissue, the rate of O_2 consumption per cell (muscle fiber) increases rapidly, and this results in a fall in the partial pressure of O_2 within the cell. The partial pressure of O_2 in the tissue fluid bathing the cells, on the other hand, falls very little, and thus a much higher pressure gradient exists to move O_2 from the capillary through the tissue fluid and into the muscle fiber. Better O_2 extraction per volume of circulating blood helps supply the increased metabolic demand. The presence of greater concentrations of CO_2 in the cell further aid the gas exchange (this factor is discussed in chap. 9).

According to the work of Reeves and others (28), it seems that the increased O_2 extraction plays the greater part in making the adjustment to mild exercise (two to three times the resting level), but that in moderate to heavy exercise the increase in cardiac output and the improved flow through tissues is more important.

In general, the blood flow through muscle increases in proportion to the metabolic demand and does not appear to be a limiting factor in high level performance (27).

The type of activity is an important factor not only in the circulation through large vessels, but also in the blood flow within the muscle itself. The classic early work of Barcroft and Millen (4) on the plantar flexors of the foot demonstrated an increased flow (*hyperemia*) when muscular contractions were 0.1 maximum or less, but a cessation of flow when the contractions were above 0.3 maximum. More recent work (33, 34) has shown that isometric contraction of the forearm muscles and elbow flexors occludes the blood flow in these muscles also, but requires 0.6 maximal contraction strength. Thus it would seem that muscles held unnecessarily in tension during athletic activity may suffer the effects of ischemia with its accompanying pain and loss of endurance.

According to Folkow and co-workers (12), the optimal rhythm for maximizing muscle blood flow is one contraction of about 0.3 second duration per second, a rhythm which is often spontaneously chosen by human subjects during bicycling, running, and swimming. The advantage from this activity pattern seems to be that it enables the muscle pump to keep the local mean venous pressure at a minimum during the relaxation phase, thus increasing the effective perfusion pressure.

There are also interesting differences in hemodynamics from sport to sport. For example, in comparing the responses to swimming and running, it was found that arterial blood pressure was higher in swimming than running by 20% at submaximal work loads and 16% at maximal, although cardiac output was lower in swimming as the result of lower heart rates (18).

As would be expected, training and conditioning have an effect on the level of blood flow through the muscles involved. Rohter and others (32) demonstrated an increase of almost 60% in blood flow through the forearm flexors in swimmers after five weeks of training, compared with controls. This work has been confirmed by recent studies by Rochelle and associates (31) (see fig. 7.7). It has also been shown that isometric training can increase exercise blood flow, at least when measured with a contraction of not over 50% of maximal (8).

It is tempting to explain this improved blood flow that results from training on the basis of the older and often referred-to work that showed increased capillarization to result from training in animals (26). However, this work depended on staining the red blood cells, and accurate counts probably cannot be made in this fashion. More recent work in which the capillary basement

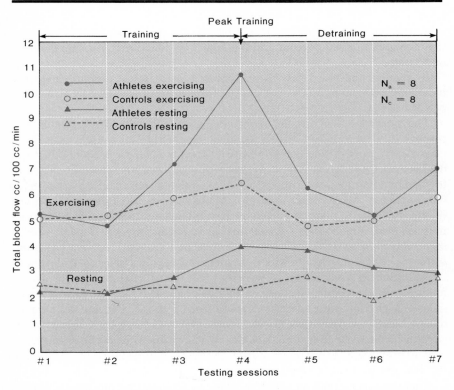

Figure 7.7 Mean resting and exercise muscle blood flows for athletes (swimmers) and controls during a twenty-week training and detraining program. Testing sessions 1, 2, 3, 4, 5, 6, and 7 correspond to the 1st, 4th, 7th, and 10th weeks of training, and 4th, 7th, and 10th weeks of detraining, respectively. (From Rochelle, R.H., et al. *Med. Sci. Sports* 3:122–29, 1971.)

Basic Physiology Underlying the Study of Physiology of Exercise

membrane itself is stained has shown no increase in the number of capillaries (17). But since the muscle fibers increased in size (human subjects), the training resulted in a more favorable capillary to fiber ratio, 1.5 to 1.0 compared to 1.0 to 1.0 in the untrained. It is possible that this may be related to the improved blood flow found to result from training.

In general, the immediate short-term effects of exercise are an increase of blood flow through muscles that are active in light, rhythmic activity, and a decrease during heavy, sustained contraction, which is followed by a period of increased flow called *reactive hyperemia*. The long-term effect of exercise almost certainly involves an improved perfusion of the active tissues.

Blood and Fluid Changes during Exercise

The total body fluid is composed of four components:

1. Blood plasma
2. Interstitial fluid
3. Intracellular fluid
4. Miscellaneous components

Only the first three have importance to exercise physiology. Because the capillary wall is freely permeable to most of the substances in the plasma (except the plasma protein), the plasma and interstitial fluid constantly mix and are very similar in makeup. Therefore the two together are referred to as extracellular fluid.

To maintain normal osmotic pressures in intracellular and extracellular fluids, water diffuses in the direction needed through the cell membrane. Thus in excessive sweating as in heavy exercise, water that is lost in sweat comes directly from an extracellular component. This raises the concentration of nondiffusible substances of the interstitial fluid, which results in transfer of water across the cell membranes of the various tissue cells from intracellular to extracellular components, thus dehydrating the tissues as well. Obviously, ingestion of water causes a reverse reaction, in which tissue hydration is restored toward normal.

Hemoconcentration. Moderate to heavy exercise bouts result in a shift of fluid from the plasma to the interstitial fluid. This, in turn, results in higher hemoglobin and serum protein concentrations, and in a higher proportion of cellular elements in the plasma. If the exercise bouts involve heavy work in a hot environment, dehydration contributes to exaggerate this phenomenon. Weight loss in wrestlers by dehydration has been reported as high as 8.8 lb, and in marathon runners it may be as high as 7% of body weight. A dehydration of 2% may cause deterioration of performance in some individuals (1).

Erythrocyte Count. The hemoconcentration discussed above results in a considerable rise in the cell count per cubic millimeter. In the normal male, the count is ordinarily 5.5 million per cubic millimeter (female, 4.8 million per cubic millimeter). These values can be raised by as much as 20%–25% in vigorous exercise. Whether this rise is due entirely to hemoconcentration, or in part to release of stored cells, is still unclear. In some animals the spleen releases erythrocytes to the general circulation during exercise, but this does not seem to be the case in man.

Blood and Fluid Changes from Training

Blood Volume. Evidence indicates that blood volume first falls, then rises, above normal during a four- to nine-week training period in dogs (9). The increased blood volume persisted about four weeks after cessation of training. Kjellberg and others (19) found similar results with men and women: blood volumes were found to be 10%–19% higher after training than before. They also found blood volume was as much as 41%–44% higher in an athletically trained group than in a comparable untrained group.

Changes in the opposite direction occur with prolonged bed rest. Thus it would seem that blood volume varies proportionately with the amount of physical activity. Increased blood volume would be a considerable advantage in heavy exercise because a circulatory demand may exist for purposes of heat dissipation at the same time that active muscle tissues also demand a greatly increased blood volume. Under these conditions, increased blood volume would help assure an adequate venous return to the heart.

Hemoglobin. Increases of *total hemoglobin* have been found in dogs and in humans as the result of physical conditioning (9, 19). These increases seem to parallel the increased blood volume, so that no increase in *hemoglobin concentration* (hemoglobin per unit blood volume) seems to occur. Indeed Dill and associates (11) found hemoglobin concentration to be 4% *lower* in highly trained runners than in controls. They concluded that endurance athletes have thin blood, but so much of it that their total hemoglobin exceeds that of nonathletes.

Alkaline Reserve. Alkaline reserve may be defined as the buffering capacity of the blood. Many investigators have shown an increased ability to tolerate acid metabolites (mainly lactic acid) after a period of training. Thus it would seem a likely hypothesis that this increased tolerance might be due to an increase in buffering ability of the blood. Robinson and Harmon (30) have shown, however, that the alkaline reserve did not change during a period of training in which large increases in ability to carry O_2 debt were observed.

Postural Effects on Circulation. Posture during physical activity varies from the usual, upright position to sitting while rowing, to the prone or supine position in swimming, to the head-down position in gymnastics. Changes in posture might be expected to exert effects upon the circulation, and these effects have been demonstrated.

The responses to exercise in supine and sitting (bicycle ergometer) and standing (treadmill) positions have been studied by many investigators. In general, the cardiac output is about 2 liters/min greater in supine position than in the upright positions. This is due to the greater stroke volume supine. A higher arteriovenous O_2 difference in the upright position compensates for the lower cardiac output and stroke volume (7).

Tilttable changes of a subject's position, from standing upright to the head-down position, produced a *vagal rebound phenomenon* (20) in which the subject's heart rates slowed by an average of fifteen beats per minute (fig. 7.8). This can be explained by the fact that in upright posture the peripheral resistance is maintained by sympathetic nervous activity that reflexly maintains a state of vasoconstriction in the arterioles of the lower extremities.

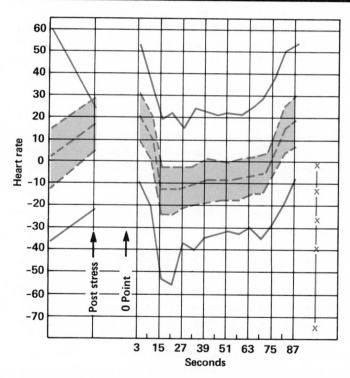

Figure 7.8 The mean change in heart rate from the baseline level for 215 subjects is plotted against time. The 0 point is a six-second period just prior to the head down tilt. Heart rate is plotted for each six-second interval therafter. The increase in heart rate after the sixty-second period is following rotation to the feet down position. The shaded area is the standard deviation of the mean. The upper and lower graphs represent the extremes, the high and low values for the group. (From Lamb, L.E., and Roman, J. *Aerospace Med.* 32:473, 1961.)

When the body is tilted head downward, the redistribution of blood due to the changed direction of the force of gravity brings about greater stimulation of the carotid sinus and central nervous system receptors which induces vagal efferent stimulation with cardioinhibitory responses. Suprisingly, no relationship was found between the reaction to head down tilting and physical fitness (measured by endurance in treadmill walking).

The circulatory reserves that can be called upon also seem to vary with body position. As soon as an individual assumes the standing position, his O_2 extraction (as measured by the difference between arterial and venous oxygen levels of the working muscle) has already greatly increased. His circulatory reserve therefore depends largely upon the factors that can increase the flow rate. In the supine position, mild demands of exercise can be satisfied by increased O_2 extraction before any increase in blood flow is demanded.

Cooling Out after Heavy Exercise. It has been the common practice in athletic events that involve large circulatory adjustments—distance running, etc.—to *cool out* at the end of the competitive effort by jogging for a few minutes. This procedure rests upon sound physiological principles, and should be encouraged. If this cooling out is not done, venous return to the heart—which has been largely supported by the muscle pump—drops too abruptly, and blood pooling may occur in the extremities. This, in turn, may result in shock, or at least in hyperventilation, which causes lower levels of CO_2 and muscle cramps.

SUMMARY

1. Blood flow is the result of a difference in pressure between two points in the circulation. This difference in pressure is called a *pressure gradient*.
   a. Other things being equal, the velocity of blood flow through any given type of blood vessel varies inversely with the cross-sectional area of the total vessels of that type.
   b. A simplification of Poiseuille's law expresses the dynamics of the circulation as follows:

   $$\text{Volume of blood flow} = \frac{\text{Pressure} \times (\text{Vessel radius})^4}{\text{Vessel length} \times \text{Viscosity}}$$

   c. Blood pressure must always be referred to a specific point in the body to account for the effects of hydrostatic pressure.
2. There are at least three routes for blood flow through the microcirculation for serving differing physiological conditions: (a) *thoroughfare* or *preferential channels* that serve the metabolic needs of resting tissue; (b) *true capillaries,* which open to serve the needs of an increased metabolic activity; and (c) *arteriovenous anastomoses* that probably dissipate heat.

3. Control of blood distribution is mediated by the sympathetic nervous system and chemical regulation. *Adrenergic fibers* of the sympathetic system bring about vasoconstriction of inactive tissues, and *cholinergic fibers* of the same system cause vasodilatation of the active tissue, which is augmented by chemical activity of the metabolites after exercise is well under way.

4. Approximately 72% of the resistance of the circulatory system at rest arises in the arterial system (largely in the arterioles) and only 28% in the venous system. In exercise, the total resistance may be reduced by at least 50%.

5. Blood pressure is subject to the effects of many variables, such as age, sex, emotional state, time of day, nutritional state, and posture.

6. Blood pressure usually shows a rise during exercise in systolic and mean values, but no change or a very small rise in diastolic pressure. The effects vary with the type of exercise.

7. The systemic arterial pressure is set not by the general level of the work load sustained by the total body, but rather by the pressure required to perfuse that muscle or muscle group that is working under greatest tension.

8. During exercise, the number of open capillaries is approximately double the number observed during rest.

9. The training and conditioning of athletes has been shown to bring improved blood flow to the active muscles at the peak of training. These changes are reversible, and show a detraining effect within three weeks of the end of training.

10. Exercise that results in heavy sweating causes a loss of intracellular as well as extracellular fluid, which must be replaced during or after the workout to maintain normal tissue hydration and electrolyte balance.

11. Total blood volume responds to training with significant increases and also declines in individuals confined to the inactivity of bed rest.

12. The response of the circulatory system to exercise varies with posture. Even at rest, changes in posture are reflected in predictable and typical circulatory response patterns.

13. *Cooling out* after heavy exercise is necessary to prevent blood pooling, which may result in muscle cramps, or even shock.

REFERENCES

1. Ahlman, K., and Karvonen, M.J. Weight reduction by sweating in wrestlers and its effect on physical fitness. *J. Sports Med. Phys. Fitness* 1:58–62, 1962.

2. Astrand, P.O.; Ekblom, B.; Messin, R.; Saltin, B.; and Stenberg, J. Intraarterial blood pressure during exercise with different muscle groups. *J. Appl. Physiol.* 20:253–56, 1965.

3. Astrand, I.; Guharay, A.; and Wahren, J. Circulatory responses to arm exercise with different arm positions. *J. Appl. Physiol.* 25:528–32, 1968.

4. Barcroft, H., and Millen, J.L.E. The blood flow through muscle during sustained contraction. *J. Physiol.* 97:17–31, 1939.

5. Barcroft, H., and Swan, H.J.C. *Sympathetic Control of Human Blood Vessels.* London: E. Arnold, 1953.

6. Bazett, H.C. A consideration of the venous circulation. In *Factors Regulating Blood Pressure,* eds. B.W. Zweifach and E. Shorr. New York: Josiah Macy, Jr., Foundation, 1949.

7. Bevegard, B.S., and Shepherd, J.T. Regulation of the circulation during exercise in man. *Physiol. Rev.* 47:178–213, 1967.

8. Byrd, R.J., and Hills, W.L. Strength, endurance and blood flow responses to isometric training. *Res. Q.* 42:357–61, 1971.

9. Davis, J.E., and Brewer, N. Effect of physical training on blood volume, hemoglobin, alkali reserve, and osmotic resistance of erythrocytes. *Am. J. Physiol.* 113:586–91, 1935.

10. deVries, H.A., and Adams, G.M. Total muscle mass activation v. relative loading of individual muscles as determinants of exercise response in older men. *Med. Sci. Sports* 4:146–54, 1972.

11. Dill, D.B.; Braithwaite, K.; Adams, W.C.; and Bernauer, E.M. Blood volume of middle distance runners; effect of 2300 m altitude and comparison with nonathletes, *Med. Sci. Sports* 6:1–7, 1974.

12. Folkow B.; Gaskell P.; and Waaler B.A. Blood flow through limb muscles during heavy rhythmic exercise, *Acta Physiol. Scand.* 80:61–72, 1970.

13. Freyschuss, V., and Strandell, T. Circulatory adaptation to one- and two-leg exercise in supine position. *J. Appl. Physiol.* 25:511–15, 1968.

14. Hamilton, W.F.; Woodbury, R.A.; and Harper, Jr., H.T. Arterial, cerebrospinal and venous pressures in man during cough and strain. *Am. J. Physiol.* 141:42–50, 1944.

15. Henry, J.P., and Cassel, J.C. Psychosocial factors in essential hypertension. *Am. J. Epidemiol.* 90:171–200, 1969.

16. Henschel, A.; De la Vega, F.; and Taylor, H.L. Simultaneous direct and indirect blood pressure measurements in man at rest and work. *J. Appl. Physiol.* 6:506–12, 1954.

17. Hermansen, L., and Wachtlova, M. Capillary density of skeletal muscle in well-trained and untrained men. *J. Appl. Physiol.* 30:860–63, 1971.

18. Holmer, I.; Stein, E.M.; Saltin, B.; Ekblom, B.; and Astrand, P.O. Hemodynamics and respiratory responses compared in swimming and running. *J. Appl. Physiol.* 37:49–54, 1974.

19. Kjellberg, S.R.; Rudhe, V.; and Sjostrand, T. Increase of the amount of hemoglobin and blood volume in connection with physical training. *Acta Physiol. Scand.* 19:146–51, 1949.

20. Lamb, L.E., and Roman J. The head-down tilt and adaptability for aerospace flight. *Aerospace Med.* 32:473–86, 1961.

21. Lind, A.R., and McNicol, G.W. Circulatory responses to sustained handgrip contractions performed during other exercise, both rhythmic and static. *J. Physiol.* 192:595–607, 1967a.

22. ———. Muscular factors which determine the cardiovascular responses to sustained and rhythmic exercise. *Can. Med. Assoc. J.* 96:706–13, 1967b.

23. ———. Cardiovascular responses to holding and carrying weights by hand and by shoulder harness. *J. Appl. Physiol.* 25:261–67, 1968.

24. Martin, E.G.; Wooley, E.C.; and Miller, M. Capillary counts in resting and active muscle. *Am. J. Physiol.* 100:407–16, 1932.

25. Nagle, F.J.; Naughton, J.; and Balke, B. Comparison of direct and indirect blood pressure with pressure-flow dynamics during exercise. *J. Appl. Physiol.* 21:317–20, 1966.

26. Petren, T.; Sjostrand, T.; and Sylven, B. Der Einfluss des Trainings auf die Häufigkeit der Capillaren in Herz und Skeletmuskulatur. *Arbeitsphysiologie* 9:376–86, 1936.

27. Pirnay, F.; Marechal, R.; Radermecker, R.; and Petit, J.M. Muscle blood flow during submaximum and maximum exercise on a bicycle ergometer. *J. Appl. Physiol.* 32:210–12, 1972.

28. Reeves, J.T.; Grover, R.F.; Filley, G.F.; and Blount, S.G. Circulatory changes in man during mild supine exercise. *J. Appl. Physiol.* 16:279–82, 1961.

29. ———. Cardiac output response to standing and treadmill walking. *J. Appl. Physiol.* 16:283–88, 1961.

30. Robinson, S., and Harmon, P.M. The lactic acid mechanism and certain properties of the blood in relation to training. *Am. J. Physiol.* 132:757, 1941.

31. Rochelle, R.H.; Stumpner, R.L.; Robinson, S.; Dill, D.B.; and Horvath, S.M. Peripheral blood flow response to exercise consequent to physical training. *Med. Sci. Sports* 3:122–29, 1971.

32. Rohter, F.D.; Rochelle, R.H.; and Hyman, C. Exercise blood flow changes in the human forearm during physical training. *J. Appl. Physiol.* 18:789–93, 1963.

33. Royce, J. Isometric fatigue curves in human muscle with normal and occluded circulation. *Res. Q.* 29:204–12, 1958.

34. Start, K.B., and Holmes, R. Local muscle endurance with open and occluded intramuscular circulation. *J. Appl. Physiol.* 18:804–7, 1963.

35. Stegall, H.F. Muscle pumping in the dependent leg. *Circ. Res.* 19:180–90, 1966.

36. Zweifach, B.W. Basic mechanisms in peripheral vascular homeostasis. In *Proceedings of Third Conference on Factors Regulating Blood Pressure.* New York: Josiah Macy, Jr., Foundation, 1949.

37. ———. General principles governing the behavior of the microcirculation. *Am. J. Med.* 23:684–96, 1957.

38. ———. Structural and functional aspects of the microcirculation in the skin. In *The Microcirculation,* eds. S.R.M. Reynolds and B.W. Zweifach, pp. 144–52. Urbana: The University of Illinois Press, 1959.

39. Zweifach, B.W., and Metz, D.B. Selective distribution of blood through the terminal vascular bed of mesenteric structures and skeletal muscle. *Angiology* 6:282–90, 1955.

8

The Lungs
and External Respiration

Each cell of every tissue in the human body depends on oxidative reactions to provide the energy for its metabolism. In very simple biological organisms, each cell is in contact with the external environment, and thus derives its supply of oxygen directly. In the human organism, the vast majority of tissues are not in direct contact with the external environment, and for this reason a specialized respiratory system is necessary to provide the oxygen for their metabolic demands.

The respiratory process can be broken down into three component functions:

1. Gas exchange in the lungs, in which the lung capillaries take up oxygen and give up much of their carbon dioxide.
2. Gas transport and distribution from the lungs to the various tissues by the blood.
3. Gas exchange between the blood and the tissue fluids bathing the ultimate consumers, the cells.

The first process is also referred to as *external respiration* or *pulmonary ventilation,* the third process is referred to as *internal* or *tissue respiration,* and the second process, *gas transport,* is the function of the cardiovascular system. This chapter is concerned with the first process, external respiration. The respiratory function of the cardiovascular system and internal respiration are discussed in chapter 9.

Anatomy of External Respiration

It will be recalled from elementary physiology and anatomy that the flow of air proceeds through the nose (or mouth) into the nasal cavity, where it is warmed, humidified, and agitated by striking the *turbinates.* From the nasopharynx, air is conducted past the *glottis,* where the pharynx separates into the *trachea* for air conduction and the *esophagus* for the passage of food. The trachea splits into the two chief *bronchi;* one goes to the left lung and the other goes to the right lung (fig. 8.1).

The lungs can be considered a system of branching tubes that perform two major functions: conduction of air and respiration. The conductive portion of the system proceeds from the chief bronchi which branch in the lung root. Subsequent branching of the conductive pathway occurs within the lungs and results in smaller bronchi which in turn branch into smaller tubes called *bronchioles.* Throughout the repeated branchings, each branching results in a larger total cross-sectional area, as was also the case in the circulatory system. The bronchioles eventually branch into the *terminal bronchioles,* the last units of the conductive system which are from 0.5 to 1.0 mm in diameter.

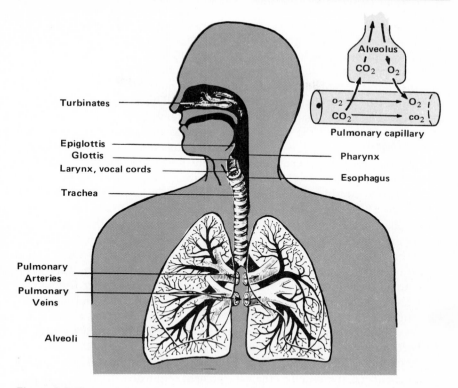

Figure 8.1 The respiratory system, showing the respiratory passages and the function of the alveolus to oxygenate the blood and to remove carbon dioxide. (From Guyton, A.C. *Function of the Human Body*, 1959. Courtesy of W.B. Saunders Company, Philadelphia.)

Each terminal bronchiole divides into two *respiratory bronchioles,* which in turn may divide once more (or even twice), with these divisions forming *alveolar ducts.* The alveolar ducts may or may not branch, but they eventually terminate in thin-walled sacs called *alveoli* (singular: *alveolus*). These alveoli, with their supporting structures, are highly vascularized. Thus the *lung unit* has been considered to consist of the alveolar duct and its subdivisions (alveoli), together with the blood and lymph vessels and the nerve supply. Within the lung unit, only two very thin endothelial layers separate the air in the system from the blood in the capillaries, thus allowing for very efficient diffusion of gasses. Furthermore, the total cross-sectional area available for diffusion has been estimated to be between 500 and 1,000 sq. ft.

Mechanics of Lung Ventilation

The laws governing fluid flow also apply to gasses. It has been pointed out that the flow of blood in the vascular system is brought about through differences in pressure, called a pressure gradient, and the flow of respiratory gas similarly depends upon a pressure gradient between the air in the lungs and the ambient (outside) air. For air to flow into the lungs, the pressure within must be lower than atmospheric pressure. This lowering of pressure in the lungs is brought about by the descent of the diaphragm (contraction of the muscle fibers) and the action of the external and anterior internal intercostal muscles in raising the ribs. Campbell (6) has shown by X-ray technique that the diaphragm descends approximately 1.5 cm in normal, quiet respiration, and as much as 6 to 10 cm during maximal breathing. Thus the volume of the lungs is considerably increased during inspiration, largely by virtue of elongation. This increase in volume results in a temporary lowering of pressure within the lungs, so that a pressure gradient exists, with the ambient air having the higher pressure and thus moving into the lungs during inspiration.

In the respiratory cycle, the inspiration phase is the active phase and is brought about by the active contraction of the ordinary muscles of respiration, the diaphragm and the intercostals. The expiration phase under resting conditions is largely due to the elastic recoil of these muscles and associated structures as they snap back to their resting length. Thus the elastic recoil creates a higher-than-atmospheric pressure within the lung which results in the necessary pressure gradient for moving the air out in expiration.

Ventilation of the alveoli, where the greatest part of the diffusion process occurs, has been attributed to the enlargement of the alveolar ducts in length and width without a concomitant enlargement of the alveoli themselves. However, it appears more likely that the alveoli participate proportionately in the overall enlargement during inspiration, this enlargement being about twofold for the alveolar duct and the alveolus. This twofold increase in volume was calculated to result in a 70% increase in alveolar area for diffusion.

The description so far holds only for normal resting breathing conditions. During exercise, metabolic demands are greater, and rate and depth of breathing are increased. The increased depth of respiration is brought about by the *accessory muscles* of breathing. In inspiration, greater volume is obtained by activity of the *scalene* and the *sternocleidomastoid* muscles, through their help in lifting the ribs. In the expiratory phase of heavy exercise, the passive elastic recoil of the ordinary muscles of breathing is greatly aided by the active contraction of the abdominal muscles. The abdominal muscles serve two important mechanical functions: (1) raising the intra-abdominal pressure which results in greater intrathoracic pressure to aid in expiration, and (2) drawing the lower ribs downward and medially. The lateral muscles (oblique and transverse) are more important than the rectus abdominis (6).

Nomenclature for the Lung Volumes and Capacities

Because respiratory physiology had been plagued by an ambiguity of terms, a group of American physiologists agreed in 1950 to standardize terms and definitions. The result of this agreement is given in table 8.1 and is illustrated in figure 8.2. Table 8.2 provides illustrative values for the lung volumes in healthy, recumbent subjects.

Table 8.1
Lung Volumes and Capacities

A. **Volumes.** There are four primary volumes, which do not overlap.
  1. Tidal volume, or the depth of breathing, is the volume of gas inspired or expired during each respiratory cycle.
  2. Inspiratory reserve volume is the maximal amount of gas that can be inspired from the end inspiratory position.
  3. Expiratory reserve volume is the maximal volume of gas that can be expired from the end expiratory level.
  4. Residual volume is the volume of gas remaining in the lungs at the end of a maximal expiration.
B. **Capacities.** There are four capacities, each of which includes two or more of the primary volumes.
  1. Total lung capacity is the amount of gas contained in the lung at the end of a maximal inspiration.
  2. Vital capacity is the maximal volume of gas that can be expelled from the lungs by forceful effort following a maximal inspiration.
  3. Inspiratory capacity is the maximal volume of gas that can be inspired from the resting expiratory level.
  4. Functional residual capacity is the volume of gas remaining in the lungs at the resting expiratory level. The resting end expiratory position is used here as a baseline because it varies less than the end inspiratory position.

(From *The Lung,* 2d ed., by Julius H. Comroe, Jr., et al. Copyright © 1962, Year Book Medical Publishers, Inc. Used by permission of Year Book Medical Publishers, Inc.)

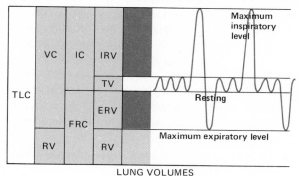

LUNG VOLUMES

Figure 8.2 Lung volumes. (From *The Lung,* 2d ed., Julius H. Comroe, Jr., et al. Copyright © 1962, Year Book Medical Publishers, Inc. Used by permission of Year Book Medical Publishers, Inc.)

Respiratory Control

The question of how pulmonary ventilation is controlled has been the subject of much physiological research. It is obvious that the rate and depth of respiration must be so controlled as to maintain homeostasis in the face of varying metabolic demands, ranging from rest to the very vigorous exercise of competitive athletics. However, the physiological mechanisms by which this homeostasis is brought about are, as yet, not so obvious.

The nerve cells responsible for the automatic and rhythmic innervation of the muscles of respiration lie in the reticular formation of the medulla. As a group, these nerve cells are referred to as the respiratory center.* The best available evidence seems to indicate that this center possesses automatic rhythmicity, which would account for the phases of inspiration and expiration, with expiration being the result of inactivity of the respiratory center.

The breathing frequency (f) and depth (tidal volume or V_T) are adjusted to metabolic demands for oxygen by a complex of factors, some of which act directly upon the nerve cells of the respiratory center (chemosensitive cells of the medulla), and some of which operate reflexly through the aortic and carotid bodies which are sensitive chemoreceptors. The six most important factors are as follows.

Carbon Dioxide and Subsequent pH Changes. In resting humans, rises in arterial carbon dioxide tensions have been shown to be a very potent stimulant for the respiratory processes. Lambertsen and others (24) have shown that approximately 45% of this respiratory drive was due to the lowered pH

*Evidence is accumulating that the respiratory center is probably not as discrete an entity as had been thought. However, for the purposes of this text, the term seems to have utility and will be retained to eliminate unnecessary discussions of neurological concepts that are as yet not completely elucidated.

brought about by the increased CO_2. The other 55% of respiratory drive may be due to extravascular pH changes due to CO_2, or to direct action of the CO_2 itself. The effect of the CO_2 and pH factors seems to be mediated directly through chemosensitive receptors in the medulla.

Oxygen (Anoxia). A low arterial level of oxygen has also been demonstrated to bring about greater ventilatory activity; however, it had been thought that the level of O_2 had to drop from a normal arterial partial pressure (PaO_2) of 100 mm Hg to about 60 mm Hg to have any effect. Hornbein and others (21) have shown that very small changes of 6 to 7 mm Hg are effective if they are sudden changes, and also that during exercise, responses are elicited to a drop in PaO_2 of 6 to 7 mm Hg. They suggest that this difference in response between the resting and exercise states may be due to increased sympathetic activity during exercise that decreases blood flow to the chemoreceptors of the carotid and aortic bodies where this reflex arises.

Proprioceptive Reflexes from Joints and Muscles. Considerable evidence exists that movement per se, independent of any other change, can have a reasonably large effect in bringing about the increased ventilation that occurs during exercise. This effect has been best demonstrated for the knee joint, but it undoubtedly is also operative at other joints and arises from receptors in the joints and muscle tissues. Recent work (22) has shown that not only can the afferents from stretch receptors elicit a ventilatory response, but even the nonmedullated C fibers can do so. However, the importance of neurogenic factors in general has been challenged by the work of Beaver and Wasserman (4) who have shown that the rise and fall of ventilation is not generally so precipitous as had been believed; consequently, the neurogenic factor is no longer necessary to explain a response which was thought to be too rapid for other than a reflex mechanism. Further work in their laboratory (7) has shown that variation in bicycle pedaling rate produces no change in ventilatory response. This finding is, of course, inconsistent with any appreciable role for proprioceptive reflexes in the control of breathing during exercise.

Temperature. As body temperature goes up, the ventilation rate goes up, in direct proportion. Since the body temperature rises from 1° to 5° F during exercise, this is one of the factors involved. However, it cannot be responsible for the early and large ventilatory response to exercise because the temperature rise requires too much time (twenty to thirty minutes for any appreciable increase). It may be noted, in passing, that respiration also drops, in direct proportion to temperatures lower than nornal, until death ensues from respiratory failure.

Cerebral Factors. An increase in the ventilation rate is frequently observed in anticipation of exercise before any of the aforementioned factors can have an effect. This increase is attributed to cerebral innervation aroused psychically in the forebrain.

Hering Breuer Reflexes. In 1868, Hering and Breuer reported the discovery of stretch receptors in the lungs whose afferent pathways traveled by way of the vagus nerve. These reflexes are named for them and are two; one is the *inhibito-inspiratory* reflex and the other is the *excito-inspiratory* reflex. The inhibito-inspiratory reflex tends to terminate inspiration that is larger than normal by firing inhibitory impulses to the respiratory center. The excito-inspiratory reflex functions only in very deep expiration, or in collapse of the lung.

Control of respiratory activity is undoubtedly brought about by a combination of the above factors in complex interactions that have not yet been completely elucidated.

Obviously, the muscular movements of breathing are also under voluntary control. Rate and depth can be changed at will, and the breath can be held for varying periods of time.

It is also interesting to note that vibration of the human body—as occurs on trucks, tractors, and high-speed, low-flying aircraft—results in hyperventilation (increase over normal ventilation). Exactly how this hyperventilation is brought about is not yet known.

Importance of Breathing Pattern

Rate versus Depth. Lung ventilation rate (minute volume of breathing) is the result of two variable factors: *rate* and *depth,* or tidal volume (V_T). There are also two types of resistance to be overcome for the respiratory muscles: the *elastic resistance,* met in stretching the lungs and muscular and connective tissues of the thorax, and *airway resistance,* met by the movement of air inflowing through the small tubes of the lungs. The elastic resistance increases with increasing tidal volume for any given lung ventilation rate because the elastic tissues are stretched farther in deeper breathing. On the other hand, the airway resistance does not increase with rate or with depth if the minute volume of breathing is held constant. On the face of it then, it would seem to be more efficient to breath at a high rate and shallow depth; however, this is not so because of the *anatomical dead space,* which is defined most simply as the volume contained within the conducting portion of the airways of the lung where no gas exchange occurs.

Figure 8.3 illustrates the importance of the anatomical dead space in helping to determine the optimum rate and tidal volume of breathing. It shows that although the minute volume is the same 8,000 ml and the dead space is 150 ml in all three cases, if the tidal volume is small and the rate is high *(A),* the dead-space air represents 4,800/8,000 of the total minute volume. In *C,* where the rate is low and the tidal volume is high, the dead-space air represents only 1,200/8,000 of the total minute volume.

The importance of the dead-space air is perhaps more evident if we consider the extreme case: where the dead-space air is 150 ml, the rate is increased, and the depth decreased, until the tidal volume is also 150 ml. In this case, of course, no real alveolar ventilation can occur because the dead-space air would simply be moved back and forth between conductive airways and the alveoli.

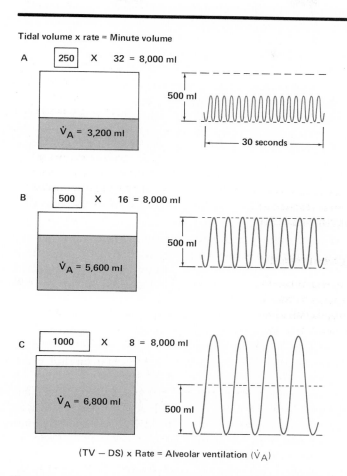

Tidal volume x rate = Minute volume

A 250 X 32 = 8,000 ml

$\dot{V}_A$ = 3,200 ml

500 ml

30 seconds

B 500 X 16 = 8,000 ml

$\dot{V}_A$ = 5,600 ml

500 ml

C 1000 X 8 = 8,000 ml

$\dot{V}_A$ = 6,800 ml

500 ml

(TV — DS) x Rate = Alveolar ventilation ($\dot{V}_A$)

Figure 8.3 Area of each small block represents tidal volume (250, 500, or 1,000 ml). Total area of each large block (shaded and unshaded areas) = minute volume of ventilation; in each case it is 8,000 ml. Shaded area of each block represents volume of alveolar ventilation per minute; this varies in each case since Alveolar ventilation/min = (Tidal volume — Dead space) X Frequency. A dead space of 150 ml is assumed in each case, although actually the dead space would increase somewhat with increasing tidal volume. *Right,* spirographic tracings. (From *The Lung,* 2d ed., Julius H. Comroe, Jr., et al., Copyright © 1962, Year Book Medical Publishers, Inc. Used by permission of Year Book Medical Publishers, Inc.)

Thus it might be predicted that for each individual, with varying degrees of elastic resistance, airway resistance, and volume of dead space, an optimum rate and depth of breathing exists, and this has proven essentially true. Recent evidence (30) suggests that in near maximal exercise the rate commonly adopted by the athlete (30 to 35 breaths per minute) is marginally more efficient than either slower or faster rates. This relatively slow breathing pattern probably minimizes the O_2 consumption of the chest muscles and is also most effective in terms of gas exchange.

Under resting conditions, the rate of breathing varies considerably, from six or seven to twenty respirations per minute; during vigorous exercise, it may go as high as fifty to sixty per minute. The tidal volume is roughly 500 ml for an average man and may increase to approximately 2,500 ml in high intensity, short duration (two to three minutes) exercise. The total lung ventilation averages about 7.0 liters/min for healthy male adults at rest and may go up to 150 liters/min or more in heavy exercise.

Effect of Type of Exercise on Breathing Pattern. The fact that various sports activities result in considerably different respiratory responses is not surprising. To begin with, the rhythm of exercise is likely to affect the rhythm of breathing, even in such sports as running and cycling (5). In some sports such as swimming (crawl stroke), the rate of breathing is totally dictated by the stroke rhythm, and even tidal volume is probably constrained to some extent. In a comparison of running and swimming it has been shown that, at maximum effort, swimming resulted in a minute ventilation of 111 liters/min compared with 154 liters for maximal running. Since arterial O_2 saturation remained similar, better O_2 extraction must have compensated for the low ventilation in swimming (20).

Diaphragmatic versus Costal Breathing. At rest, there are apparently no major differences with respect to either sex or age in the relative contribution of the diaphragm and the intercostal muscles to the movement of the tidal volume (29). Most normal subjects are abdominal breathers when supine and thoracic when upright.

In moderate and heavy exercise tidal volume increases both in inspiratory and expiratory directions. Virtually all of the inspiratory increase is produced by the rib cage and most of the expiratory increase by the abdomen (diaphragm) (17). Physical education and singing instructors have often advised modification of costal breathing (chest) to abdominal breathing (greater use of the diaphragm). Campbell (6), after electromyographic research into the muscular activity of breathing, concluded that the activity pattern of the normal muscles of breathing (diaphragm and intercostals) is probably not changed by any of the breathing exercises advocated, even though externally observed movements of the thorax and abdomen may seem different. He feels, however, that the accessory muscles of breathing (abdominals and elevators of ribs) can be trained through breathing exercises but that their activity probably does not become an unconscious habit pattern. Thus a change in the breathing pattern would require constant voluntary control and attention.

Efficiency of Breathing

Lung Ventilation versus O_2 Consumption in Exercise. Between the resting state and a moderate level of exercise (approximately 2.0 liters of O_2 consumption), a very constant ratio of ventilation rate to O_2 consumption is maintained. This ratio is usually termed the *ventilation equivalent,* and is defined as the number of liters of air breathed for every 100 ml of oxygen consumed. Thus at rest the ventilation equivalent (VE) is

$$\frac{7.0 \text{ liters air breathed/min}}{275 \text{ ml } O_2 \text{ uptake/min}} \quad \text{or} \quad \frac{7.0}{2.75 \text{ (hundred ml)}} = 2.54$$

This ratio also indicates that 25.4 liters of air must be respired to achieve an O_2 uptake of 1 liter: $\frac{7.0 \text{ liters}}{.275 \text{ liters}} = 25.4$ liters

This ratio holds until the work load demands more than 2 liters of O_2 per minute, at which point the ratio grows higher; and this relationship is shown in figure 8.4. The reason for the higher ventilation equivalent at the higher work loads is that the steady state is no longer maintained, and lactic acid accumulates which acts upon the respiratory center through lowering the pH.

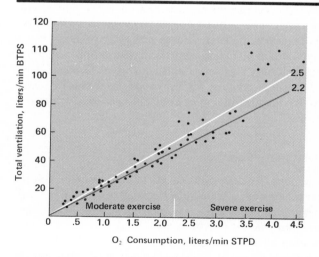

Figure 8.4 Ventilation as a function of O_2 consumption in physical exercise (walking, running, bicycling). The experimental points represent the means of 611 determinations in eighty-six subjects reported by eleven different authors. The straight lines correspond to ventilation equivalents for O_2 of 2.2 and 2.5 liters/min STPD. (From Gray, J.S. *Pulmonary Ventilation and Its Physiological Regulation,* 1950. Courtesy of Charles C Thomas, Publisher, Springfield, Ill.)

The Lungs and External Respiration

169

Oxygen Cost of Breathing. Under resting conditions, the muscular work done in breathing is relatively small; probably not much more than 1% of the resting metabolism is devoted to this function. As exercise becomes vigorous, however, the oxygen cost of breathing increases disproportionately. Probably at a point between 150 and 250 liters of ventilation, the cost of moving the air takes all of the additional O_2 provided; however, this range of ventilation is seldom achieved in the normal ventilation of even the heaviest work loads. (The experimental values have been achieved through artificially contrived hyperventilation.)

One of the important factors in the physiology of athletic training is the increased efficiency of the breathing mechanics. The ventilation equivalent discussed above is smaller in conditioned athletes, meaning that less breathing work is required to maintain a given O_2 supply. It is quite likely that high efficiency of breathing (as measured by the VE) is a very important factor in such feats as the four-minute mile.

Improving Performance by Better Breathing

It is of practical interest that efficiency of lung ventilation apparently could be improved on theoretical bases (33) by changing the breathing pattern from the normal, which approximates a sine wave, to that approaching a square wave involving an end inspiratory pause. It has been shown in experiments on dogs that such a change in breathing pattern results in an increase in the arterial O_2 pressure of 9.5%, decrease in arterial CO_2 of 8.2%, and 20.3% better alveolar ventilation (23). The only experimental evidence available on human subjects suggests that spontaneously chosen breathing patterns result in the most effective combination of O_2 transport and ventilation cost (19). However, this study included only the very lightest of exercise loads. This would appear to be an interesting area for further experimentation by scientifically inclined coaches. To acquire such new breathing patterns would not be a simple matter, however.

Training Effects on Pulmonary Function

In a recent review of pulmonary adaptations to acute and chronic exercise (training), Dempsey and others (13) pointed out that neither they nor other investigators had been able to demonstrate a training effect on the lung tissue itself in terms of improved diffusion capacity.

However, it has long been known that the VE for O_2 under exercise conditions decreases as the result of training. But this improved efficiency in breathing is not the result of improvement in the lung tissue per se, but rather is the result of a reduced metabolic acidosis and thus a lessened drive to increase ventilation.

Endurance training does appear to bring about important changes in the lung volumes and capacities. Bachman and Horvath (3) found significant decreases in functional residual capacity, residual volume, and the ratio of residual volume/total lung capacity in swimmers after four months of training. Controls and a similar group of wrestlers showed no significant changes. The swimmers also showed significantly increased vital capacity which was the result of an increased inspiratory capacity. All of these changes would result in better alveolar ventilation and consequently should weigh in favor of improved athletic performance.

Genetic Effects on Respiratory Responses to Endurance Exercise. Endurance athletes compared to nonathletes have a decreased ventilatory response to hypoxia (low blood O_2 level) and to hypercapnia (high blood CO_2 level). Interestingly, the former is a genetic effect, but the latter is probably not (2, 28). In any event, the lesser sensitivity to the buildup of CO_2 acting in concert with the smaller amount of CO_2 produced in the muscle tissue (see chap. 3) result in the significantly lower VE discussed above.

Respiratory Phenomena

Stitch in the Side. Frequently, in the course of making the respiratory adjustment to a heavy exercise load such as distance running, athletes experience a rather severe, sharp pain on the lower, lateral aspects of the thoracic wall. This pain has been called a "stitch in the side," but no scientific evidence is available to explain its cause. Because it usually occurs during adjustment to new metabolic demands, it seems reasonable to postulate ischemia of either the diaphragm or intercostal muscles as the cause. Ischemia of any skeletal muscle brings the sensation of pain.

Second Wind. "Second wind," familiar to most endurance athletes, is typified by the feeling of relief upon making the necessary metabolic adjustments to a heavy endurance load. Although it is not entirely, and probably not even mainly, a respiratory adjustment, it is treated under respiratory phenomena because the major manifestation to the athlete is the changeover from dyspnea (labored breathing, shortness of breath) to eupnea (normal breathing).

Again, very little scientific evidence can be brought to bear upon this problem. Close observation seems to indicate a lack of relationship between the time of occurrence and cardiovascular adjustments. Furthermore, the respiratory adjustment is probably only a reflection (an effect rather than a cause) of metabolic adjustment to the exercise load. The most likely explanation of the cause would seem to be one that invokes a change in skeletal muscular efficiency such as might be brought about by increasing muscle temperature.

Exercise-Induced Asthma. It is now a well-established fact that exercise may induce bronchoconstriction (attack of asthma) in asthmatic subjects (16). Since physical educators frequently work with asthmatic children in corrective and adaptive physical education classes, this fact is of some importance. Interestingly, there seems to be a much lower tendency to an attack when cycling than running (1). Thus, if the supervising physician permits endurance exercise at all, then stationary bicycles would be the method of choice. The physiology underlying this interesting difference is as yet unexplained.

Unusual Respiratory Maneuvers

Hypoventilation. If lung ventilation is decreased, either voluntarily or involuntarily, without a corresponding decrease in metabolic rate, this reduction is called *hypoventilation*. It occurs only in abnormal situations, such as those involving airway obstruction. Because metabolism continues at a faster rate than lung ventilation, CO_2 accumulates and the arterial CO_2 must rise (hypercapnia).

Hyperventilation. The converse of hypoventilation is *hyperventilation,* in which the lung ventilation rate is greater than is needed for the existing metabolic rate. In this case, CO_2 is *blown off* faster than it is produced, and thus hyperventilation or forced breathing results in decreasing quantities of CO_2 in the circulorespiratory system. This lowering of blood CO_2 is called *hypocapnia*. Hyperventilation and hypocapnia have considerable interest since they occur accidentally, as the result of emotional excitement, particularly in the inexperienced athlete, and as an intentional device to increase breath-holding ability. It should be pointed out here that hyperventilation has no appreciable effect on O_2 values in the blood since blood is virtually saturated with O_2 when it leaves the lungs anyway.

That hyperventilation increases breath-holding time is shown by table 8.3, which illustrates an experiment by the author on sixteen college men and women. One minute of hyperventilation more than doubled the mean breath-holding time. In light of the earlier discussion on the significance of the dead space, it is readily understood why hyperventilation is best performed by increasing the depth rather than the rate of breathing.

There is no doubt that hyperventilation can provide a significant advantage in competitive athletics wherever breath-holding time is a factor in affecting overall performance. In swimming the crawl stroke, for example, turning the head away from the midline to breathe slows the sprint swimmer's time; hyperventilation immediately before an event allows the swimmer to go farther before breathing becomes necessary.

A physical educator or coach will occasionally witness a situation in which an inexperienced individual passes out immediately after completing a sprint run. If the subject's fingernails show cyanosis (blue color) and there is evidence of tetany, such as involuntary flexing of the fingers or toes, the faint may well have been caused by prolonged, unintentional hyperventilation due to the emotional content associated with the run (such as testing for grading purposes). In any event, the situation calls for medical attention to determine if more serious factors are involved.

Individual variations in reaction to hyperventilation are illustrated by the symptoms found in table 8.3. If overbreathing is done at a lesser depth, it can be continued for a longer period before symptoms become evident.

Breath-holding. Many athletic events are performed with the breath held, notably swimming and track sprints. The physiology of breath-holding involves respiratory, circulatory, and cardiac changes, all of which are important in the light of recent research. The most obvious changes when the breath is held are the increasing level of CO_2 and the decreasing level of O_2 in the alveolar air. These changes, of course, reflect the changes in the level of the respiratory gasses in the blood, the result of the continuing metabolism. It will be recalled that O_2 and CO_2 levels are involved in respiratory control, but the rising CO_2 level is more important in determining the length of time the breath can be held.

Table 8.3
Seconds of Breath-holding with and without Hyperventilation*(Sitting at Rest)

Subject	Without Hyper- ventilation	10 Times in 30 Sec	20 Times in 60 Sec	Dizziness at	Other Symptoms
1	35	95	110	20 breaths	none
2	63	83	93	10	cyanosis of fingers
3	80	122	183	20	cyanosis of fingers
4	40	70	152	20	none
5	45	63	105	10	tension in fingers
6	65	125	160	10	cyanosis of fingers
7	45	80	120	20	none
8	63	108	153	20	none
9	90	135	120		none
10	80	180	140		none
11	35	55	80	10	none
12	55	95	120		cyanosis of fingers
13	55	105	145	20	cyanosis of fingers
14	105	150	195	20	tension, headache
15	120	150	160		finger perspiration
16	30	60	75		tension in gastrocnemius
mean	62.9	104.8	131.9		

*Hyperventilation achieved through increasing depth by using both inspiratory and expiratory reserve volumes.

It has been shown by Craig and his associates (8,9,10) that when the partial pressure of CO_2 in the alveolar air exceeds approximately 50 mm Hg, the stimulus to breathe is so strong that the breath can no longer be held. This is called the *break point,* at which breathing recommences. It has, however, become common practice to hyperventilate in preparation for any event that requires breath-holding, and this procedure, since it allows athletes to start with a lower level of CO_2 also enables them to continue the breath-hold longer (see table 8.3). As the breath-hold continues, lower levels of alveolar O_2 are reached because of the continuing consumption for metabolic needs. When the partial pressure of O_2 in the alveolar air has been reduced from its normal value (approximately 100 mm Hg) to 25 or 30 mm Hg, cerebral function is affected and consciousness is lost. Craig's work (8) indicates this may happen in underwater swimming at distances of 114 –185 ft. Thus the combination of hyperventilation and subsequent underwater swimming constitutes a hazardous situation.

Craig has recently summarized fifty-eight cases of loss of consciousness during underwater swimming and diving in which twenty-three athletes died (11). His breath-holding experiments indicated that the time between loss of consciousness and death may be no longer than 2.5 min. Those responsible for aquatic safety must watch for swimmers who are practicing hyperventilation and underwater swimming in competition with themselves or with others.

Valsalva Maneuver. This maneuver involves a deep inspiration that is followed by attempted expiration against a closed glottis. Its physiology was described in chapter 7.

Effects of Air Pollution on Respiration

There are many different chemical entities that pollute our urban environments. The major contaminants are (1) *particulate matter* (dusts, fumes, and mists of solid particles); (2) *carbon monoxide,* or CO, the result of incomplete combustion of hydrocarbons such as gasoline; (3) *hydrocarbons* from industrial plants, etc.; (4) *sulfur oxides,* which result from burning of fossil fuels such as coal; (5) *nitrogen oxides,* which are formed when most combustibles are burned because of the large fraction of nitrogen in air; (6) *ozone,* which is the result of the sun's action on nitrogen dioxide and certain hydrocarbons. There are many other contaminants that achieve importance under certain specified and unusual circumstances.

Of greatest importance to us is ozone, because it is such a common and typical constituent of the upper atmosphere and because it is a toxic contaminant predominant in the smog of numerous urban areas. It is one of the most potent oxidizing agents in our atmosphere and can seriously disrupt biochemical and physiological functions through direct oxidative lesions in the respiratory tissues and blood.

Until recently, most experimentation with air pollutants had been carried out with exposure to varying levels of contaminant only at rest or light exercise. But the total exposure to ozone (O_3), for example, is related not only to its concentration in the air breathed, but also to ventilation, which may increase by twenty times in heavy work or exercise. Thus the important question for us is, What is the effect of smog as measured by O_3 concentration upon the exercising human?

Effect on Performance.   One of the earlier studies on this problem showed that the performance of high school cross-country runners decreased with increasing concentrations of photochemical oxidants (smog) (32). The basis for these losses in endurance performance is probably a loss of aerobic capacity. It has been shown recently that exposure to ozone concentrations approximating peak ambient levels resulted in significant reductions of 10% in aerobic capacity accompanied by a 16% loss in maximum ventilation (14).

Toxicity.   At rest, the short-term effects (less than twenty-four hours exposure) of the nitrogen oxides are mainly a loss in visual dark adaptation and increases in airway resistance. The effects of chronic exposure are more serious and include a lessened resistance to infection in the respiratory system, ciliary loss, alveolar cell disruption, and obstruction of respiratory bronchioles, at least in animal studies and quite probably in the human as well (26).

At an exercise in which normal young males were exposed to ozone concentrations below current smog-alert levels, all subjects clearly demonstrated some signs of toxicity while working at 45%–65% of maximum VO_2. The major ventilatory effects were a trend toward shallower breathing, leading to a 25% increase in breathing frequency and a 30% decrease in mean tidal volume. DeLucia and Adams (12) also showed distinct differences in individual sensitivity to ozone. Their two most sensitive subjects were unable to complete one hour at 65% maximum exercise.

Silverman and colleagues (31) have shown that ozone exposure at concentrations that may be encountered in some North American cities is associated with significant decrements in breathing function and that the addition of even light exercise increased these losses of function. They concluded that currently encountered ambient levels are too high and may provoke physiological responses even in healthy, normal individuals.

On the other hand, Hackney and his colleagues (18) have found that some human subjects seem to adapt to severe ambient ozone exposure, at least to the extent that obvious acute respiratory effects are prevented. Whether this is also true for the chronic effects of smog exposure is not yet known.

In any event, the only dose-response data available at the present time suggest that for subjects engaged in vigorous exercise, 0.30 ppm of ozone is an unacceptably high level of exposure (15). On the basis of these data, it appears that physical educators, coaches, and school administrators in urban areas exposed to smog would be well advised to set standards along these lines, followed by day-to-day monitoring of ozone level. Classes and team exercises involving vigorous activity should be discontinued, or activities should be made less vigorous when ozone levels approach 0.30 ppm.

Smoking—Self-Induced Air Pollution

Although smoking has long been indicted as a cause of respiratory problems in athletes and has customarily been forbidden for members of athletic teams, the scientific evidence has been meager until recently. Nadel and Comroe (25), in a well-controlled study, demonstrated that fifteen puffs of cigarette smoke in five minutes caused an average decrease in airway conductance of 31% in thirty-six normal subjects. This finding was highly significant and was found in both smokers and nonsmokers. Changes occurred as early as one minute after smoking began and lasted from ten to eighty minutes (mean was equal to thirty-five minutes). These investigators attributed the changes to inhalation of submicronic particles rather than nicotine or oxides of nitrogen.

In light of the earlier discussion on the cost and efficiency of breathing, it is readily seen that smoking, which can increase airway resistance by 31% under resting conditions, can be a very great detriment under conditions of maximum ventilation to an athlete.

Indeed, this increased resistance to breathing may usurp as much as 5%-10% of the total O_2 supply at heavy exercise work loads such as those encountered in endurance-type athletic activities (27). This is one form of air pollution we can and should remove immediately. While other forms of air pollution are difficult to control, this self-induced air pollution is completely within our control.

SUMMARY

1. The total respiratory process consists of three component functions: (a) gas exchange in the lungs, (b) gas transport to the tissues by the blood, and (c) gas exchange between the blood and the tissue fluids bathing the cells.
2. The lungs consist of two systems: (a) a conductive system, whose smallest components are the *terminal bronchioles,* and (b) a respiratory system, whose function is performed largely by the *alveoli.*
3. Air flow into and out of the lungs depends on differences of pressure between the ambient air and the air within the lung. This pressure gradient is brought about by the muscular activity of the diaphragm and intercostals in normal resting breathing.
4. Respiration during vigorous exercise brings accessory muscles into play. Inspiration is aided by action of the *sternomastoids* and *scalenes.* Expiration is aided by the abdominal group.

5. Respiratory control is brought about by the interaction of several factors, acting either directly or reflexly on the respiratory center in the medulla: (a) CO_2 rise and consequent pH depression; (b) anoxia; (c) proprioceptive reflexes from the joints and muscles; (d) body temperature rise; (e) cerebral factors; (f) Hering-Breuer reflexes.

6. Every individual has an optimal combination of rate and depth of breathing for greatest efficiency. In normal individuals, increases in depth are more effective because of the lessening effect of the anatomical dead space as depth is increased.

7. Various types of athletic activity require considerably different breathing patterns. Where minute ventilation is restricted by the nature of the movement, as in swimming, better O_2 extraction appears to compensate.

8. Under steady-state conditions, approximately twenty-five liters of air are required to furnish one liter of O_2 to the tissues. The ratio goes even higher during overload conditions.

9. The work of breathing is relatively small at rest. Under vigorous exercise conditions, however, a point is reached at which all of the extra O_2 made available by increased breathing is necessary to supply the needs of the respiratory muscles.

10. Theoretical considerations suggest the possibility for improving performance by optimizing the mechanics of breathing, but presently available evidence relating to application is meager.

11. Training results in more efficient breathing in that less ventilation is required per liter O_2 consumption. This appears to be more a matter of improving various lung volumes and capacities rather than bringing about changes in the lung tissue per se.

12. *Hyperventilation* has important physiological advantages in increasing breath-holding time. On the other hand, it also brings about possibly hazardous conditions of which every physical educator, coach, and athlete should be aware.

13. The effects of air pollution upon respiration are well documented and significantly affect performance, as well as health.

14. Scientific evidence indicates that smoking has a deleterious effect not only on the circulatory system, but also on the respiratory system, and on performance as well as health.

REFERENCES

1. Anderson, S.D.; Connolly, N.M.; and Godfrey, S. Comparison of bronchoconstriction induced by cycling and running. *Thorax* 26:396–401, 1971.
2. Arkinstall, W.W.; Nirmel, K.; Klissouras, V.; and Milic-Emili, J. Genetic differences in the ventilatory response to inhaled CO_2. *J. Appl. Physiol.* 36:6–11, 1974.
3. Bachman, J.C., and Horvath, S.M. Pulmonary function changes which accompany athletic conditioning programs. *Res. Q.* 39:235–39, 1968.

4. Beaver, W.L., and Wasserman, K. Transients in ventilation at start and end of exercise. *J. Appl. Physiol.* 25:390–99, 1968.

5. Bechbache, V.; Bechbache, R.R.; and Duffin, J. The entrainment of breathing frequency by exercise rhythm. *J. Physiol.* 272:553–61, 1977.

6. Campbell, E.J.M. *The Respiratory Muscles and the Mechanics of Breathing.* Chicago: Year Book Medical Publishers, Inc., 1958.

7. Casaburi, R.; Whipp, B.J.; Wasserman, K.; and Koyal, S.N. Ventilatory and gas exchange responses to cycling with sinuisoidally varying pedal rate. *J. Appl. Physiol.* 44:97–103, 1978.

8. Craig, A.B., Jr. Causes of loss of consciousness during underwater swimming. *J. Appl. Physiol.* 16:583–86, 1961.

9. Craig, A.B., Jr.; Halstead, L.S.; Schmidt, G.H.; and Schnier, B.R. Influences of exercise and oxygen on breath-holding. *J. Appl. Physiol.* 17:225–27, 1962.

10. Craig, A.B., Jr.; and Babcock, S.A. Alveolar CO_2 during breath-holding and exercise. *J. Appl. Physiol.* 17:874–76, 1962.

11. Craig, A.B., Jr. Summary of 58 cases of loss of consciousness during underwater swimming and diving. *Med. Sci. Sports* 8:171–75, 1976.

12. DeLucia, A.J., and Adams, W.C. Effects of O_3 inhalation during exercise on pulmonary function and blood biochemistry. *J. Appl. Physiol.* 43:75–81, 1977.

13. Dempsey, J.A.; Gledhill, N.; Reddan, W.G.; Forster, H.V.; Hanson, P.G.; and Claremont, A.D. Pulmonary adaptation to exercise: effects of exercise type and duration, chronic hypoxia and physical training. *Ann. N.Y. Acad. Sci.* 301:243–61, 1977.

14. Folinsbee, L.J.; Silverman, F.; and Shephard, R.J. Decrease of maximum work performance following ozone exposure. *J. Appl. Physiol.* 42:531–36, 1977.

15. Folinsbee, L.J.; Drinkwater, B.L.; Bedi, J.F.; and Horvath, S.M. The influence of exercise on the pulmonary function changes due to exposure to low concentrations of ozone. In *Environmental Stress: Individual Human Adaptations,* eds. L.J. Folinsbee, et al. New York: Academic Press, 1978.

16. Freedman, S.; Tattersfield, A.E.; Pride, N.B. Changes in lung mechanics during asthma induced by exercise. *J. Appl. Physiol.* 38:974–82, 1975.

17. Grimby, G.; Bunn, J.; and Mead, J. Relative contributions of rib cage and abdomen to ventilation during exercise. *J. Appl. Physiol.* 24:159–66, 1968.

18. Hackney, J.D.; Linn, W.S.; Mohler, J.G.; and Collier, C.R. Adaptation to short-term respiratory effects of ozone in men exposed repeatedly. *J. Appl. Physiol.* 43:82–85, 1977.

19. Hanson, P.G.; Lin, K.H.; and McIlroy, M.B. Influence of breathing pattern on oxygen exchange during hypoxia and exercise. *J. Appl. Physiol.* 38:1062–66, 1975.

20. Holmer, I.; Stein, E.M.; Saltin, B.; Ekblom, B.; and Astrand, P-O. Hemodynamic and respiratory responses compared in swimming and running. *J. Appl. Physiol.* 37:49–54, 1974.

21. Hornbein, T.F.; Roos, A.; and Griffo, Z.J. Transient effect of sudden mild hypoxia on respiration. *J. Appl. Physiol.* 16:11–14, 1961.

22. Kalia, M.; Senapati, J.M.; Parida, B.; and Panda, A. Reflex increase in ventilation by muscle receptors with nonmedullated fibers (C fibers). *J. Appl. Physiol.* 32:189–93, 1972.

23. Knelson, J.H.; Howatt, W.F.; and DeMuth, G.R. Effect of respiratory pattern on alveolar gas exchange. *J. Appl. Physiol.* 29:328–31, 1970.

24. Lambertsen, C.J.; Semple, S.J.G.; Smyth, M.G.; and Gelfand, R. "H+ and pCO_2 as chemical factors in respiratory and cerebral circulatory control." *J. Appl. Physiol.* 16:473–84, 1961.

25. Nadel, J.A., and Comroe, J.H., Jr. Acute effects of inhalation of cigaret smoke on airway conductance. *J. Appl. Physiol.* 16:713–16, 1961.

26. National Academy of Sciences. *Medical and Biological Effects of Environmental Pollutants: Nitrogen Oxides.* Washington D.C.: The Academy, 1977.

27. Rode, A., and Shephard, R.J. The influence of cigarette smoking upon the oxygen cost of breathing in near maximal exercise. *Med. Sci. Sports.* 3:51–55, 1971.

28. Scoggin, C.H.; Doekel, R.D.; Kryger, M.H.; Zwillich, C.W.; and Weil, J.V. Familial aspects of decreased hypoxic drive in endurance athletes. *J. Appl. Physiol.* 44:464–68, 1978.

29. Sharp, J.T.; Goldberg, N.B.; Druz, W.S.; and Danon, J. Relative contributions of rib cage and abdomen to breathing in normal subjects. *J. Appl. Physiol.* 39:608–18, 1975.

30. Shephard, R.J., and Bar-or, O. Alveolar ventilation in near-maximum exercise. *Med. Sci. Sports* 2:83–92, 1970.

31. Silverman, F.; Folinsbee, L.J.; Barnard, J.; and Shephard, R.J. Pulmonary function changes in ozone-interaction of concentration and ventilation. *J. Appl. Physiol.* 41:859–64, 1976.

32. Wayne, W.S., and Wehrle, P.F. Oxidant air pollution and school absenteeism. *Arch. Environ. Health* 19:315–22, 1969.

33. Yamashiro, S.M., and Grodins, F.S. Optimal regulation of respiratory airflow. *J. Appl. Physiol.* 30:597–602, 1971.

9

Gas Transport
and Internal Respiration

Thus far we have discussed the mechanical factors involved in breathing and their physiological controls; we will now consider the processes that are necessary for bringing about the ultimate goal of tissue respiration. Three processes intervene between lung ventilation and actual tissue respiration: (1) diffusion of O_2 across two very thin membranes—the wall of the alveolus and the wall of the capillary; (2) transport of O_2 via the blood to the capillary bed of the active tissues; (3) diffusion of O_2 across the capillary wall to the tissue fluids that bathe the actual consumers, the metabolizing cells. As O_2 is unloaded from the blood, CO_2 is being taken on for the return trip to the right heart and back to the lungs.

At this point we must consider the nature of the diffusion process and some of the laws that govern it. Recall that fluids flow from point to point only because of differences in pressure called *pressure gradients*. This is equally true for gasses. For this reason physiologists usually refer to respiratory gasses in terms of pressure rather than in terms of concentration (percentage, etc.).

Table 9.1 gives percentages and approximate pressures of O_2 exerted in a standard atmosphere. It should be noted that the percentage concentration of O_2 is the same at 40,000 feet as it is at sea level; however, since one becomes unconscious in a matter of seconds at 40,000 feet altitude due to O_2 lack, it is obvious that percentage has little meaning in this situation. The atmospheric pressure of O_2, on the other hand, tells the story rather well.

Properties of Gasses and Liquids

Basic to all gas laws is the molecular theory that all gasses are composed of molecules that are constantly in motion at very high velocities. A gas has no definite shape or volume and conforms to the shape and volume of its container. Its pressure is the result of the constant impacts of its many molecules upon the walls of the container. Obviously, the pressure of a gas is increased by confining it in a smaller volume or by increasing the activity of each molecule. Because a rise in temperature increases the velocity of molecular movement, heat also results in producing increased pressure.

Liquids, on the other hand, are composed of molecules that are much closer together, and this closeness results in their having a definite, independent volume that varies little with temperature or with the size and shape of the container.

For the student to understand respiratory physiology (a very important part of exercise physiology), knowledge of the laws that govern behavior of gasses is essential.

Boyle's Law. This law states that if temperature remains constant, the pressure of a gas varies inversely with its volume. If, for example, we decrease the volume by one-half, the pressure will be doubled.

Table 9.1
Percentage and Partial Pressures of O_2 by Altitude

Altitude	Atmospheric Pressure (mm Hg)	Percent O_2	Approximate Pressure Exerted by O_2 in the Atmosphere (mm Hg)
Sea level	760	20.93	159
10,000	523	20.93	109
20,000	349	20.93	73
30,000	226	20.93	47
40,000	141	20.93	29

Table 9.2

Composition of Atmospheric Air and the Consequent Partial
Pressures of Respiratory Gasses

Gas	Percent in Dry Atmosphere	Partial Pressure in Dry Atmosphere	Partial Pressure in Alveolar Air	Partial Pressure in Mixed Venous Blood	Diffusion Gradient
Total	100	760	760	705	
H_2O	0	0	47	47	
O_2	20.93	159	100	40	60
CO_2	0.03	0.2	40	45	5
N_2	79.04	600.8	573	573	

Gay-Lussac's Law. If its volume remains constant, the pressure of a gas increases directly in proportion to its (absolute) temperature.

Law of Partial Pressures. In a mixture of gasses, each gas exerts a *partial pressure,* proportional to its concentration. Thus in atmospheric air with a total pressure of 760 mm Hg, O_2—which makes up 20.93%—has a partial pressure of 159 mm Hg: $20.93/100 \times 760$ mm Hg = 159 mm Hg.

Henry's Law. The quantity of a gas that will dissolve in a liquid is directly proportional to its partial pressure, if temperature remains constant.

Composition of Respiratory Gasses. The atmospheric air is composed mainly of nitrogen (N_2), oxygen (O_2), and carbon dioxide (CO_2); there are also rare gasses (argon, krypton, etc.), but these are ordinarily lumped together and included with the nitrogen fraction.

Table 9.2 illustrates the salient features of the respiratory gas exchange. It will be noted that the partial pressures of dry atmospheric air are proportional to the percentages as per the law of partial pressures; however, the alveolar air is saturated with water vapor, which contributes a partial pressure of 47 mm Hg at body temperature. The partial pressure of O_2 in the lungs, then, would be 20.93% of 713 (760−47) mm, or approximately 149 mm Hg—if we could completely exchange the air in the lungs. This, of course, is

Gas Transport and Internal Respiration

183

impossible because alveolar air in the lungs is a mixture of atmospheric air with air that has already participated in the respiratory exchange. For this reason, it will be noted in Table 9.2 that the actual partial pressure of O_2 in the alveolar air is 100 mm instead of the 149 mm that would be present if there were no dead space and if the lung collapsed to empty itself completely at each breath.

The importance of all this lies in the *diffusion gradients* for O_2 and CO_2 that, after all, ultimately determine the amount of gaseous exchange taking place. The diffusion gradient for O_2 is some 60 mm Hg, for CO_2 only 5 to 6 mm Hg. Since these figures obtain for normal, healthy individuals, the diffusion gradient for CO_2 is obviously sufficient to maintain the necessary homeostatic relations for CO_2 levels. This can be explained by the fact that diffusion of gasses across a membrane depends not only upon the gradient, but also upon the ease with which a particular gas can penetrate the membrane. This, in turn, depends upon the solubility of the gas in the membrane (largely water). Because the solubility of CO_2 in water is some twenty or more times that of O_2, this explains the need for a greater diffusion gradient for O_2.

Acids, Bases, and pH. Acids may be defined as compounds that yield positively charged hydrogen ions (H^+) in solution—and, conversely, bases as compounds that yield negatively charged hydroxyl ions (OH^-) in solution. A very convenient yardstick for measuring and describing degrees of acidity or alkalinity has been set up: pH, which is the negative logarithm of the hydrogen ion concentration in gram molecular weight. This may be visualized as follows:

Strongest Acid		Strongest Base
← Increasing Acidity	Neutrality	Increasing Alkalinity →

pH = 1 2 3 4 5 6 7 8 9 10 11 12 13 14

Since this is a logarithmic scale, a very small change of pH makes a considerable change in acidity or alkalinity, and therefore pH is given to at least one and usually two decimal places. For example, the extreme fluctuations of the pH of normal blood lie within pH values of 7.30 to 7.50. The extreme values in illness have been known to go as low as 6.95 and as high as 7.80. However, in healthy subjects, recent work has shown (12) that surprisingly low values, down to 6.80 can result from heavy anaerobic exercise. Such *short-term* pH changes appeared to be well tolerated in *healthy subjects*. For a more complete treatment of pH and acid-base balance, the student is referred to a text in physiological chemistry.

Gas Transport by the Blood

Oxygen. Although chemical analyses have demonstrated that blood is capable of carrying only about 0.2 volume percent of O_2 (0.2 ml of O_2 per 100 ml of blood) in solution at normal atmospheric pressures, it actually transports 20 volume percent of O_2, 100 times as much as will dissolve in physical solution.

The reason for this great discrepancy is the presence of hemoglobin in the erythrocytes. Hemoglobin is an iron-bearing pigment, consisting of *heme* (which contains iron) and *globin* (which is a protein). Hemoglobin has the unique characteristic of combining with O_2 quickly and reversibly, and without the necessity for help from enzyme reactions. This is not true of typical oxidation reactions in which the O_2, once combined, is separated only with difficulty. Therefore the term *oxygenation* is used for the process, rather than *oxidation,* and the reaction can be described thus:

$$\text{Hemoglobin} \quad + \quad \text{Oxygen} \rightleftarrows \text{Oxyhemoglobin}$$
$$\text{Hb} \qquad\qquad + \quad O_2 \quad \rightleftarrows \quad HbO_2$$

As is the case with all reversible reactions, if the end product on the right, HbO_2, is constantly removed, as in the lungs where oxygenated blood moves off to the tissues, the reaction continues to proceed from left to right. On the other hand, in the tissue exchange the O_2 is carried off by tissue fluids to the cells, and the reaction proceeds to the left.

If the organism depended on dissolved O_2, an all-out effort by the cardiovascular and respiratory systems could not meet the O_2 needs of the resting metabolism, let alone the metabolism of exercise.

Carbon Dioxide. CO_2, which is the constant end product of metabolism in the cell, diffuses across the cell membrane into the tissue fluid, thence across the capillary wall into the blood plasma, where a small portion of it is transported. The larger proportion, probably 90%-95%, diffuses from the plasma into the erythrocytes. It is transported by the erythrocyte in three forms: (1) in combination with hemoglobin as carbamino-hemoglobin; (2) as bicarbonate, HCO_3; and (3) as dissolved CO_2, a portion of which ionizes into carbonic acid, H_2CO_3.

The exchange of CO_2, at the lungs and at the tissues, involves the following reversible reaction.

$$\text{In the Lungs} \rightarrow$$
$$HCO_3^- + H^+ \rightleftarrows H_2CO_3 \rightleftarrows H_2O + CO_2$$
$$\leftarrow \text{In the Tissues}$$

This is ordinarily a slow reaction; however, an enzyme, *carbonic anhydrase,* catalyzes these reactions so that they can go to completion before the blood leaves the lung or tissue capillaries.

Gas Transport and Internal Respiration

Internal Respiration

Much of the story concerning internal respiration, or gas exchange in the tissues, has already been told. Two other factors, however, are basic to an understanding of the respiratory exchange.

O_2 Dissociation Curve. It is an interesting biochemical fact that the loading of CO_2 into the blood at the tissues considerably aids the unloading of O_2 from blood to tissues. The reverse is also true in the lungs: the unloading of CO_2 in the lungs aids the loading of O_2 into the blood.

These facts are best illustrated by the O_2 dissociation curve (fig. 9.1). If one places a straightedge vertically along the line representing a partial pressure of O_2 of 30 mm Hg (that of the tissues), the difference in hemoglobin saturation level between the points where the 40 mm CO_2 curve (blood CO_2 level) and the 80 mm CO_2 curve (active tissue level) cross this vertical line represents the difference of O_2 that the hemoglobin can hold.

This amount of O_2—in this case from approximately 58% to approximately 42% (16%)—is driven off by changing CO_2 levels in the tissues. In other words, increasing the CO_2 level of the blood from its arterialized level of 40 mm Hg to 80 mm Hg, which probably represents the temporary local changes in blood level at the tissues, results in driving off 16% of the total O_2 load.

Another point regarding the O_2 dissociation curve is of very practical interest on the effect of altitude on human respiration. It should be noted that the curve for 40 mm of CO_2, which represents the mixed venous blood (typical of the body as a whole, but not of any localized tissue site), is not very steep from 100 mm to approximately 60 mm of O_2 partial pressure. The upper figure is typical of alveolar O_2 tension at sea level, and the lower figure represents alveolar O_2 tension at about 15,000–16,000 feet above sea level, the level at which resting humans (pilots, and others) begin having serious symptoms due to the lack of O_2. (Military regulations require the use of oxygen above 10,000 feet to provide a safety margin that allows for individual variance.) If this O_2 dissociation curve were a straight diagonal line, physical impairment would probably commence at about 90 mm of O_2 tension instead of at 60 mm, or at an altitude of about 6,000 feet.

Another salient feature of the O_2 dissociation curve is that it is steep (close to vertical) when the partial pressures of O_2 are low. This fact means, of course, that small changes in partial pressure of O_2 on this part of the curve make large changes in the amount of O_2 the hemoglobin can hold, thus making large exchanges of respiratory gas efficient when the need is greatest.

Coefficient of O_2 Utilization. This term can be defined as the proportion of O_2 transported by the blood that is given off to the tissues. Since 99% of the transported O_2 is bound to hemoglobin, this story also can be related in terms of the O_2 dissociation curve. During resting conditions at sea level, the hemoglobin leaving the lungs is at least 95% saturated with O_2. After leaving the capillary bed of resting tissue, it is still at least 70% saturated. (In figure 9.1, use the 40 mm CO_2 curve and note its intersections at the 100 mm and the 40 mm O_2 tension lines.) Thus the hemoglobin has given off 23/98 of its O_2, or 23%, in resting conditions.

In exercise, this situation becomes much more favorable. The hemoglobin leaving the lungs is still approximately 95% saturated, but after leaving active muscle tissue, it may approach zero saturation. Thus the coefficient of oxygen utilization may increase from three to four times in exercise. It should be reiterated that this increase is facilitated by the steepness of the O_2 dissociation curve at the lower O_2 tensions (as was mentioned before).

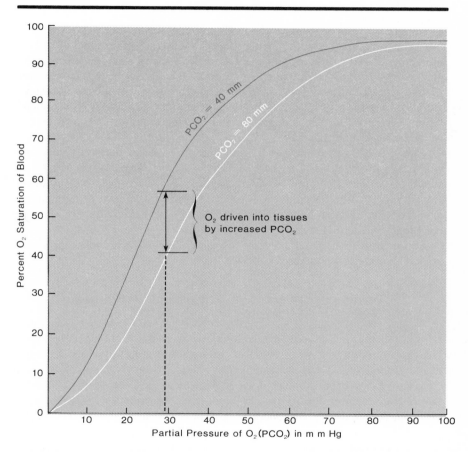

Figure 9.1 The oxygen dissociation curve for human blood. Drawn from the data of J.W. Severinghaus, *J. Appl. Physiol.* 21:1108, 1966, and R.M. Winslow, et al., *J. Appl. Physiol.* 45:289, 1978.

At this point it may be well to note the combination of factors that contribute to supplying the increased oxygen demands of exercising muscle tissue. First, recall that cardiac output can increase about six times its resting value. Second, in combination with an increased utilization coefficient of three to four times, this means a possible increase of at least eighteen times the resting value. Third, in respect to the local situation at any given active muscle, this may be multiplied by another factor of two, due to the approximate doubling of the number of open capillaries (discussed earlier). Thus a total increase of at least thirty-six times the resting oxygen supply is possible at any active muscle group.

Regulation of Acid-Base Balance

Even under resting conditions, the acid-base equilibrium of the body fluids is constantly challenged by the formation of CO_2 as the end product of cellular metabolism. Furthermore, when exercise work loads become severe, lactic acid is also formed, constituting an additional influence that tends to drive pH downward. Illness brings about other acidifying or alkalinizing influences. Because long-term pH changes beyond the range of 7.30 to 7.50 are inconsistent with good health, it is obvious that the human organism must be able to control acid-base balance.

Two processes are involved in this. The first line of defense against pH changes is the combination of three *buffer systems* that serve to absorb the shock, as it were, of sudden changes. Ultimately, however, *physiological changes* have to be brought about to maintain the organism in homeostasis over the longer period of time, and these physiological changes mainly involve the lungs and the kidneys.

Buffer Systems. A buffer system consists of a weak acid and a salt of that acid. The system functions as follows.

HL		NaHCO$_3$		NaL		H$_2$CO$_3$
Lactic	+	Sodium	$\rightarrow$	Sodium	+	Carbonic
acid		bicarbonate		lactate		acid

In this schema, lactic acid (a relatively strong acid) combines with sodium bicarbonate (salt of a weak acid) to form sodium lactate (which no longer has acid tendencies) and carbonic acid (which is a very weak acid). Thus a strong acid has been exchanged for a weak acid, and the tendency for the lactic acid to lower the pH of the blood has been greatly lessened by the buffering action of the carbonic acid bicarbonate system.

Table 9.3 illustrates what happens when a strong acid, HCl, is added to a water solution that contains the bicarbonate buffer system. It should be noted that when the ratio of H_2CO_3 and $NaHCO_3$ is favorable (about 1:5), the addition of the first 10 gm of acid brings about almost no change in

Basic Physiology Underlying the Study of Physiology of Exercise

relative acidity or pH. When the buffer ratio becomes less favorable, however, the addition of the same quantity of acid (300-310 gm) more than doubles the relative acidity, and the pH changes from 6.00 downward to 5.66. When the bicarbonate is used up, there is no longer any buffering, and the addition of 10 gm of HCl (bringing the HCl from 320 to 330 gm) brings about a *sixfold* increase in relative acidity and, of course, a much larger drop in pH. The effectiveness of the buffer system depends on the ratio of the acid to the salt.

Blood must now be considered as two fluids that need buffering: the plasma and the fluid within the erythrocytes. In the plasma, the acids to be buffered are largely *fixed acids,* so called because they are not subject to rapid excretion. They are, in general, stronger acids, such as hydrochloric, phosphoric, sulfuric, and lactic acids (the stronger acids occur in very small quantities). The most important buffers in the plasma are

$$\frac{H_2CO_3}{NaHCO_3} \quad \text{and} \quad \frac{H \text{ protein}}{Na \text{ proteinate}}$$

The blood proteins can act as buffer systems because, at the pH of blood, they behave as weak acids and react with the base to form a salt. In the intracellular fluids of the erythrocytes, on the other hand, the major acid to be buffered is the carbonic acid that results from the respiratory exchange.

Table 9.3
Effect of Buffering in Combating Acidifying Effect of Adding Hydrochloric Acid to a Solution

HCl (gm) Added	Buffer Ratio $H_2CO_3/NaHCO_3$	H^+ Concentration	pH	Relative Acidity
0	2.27:11.9	0.000000057N	7.24	.57
10	2.27:11.5	0.000000059	7.23	.59
50	2.27:10.0	0.000000068	7.13	.68
100	2.27: 8.2	0.000000083	7.08	.83
150	2.27: 6.3	0.000000108	6.97	1.08
200	2.27: 4.4	0.000000154	6.81	1.54
250	2.27: 2.6	0.000000260	6.59	2.60
300	2.27: 0.68	0.000001000	6.00	10.
310	2.27: 0.31	0.000002200	5.66	22.
318		0.000260000	3.59	260.
320		0.000450000	3.35	450.
330		0.002700000	2.57	2,700.

Based on data from Henderson L. J., in Best, C.H., and Taylor, N.B. *The Physiological Basis of Medical Practice,* 1943, p. 170. Courtesy of Williams & Wilkins Company, Baltimore.

The buffer systems mainly responsible for this are hemoglobin and oxyhemoglobin, each of which can act as a weak acid or as a potassium salt as follows:

$$\frac{H\ Hb\ O_2}{K\ Hb\ O_2} \quad \text{and} \quad \frac{H\ Hb}{K\ Hb}$$

Oxyhemoglobin	Hemoglobin
system	system

There are also other, less important buffer systems in both plasma and cells, such as the acid and basic sodium and potassium phosphates.

Physiological Regulation of Acid-Base Balance. Although the buffering systems can resist fast changes in pH, a change does occur with the addition of an acid (or base) to the body fluids (table 9.3). These changes are corrected by two physiological mechanisms for excretion of the acid (or base): changes in respiratory function and changes in kidney function.

The function of the lungs in regulating acid-base balance is another example of the body's servomechanisms. For example, if the breath is held, the CO_2 resulting from metabolism accumulates. This has the effect of pouring acid into the tissue fluids and blood: $CO_2 + H_2O \rightarrow H_2CO_3$. This decrease in pH is interpreted as the *error signal* by the respiratory center in the medulla, which corrects the situation by increasing the rate and depth of ventilation (if possible). When the breath is held, the error signal simply grows larger and larger, until the urge to breathe overcomes the willpower to hold the breath. The rate and depth will then be greater than normal, until equilibrium has been reestablished by *blowing off* more CO_2 than is being formed: $H_2CO_3 \rightarrow CO_2 + H_2O$.

Thus the importance of the respiratory function in acid-base regulation lies in the fact that the decomposition products of carbonic acid are volatile and can be readily blown off by the lungs.

In hyperventilation, on the other hand, more CO_2 is blown off than is formed and consequently a change in the ratio of H_2CO_3 to $NaHCO_3$ is brought about. The normal ratio is 1:20. If H_2CO_3 and $NaHCO_3$ are increased or decreased proportionately, no change in ratio occurs, and consequently there is no change in pH. In the case of hyperventilation, however, the H_2CO_3 decreases disproportionately because the $NaHCO_3$ remains the same, and thus a relative increase in the base occurs with a rise in pH.

If the hyperventilation is of short duration—as in preparing for an athletic event—the increased production of CO_2 during the event corrects the situation; but if hyperventilation is long-term (one hour or more), as in fever, a secondary adjustment must be made to excrete some of the bicarbonate to keep the ratio close to 1:20. This excretion of bicarbonate is done through the kidney.

The last line of defense and the one concerned with long-term changes in acid-base equilibrium is the excretion of abnormal amounts of acid or base by the kidney to maintain all the buffer systems at the proper acid-salt ratio for maintaining normal pH. When it is no longer possible to maintain these ratios, acidosis or alkalosis ensues. In severe acidosis, death ensues as a result of coma; in severe alkalosis, death may be brought about by tetany and the resulting muscle spasm of respiratory muscles.

Acid-Base Balance As a Factor Limiting Performance

The preceding discussion makes it obvious that metabolism in general provides a constant acidifying influence. When the metabolic rate is raised to seven or eight times that of the resting level, the increase in CO_2 is proportional, but ventilation can usually keep pace to maintain acid-base equilibrium. However, when the work load goes beyond aerobic capacity, lactic acid becomes the end product of metabolism, instead of CO_2. This is a much stronger acid, and it cannot be excreted quickly by respiration as could the CO_2. It has already been pointed out that under conditions of heavy anaerobic exercise the pH can drop as low as 6.80 (12).

This line of reasoning would seem to indicate that the body's ability to buffer fixed acids (such as lactic acid) should play a large part in determining the end point of anaerobic activity. Because these fixed acids are largely buffered by the bicarbonate system, the combining power of the plasma bicarbonate has been referred to as the *alkaline reserve*. Although this concept rests on a reasonably sound rationale, it is an oversimplification of the complex biochemical interactions involved in acid-base regulation.

Changes in Lung Diffusion in Exercise

The diffusion of oxygen from the alveoli to the pulmonary capillaries increases in virtually direct proportion to the effort involved in the exercise, as measured by O_2 consumption. This relationship has been demonstrated to a level at least seven or eight times that of the resting metabolism (19). In the absence of disease processes, it is very unlikely that pulmonary diffusing capacity for O_2 is a limiting factor in exercise.

The reasons for increased pulmonary diffusion are as yet not completely understood, but the increased ventilation of exercise does not seem to be a necessary factor. It has been shown that similar increases in pulmonary diffusion can occur during exercise when the ventilation is voluntarily held to resting levels (14). It seems most likely that the increase in pulmonary diffusion is the result of increased pulmonary capillary blood volume during exercise brought about by the opening of previously unopened capillaries—as discussed in the chapter on circulation.

There is evidence that highly trained athletes demonstrate better pulmonary diffusion under maximal (11) and submaximal work rates (3) than do nonathletes. This effect may be an innate, inherited characteristic of champion athletes, but it can also be brought about by the effects of a rigorous training program (11).

The work of Mostyn and others (10) seems to indicate that championship-level swimmers are unusually high in pulmonary diffusing capacity. Their experiments would indicate this is due to a larger-than-normal pulmonary capillary blood volume; however, a change in breathing pattern can bring about significant increases in normal nonswimmers. When subjects breathed during exercise with a *held inspiration* maneuver in which they took a fast inspiration to full capacity (one second), held the breath at full inspiration for about seven seconds, then exhaled as fast as possible, they showed a significant improvement in pulmonary diffusion. Obviously, this type of breathing is very similar to that imposed upon competitive swimmers by their environment.

Thus it would seem that in activities in which the respiratory exchange is made under unfavorable conditions such as swimming or mountain climbing, or other activities in which the ambient atmospheric pressure of O_2 is low, the held inspiration type of breathing might provide an advantage in increasing pulmonary diffusing capacity. It is conceivable that this procedure might be of practical use to athletes who must compete at higher-than-normal elevations.

Use of Oxygen to Improve Performance

It is not uncommon to see a coach administer O_2 to his athletes. This has been done by track and swimming coaches immediately before the event, and by many other coaches to hasten the recovery process—for example, between halves of a football or basketball game. Let us consider first the theoretical aspects of this procedure to determine its likely validity, and then take note of the applied research.

We must remember that the arterial blood that leaves the lungs is saturated to the extent of 95%-98%, and this degree of saturation does not seem to be changed in vigorous exercise in a normal subject at sea level. (All of this discussion will apply only to sea-level atmosphere.) If alveolar air increases its partial pressure of O_2 by a factor of three (not unreasonable in 70%-100% mixtures), O_2 transport would conceivably be enhanced by two factors.

1. Hemoglobin saturation would increase to a maximum of 95%-100%. Because blood normally carries twenty volume percent O_2 or 20 ml of O_2 in 100 ml of blood (in round figures), it would increase by 5% (0.05 $\times$ 20), for an increase of 1 ml/100 ml of blood.

2. The O_2 in solution would increase in proportion to the increased partial pressure of O_2 (Henry's law). The amount in solution is ordinarily about 0.2 ml/100 ml of blood. This, then, would become about 0.6 ml/100 ml of blood, or an increase of 0.4 ml. The total potential advantage accruing in O_2 transport would therefore be something like 1.4 ml, or 1.4/20, or about 7%.

It must be realized, however, that this discussion is predicated upon the breathing of O_2 *during* the athletic performance. If it is breathed before the performance, the O_2 would have to be stored to benefit the performance, unless subjects were able to use O_2 right up to the start of the performance and then hold their breath during the performance. In this unlikely event, storage of O_2 might be thought to occur in the sense of a higher partial pressure of O_2 in the lungs at the start of the event.

Now let us review the research in this area under three categories: (1) O_2 before exercise, (2) O_2 during exercise, and (3) O_2 during recovery.

O_2 before Exercise. In a well-controlled study by Miller (9) in which he administered O_2 before, during, and after a treadmill exercise, no changes in heart rate, blood pressure, blood lactate, or endurance were found when the O_2 was administered before or during recovery from the exercise. He was able to show a psychological effect from breathing air that had been marked *oxygen,* and this very likely is also the explanation for earlier studies that had shown improved performance after O_2 breathing.

O_2 during Exercise. Available research seems to agree upon the value of O_2 administration during exercise. Miller found a significant decrease in blood lactate and an increase in running time to exhaustion when O_2 was administered during the treadmill run. Elbel and others (4) found significantly slower heart rates when O_2 was administered during exercise. Bannister and Cunningham (2) set exercise loads on a treadmill so that their four subjects would run to exhaustion in seven to ten minutes while breathing air. While breathing 66% O_2, all subjects were able to maintain these work loads for longer periods, and three of the four subjects went beyond twenty-three minutes. They also compared the effects of 21%, 33%, 66%, and 100% O_2, and it is of interest that best results were obtained with 66%. Subjective reports of euphoria were gotten only from the 66% mixture.

O_2 during Recovery. Miller found no improvement in recovery for any of the measurements made as the result of breathing O_2 during recovery. Elbel (4) found no significant changes in O_2 debt repayment as the result of breathing O_2 during recovery, but they found significantly slower heart rates. Their heart rate findings could be construed as hastening the recovery process, but the differences were very small, one to five beats per minute.

To summarize the case for breathing O_2 to improve performance, it seems unlikely that any *physiological* changes are brought about by breathing it before the event and (at best) only very small improvements in the recovery rate. Improved performance undoubtedly can result if O_2 is breathed *during* the event, but this can have no practical importance in athletics.

On the other hand, *psychological* improvement in performance can be brought about through suggestion (in the use of oxygen). Also, if an athlete is conditioned to the use of O_2, accidental deprivation at an important meet or game could result in a calamitous decrement in performance.

What Sets the Limits of Aerobic Capacity?

Aerobic capacity—the maximum O_2 consumption—depends on the transport of O_2 from the atmosphere to the mitochondria of the muscle cells. This transit involves basically four processes: (1) lung ventilation, or more precisely alveolar ventilation; (2) the interaction between pulmonary diffusion and blood transport; (3) blood transport; and (4) the interaction between tissue diffusion and blood transport.

R. J. Shephard (16) has suggested the use of the *O_2 conductance equation* to estimate where the bottleneck to O_2 transport may lie on theoretical bases. In this approach each of the four links in the O_2 transport chain is treated as a conductance. Conductances can be added as reciprocals to give the sum of series conductances, so that the total conductance, which is equivalent to aerobic capacity or *maximal O_2 consumption,* can be predicted in simplified form as follows:

$$\frac{1}{Uo_2} = \frac{1}{A} + \frac{1}{B} + \frac{1}{C} + \frac{1}{D}$$

where Uo_2 is the total conductance and A, B, C, D represent the four phases in gas transport.

When realistic estimates for maximal conductance are substituted for each of the four links in the chain, the equation looks like this:

$$\frac{1}{Uo_2} = \frac{1}{90} + \frac{1}{176} + \frac{1}{30} + \frac{1}{681}$$

It can readily be seen that the term C, which represents blood transport, has the greatest effect in setting the limit for maximal aerobic capacity. All of these data apply only to normal young subjects at sea level. Thus the theoretical application of the O_2 conductance equation suggests that the most important determinants of aerobic capacity are the interaction of cardiac output and hemoglobin level, which together determine the level of blood transport. As Shephard points out, however, this still leaves us with the question of the relative contributions of the cardiac musculature and of the venous return to this ceiling of performance.

Furthermore, in the last few years during which muscle energetics at the cellular level have come under closer investigation, it no longer appears so certain that human endurance performance (as measured by aerobic capacity, $\dot{V}O_2$ max) is limited by O_2 transport. Considerable evidence now suggests that endurance performance may be limited by the ability of active muscle tissue to utilize O_2, rather than by the delivery of O_2 to it.

Two lines of research have been most popular in attempts to resolve this rather basic question. The first research paradigm is based on the concept that adding more active muscle mass should increase the $\dot{V}O_2$ max, if the limits are set by O_2 utilization. On the other hand, if aerobic capacity is limited by O_2 transport, there should be no increase by adding muscle mass. In such studies where arm work has been added to leg work on the bicycle ergometer, some investigators have found that arm plus leg work elicited greater $\dot{V}O_2$ max than leg work alone (5,6,15,18). Other investigators have been unable to demonstrate any difference between arm plus leg work vs. leg work only (1,17).

Evidence suggests that the divergent results can be explained on the basis of interindividual difference. Those subjects having high fitness levels for leg work show no increase by adding muscle mass because the high O_2 utilization of well-trained legs may reach the capacity for O_2 delivery. Less well conditioned subjects can increase $\dot{V}O_2$ max because the capacity of the O_2 transport system exceeds the demand of the poorly trained leg muscles (13). This is an attractive hypothesis to explain the divergent research findings, but unfortunately it rests on findings from only three subjects.

The second research paradigm is based on the response of $\dot{V}O_2$ max when subjects breathe greater than normal values of O_2 (hyperoxia), either by breathing gasses with a greater than normal percentage of O_2 (20.93) or by breathing air at greater than sea-level pressure. Again, there is disagreement in the experimental results. Some investigators have found an increased $\dot{V}O_2$ max or improved performance resulting from hyperoxia (1,8,20,21,22), which suggests that the limit is set by O_2 transport. If increasing the O_2 offered to the tissues increases O_2 consumption, then obviously O_2 utilization could not have been limiting. But other investigators have found no increase in $\dot{V}O_2$ max as a result of hyperoxia (7,9).

In any event, the hyperoxic paradigm is not very convincing even when an increased $\dot{V}O_2$ max is found, because the CO_2 produced does not increase under these conditions, and therefore it is unlikely to represent a real increase in metabolic activity (see chap. 3). Furthermore, Welch and associates (20) have recently shown that during exercise under hyperoxic conditions, leg muscle blood flow is reduced, and therefore the O_2 available to the active leg muscles is not really increased.

While this question of where the limits are set is not finally resolved, the available evidence suggests that in activities involving large muscle masses (such as running), aerobic capacity is probably limited by cardiac output. Activities involving smaller muscle masses, as in arms only or legs only sports, peripheral factors such as muscle blood flow and muscle O_2 utilization probably set the limit. It is also likely that interindividual differences in cardiovascular dimensions and functional capacities of the various systems involved determine where the limits are set for any given individual.

SUMMARY

1. The important factor governing the diffusion of respiratory gasses is the difference in pressure between two points, called a *diffusion gradient.*
2. If atmospheric pressure is much below sea-level pressure, the percentage concentration of a gas such as O_2 is meaningless, because its partial pressure may be insufficient to bring about sufficient diffusion.
3. At sea level, the partial pressures of O_2 are approximately 100 and 40 mm Hg in the alveoli and venous blood, respectively, yielding a diffusion gradient of 60 mm Hg. For CO_2, the corresponding figures are 45 and 40 mm in venous blood and alveoli, respectively, with a diffusion gradient of 5 mm Hg.
4. CO_2 requires a smaller diffusion gradient than O_2 because of its greater solubility in water, and hence its greater solubility in the alveolar and endothelial tissues.
5. Acidity and alkalinity are expressed in pH units: 7.00 represents neutrality; higher values represent alkalinity, and lower values represent acidity.
6. Ninety-nine percent of the oxygen transport of the blood is accomplished by combination with hemoglobin; about 1% is carried in physical solution.
7. Carbon dioxide transport is accomplished largely within the erythrocytes and in three forms: (a) carbamino-hemoglobin, (b) bicarbonate, and (c) dissolved CO_2.
8. The shape of the O_2 dissociation curve is such that gaseous exchange is greatly expedited at the lungs and at the tissues.
9. At rest, hemoglobin enters the tissues about 95% saturated and leaves about 70% saturated. This difference is called the *coefficient of O_2 utilization.* In exercise, comparable percentages may be 95% and close to 0%, thus effecting a great increase in O_2 utilization.
10. Acid-base balance is maintained throughout the body by buffer systems. Each is comprised of a weak acid and a salt of the acid, and functions by converting strong acids to weak acids and neutral salts.

11. When the capacity for buffering is stressed, acids and bases are excreted by the lungs and kidneys as a secondary line of defense against pH changes that may be inconsistent with the welfare of the organism.

12. As the rate of metabolism increases in exercise, the rate of diffusion for O_2 increases proportionately. This is probably brought about by the better perfusion of the lung capillaries with blood.

13. Use of O_2 to improve performance rests on sound scientific evidence only when it is used during the exercise period. A small improvement in rate of recovery seems to be possible, but this awaits corroborative evidence.

14. Both theoretical and experimental data suggest that the limits for maximal O_2 transport *(aerobic capacity)* for healthy young subjects exercising at sea level are set by the interaction of cardiac output and hemoglobin level when large muscle masses are involved. In exercise involving small muscle masses, muscle O_2 utilization and muscle blood flow may become critical.

REFERENCES

1. Astrand, P-O., and Saltin, B. Maximal oxygen uptake and heart rate in various types of muscular activity. *J. Appl. Physiol.* 16:977–81, 1961.

2. Bannister, R.G., and Cunningham, D.J. The effects on the respiration and performance during exercise of adding oxygen to the inspired air. *J. Physiol.* 125:118, 1954.

3. Bannister, R.G.; Cotes, J.E.; Jones, R.S.; and Meade, F. Pulmonary diffusing capacity on exercise in athletes and nonathletic subjects. *J. Physiol.* 152:66–67, 1960.

4. Elbel, E.R.; Ormond, D.; and Close, D. Some effects of breathing O_2 before and after exercise. *J. Appl. Physiol.* 16:48–52, 1961.

5. Gleser, M.A.; Horstman, D.H.; and Mello, R.P. The effect on VO_2 max of adding arm work to maximal leg work. *Med. Sci. Sports* 6:104–7, 1974.

6. Hermansen, L. Oxygen transport during exercise in human subjects. *Acta Physiol. Scand.* [Suppl.] 399, 1973.

7. Kaijser, L. Oxygen supply as a limiting factor in physical performance. In *Limiting Factors of Human Performance,* ed. J. Keul. Stuttgart: Georg Thieme. 1973.

8. Margaria, R.; Camporesi, E.; Aghemo, P.; and Sassi, G. The effect of O_2 breathing on maximal aerobic power. *Pfluegers Arch.* 336:225–35, 1972.

9. Miller, A.T., Jr. Influence of oxygen administration on cardiovascular function during exercise and recovery. *J. Appl. Physiol.* 5:165–68, 1952.

10. Mostyn, E.M.; Helle, S.; Gee, J.B.L.; Bentivoglio, L.G.; and Bates, D.V. Pulmonary diffusing capacity of athletes. *J. Appl. Physiol.* 18:687–95, 1963.

11. Newman, F.; Smalley, B.F.; and Thomson, M.L. Effect of exercise, body and lung size on CO diffusion in athletes and nonathletes. *J. Appl. Physiol.* 17:649–55, 1962.

12. Osnes, J.B., and Hermansen, L. Acid-base balance after maximal exercise of short duration. *J. Appl. Physiol.* 32:59–63, 1971.

13. Reybrouck, T.; Heigenhauser, G.F.; and Faulkner, J.A. Limitations to maximum oxygen uptake in arm, leg and combined arm-leg ergometry. *J. Appl. Physiol.* 38:774–79, 1975.

14. Ross, J.C.; Reinhart, R.W.; Boxell, J.F.; and King, L.H., Jr. Relationship of increased breath-holding diffusing capacity to ventilation in exercise. *J. Appl. Physiol.* 18:794–97, 1963.

15. Secher, N.H.; Ruberg-Larson, N.; Binkhorst, R.A.; and Bonde-Peterson, F. Maximal oxygen uptake during arm cranking and combined arm plus leg exercise. *J. Appl. Physiol.* 36:515–18, 1974.

16. Shephard, R.J. The validity of the oxygen conductance equation. *Int. Z. Angew. Physiol.* 28:61–75, 1969.

17. Stenberg, J.; Astrand, P-O.; Ekblom, B.; Royce, J.; and Saltin, B. Hemodynamic response to work with different muscle groups, sitting and supine. *J. Appl. Physiol.* 22:61–70, 1967.

18. Taylor, H.L.; Buskirk, E.; and Henschel, A. Maximal oxygen uptake as an objective measure of cardio-respiratory performance. *J. Appl. Physiol.* 8:73–80, 1955.

19. Turino, G.M.; Bergofsky, E.H.; Goldring, R.; and Fishman, A.P. Effect of exercise on pulmonary diffusing capacity. *J. Appl. Physiol.* 18:447–56, 1963.

20. Welch, H.G.; Bonde-Petersen, F.; Graham, T.; Klausen, K.; and Secher, N. Effects of hyperoxia on leg blood flow and metabolism during exercise. *J. Appl. Physiol.* 42:385–90, 1977.

21. Wilson, G.D., and Welch, H.G. Effects of hyperoxic gas mixtures on exercise tolerance in man. *Med. Sci. Sports* 7:48–52, 1975.

22. Wilson, B.A.; Welch, H.G.; and Liles, J.N. Effects of hyperoxic gas mixtures on energy metabolism during prolonged work. *J. Appl. Physiol.* 39:267–71, 1975.

10

Exercise Metabolism

The essence and the uniqueness of the study of physiology of exercise lie in its concern with physiological mechanisms in operation not during rest, but while the organism is stressed by physical activity. This physical activity may be work, physical education activity, athletics, or informal play. By observing the stress of vigorous physical activity, the exercise physiologist gains insight into physiology that is withheld when the organism is at rest. Questions pertaining to how well various organic systems can function under stress can be answered only in terms of the functional capacity an individual has in respect to his cardiovascular system, his respiratory system, his heat dissipation system, etc. An individual may show no cardiac abnormality in a physician's diagnostic examination, but this obviously does not tell us anything about his cardiac capacity for running a good time in the 440 or the mile. Functional tests are needed here, and these are in the domain of the exercise physiologist. To make these functional tests meaningful to other professions the exercise physiologist uses a vocabulary of terms that is taken from physics as well as physiology.

Definition of Terms

Work. Work is defined by the physicist as the product of force times the distance through which that force acts: $W = F \times D$. For example, if a man lifts a weight of 100 pounds to a height of 3 feet, he has done work of 100 pounds $\times$ 3 feet, or 300 ft-lb of work. In the metric system (which is commonly used internationally in exercise physiology, as in other sciences), the same individual if he weighed 100 kg and climbed up to stand on the 3-meter diving board would have performed 100 kg $\times$ 3 meters, or 300 kg-m of work. It will be noted that no mention has been made of the time it took to do the work, in either case, and this is not a relevant factor in the concept of work. The same amount of work is performed regardless of how long it takes.

Unfortunately, the physicist's definition leaves something to be desired when the muscular activity is isometric, as in the case where one "merely holds 100 pounds motionless"; here, since the distance is zero, the work must also be zero. However, other methods are available by which the effort involved can be evaluated.

Power. If two individuals can each lift 100 pounds a distance of 3 feet, but one does it twice as fast as the other, we have introduced the concept of power. The man who does it twice as fast is twice as powerful. Power is usually defined in terms of horsepower, just as in rating an automobile engine.

$$1 \text{ hp} = 33,000 \text{ ft-lb/min} = 550 \text{ ft-lb/per sec}$$

It should be noted that *power* is distinct from *strength;* power is composed of strength and speed. An athlete's power in an event (such as the shot put) thus can be increased either by improving the strength of the muscles involved—as by heavy resistance training—or by improving the speed with which the movement is made.

As an illustrative example of the calculation of horsepower, let us consider an individual performing the Harvard Step Test. In this test of physical fitness one lifts one's body weight (150 pounds, let us say) onto a 20-inch-high bench 30 times per minute. Thus the work done per minute would be as follows:

$$W = 150 \text{ lb} \times 1\text{-}\tfrac{2}{3} \text{ ft (20 in)} \times 30 = 7{,}500 \text{ ft-lb}$$

In terms of power, the subject has worked at the level of:

$$\frac{7{,}500 \text{ ft-lb/min}}{33{,}000 \text{ ft-lb/min}} = 0.227 \text{ hp}$$

A very powerful person can produce as much as three or four horsepower, but only for very short periods of time (five to ten seconds).

Two other methods for expressing the magnitude of an exercise load are commonly used. Since many of the bicycle ergometers now in use are electrical devices, we often refer to the exercise load in *watts.*

For example:

$$1 \text{ w} = 6.12 \text{ kg-m/min or approx. 6 kg-m/min.}$$
$$50 \text{ w} = \text{approximately 300 kg-m/min.}$$
$$746 \text{ w} = 1 \text{ hp}$$

From a more physiological viewpoint the exercise load is often expressed in terms of METs, which is simply a number expressing the ratio of the exercise metabolic load to resting metabolic rate. Resting metabolic rate is approximately 3.5 ml of O_2 consumption per minute/per kilogram body weight. Thus the METs required for a given exercise load are calculated as follows:

$$\text{METs} = \frac{O_2 \text{ required for exercise}}{O_2 \text{ required for rest}}$$

$$= \frac{VO_2 \text{ at exercise in ml/kg/min}}{3.5 \text{ ml/kg/min}}$$

Expressing the intensity of an exercise load in foot pounds per minute, kilogram-meters per minute, or in watts results in a measure of power output, whereas METs are a measure of energy consumption.

As Knuttgen (26) has recently pointed out, exercise physiologists often refer to exercise loads measured in power units as "work loads." This is technically incorrect and can lead to imprecise thinking, with its attendant ambiguities. As we pointed out above, $Work = Force \times Distance$, while $Power = Force \times Distance \div Time$—two very different dimensions.

Energy. Energy is defined as the capacity for doing work and can be expressed in the same units. Energy can be stored, in which case it is *potential energy,* and the energy involved in the production of work is *kinetic energy.* Because humans are ultimately dependent on food for their energy, it is obvious that energy can be transformed from one form to another. This is done in accord with the *law of conservation of energy,* which states that in the conversion of energy from one form to another, energy is neither created nor destroyed. The energy in food is chemical energy, and it is converted into mechanical and heat energy by the muscles in bringing about movement and doing work.

The energy of food, which produces work, and the work itself can also be described in terms of calories (or kilocalories). One kilocalorie represents the heat required to raise the temperature of 1 kg of water 1° C. To convert heat units to mechanical units:

$$1 \text{ kcal} = 3,087 \text{ ft-lb} = 427 \text{ kg-m}$$

Thus a one-ounce chocolate bar that contains 150 kilocalories can theoretically produce energy for some 463,000 ft-lb of work, or enough to keep a 150-pound person doing the Harvard Step Test for more than one hour at 100% efficiency. (However, the actual efficiency of such an exercise is not over 25%.)

Efficiency. Efficiency is usually defined as the percentage of energy input that appears as useful work. Thus if a person requires 4,000 kcal input of energy to perform a muscular activity that represents 1,000 kcal, efficiency would be 25%.

Most of the experimentation in this area indicates that muscular performance under favorable circumstances achieves a mechanical efficiency of 20%-25%. However, recent evidence suggests that efficiency may be much higher in running, (see chap. 19, Efficiency of Muscle Activity). It must be realized that the other energy, which does not appear as work, is not lost; it appears as heat, is dissipated, and tends to raise the body temperature during exercise.

Methods for Standardizing and Measuring Exercise Loads

To make meaningful measurement of the physiological processes during exercise, the exercise work load must be set up in such fashion as to be measurable and repeatable, and it should require little skill. Much of our athletic activity does not lend itself well to these requirements. Measuring the energy or work output of a football player, for instance, would be most difficult because bursts of activity are interspersed with variable periods of relative inactivity, such as huddles. Furthermore, the work done varies from moment to moment.

 Three methods for establishing a standard measurable work load are in common use. Each has its advantages and disadvantages.

Bench-stepping. Subjects lift their weight to a known height (the height of the bench), and the rate can be easily set with a metronome. This activity requires minimum skill and lends itself well to large groups, but it is subject to several sources of inaccuracy. First, subjects, particularly when they are tired, tend not to straighten their bodies at the hip and knee joints, and consequently have not lifted their center of gravity to the full height of the bench. Second, they are doing positive work (stepping up) and negative work (stepping down). Negative work, a relatively recent concept, requires considerably less energy expenditure than positive work, but it is difficult to assess.

Treadmill. Treadmills consist of a motor-driven conveyor belt that is large and strong enough for subjects to walk and run on. These devices are usually constructed so that the speed of the belt and the incline are adjustable. Use of the treadmill is advantageous in using a skill with which everyone is familiar (walking or running). Furthermore, it seems to bring about a slightly better involvement of large muscle masses than any other device since the arms and shoulders can and do enter into the activity.

It has two major disadvantages. First, the subject's movements make instrumentation somewhat difficult; second, and more important, the units of work must be stated in arbitrary fashion—as running at 7 mph on a 10% incline—because much of the work is done in a horizontal direction, and this does not allow evaluation in the standard units of foot-pounds, kilogram-meters, or watts.

Bicycle Ergometer. This instrument is a stationary bicycle whose front or back wheel is driven by subjects pedaling (fig. 10.1). The resistance against which subjects pedal is provided by a frictional band or by electromagnetic braking. The work load can be quickly and easily adjusted by changing the tension (and hence the frictional load) of the brake band or the electromagnetic load across the generator. Work is calculated easily from a scale reading, which provides the frictional resistance (force), and from a counter that records the number of times the wheel has turned and thus allows calculation of distance: $D = 2\pi r \times N$. The wheel's circumference, $2\pi r$, is the distance traveled by any point on the wheel in one revolution; N is the number of revolutions during the work period. Then, since $W = FD$, the total work done may be expressed as $W = F (2\pi r \times N)$.

This piece of equipment has several advantages. First, it is relatively inexpensive or it can be built in most school workshops from a discarded bicycle. Second, the subject's upper body is relatively motionless, and thus instrumentation for electrocardiograph leads, etc., is greatly facilitated. Third, and most important, the exercise load is expressed in standard units of work, foot-pounds or kilogram-meters, and thus work comparisons may be made more easily with the bicycle ergometer than with the treadmill.

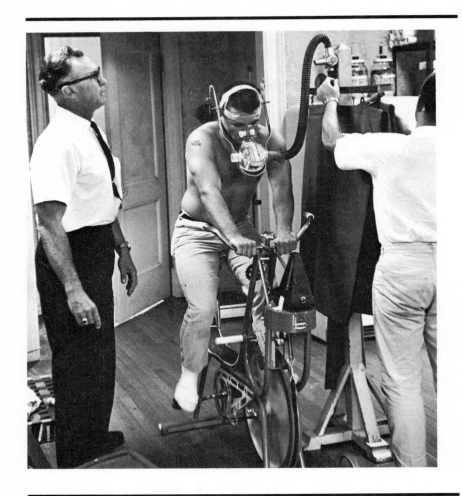

An international team of work physiologists compared the maximal O_2 consumption measured by the three ergometric methods in the same twenty-four healthy young subjects. They found that the terminal pulse rates and arterial lactate levels two minutes after exercise are very similar for step, bicycle, and treadmill exercise (3). However, the maximal O_2 uptake in the treadmill test was 7% greater than that in the bicycle test, while the step test values were intermediate between the treadmill and bicycle ergometer values. Many other investigators have found similar results. The difference does not appear to be due to the different leg forces involved (20). However, recent evidence suggests that the difference might be minimized by using the handlebars in the typical racing position, which allows the participation of a greater muscle mass, at least in the skilled cyclist (14).

Basic Physiology Underlying the Study of Physiology of Exercise

Figure 10.1 Figure 10.1a (left) shows gas collection by the "classic" Douglas bag method using a simple frictional bicycle ergometer. Figure 10.1b (above) shows the author's laboratory as it is currently set up using semiautomated electronic methods for gas collection and an electrically controlled and programmed bicycle ergometer.

The consensus of an international group of experts reported to the World Health Organization (WHO) was that the "order of preference of exercise tests is considered to be as follows: upright bicycle ergometer, step test, and treadmill" (39).   *CONCLUSION*

It must be recognized in setting work loads on the bicycle ergometer that equal work loads can have very different physiological effect if pedal frequency is allowed to vary (7, 18). Highest values of maximum O_2 consumption are gotten using between 60 and 70 rpm (17).

It is also important to consider the effects of learning and habituation upon the performance, particularly when measurements such as heart rate at submaximal work loads are of concern. Most subjects experience anxiety in a new test situation, and this of course affects observed heart rate. There is also a learning effect that produces small increases in efficiency. These effects must be controlled or balanced out in any test-retest experiment, whether the exercise is on a bicycle ergometer, treadmill, or stepping bench. It has also been shown that there is considerable specificity of training effect such that bicycle training results are best measured on a bicycle, although run training seems to be reflected in all tests (32, 33).

Methods for Measuring Energy Consumption

Direct Calorimetry. Because the human organism is essentially a heat engine, the direct approach to measurements of energy consumption would be to measure the heat produced by an individual's metabolic processes. This has been done in specially constructed chambers, where all metabolic heat is accumulated by the air and walls of the chamber and changes in their temperature are used to calculate the energy output. This method is called *direct calorimetry;* however, the equipment is expensive and difficult to use, and consequently is seldom used in exercise physiology.

Indirect Calorimetry. Because all of the body's metabolic processes utilize oxygen and produce carbon dioxide (either during activity or immediately after), the energy output is directly related to the quantity of these respiratory gasses. The gasses can be collected from the expired air and measured. This is a much simpler process than direct calorimetry, and it is therefore commonly used in exercise physiology. There are two methods for accomplishing indirect calorimetry: the *closed-circuit* and the *open-circuit* methods.

In the closed-circuit method illustrated in figure 10.2, the subject inspires from a face mask that is connected to an oxygen chamber, which is charged from an O_2 cylinder. Expired air is conducted back to the oxygen chamber by way of a soda lime cannister, where the CO_2 produced is absorbed. Thus only the O_2 that remains after the respiratory exchange is returned to the oxygen chamber, and the changes in the volume of the O_2 that remains in the chamber, are recorded from breath to breath. Each peak in the kymogram in figure 10.2 represents one respiration and, by measuring the downward slope of the bottom points of the record per unit time, the value of O_2 consumed can be calculated.

This method has the advantage of simplicity, but its accuracy is not much better than plus or minus 10% of the true value. Furthermore, no value for the CO_2 produced is obtained, and consequently the respiratory quotient (to be discussed below) must be estimated.

In the classic open-circuit method, the subject inspires directly from the atmospheric air and expires into a rubberized canvas bag called a *Douglas bag* (fig. 10.3). After an exercise period, during which gas collection is accurately timed, samples of the expired gas are taken from the bag for analysis, and the volume of expired gasses is measured by a gas meter similar to that used for metering the gas used in a home (fig. 10.3). The concentrations of

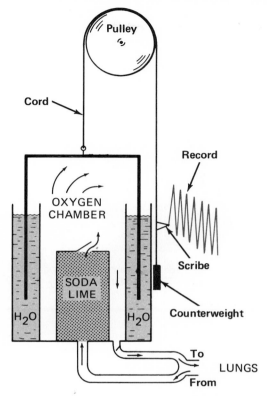

Figure 10.2 Apparatus for indirect determination of heat production by measuring oxygen consumption. The subject breathes into and from the oxygen chamber through the tubes at the bottom. The slope of the curve recorded by the movements of the upper cylinder is a measure of the rate at which oxygen is used by the subject. Arrows indicate direction of oxygen movement as the subject breathes. (From Carlson, A.J., and Johnson, R.E. *The Machinery of the Body,* 1953. Courtesy of The University of Chicago Press, Chicago.)

O_2 and CO_2 in the atmosphere are very constant—20.93% and 0.03%, respectively—and on the assumption that the remaining gasses (79.04%), lumped together as N_2 do not enter into physiological reactions, the volume of inspired air can be calculated from the volume of expired air as follows:

$$\frac{\text{Volume inspired}}{\text{Volume expired}} = \frac{\text{Percent of } N_2 \text{ in expired air}}{79.04.}$$

or

$$\text{Volume inspired} = \frac{\text{Percent of } N_2 \text{ in expired air} \times \text{Volume expired}}{79.04}$$

Figure 10.3 Measuring the volume of expired respiratory gasses from the Douglas bag by wet-test gas meter. Gasses are drawn through the meter at a constant rate by use of an electrical vacuum pump.

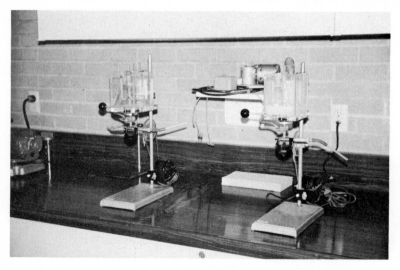

Figure 10.4 Micro Scholander gas analyzers used for estimation of the percentage of O_2 and CO_2 in the expired gas volume.

This calculation is necessary because the volume of CO_2 produced is not usually equal to the volume of O_2 consumed, and consequently the total volume of expired air also differs from the total inspired.

To calculate the O_2 consumed, one need only subtract the volume of O_2 remaining in the expired air (percent of O_2 expired times volume expired) from the volume of O_2 in the inspired air (20.93 times volume inspired). The same sort of calculation will also provide the volume of CO_2 produced during the exercise. This method is obviously somewhat more involved than the closed-circuit method, but the gain in precision is commensurate. In the open-circuit method, the error may be less than $\pm 1.0\%$, compared with $\pm 10\%$ for the closed-circuit. Furthermore, data are obtained for the percent of CO_2, which enables computation of the respiratory quotient (RQ).

Gas Analysis. The percentage of O_2 remaining and CO_2 produced in the expired air are analyzed either by biochemical or electronic methods. The classic biochemical method—the *Haldane apparatus* and *procedure*—uses the absorption of CO_2 by potassium hydroxide from a sample of known size. The change in volume of the sample by absorption of CO_2 is observed, and the proportion this change represents in the total volume of the sample represents the proportion or percent of CO_2 in the sample. The O_2 is then absorbed out by a strong reducing solution, and the same reasoning is applied. All of the gasses left in the sample at this point are considered N_2.

A more recent development of the chemical (absorptiometric) method is the *Scholander apparatus* (fig. 10.4); it utilizes the same principle described above, but is a considerably faster procedure. (Both methods are capable of a precision of better than $\pm 0.02\%$ in the range of respiratory gasses.)

Still more recently, electronic methods have been developed for measuring respiratory gasses, and they have advantages in both speed and simplicity of operation. Indeed, on-line computer analysis and breath-by-breath graphic display of exercise function tests are now possible and in use in more sophisticated laboratories. The electronic methods compare favorably with the precision of the Haldane and Scholander apparatus.

Oxygen Debt and Oxygen Deficit

In the normal, resting individual, the supply of O_2 to the tissues is sufficient for the complete breakdown of glycogen to CO_2 and H_2O, with no accumulation of lactic acid. This situation also applies when the rate of work is such that the metabolic demands can be met aerobically.

When exercise creates a metabolic need for greater O_2 than can be supplied by the cardiorespiratory processes, part of the energy of muscular activity is supplied by the anaerobic mechanism described in chapter 3, and lactic acid accumulates as the end product of metabolism. Whenever the supply of O_2 is insufficient to meet demands, an individual is said to contract an *oxygen debt* (a term coined by A. V. Hill, a pioneer in exercise physiology).

In any exercise bout there is a transition period between rest and exercise, involving a short period during which the circulatory and respiratory adjustments lag behind.

The amount by which the O_2 supply fails to meet need by virtue of this lag in the organism's adjustment to the rise in metabolic rate is called the *oxygen (O_2) deficit*. Function of the organism under O_2 deficit conditions is made possible by several energy sources not dependent on O_2 transport. Most important are (1) the splitting of ATP and CP; (2) anaerobic breakdown of glycogen (glycolysis) to lactic acid; and (3) use of O_2 bound to muscle myoglobin and blood O_2 stores as shown in figure 10.5.

O_2 debt is the amount of O_2 taken up in excess of the resting value during the recovery period. This is ordinarily larger than the O_2 deficit, since any anaerobic metabolism beyond the initial lag must be repaid here as well as the O_2 deficit.

In light exercise, where a steady level of O_2 consumption is attained as shown in figure 10.5, the O_2 debt may be due entirely to the O_2 deficit at the beginning of exercise. At the end of exercise then, there must be a recovery period during which the O_2 debt is repaid. This is why the heart and the ventilation rates remain elevated after exercise ceases.

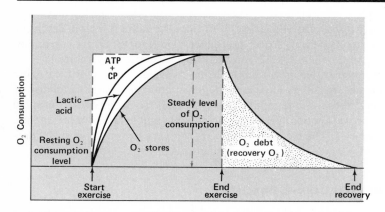

Figure 10.5 Diagram of the relationship of O_2 consumption and time before, during, and after submaximal exercise. The crosshatched area represents O_2 deficit which depends on at least three factors: (1) breakdown of high energy phosphates ATP + CP; (2) glycolysis to form lactic acid; and (3) use of O_2 stores such as oxymyohemoglobin and O_2 of the venous blood. The stippled area represents the O_2 debt (the O_2 used during recovery). All values of O_2 consumption are measured above the resting value as a baseline.

Basic Physiology Underlying the Study of Physiology of Exercise

When an exercise represents a true overload in which a steady state cannot be achieved, the duration of the effort (or the level of performance) is limited by the athlete's ability to sustain an O_2 debt. In maximum work load situations where energy supply is predominately from anaerobic sources, duration is limited to one to two minutes and recovery may take forty-five minutes or even longer.

To illustrate the O_2 debt concept, let us consider a typical experiment on the bicycle ergometer.

It is clear that such an individual had a metabolic demand for 4 liters of O_2 per minute, 3 liters of which were provided by aerobic mechanisms and 1 liter of which was provided by anaerobic mechanisms. This constituted an O_2 debt that was repaid during the recovery period by consumption of more O_2 per minute than the resting state would have demanded, until the O_2 debt was paid. The length of time for monitoring recovery O_2 consumption is determined by observations of heart rate and minute-by-minute O_2 consumption. When these values have returned to their preexercise values, recovery is complete. Girandola and Katch (15) have shown that training results in greater VO_2 *during* exercise and therefore a lower O_2 debt.

Time	Activity	Total Liters of O_2 Consumed	Liters of O_2 Consumed per Minute
8:00–8:05	Resting (seated on bicycle)	1.50	.30
8:05–8:10	Riding (@ 1,500 kg-m/min)	16.50	3.30
8:10–8:40	Recovery (seated on bicycle)	14.00	.47

Calculations (in liters)

1. Total gross O_2 cost $=$ O_2 during $+$ O_2 recovery
    $=$ 16.50 liters $+$ 14.00 liters $=$ 30.50 liters
2. Total net O_2 cost $=$ O_2 during $+$ O_2 recovery $-$ O_2 for equivalent period of rest (35 min)
    $=$ 16.50 liters $+$ 14.00 liters $-$ (35 $\times$ 0.30) $=$ 20.0 liters
3. Net O_2 cost per min of ride $= \dfrac{20.00 \text{ liters}}{5 \text{ min}} =$ 4.00 liters/min

4. Net O_2 intake during ride $=O_2$ during $-$ O_2 for equivalent rest period
    $=$ 16.50 $-$ (5 $\times$ 0.30) $=$ 15.00 liters
5. Net O_2 intake during per min $= \dfrac{15.00 \text{ liters}}{5 \text{ min}}$ 3.00 liters/min

6. O_2 debt incurred per min $=$ net O_2 cost per min of ride $-$ net O_2 intake during per min
    $=$ 4.00 liters $-$ 3.00 liters $=$ 1.00 liters/min

7. Total O_2 debt incurred $=$ 5 $\times$ 1.00 liters $=$ 5.00 liters

Lactacid and Alactacid O₂ Debt and New Concepts

The older literature had established that, under overload conditions (work loads greater than aerobic capacity), for every liter of O_2 debt, the lactic acid level increased by 7 gm. However, the classic work of Margaria, Edwards, and Dill (28) had shown that for the first 2.5 liters of O_2 debt no increase in lactate could be demonstrated. On this basis, O_2 debt was thought to have two components: *lactacid,* which was represented by proportional increases in blood lactate, and *alactacid,* for which no lactate increase was found. Furthermore, Margaria and colleagues also demonstrated a great difference in the repayment of these two components of the O_2 debt. The alactacid debt was repaid at a rate approximately thirty times faster than the lactacid debt. Thus the fast component (alactacid) was ascribed to replacement of O_2 and energy stores, and the slow or lactacid component was thought to be used to remove lactate from the blood.

While the two components of O_2 debt with respect to rate of repayment are well verified, unfortunately the *lactacid-alactacid* explanation of the physiology involved has proved to be an oversimplification. It is now clear that many processes besides the elimination of lactate may be involved in the delayed return of O_2 uptake to the resting value after cessation of exercise (which we call O_2 debt):

1. During exercise, O_2 stores of the body are greatly reduced, and part of the recovery O_2 is used to
   a. restore muscle myoglobin to resting values.
   b. restore venous oxyhemoglobin levels.
   c. replenish O_2 dissolved in tissue fluids.
2. The rise in body temperature resulting from vigorous exercise creates a demand for more O_2.
3. Neither heart rate nor cardiac output return immediately to resting values, and thus excess O_2 is required for cardiac metabolism.
4. What is true for heart rate and cardiac output is also true for pulmonary function.
5. The output of catecholamines is probably still above resting values, and this augments O_2 consumption.
6. The high energy phosphate breakdown (see fig. 10.6) must be reversed at a considerable cost of O_2 consumption (25).

To further complicate the relationship of lactate level to O_2 debt as measured by excess recovery O_2, Rowell and co-workers (34) have shown that even during moderate exercise, as much as 50% of the lactate production may be removed by hepatic splanchnic tissues. It has also been shown that the nonexercising muscles may be consuming lactate even as the exercising muscles are producing it (2). Obviously, this precludes any neat 7 gm lactate per liter O_2 debt relationship as had been suggested by earlier workers. Thus a lactate buildup in the blood would be unlikely at any but the heaviest work loads, as shown in figure 10.7. Of course, no proportionality between blood lactate and level of anaerobic metabolism could be expected.

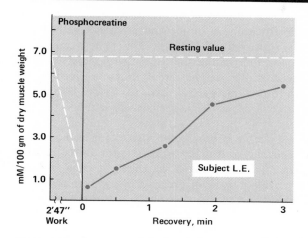

Figure 10.6 Phosphocreatine concentration (mM per 100 gm dry muscle) before and after a maximal work load. (From Hermansen, L. *Med. Sci. Sports* 1:32, 1969.)

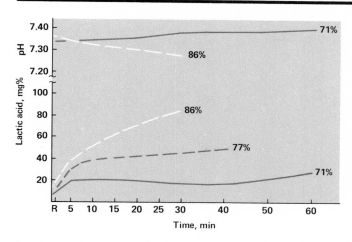

Figure 10.7 Mean lactic acid values and pH values over duration of runs at 82%-89% VO_2, max 74%-79% VO_2 max, and 67%-74% VO_2 max. (From Nagle, F. *Med. Sci. Sports* 2:185, 1970.)

Training Effect on Anaerobic Metabolism and O_2 Debt

In football, baseball, basketball, and probably most athletic activities that find favor in our country, one important determinant of success is *anaerobic capacity,* the ability to get moving quickly for short distances. Relatively few sports events in the USA require a sustained effort longer than thirty to sixty seconds, and consequently the vast majority of events depend upon anaerobic capacity, which has been given very little attention by researchers in physiology of exercise.

Until recently, we did not even have a simple means for measurement of this important parameter. Fortunately, a simple and practical method has been developed by Margaria, Aghemo, and Rovelli (29). The test consists of measuring the vertical component of the maximal speed with which an individual can run up an ordinary staircase. The test is fully described in the laboratory manual designed to accompany this text (13). More recently, a bicycle ergometer test of anaerobic capacity has been developed by Katch and co-workers (21).

Figure 10.8 shows the very sizable improvements that occur in the ability to contract an O_2 debt as the result of training swimmers for 100- and 200-meter swimming events (16). Figure 10.9 shows the effects of swim training on the lactate concentration after maximum exercise. These results, which show the trainability for anaerobic events, have been supported by Cunningham and Faulkner (11) who also found an improvement in O_2 debt of 9% and postexercise blood lactate of 17% after training with interval sprints.

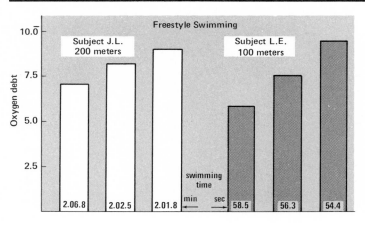

Figure 10.8 O_2 debt in relation to the actual time of performance. Note the increasing O_2 debts with improvement in performance (lower times). (From Hermansen, L. *Med. Sci. Sports* 1:37, 1969.)

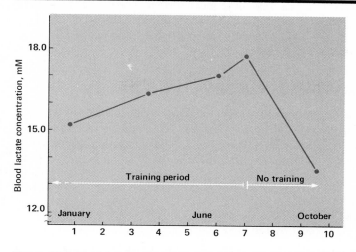

Figure 10.9 Blood lactate concentration (peak value) after maximal exercise (100 meters swimming) during the training period and after 2½ months with no training. (From Hermansen, L. *Med. Sci. Sports* 1:37, 1969.)

Intermittent Work (Interval Training)

At this point it is of interest to examine the work of the Astrands, Christensen, and their co-workers (5, 6, 8); it may bear on our discussion of O_2 debt, and it seems to have large implications for the planning of training regimens. In one experiment, for example, a well-trained subject was able to work for thirty minutes at a very high work load (4.4 liters of O_2 per minute) by alternating five seconds of work with five seconds of rest. This appeared to be something like a steady state in that very little lactic acid accumulated. Even for a highly trained athlete, a work load of that size done continuously would result in a large O_2 debt.

In another experiment a subject alternately ran ten seconds and rested five seconds, for a total of thirty minutes and 6.67 km. His O_2 intake for the thirty-minute period averaged 5.0 liters/min (his maximum capacity was 5.6 liters/min). His actual O_2 uptake for the twenty minutes of running was 101 liters, and his uptake during the accumulated rest periods was 49 liters. Subtracting a resting O_2 consumption of 4.0 liters for the ten minutes of rest from the total 49 liters leaves a tremendous O_2 debt of 45 liters, which had been eliminated in some fashion during the five-second rest periods. The largest O_2 debts ordinarily reported after continuous work vary from 15 to 20 liters of O_2. These experimenters suggest that O_2 is stored as oxymyohemoglobin in the muscles during the rest periods to support metabolism during the work

Exercise Metabolism   215

periods without resort to the anaerobic mechanisms. This is an interesting possibility and provides a scientific basis for interval training of athletes. Indeed, evidence is now available that intermittent exercise can train both aerobic and anaerobic capacity (22).

Maximal O_2 Consumption as a Measure of Physical Fitness

The maximal O_2 consumption (VO_2 max) for any individual is a good criterion of how well various physiological functions can adapt to the increased metabolic needs of work or exercise. At least the following functions are involved and contribute to the magnitude of an athlete's ability to maintain a steady state:

1. Lung ventilation
2. Pulmonary diffusion
3. O_2 and CO_2 transport by the blood
4. Cardiac function
5. Vascular adaptation (vasodilatation of active tissues and vasoconstriction of inactive tissues)
6. Physical condition of the involved muscles

Increasing Work Load. The method for measuring maximal O_2 consumption (aerobic capacity) involves working subjects at ever-increasing work loads, during each of which steady state level O_2 consumption is measured (usually by open-circuit spirometry). When an increase in work load fails to elicit a significant increase in O_2 consumption, the highest value attained represents the maximum O_2 consumption, as shown in figure 10.10.

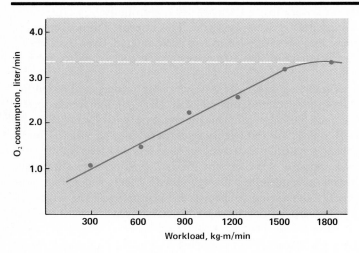

Figure 10.10 Diagram showing the O_2 consumption work load relationship when the subject was tested at repeated five-minute workouts with rest intervals. The maximum O_2 consumption is indicated by the dash line at 3.30 liters/min.

Any of the ergometric methods discussed earlier can be used. The test can be administered either on a continuous or discontinuous protocol. In the discontinuous method, subjects are worked at each load until a steady O_2 level is attained (at least five minutes), with rest periods between work loads sufficient to allow recovery. Unfortunately, this method usually involves at least two or three visits to the laboratory and requires a great deal of technician and lab time in addition. Fortunately, the results obtained by the continuous method, where the load is increased every minute or every two minutes, are closely comparable to those obtained in the discontinuous method (2). This procedure can be consummated in one half-hour visit to the laboratory. The error of the measurement of aerobic capacity has been reported to be about 2.5% (36).

Saltin and Astrand (35) tested ninety-five male and thirty-eight female members of the Swedish national teams and found the mean maximal O_2 uptake for the best fifteen males to be 5.75 liters/min and the best ten females 3.6 liters/min. The highest values found were 6.17 liters/min and 4.07 liters/min for the male and female, respectively. It is of interest that the highest values were achieved by the cross-country ski team.

A clever experiment was designed by Klissouras (23) to determine to what extent aerobic capacity is determined genetically. He tested fifteen monozygous and ten dizygous twins and found that in these young boys aged seven through thirteen the variability in aerobic capacity was determined 93% by genetic factors. In a subsequent experiment (24) in which comparisons were made of a trained and an untrained monozygous twin, the trained twin was found to be superior in aerobic capacity by 37%; but the absolute value after training was still only in the average category, leading to the conclusion that while training can bring about substantial improvement, the ceiling is set by genetic factors.

O_2 Pulse. The measurement of maximal O_2 consumption requires a willingness on the part of the subject to work to exhaustion, plus a sufficiently conditioned musculature to fully load the O_2 transport systems. In sedentary middle-aged or older adults neither of these conditions is apt to be satisfied. Furthermore, exhaustive physical tests are not completely without hazard for sedentary older populations where unrecognized heart disease may complicate matters. Under such conditions, considerable information can be derived from measuring O_2 pulse at a standardized submaximal level that can be attained by all members of the group to be tested. O_2 pulse is derived by simply dividing O_2 consumption by the heart rate at the time of measurement, thus giving the dimension of O_2 transport per heartbeat.

It has been shown that at any given work rate the subject with the greatest maximum work capacity (aerobic capacity) has the highest O_2 pulse, and conversely the lowest work capacity is associated with the lowest O_2 pulse (37). O_2 pulse under exercise conditions is, of course, largely determined by

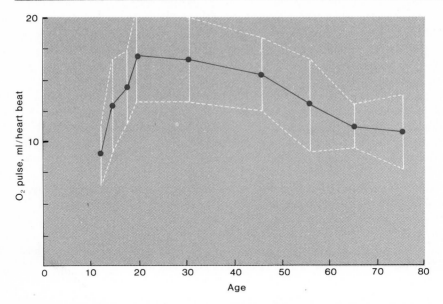

Figure 10.11 Maximum O_2 pulse as a function of age. Mean $\pm$ S.E.M. (Drawn from the data of Hollmann, W., Internationales Seminar für Ergometric, 1965, p. 186. Inst. für Leistungsgrenzen, Berlin.)

stroke volume and arteriovenous O_2 difference (31). Norms for this measurement are shown in fig. 10.11.

Respiratory Quotient

The relationship of CO_2 produced to O_2 consumed—RQ, the *respiratory quotient*—is an important physiological concept because it provides information that tells which foodstuff is being used for energy supply, if the subject is resting or in a steady state of moderate exercise.

$$RQ = \frac{\text{Volume } CO_2 \text{ produced}}{\text{Volume } O_2 \text{ consumed}}$$

Thus, if carbohydrate is completely oxidized to CO_2 and H_2O, the relationship can be described as:

$$C_6H_{12}O_6 + 6O_2 \rightarrow 6CO_2 + 6H_2O$$

And it follows that $RQ = \dfrac{6CO_2}{6O_2} = 1.00$, if one volume of CO_2 is produced for each volume of O_2 consumed. If fat is used as a source of energy, however, the ratio is somewhat different. The fats and oils of our foods are

Basic Physiology Underlying the Study of Physiology of Exercise

largely mixtures of palmitin, stearin, and olein. These substances are of similar chemical compositions, and their oxidation can be simplified as follows.

$$2C_{51}H_{98}O_6 + 145O_2 \longrightarrow 102CO_2 + 98H_2O$$
$$RQ = \frac{102}{145} = 0.70$$

Since the exact structure of the extremely large protein molecules has not yet been completely elucidated, the RQ for protein metabolism is estimated from known amino acid structures as approximately 0.80. However, protein plays a very small part in energy metabolism, and because its participation can be closely estimated from urine analysis, it is not important for the present discussion.

Consequently, if a subject is in a steady state, a reasonably valid deduction of the foodstuff being oxidized can be made on the basis of the observed value of the RQ. For example, if analysis of the respired gasses yielded an RQ of 1.00, the subject could be considered to be utilizing only carbohydrate for energy, and a value between 0.70 and 1.00 would indicate a mixture of fat and carbohydrate being burned. The exact amounts of each of the latter are shown in table 10.1.

Table 10.1
The Caloric Equivalents of Oxygen and Carbon Dioxide for Nonprotein Respiratory Quotients

Nonprotein respiratory quotient	Kcal/liter	
	Oxygen	Carbon dioxide
0.70	4.686	6.694
0.72	4.702	6.531
0.74	4.727	6.388
0.76	4.752	6.253
0.78	4.776	6.123
0.80	4.801	6.001
0.82	4.825	5.884
0.84	4.850	5.774
0.86	4.875	5.669
0.88	4.900	5.568
0.90	4.928	5.471
0.92	4.948	5.378
0.94	4.973	5.290
0.96	4.997	5.205
0.98	5.022	5.124
1.00	5.047	5.047

From Carpenter, T.M.: *Tables, Factors, and Formulas for Computing Respiratory Exchange and Biological Transformation of Energy,* 3d ed. Washington: Carnegie Institute of Washington, 1939.

Exercise Metabolism

However, during exercise in which a steady state is not attained, RQ does not truly reflect the percentage of fat and carbohydrate utilized. If the work load is partially anaerobic, lactic acid accumulates, and this causes a temporary metabolic acidosis, which is compensated for by a respiratory alkalosis brought about by *blowing off* more CO_2 than is forming metabolically. This compensatory hyperventilation has been found to be closely related to excess lactate formed during anaerobic exercise (19). In heavy exercise, it is common to find RQs above 1.00 and as high as 1.30. On the other hand, during recovery the RQ is usually below .70, showing that lactate is being removed from the circulation and is bringing about a temporary alkalosis that is compensated by hypoventilation in recovery.

Anaerobic Threshold

As was pointed out earlier in this chapter, maximal O_2 consumption (VO_2 max) has long been recognized as an important determinant of performance in events requiring endurance. It has recently been shown that an even more important determinant with respect to distance running is the fraction of VO_2 max that can be utilized over the distance without incurring a large buildup in blood lactate.

Costill (9) has shown that highly trained distance runners are capable of utilizing more than 90% of their VO_2 max for twenty-five to thirty minutes with only a moderate accumulation of blood lactate. Later work from the same laboratory showed a correlation of -0.91 between performance in a ten-mile race and VO_2 max, while the correlation with percentage of VO_2 max utilized at a given speed was even greater, $-.94$ (10). Similar data have been presented for a 1½-mile run (30).

It has also been pointed out by Londeree and Ames (27) that since VO_2 max is largely determined by hereditary factors, one cannot use it as a measure of training status. For a given VO_2 max it is impossible to know whether the subject has a lot of inherited ability and is out of shape, or has little ability and is in good shape. Their work showed that training status is best reflected in the percentage of VO_2 max and maximal heart rate that can be maintained without greatly exceeding resting blood lactate values. However, these findings do not preclude the fact that *changes* in VO_2 max can serve as excellent indices of *change* in condition.

For all of the above reasons, the concept of an anaerobic threshold (AT) has become important in exercise physiology. The anaerobic threshold is defined as the level of work, or O_2 consumption, just below that at which metabolic acidosis and associated changes in gas exchange occur (38). That is, when the exercise load is ever increasing, at some point the O_2 demand exceeds the O_2 supply, and at this point (the AT) energy release from anaerobic metabolism increases, with a subsequent increase in lactate formation.

Wasserman and his colleagues have developed noninvasive methods for estimating AT, using the point of a nonlinear increase in ventilation, CO_2 production, and a sudden increase in RQ and other gas exchange parameters (38). Figure 10.12 shows the application of these concepts by Davis and others (12) using instrumentation typically used in well-equipped exercise physiology labs. The AT can be estimated from gas exchange without the drawing of blood for lactate concentration analysis.

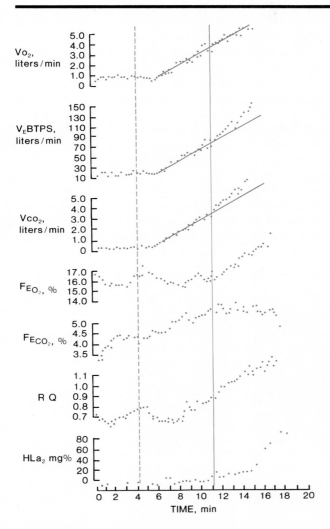

Figure 10.12 Measurements of respiratory gas exchange and venous blood lactate during an incremental leg cycling test. Solid, thin vertical line denotes anaerobic threshold. Dashed vertical line indicates onset of incremental work. Exercise duration was sixteen minutes. Lactate levels during four minutes of recovery are also shown. (From Davis, J.A., et al. *J. Appl. Physiol.* 41:544, 1976.)

Use of the AT concept offers many potential advantages for evaluation of work capacity in healthy athletes, as well as for clinical evaluation of cardio-vascular-respiratory disease in the general public.

Negative Work

So far, the main concern has been with exercise loads involving predominantly concentric contraction in which the muscle shortens to do work; this can be called *positive work.*

But effort is also involved in the muscular activity of resisting lengthening, as in eccentric contraction, and the physicist's definition of work can no longer be applied. It has become common usage to refer to the work done by eccentric contraction as *negative work,* and to compute it as if it were positive: $W = F \times D$. However, the work calculated in this fashion cannot be used interchangeably with the work of positive work; the energy involved per unit work is quite different.

This was ingeniously demonstrated by Abbott, Bigland, and Ritchie (1) who coupled two bicycle riders in opposition, one of whom pedaled concentrically while the other pedaled eccentrically. Although the forces developed exactly balanced each other, the O_2 consumption of the subject doing positive work was 3.7 times higher. Asmussen (4) confirmed this, and plotted negative and positive work loads versus O_2 consumption (fig. 10.13). Although negative work was linearly related to O_2 consumption, the ratio of the slope lines was 7.4 for their rate of pedaling. For varying rates of pedaling, the ratio of O_2

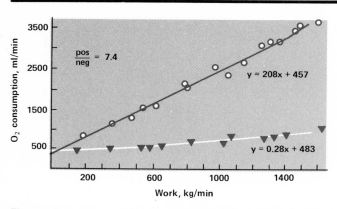

Figure 10.13 The oxygen consumption in milliliters per minute of a man bicycling "uphill" and "downhill" on a motor-driven treadmill plotted against the rate of work in kilograms per minute. (From Asmussen, E. In *Ergonomics Society Symposium on Fatigue,* 1953. Courtesy of H.K. Lewis, London.)

Basic Physiology Underlying the Study of Physiology of Exercise

consumption for positive/negative work varied from three to nine, the ratio increasing with the speed of pedaling. Thus it would seem that positive work is from three to nine times more costly in terms of energy expenditure than negative work.

SUMMARY

1. Some of the terms needed for discussing exercise metabolism intelligently are as follows: (a) *work:* $W = F \times D$; (b) *power* (in hp): 1 hp = 33,000 ft-lb/min; (c) *energy* (in kcal): 1 kcal = 3,087 ft-lb; (d) *efficiency* (in ratio or percentage):

$$\text{Efficiency} = \frac{\text{Work output}}{\text{Energy input}}$$

2. Three methods are commonly used for setting up standard work loads: *bench-stepping, treadmill,* and *bicycle ergometer.*
3. The two methods for measuring energy expenditures are *direct* and *indirect calorimetry;* the latter uses either the *closed-circuit* or the *open-circuit* method.
4. Muscular work can be performed *aerobically,* if the energy source is completely oxidized to CO_2 and H_2O, or *anaerobically,* if the biochemical breakdown of glycogen ends at the lactic acid stage.
5. When exercise begins, there is a lag in the response of the O_2 transport systems. The amount by which the O_2 supply lacks being adequate until the O_2 transport catches up with the demand is called O_2 *deficit.*
6. O_2 *debt* is defined as the amount of O_2 taken up in excess of the resting value during the recovery period. This constitutes the repayment of the O_2 deficit plus any anaerobic metabolism which may have occurred.
7. Anaerobic capacity is an important determinant of success in many American athletic contests. Improvement in this parameter can be brought about by appropriate training directed to short, sprint-type activity.
8. When equal loads of continuous and intermittent work are compared, much lower stress seems to result from intermittent work. Storage of O_2 as oxymyohemoglobin has been postulated to explain this phenomenon.
9. If physical fitness is defined as physical working capacity, the best single measure of this factor is maximal O_2 consumption.
10. Under resting and steady-state exercise conditions, the *respiratory quotient* (RQ) accurately reflects the foodstuff being utilized.
11. During exercise that cannot be performed aerobically, RQ seems to reflect the percentage participation of aerobic processes.

12. The *anaerobic threshold* (AT) is defined as the level of work, or O_2 consumption, just below that at which metabolic acidosis and associated changes in gas exchange occur. This measure appears to be useful in the prediction of athletic performance in endurance events and in the assessment of exercise capacity in patients with cardiovascular-respiratory disease.

13. Work in which muscular contraction is eccentric is called *negative work*. Depending on the rate, negative work can be performed from three to nine times more economically than *positive work*.

REFERENCES

1. Abbott, B.C.; Bigland, B.; and Ritchie, J.M. Physiological cost of negative work. *J. Physiol.* 117:380–90, 1952.
2. Ahlborg, G.; Hagenfeldt, L.; and Wahren, J. Substrate utilization by the inactive leg during one-leg or arm exercise. *J. Appl. Physiol.* 39:718–23, 1975.
3. Anderson, K.L.; Shephard, R.J.; Denolin, H.; Varnauskas, E.; and Masironi, R. Fundamentals of exercise testing. Geneva: World Health Organization, 1971.
4. Asmussen, E. Experiments on positive and negative work. In *Ergonomics Society Symposium on Fatigue,* eds. W.F. Floyd and A.T. Welford. London: Lewis and Co., 1953.
5. Astrand, I.; Astrand, P-O.; Christensen, E.H.; and Hedman, R. Intermittent muscular work. *Acta Physiol. Scand.* 48:448–53, 1960a.
6. ———. Myohemoglobin as an oxygen-store in man. *Acta Physiol. Scand.* 48:454–60, 1960b.
7. Bannister, E.W., and Jackson, R.C. The effect of speed and load changes on oxygen intake for equivalent power outputs during bicycle ergometry. *Int. Z. Angew. Physiol.* 24:284–90, 1968.
8. Christensen, E.H.; Hedman, R.; and Saltin, B. Intermittent and continuous running. *Acta Physiol. Scand.* 50:269–86, 1960.
9. Costill, D.L. Metabolic responses during distance running. *J. Appl. Physiol.* 28:251–55, 1970.
10. Costill, D.L.; Thomason, H.; and Roberts, E. Fractional utilization of the aerobic capacity during distance running. *Med. Sci. Sports* 5:248–52, 1973.
11. Cunningham, D.A., and Faulkner, J.A. The effect of training on aerobic and anaerobic metabolism during a short exhaustive run. *Med. Sci. Sports* 1:65–69, 1969.
12. Davis, J.A.; Vodak, P.; Wilmore, J.H.; Vodak, J.; and Kurtz, P. Anaerobic threshold and maximal aerobic power for three modes of exercise. *J. Appl. Physiol.* 41:544–50, 1976.
13. deVries, H.A. *Laboratory Experiments in Physiology of Exercise.* Dubuque, Iowa: Wm. C. Brown Company Publishers, 1971.
14. Faria, I.; Dix, C.; and Frazer, C. Effect of body position during cycling on heart rate, pulmonary ventilation, oxygen uptake, and work output. *J. Sports Med. Phys. Fitness* 18:49–56, 1978.

15. Girandola, R.N., and Katch, F.I. Effects of physical conditioning on changes in exercise and recovery O_2 uptake and efficiency during constant-load ergometer exercise. *Med. Sci. Sports* 5:242–47, 1973.

16. Hermansen, L. Anaerobic energy release. *Med. Sci. Sports* 1:32–38, 1969.

17. Hermansen, L., and Saltin, B. Oxygen uptake during maximal treadmill and bicycle exercise. *J. Appl. Physiol.* 26:31–37, 1969.

18. Hess, P., and Seusing, J. Der Einfluss der Tretfrequenz und des Pedaldruckes auf die Sauerstoff aufnahme die Untersuchungen am Ergometer. *Int. Z. Angew. Physiol.* 19:468–75, 1963.

19. Issekutz, B., Jr., and Rodahl, K. Respiratory quotient during exercise. *J. Appl. Physiol.* 16:606–10, 1961.

20. Katch, F.I.; McArdle, W.D.; and Pechar, G.S. Relationship of maximal leg force and leg composition to treadmill and bicycle ergometer maximum oxygen uptake. *Med. Sci. Sports* 6:38–43, 1974.

21. Katch, V.; Weltman, A.; Martin, R.; and Gray, L. Optimal test characteristics for maximal anaerobic work on the bicycle ergometer. *Res. Q.* 48:319–27, 1977.

22. Keul, J., and Doll, E. Intermittent exercise: metabolites, pO_2 and acid-base equilibrium in the blood. *J. Appl. Physiol.* 34:220–25, 1973.

23. Klissouras, V. Heritability of adaptive variation. *J. Appl. Physiol.* 31:338–44, 1971.

24. ———. Genetic limit of functional adaptability. *Int. Z. Angew Physiol.* 30:85–94, 1972.

25. Knuttgen, H.G., and Saltin, B. Muscle metabolites and oxygen uptake in short term submaximal exercise in man. *J. Appl. Physiol.* 32:690–94, 1972.

26. Knuttgen, H.G. Force, work, power and exercise. *Med. Sci. Sports* 10:227–28, 1978.

27. Londeree, B.R., and Ames, S.A. Maximal steady state versus state of conditioning. *Eur. J. Appl. Physiol.* 34:269–78, 1975.

28. Margaria, R.; Edwards, H.T.; and Dill, D.B. The possible mechanisms of contracting and paying the O_2 debt and the role of lactic acid in muscular contraction. *Am. J. Physiol.* 106:689–715, 1933.

29. Margaria, R.; Aghemo, P.; and Rovelli, E. Measurement of muscular power (anaerobic) in man. *J. Appl. Physiol.* 21:1662–64, 1966.

30. Mayhew, J.L., and Andrew, J. Assessment of running performance in college males from aerobic capacity percentage utilization coefficients. *J. Sports Med. Phys. Fitness* 15:342–46, 1975.

31. Musshoff, K.; Reindell, H.; Stein, H.; and Konig, K. Die Sauerstoff aufnahme pro Herzschlag (O_2 puls) als Funktion des Schlagvolumens der Arterio-venosen Differenz Des Minuten volumens und Herzvolumens. *Z. Kreislaufforschung* 48:255–77, 1959.

32. Pechar, G.S.; McArdle, W.D.; Katch, F.I.; Magel, J.R.; and DeLuca, J. Specificity of cardiorespiratory adaptation to bicycle and treadmill training. *J. Appl. Physiol.* 36:753–56, 1974.

33. Roberts, J.A., and Alspaugh, J.W. Specificity of training effects resulting from programs of treadmill running and bicycle ergometer riding. *Med. Sci. Sports* 4:6–10, 1972.

34. Rowell, L.B.; Kraning, K.K.; Evans, T.O.; Kennedy, J.W.; Blackmon, J.R.; and Kusumi, F. Splanchnic removal of lactate and pyruvate during prolonged exercise in man. *J. Appl. Physiol.* 21:1773–83, 1966.

35. Saltin, B., and Astrand, P-O. Maximal oxygen uptake in athletes. *J. Appl. Physiol.* 23:353–58, 1967.

36. Taylor, H.L.; Buskirk, E.; and Henschel, A. Maximal oxygen intake as an objective measure of cardiorespiratory performance. *J. Appl. Physiol.* 8:73–80, 1955.

37. Wasserman, K.; Van Kessel, A.L.; and Burton, G.G. Interaction of physiological mechanisms during exercise. *J. Appl. Physiol.* 22:71–85, 1967.

38. Wasserman, K.; Whipp, B.J.; Koyal, S.; and Beaver, W.L. Anaerobic threshold and respiratory gas exchange during exercise. *J. Appl. Physiol.* 35:236–43, 1973.

39. World Health Organization. Exercise tests in relation to cardiovascular function. WHO Technical Report series no. 388. Geneva, Switzerland, 1968.

Part 2

Physiology Applied to Health and Fitness

11

Health Benefits: Prophylactic and Therapeutic Effects of Exercise

The Cardiovascular System and Exercise
Lipid Metabolism and Exercise
Pulmonary Function Effects
Oxygen Transport Effects
Effects on Bones, Joints, and Connective Tissue
Effects of Exercise on Cancer
Exercise and Resistance to Disease
The "Tranquilizer Effect"
Effect of Exercise on Psychiatric State

The nature of the illnesses that beset our American population has in recent years undergone a transition from a predominance of infectious diseases to the present predominance of degenerative diseases. This change represents an implicit compliment to the medical profession for its contributions, both in research and clinical practice, toward the virtual control and the imminent eradication of a large portion of the formerly dreaded infectious scourges.

The increase of such degenerative diseases as cardiovascular accidents (heart attacks and strokes), hypertension, neuroses, and malignancies offers a challenge not only to medicine but to physical education as well. It seems that as improvements in medical science allow us to escape decimation by such infectious diseases as tuberculosis, diphtheria, poliomyelitis, we live longer, only to fall prey to the degenerative diseases at a slightly later date. Whether this involvement with the degenerative problems is a necessary concomitant of our living longer or the result of our simultaneous change in life style cannot yet be answered.

Along with our newly acquired control over the infectious diseases, we have made at least three other changes that seem likely to affect adversely our physical and mental well-being:

1. We have learned to produce more food than we need, and we eat commensurately.
2. We have learned to control our environment with very little expenditure of physical energy.
3. We have so constituted our society that most of us are subjected to unusual stresses, for which our biological responses are inadequate or, indeed, deleterious.

Our grandfathers labored hard and long physically in agriculture or in industry. We now use automobiles to go to the corner drugstore. We have power lawn mowers, automatic washers, and dishwashers for the adults in the house, and the children get motor scooters at the earliest possible age. On weekends we stage athletic spectacles in which our population gets its exercise vicariously by watching its hired athletes work out. The results of this sedentary life style appear to be the growth of degenerative diseases and an increasing involvement with neuroses and psychoses for which our grandparents just did not have time.

No one advocates a return to the long, tedious drudgery of manual work, but we cannot deny there is a need to learn how to adjust in better fashion to our newly found leisure time. The *fun* of exercise, sport, and physically vigorous recreation must replace the *tedium* of hard work that kept our grandfathers physically fit. Herein is the challenge to concerted effort by the medical and physical education professions.

The importance of life style to the health and illnesses of a lifetime has been suggested in the past, but the hard-core scientific evidence has only recently become available. In a survey of 6,928 adults of Alameda County, California (5, 6), individual health practices were related to health and also to mortality statistics. The health practices surveyed included (1) smoking, (2) weight in relation to desirable standards, (3) use of alcohol, (4) hours of sleep, (5) breakfast eating, (6) regularity of meals, and (7) physical activity. It was found that the average life expectancy of men age forty-five who reported six or seven "good" practices was more than eleven years more than that of men reporting fewer than four. For women, the difference in life expectancy was seven years. It was also found that the good health practices were reliably associated with positive health and that the relationship of the different health practices was cumulative: those who followed all of the good practices were in better health, even though older, than those who failed to follow them. This association was found to be independent of age, sex, and economic status.

The Cardiovascular System and Exercise

Physical Activity and Coronary Heart Disease. Since WW II, several large-scale statistical surveys have been conducted to evaluate the relationships between activity level and coronary heart disease (CHD).

Probably the most widely known study was conducted by Morris and associates (37) on bus drivers and conductors of the London Transport Authority. They found, among 31,000 drivers and conductors, that the drivers suffered significantly more coronary heart disease than the conductors. Since the drivers might be considered sedentary, while the conductors (of double-deck busses) did considerable walking and stair-climbing, it would seem that men in active jobs suffer less CHD. However, we cannot deduce from this that the exercise involved was the causative factor, because it is possible that coronary-prone people selected the driver jobs.

Taylor and his colleagues at the Laboratory of Physiological Hygiene at the University of Minnesota conducted a similar study on American railway employees (53). They found that in 191,609 man-years of risk and 1,978 reported deaths, the age-adjusted deaths for arteriosclerotic heart disease were 5.7, 3.9, and 2.8 for clerks, switchmen, and section men, respectively. Since the clerks' jobs were sedentary, the switchmen's moderately active, and the section men's very active, these data support those of the Morris group.

In the Framingham, Massachusetts, study a team of investigators from the U.S. Public Health Service classified men by habitual level of physical activity (29). In the ten years following the physical activity assessments, 207 men developed some manifestation of a coronary attack, and those who had been classified as most sedentary in each age group had an incidence almost twice that of the group who were at least moderately active.

Brunner (9) studied 5,279 men and 5,229 women in the Israeli kibbutzim. The kibbutzim are run as communes and consequently allow comparison of physical activity effects uncontaminated by the effects of income, diet, and other factors, since all members regardless of the nature of their work have the same income and eat in the same communal dining room. He found the incidence of the anginal syndrome, myocardial infarction, and fatalities due to CHD was 2.5 to 4 times higher in sedentary than in physically active workers. Cooper and associates (14), after study of some 3,000 men at the Aerobics Center in Dallas, concluded that physical fitness is related to lower coronary risk factors.

Many more studies could be cited to support the need for physical activity as a prophylactic measure against CHD. A good review of the literature on this subject is available (20).

Exercise and Coronary Circulation. Eckstein (19) operated on 117 dogs to produce various degrees of narrowing of the circumflex coronary artery. This simulated the narrowing brought about in coronary disease by deposition of cholesterol in the intima of the coronary arteries. When the dogs with constricted arteries were exercised, coronary blood flow capacity increased significantly by virtue of increased collateral circulation.

Raab (41, 42) provided considerable data on the importance of the autonomic control of the heart. The rate and metabolism of the heart are established as the result of a balance between the parasympathetic system (vagus nerve) and the sympathetic system (accelerator nerve). This balance is established in the midbrain and is mediated through release of neurohormonal (chemical) transmitters. The sympathogenic effects are brought about by the catecholamines, epinephrine and norepinephrine, and the vagal effects are brought about through acetylcholine.

In general, athletic training brings about vagal preponderance, as indicated by the slower heart rate in the athlete both at rest and under any given work load. A state of nervous excitement (emotional upset) causes a sympathogenic preponderance. The sympathogenic catecholamines were shown (by Raab) to have undesirable effects on the myocardium, such as an excessive increase in O_2 consumption. Raab and others (42) demonstrated that the combination of coronary constriction (as in atherosclerosis) and a sympathogenic supply of catecholamines brings about the typical ECG changes of coronary disease. He felt that this neurohormonal imbalance (sympathetic preponderance) is caused jointly by "(1) hypothalamic-stimulating emotional socioeconomic pressures, and (2) a deficiency of vagal and sympathoinhibitory counterregulation resulting from lack of physical exercise."

In a recent experiment by Heusner and co-workers (23), it was shown that similar myocardial damage can be produced by (1) epinephrine injection, (2) anoxia, (3) severe emotional stress, and (4) *severe* exercise. Most interestingly they also showed that appropriate physical conditioning can protect against the stressor effect of extreme anxiety or emotional stress.

Results from two different laboratories agree in showing increased coronary tree size to result from physical conditioning in rats (52, 54). Stevenson and associates extended their investigation to include the question of the effect of intensity and frequency upon the increase in coronary vessel size. They found that moderate exercise (twice weekly) had a more beneficial effect than extremely severe exercise (four hours per day, four days per week) (52).

More recent work on rats has confirmed the growth of capillary blood vessels in the myocardium as the result of endurance training, but only if the training is heavy enough to bring about cardiac hypertrophy (32).

Exercise and the Peripheral Circulation. The effect of exercise training on the capillary bed of skeletal muscles is not quite so clear. While older work had suggested an increase in the capillary density of exercising skeletal muscles in experimental animals, this has not been confirmed by more modern methods such as electron microscopy (22, 32). In a study comparing well-trained and untrained humans, no difference in capillary density was found either, but the muscle fibers were larger in the well-trained subjects, and consequently the number of capillaries per fiber was increased (22).

Clarke (12) reports on work done at the Longevity Research Institute in Santa Barbara, California, on patients with severe peripheral vascular disease. Walking exercise over a six-month period resulted in a dramatic improvement of 300% in treadmill walking distance, which strongly suggests an improvement in peripheral circulation for these patients for whom other therapies have not been encouraging. A combination of walking with rigorous dietary control was even more effective.

Exercise Effect on Blood Pressure. One of the better experiments in the exercise and blood pressure area was performed by Boyer and Kasch (7), who found that six months of participation in their San Diego State fitness program resulted in decreases of systolic pressure of almost 12 mm Hg and 13 mm Hg in diastolic values in twenty-three hypertensive men.

It has also been shown that a daily exercise program can prevent the hypertension and atherosclerosis which usually develop in rats on a high salt diet (10).

Perhaps the most impressive evidence for the exercise effect on blood pressure is that provided by Montoye and co-workers (33) as a result of the Tecumseh Project in which they studied approximately 1,700 men. They found that the more active men had significantly lower systolic and diastolic blood pressure on the average, regardless of age.

Changes in the Blood Accompanying Stress and Exercise. Thus far we have discussed the effects of stress and exercise upon the blood vessels and muscle tissue of the heart; however, another factor is thought to be of considerable importance in the etiology of heart disease: the *physicochemical properties* of the blood that courses through these coronary vessels. It is obvious that changes in the blood that lead to quicker coagulation or clotting time

—clot

might also be more likely to result in thrombus formation, the plugging of a coronary artery in a heart attack.

The work of Schneider and Zangari (47) demonstrated the effects of stress upon the blood. Anxiety, tension, fear, anger, and hostility were associated with shorter clotting times, increased viscosity, and blood pressure. It was suggested that this pattern was appropriate as a protective reaction when the organism was under attack because excessive blood loss would be prevented by the shortened clotting time and O_2 transport would be enhanced by the increased viscosity. If this pattern were used chronically, however—as seemed to be the case in their hypertensive subjects—it could prove detrimental by favoring intravascular thrombosis and by increasing the work of the heart (because of the increased viscosity of the blood). Their work has since been corroborated by Friedman (21) and others.

Since moderate exercise has been shown to have significant tranquilizer effects, it is reasonable to believe that these undesirable effects of stress might be relieved to some extent by appropriate exercise.

Another important effect of exercise training is the reduction of serum uric acid (SUA) (15). Recent reports from the Tecumseh Project involving over 1,200 men showed that physically active men had significantly lower SUA levels than sedentary men, (34). SUA is important because of its causal relationship to gout and its suggested relationship to coronary artery disease.

Increases of *total blood volume* (TBV) that had been reported earlier were considered equivocal because of methodological questions, but recent evidence has now been presented to confirm the fact that endurance training can increase the blood volume by as much as 6% (39). The increased TBV was found to be due to increased plasma volume; red cell volume did not change significantly. These facts are important to cardiac function because of the implications for improved venous return without increased flow resistance.

Lipid Metabolism and Exercise

The lipids include the *typical fats,* which are esters of fatty acids and glycerol (triglycerides), and *sterols* such as cholesterol (as well as other categories). Our interest in these members of the lipid family stems from the fact that both are found in the deposits that narrow the lumen of arteries in atherosclerosis. The physician's concern with the blood cholesterol level as a predisposing factor to heart disease is well known, and it is based on statistical evidence that indicates a strong relationship (though possibly not causal) between cholesterol level and heart disease. For these reasons, the effect of exercise on blood triglyceride and cholesterol is of interest.

Cholesterol. Cholesterol is transported in the blood in combination with special proteins to form lipoprotein. Three different lipoproteins are responsible for this transport, and they are differentiated by their densities: (1) alpha lipoprotein, also called high density lipoprotein (HDL) because it has the greatest density; (2) beta lipoprotein, also called low density lipoprotein (LDL) because of its lower density; and (3) pre-beta lipoprotein, also called very low density lipoprotein (VLDL) because it is the least dense. It is now known that only about 17% of the cholesterol in fasting plasma is carried by the HDL, whereas the remainder is carried by the LDL and VLDL. Furthermore, there is good agreement among several epidemiologic studies that ischemic heart disease is associated with high levels of LDL and VLDL, whereas high levels of HDL appear to be protective against coronary artery disease and seem to be related to longevity. There seems to be little doubt that LDL and VLDL are the transport mechanism by which cholesterol is transported from the periphery into the smooth muscle cells of the arteries where it collects, plugging the arteries and causing the disease we call atherosclerosis. On the other hand, the HDL works against this atherosclerotic process, either by (1) resisting the movement of LDL cholesterol into the arterial wall, or (2) promoting the efflux of cholesterol from the tissues to the liver where it is broken down and excreted. The mechanism of the HDL protective effect is not yet completely clear, and it is possible that both mechanisms are operative. In any event the HDLs are the "good guys" and the LDL and VLDLs are the "bad guys." This is important to us because there is now considerable evidence that exercise is one of the important means by which HDL can be increased (2,43,45).

Triglycerides. Holloszy and colleagues (24) have shown that six months of physical conditioning by calisthenics and distance running reduced serum triglycerides by 40%. However, this effect appeared to last only about two days. Thus it may be inferred from their work that serum triglycerides can be maintained at a significantly lower level by exercise, but the exercise must be done at least every other day. More recent work has supported their data and has also shown that exercise is effective in correcting certain abnormalities of fat metabolism (40).

Pulmonary Function Effects

Until recently, the only beneficial effects of physical conditioning upon pulmonary function seemed to be improvements in static lung volumes and capacities such as functional residual capacity, residual volume, vital capacity, and the ratio of residual volume to total lung capacity (4), and in lung diffusion (25).

Although the muscles involved in breathing are skeletal muscles, no one had thought to investigate the trainability of these muscles until very recently. Leith and Bradley (31) have now shown that in normals the respiratory muscles respond to strength and endurance training just as we might expect—i.e., by increases in these measures of approximately 55% and 14%, respectively. Subsequently, it has been shown that cystic fibrosis patients improve their respiratory muscle endurance by even greater percentages and that the training can be accomplished equally well by specific breathing exercise or by upper body endurance exercise (30).

A considerable volume of literature has developed with respect to the effects of exercise upon chronic obstructive lung disease. To sum up the evidence, there is no reason to believe that exercise can restore destroyed alveolar tissue. However, the improvements in respiratory muscle strength and endurance bring about a much improved exercise or work tolerance and for some individuals considerable symptomatic relief.

Oxygen Transport Effects

The benefits to the systems that determine the capacity for oxygen transport have been supported by experiments too numerous to cite. In general, improvements in maximal O_2 consumption from appropriate physical conditioning have been shown to occur in all ages and both sexes. Improvements reported have ranged from 5% to 30%. Differences in training effect would be expected according to the fitness level at the start of an experiment and the intensity-duration-frequency characteristics of the training regimen.

Effects on Bones, Joints, and Connective Tissue

It is well known that disuse of the skeletal system results in its atrophy with the eventual development of osteoporosis. Loading of the bones of the skeletal framework is necessary for normal bone metabolism, which consists of both anabolic and catabolic processes as that occurring in other living tissues. In bone, both mineral and organic metabolic processes are involved, and a study using swine has provided evidence that the stress of exercise exerts a conservatory influence on both the mineral and the collagenous (organic) components of bone (3). Smith and Reddan (50) have shown similar results in man.

Recent work from several independent investigations has shown the importance of physical conditioning on connective tissues such as ligaments. It was shown in rats that the strength of ligaments of the knee joint improves with physical activity (1,59). Tipton and co-workers have performed a series of experiments in which the same ligament strength results were shown in large animals (dogs), and in addition they demonstrated that the collagen content and fiber bundle size were significantly greater in trained dogs (55). Interestingly, they also showed the beneficial exercise effects on ligament strength after surgical repair. This, of course, has important implications for the post-surgical treatment of knee injuries in athletes. In another investigation Tipton and colleagues questioned whether these effects were hormonal, since connective tissues are known to be responsive to hormonal effects. They found that the mechanical stresses of training can act independently of the hormonal effect, which was verified. An excellent review of this work is now available (56).

In the author's laboratory, Chapman (11) showed that the resistance to movement in a joint can be significantly reduced both in the old and in the young by appropriate exercise.

Effects of Exercise on Cancer

As long ago as 1921, it was reported from a study of 86,838 men in Minnesota (49) that the death rate from cancer was roughly inversely proportional to the amount of muscular effort required on the job.

Rigan (44) has summarized research reports for the years 1920 until 1963 that relate to the effects of exercise on cancer. Evidence was reported to show an inverse relationship between physical activity and the cancer death rate in men. In various animal experiments over these forty years, the following observations were reported.

1. Caloric restriction inhibited the growth rate of malignancies.
2. In mice, two hours of daily exercise reduced the incidence of mammary gland carcinoma.
3. Three studies indicated a retarded tumor growth rate in exercised mice.
4. Tumor growth rate was reported to be inhibited in rats when they were injected with saline solution that had bathed excised rat muscle fatigued by exercise.

More recent evidence has been supplied by Colacino and Balke (13) who found only half the number of tumors in exercising mice compared with controls when both groups had been subjected to carcinogenic agents. Since there were no differences in food intake or weight, their study supports the older work with respect to the effect of exercise on tumor growth.

It must be emphasized that none of these studies provides evidence of a direct cause-and-effect relationship between exercise and cancer. Exercise is *not* being advanced as a panacea for cancer prevention; however, if a life of hygienic exercise can make a contribution—no matter how small in a statistical sense—to the prevention of this dread disease, this information (even though causal relationships are not scientifically valid) may be extremely important.

Exercise and Resistance to Disease

Advocates of the vigorous life have long felt that physical training and conditioning should provide a degree of resistance to infectious disease, but the absence of scientific evidence on which to base such a theory has been most frustrating. There is still no definitive evidence from which valid conclusions can be drawn, but the work of Zimkin (58) and his colleagues suggests that this is a fruitful area for further research. Their work used Selye's general adaptation syndrome as a point of departure, and they hypothesized that the stage of resistance was nonspecific as to type of stress, and would therefore have prophylactic significance.

They tested this hypothesis by developing a *stage of resistance*, using exercise as the stressor. It was found that physical exercise materially increased the body's resistance to infection. They reported experiments that showed agglutination titre and phagocytic activity (both advantageous in the prevention of illness) were increased much more in trained than in untrained animals.

Very important, they found that the training effects required optimum intensity and duration of work. Too much work not only failed to produce the desired resistance, but also resulted in decreased resistance in some cases.

It is difficult to evaluate this work because no statistical treatments are offered, and in almost all cases the original work is in Russian. A very interesting field of research is suggested, however, and corroboration of their results could have large implications for physical education and for medicine.

The "Tranquilizer Effect"

The importance of the ability to achieve neuromuscular relaxation is emphasized by the many references to this topic in popular literature. More than $300 million are spent yearly in the United States on tranquilizers to quiet jangled nerves. Many books and articles have been written on the subject, and physical educators often voice the opinion that a good workout can relieve nervous tension. Until recently, however, little scientifically acceptable evidence had been submitted that relates exercise and relief of residual neuromuscular tension.

State of Neuromuscular System Related to Anxiety and Tension. Overwhelming evidence supports the concept of a neuromuscular manifestation of various psychologically induced *anxiety* and *tension* states. The classic work, and much of the evidence, has been provided by Edmund Jacobson (26, 27, 28), who was the first to recognize this relationship and to apply it to the need for making objective measurements of previously unmeasurable symptoms. Thus, by making electronic measurements of the activity of the skeletal muscles (electromyography or EMG), it was possible to gain an objective insight into subjects' emotional states and nervousness. Later investigators have supported the work of Jacobson, and, on the basis of the work of Sainsbury and Gibson (46), and Nidever (38), it would appear that sampling even one or two representative muscles in the resting state can provide good evidence on the state of the entire organism at any given moment. Indeed, work in the author's laboratory has shown a correlation of 0.58 between resting oxygen consumption (total body) and the EMG activity in one muscle group, the right elbow flexors (18). Significant relationships have been shown to exist between these EMG measurements on selected skeletal muscles and such clinical states as headache, backache, mental activity, and emotional states.

Exercise and Relaxation. The earliest objective work relative to exercise and relaxation was done by Jacobson (26), who compared the ability of college athletes with normal subjects who were untrained in relaxation technique. He found that the athletes could relax more quickly and completely than the untrained controls; however, controls who were trained in the art of relaxation were superior to the athletes (as a group). Obviously, this experiment does not tell us whether an athletic program contributes to relaxation ability or whether more relaxed persons take up athletics.

In the author's laboratory, twenty-nine young, healthy subjects were studied for EMG changes after five minutes of bench-stepping as standard exercise. Neuromuscular tension was decreased significantly in the experimental situation, dropping 58% in electrical activity one hour after exercise. No significant change was seen in the same subjects on the control day. A chronic effect of conditioning was also shown (16). More recent work in the laboratory was directed toward comparison of the exercise effect with that of a recognized tranquilizer drug, *meprobamate* (17). To make the experiment more sensitive, older people with complaints of nervous tension acted as subjects. EMG measurements were made before and after (immediately, thirty minutes, and sixty minutes after) each of the five following treatment conditions:

1. Meprobamate 400 mg (normal dosage)
2. Placebo, 400 mg lactose
3. Fifteen minutes of walking-type exercises at a heart rate of 100
4. Fifteen minutes of the same exercise at a heart rate of 120
5. Resting control

Conditions 1 and 2 were administered double blind. It was found that exercise at a heart rate of 100 lowered electrical activity in the musculature by 20%, 23%, and 20% at the first, second, and third posttests, respectively. These changes were highly significant ($P < .01$). Neither meprobamate nor placebo treatments were significantly different from controls. Exercise at the higher heart rate was only slightly less effective, but the data were more variable and approached, but did not achieve, significance.

Similar results have been found by Morgan and Horstman (36), who showed a reduction in anxiety in normals as well as in clinically anxious individuals, and by Sime (48), who found brief, mild exercise to have a potent effect in reducing the physiological response to an acute stressor.

The data suggest that the exercise modality should not be overlooked when a tranquilizer effect is desired, since in single doses, at least, appropriate exercise has a significantly greater effect than does one of the most frequently prescribed tranquilizer drugs, meprobamate. It must be added that exercise has no undesirable side effects, whereas tranquilizer drugs used in sufficient repeated dosage to bring about the same effect must also impair motor coordination, reaction time, etc., with subsequent hazards involved in driving an automobile and any other activity requiring normal reactions.

Effect of Exercise on Psychiatric State

Only very recently has there been any interest in the effects of exercise and fitness level upon mood and psychiatric state. Morgan and Horstman (36) were among the first to show the important effect of exercise therapy in bringing about significant reductions in depression. Current work by Brown and others (8) at the University of Virginia also shows significant improvements in various depressive disorders from ten weeks of jogging.

A study by Young and Ismael (57) has shown an interesting relationship between fitness and emotional stability in middle-aged men, and Stamford and associates (51) found both physiological and psychiatric improvement in institutionalized elderly mental patients as a result of daily exercise.

It would seem that the wisdom of the ages which suggested that vigorous exercise makes you feel good, now rests on laboratory evidence as well.

SUMMARY

Although much evidence has been furnished that supports the value of exercise as a prophylactic and therapeutic measure, exercise is not a panacea. In none of the areas we have discussed in this chapter is the evidence final and conclusive, but we can confidently say that the evidence indicates the desirability of the vigorous life in maintaining optimum levels of health and well-being. Every physical educator should be dedicated to this principle, both in his personal and in his professional life. In no other way can the youth of this nation be led into the full life that only vigorous activity can bring.

REFERENCES

1. Adams, A. Effect of exercise on ligament strength. *Res. Q.* 37:163–67, 1966.
2. Altekruse, E.B., and Wilmore, J.H. Changes in blood chemistries following a controlled exercise program. *J. Occup. Med.* 15:110–13, 1973.
3. Anderson, J.J.B.; Milin, L.; and Crackel, W.C. Effect of exercise on mineral and organic bone turnover in swine. *J. Appl. Physiol.* 30:810–13, 1971.
4. Bachman, J.C., and Horvath, S.M. Pulmonary function changes which accompany athletic conditioning programs. *Res. Q.* 39:235–39, 1968.

5. Belloc, N.B. Relationship of health practices and mortality. *Prev. Med.* 2:67–81, 1973.

6. Belloc, N.B., and Breslow, L. Relationship of physical health status and health practices. *Prev. Med.* 1:409–21, 1972.

7. Boyer, J.L., and Kasch, F.W. Exercise therapy in hypertensive men. *J. Am. Med. Assoc.* 211:1668–71, 1970.

8. Brown, R.S.; Ramirez, D.E.; and Taub, J.M. The prescription of exercise for depression. Paper read at ACSM meeting, May 24, 1978, Washington. D.C.

9. Brunner, D. The influence of physical activity on incidence and prognosis of ischemic heart disease. In *Prevention of Ischemic Heart Disease,* ed. W. Raab, pp. 236–43. Springfield: Charles C Thomas, 1966.

10. Buuck, R.J. Effect of exercise on hypertension. Abstracted in *Med. Sci. Sports* 10:37, 1978.

11. Chapman, E.A.; deVries, H.A.; and Swezey, R. Joint stiffness: effects of exercise on young and old men. *J. Geront ol.* 27: 218–21, 1972.

12. Clarke, H.H. Diet and exercise related to vascular disease. *Phys. Fitness Res. Digest* 6:11–17, 1976.

13. Colacino, D., and Balke, B. Tumor reduction in endurance trained mice. Paper read at ACSM Meeting, May 1972, Philadelphia.

14. Cooper, K.H.; Pollock, M.L.; Martin, R.P.; White, S.R.; Linnerud, A. C.; and Jackson, A. Physical fitness levels vs. selected coronary risk factors. *J.Am. Med. Assoc.* 236:166–69, 1976.

15. Cronau, L.H.; Rasch, P.J.; Hamby, J.W.; and Burns, H.J. Effects of strenuous physical training on serum uric acid levels. *J. Sports Med.* 12:23–25, 1972.

16. deVries, H.A. Immediate and long-term effects of exercise upon resting muscle action potential level. *J. Sports Med. Phys. Fitness* 8:1–11, 1968.

17. deVries, H.A., and Adams, G.M. Electromyographic comparison of single doses of exercise and meprobamate as to effects on muscular relaxation. *Am. J. Phys. Med.* 51:130–41, 1972.

18. deVries, H.A.; Burke, R.K.; Hopper, R.T.; and Sloan, J.H. Relationship of resting EMG level to total body metabolism with reference to the origin of "tissue noise." *Am. J. Phys. Med.* 55:139–47, 1976.

19. Eckstein, R.W. Effect of exercise on coronary artery narrowing and coronary collateral circulation. *Circ. Res.* 5:230–35, 1957.

20. Fox, S.M., and Boyer, J.L. Physical activity and coronary heart disease. In *Physical Fitness Research Digest,* ser. 2, no. 2, ed. H.H. Clarke. Washington, D.C.: The President's Council on Physical Fitness and Sports, 1972.

21. Friedman, Meyer. *Pathogenesis of Coronary Artery Disease.* New York: McGraw-Hill Book Co., 1969.

22. Hermansen, L., and Wachtlova, M. Capillary density of skeletal muscle on well-trained and untrained men. *J. Appl. Physiol.* 30:860–63, 1971.

23. Heusner, W.W.; Van Huss, W.D.; Carrow, R.E.; Wells, R.L.; Anderson, D.J.; and Ruhling, R.O. Exercise, anxiety, and myocardial damage. Paper read at ACSM Meeting, May 1, 1972, Philadelphia.

24. Holloszy, J.O.; Skinner, J.S.; Toro, G.; and Cureton, T.K. Effects of a 6-month program of endurance exercise on the serum lipids of middle-aged men. *Am. J. Cardiol.* 14:748–55, 1964.

25. Holmgren, A. On the variation of DL_{CO} with increasing oxygen uptake during exercise in healthy trained young men and women. *Acta Physiol. Scand.* 65:207–20, 1965.
26. Jacobson, E. The course of relaxation of muscles of athletes. *Am. J. Psychol.* 48:98–108, 1936.
27. —. *Progressive Relaxation.* Chicago: University of Chicago Press, 1938.
28. —. The cultivation of physiological relaxation. *Ann. Intern. Med.* 19:965–72, 1943.
29. Kannel, W.B.; Sorlie, P.; and McNamara, P. The relation of physical activity to risk of coronary heart disease: the Framingham study. In *Coronary Heart Disease and Physical Fitness,* eds. O.A. Larson and R.O. Malmborg, p. 256. Baltimore: University Park Press, 1971.
30. Keens, T.G.; Krastins, I.R.B.; Wannamaker, E.M.; Levison, H.; Crozier, D.N.; and Bryan, A.C. Ventilatory muscle endurance training in normal subjects and patients with cystic fibrosis. *Am. Rev. Respir. Dis.* 116:853–60, 1977.
31. Leith, D.E., and Bradley, M. Ventilatory muscle strength and endurance training. *J. Appl. Physiol.* 41:508–16, 1976.
32. Ljunggvist, A., and Unge, G. Capillary proliferative activity in myocardium and skeletal muscle of exercised rats. *J. Appl. Physiol.* 43:306–7, 1977.
33. Montoye, H.J.; Metzner, H.L.; Keller, J.B.; Johnson, B.C.; and Epstein, F.H. Habitual physical activity and blood pressure. *Med. Sci. Sports* 4:175–81, 1972.
34. Montoye, H.J.; Mikkelsen, W.M.; Metzner, H.L.; and Keller, J.B. Physical activity, fatness, and serum uric acid. *J. Sports Med. Phys. Fitness* 16:253–60, 1976.
35. Morgan, W.P.; Roberts, J.A.; Brand, F.R.; and Feinerman, A.D. Psychological effect of chronic physical activity. *Med. Sci. Sports* 2:213–17, 1970.
36. Morgan, W.P., and Horstman, D.H. Anxiety reduction following acute physical activity. Abstracted in *Med. Sci. Sports* 8:62, 1976.
37. Morris, J.N.; Heady, J.A.; Raffle, P.A.B.; Roberts, C.G.; and Parks, J.W. Coronary heart disease and physical activity of work. *Lancet* 2:1053–1111, 1953.
38. Nidever, J.E. A factor analytic study of general muscular tension. Doctoral dissertation, psychology, UCLA, 1959.
39. Oscai, L.B.; Williams, B.T.; and Hertig, B.A. Effect of exercise on blood volume. *J. Appl. Physiol.* 24:622–24, 1968.
40. Oscai, L.B.; Patterson, J.A.; Bogard, D.C.; Beck, R.J.; and Rothermel, B.L. Normalization of serum triglycerides and lipoprotein electrophoretic patterns by exercise. *Am. J. Cardiol.* 30:775–80, 1972.
41. Raab, W. Metabolic protection and reconditioning of the heart muscle through habitual physical exercise. *Ann. Int. Med.* 53:87–105, 1960.
42. Raab, W.; van Lith, P.; Lepeschkin, E.; and Herrlich, H.C. Catecholamine-induced myocardial hypoxia in the presence of impaired coronary dilatability independent of external cardiac work. *Am. J. Cardiol.* 9:455, 1962.
43. Ratliff, R.; Elliott, K.; and Rubenstein, C. Plasma lipid and lipoprotein changes with chronic training. *Med. Sci. Sports* 10:55, 1978.
44. Rigan, D. Exercise and cancer, a review. *J. Am. Osteopathic Assoc.* 62:596–99, 1963.

45. Roundy, E.S.; Fisher, G.A.; and Anderson, S. Effect of exercise on serum lipids and lipoproteins. *Med. Sci. Sports* 10:55, 1978.

46. Sainsbury, P., and Gibson, J.G. Symptoms of anxiety and tension and the accompanying physiological changes in the muscular system. *J. of Neurol. Neurosurg. Psychiatry* 17:216–24, 1954.

47. Schneider, R.A., and Zangari, V.M. Variations in clotting time, relative viscosity and other physicochemical properties of the blood accompanying physical and emotional stress in the normotensive and hypertensive subject. *Psychosom. Med.* 13:289–303, 1951.

48. Sime, W.E. A comparison of exercise and meditation in reducing physiological response to stress. *Med. Sci. Sports* 9:55, 1977.

49. Sivertsen, I., and Dahlstrom, A.W. Relation of muscular activity to carcinoma; preliminary report. *J. Cancer Res.* 6:365–78, 1921.

50. Smith, E.L., and Reddan, W. Physical activity—a modality for bone accretion in the aged. *Am. J. Roentgen, Radium Ther. Nucl. Med.* 126:1297, 1977.

51. Stamford, B.A.; Hambacher, W.; and Fallica, A. Effects of daily physical exercise on the psychiatric state of institutionalized geriatric mental patients. *Res. Q.* 45:34–41, 1974.

52. Stevenson, J.A.; Feleki, V.; Rechnitzer, P.; and Beaton, J.R. Effect of exercise on coronary tree size in the rat. *Circ. Res.* 15:265–69, 1964.

53. Taylor, H.L.; Klepetar, E.; Keys, A.; Parlin, W.; Blackburn, H.; and Puchner, T. Death rates among physically active and sedentary employees of the railway industry. *Am. J. Public Health* 52: 1697–1707, 1962.

54. Tepperman, J., and Pearlman, D. Effects of exercise and anemia on coronary arteries of small animals as revealed by the corrosion-cast technique. *Circ. Res.* 9:576–84, 1961.

55. Tipton, C.M.; James, S.L.; Mergner, W.; and Tcheng, T.K. Influence of exercise on strength of medial collateral knee ligaments of dogs. *Am. J. Physiol.* 218:894–901, 1970.

56. Tipton, C.M.; Matthes, R.D.; Maynard, J.A.; and Carey, R.A. The influence of physical activity on ligaments and tendons. *Med. Sci. Sports* 7:165–75, 1975.

57. Young, R.J., and Ismael, A.H. Relationship between anthropometric, physiological, biochemical, and personality variables before and after a four-month conditioning program for middle-aged men. *J. Sports Med. Phys. Fitness* 16:267–76, 1976.

58. Zimkin, N.V. Stress during muscular exercise and the state of non-specifically increased resistance. In *International Research in Sport and Physical Education,* eds. E. Jokl and E. Simon. Springfield, Ill.: Charles C Thomas, 1964.

59. Zuckerman, J., and Stull, G.A. Effects of exercise on knee ligament separation force in rats. *J. Appl. Physiol.* 26:716–19, 1969.

12

Physical Fitness Testing

Hardly a day goes by without a newspaper reference to physical fitness or the lack of it. We hear frequently from the medical profession that obesity is our most common disease, that we suffer from a softness brought about by our highly mechanized lives, and that our complex civilization is producing ever-increasing levels of nervous and mental disease. On the other hand, sports-writers have a field day after our Olympic successes, enthusiastically rebutting allegations of *our* lack of physical fitness because our *athlete's* demonstrated superb fitness.

Herein lies one of the greatest fallacies of American physical education. We do indeed develop outstanding athletes to represent us in international competition, but they are in no way typical of the population. It is unfortunate that the spectator sports—football, basketball, and baseball—occupy such a prominent place in physical education, for there is little opportunity to pursue them in adult life. In fact, this emphasis is probably also responsible for the neglect of our less physically talented children; our culture encourages them to be spectators rather than participants.

Despite all the interest shown in physical fitness, by laymen and professionals alike, we are not yet prepared to offer a universally acceptable definition of the term, much less an operational definition. It must be realized that not all definitions can be couched in terms of absolutes; sometimes a definition must be arbitrary and arrived at by consensus. Thus a nautical mile is based on an absolute measure, one minute of latitude, but the statute mile, which we use much more frequently, is an arbitrary 5,280 feet. It is necessary that we define physical fitness arbitrarily so that we may proceed with an operational definition, and with the most important work of all: improving physical fitness at all levels in our population.

The best possible definition of physical fitness encompasses the work that has been performed and accepted by the two professions most interested in this area: physical education and medicine. Thus physical educators have developed many fine tests, which include such items as running, jumping, throwing, pull-ups, and push-ups. These test batteries, which are categorized as tests of *motor fitness,* attempt to measure the following *elements* of physical fitness: strength, speed, agility, endurance, power, coordination, balance, flexibility, and body control.

The concept of *physical working capacity* (PWC) a measure of aerobic power has gained wide acceptance as a measure of fitness among physiologists, pediatricians, cardiologists, and other members of the medical profession. PWC may be defined as the maximum level of metabolism (power) of which an individual is capable. PWC is measured by objective and accurate means (maximal O_2 consumption), and simpler but valid methods are available for predicting PWC from the submaximal heart rate tests described below.

Much could be gained by wider use of the PWC concept in the physical education profession. First, a unification of thought between the physical education and medical professions would be most beneficial to both. Second,

PWC testing would provide a motivating factor for students in physical activity classes who are not skillful enough to compete successfully in athletics with their peer groups.

That the PWC concept is not widely used in physical education is probably due to three factors:

1. Physical educators' inability to perform these analyses.
2. Lack of facilities.
3. Classes that are too large to permit sufficient attention to individual testing.

We suggest that some of the tests described in this chapter, such as the Astrand-Ryhming nomogram, which require only one inexpensive piece of equipment, the bicycle ergometer, ought to be at least part of every corrective physical education program. It is also to be hoped that eventually an enlightened public will demand smaller size physical education classes, at which time PWC testing should become an integral part of the general program.

An individual's PWC ultimately depends on the capacity to supply oxygen to the working muscles. This, in turn, means that PWC probably evaluates, directly or indirectly, at least the following elements of physical fitness: (1) cardiovascular function, (2) respiratory function, (3) muscular efficiency, (4) strength, (5) muscular endurance, and (6) obesity. Obesity becomes a factor because the final score in maximal O_2 consumption is usually expressed in milliliters of O_2 per kilogram of body weight.

It is readily seen that PWC and motor fitness testing are needed in a well-rounded physical education curriculum. The relative importance of the elements tested by the two major components varies with the age group under consideration, and this factor will be considered later in this chapter.

Measurement of PWC by Maximum O_2 Consumption

Work by human muscular effort can be produced by aerobic and anaerobic metabolic processes as discussed earlier. However, anaerobic processes, when fully loaded, can function only for approximately forty seconds. For this reason it is really *aerobic capacity* that we measure when we measure PWC, and it is sometimes referred to in these terms.

The measurement of maximal O_2 consumption (VO_2 max) requires going not only to the exercise load that elicits VO_2 max, but at least one step beyond to assure that a true maximal value has been reached (see fig. 10.10). Although this causes little concern in tests on healthy college age or younger subjects, this discussion will deal with adult fitness testing as well, and consequently safety measures are emphasized. Where feasible it is sensible to employ the same precautions for the school age subjects, since there are occasionally unrecognized, undiagnosed cases of heart disease among them.

Table 12.1

Medical Referral Form for Participation in Graded Exercise Test and Exercise Program

---

Patient's name _____ Date_____

                      Last   First   Initial

Address_____ Age_____ Phone_____

I consider the above individual as:

        _____Normal
        _____Cardiac patient
        _____Prone to coronary heart disease
        _____Other (Explain) _____

Diagnostic Data Etiologic	Present Physical Activity	ECG	Rhythm
1. No heart disease	1. Very active	1. Normal	1. Sinus
2. Rheumatic heart disease	2. Normal	2. Dig. effect only	2. Atrial fib.
3. Congenital heart disease	3. Limited	3. Abnormal	3. Other
4. Hypertension	4. Very limited	4. Infarct	
5. Ischemic heart disease			
6. Other			

Specific cardiac diagnosis _____

_____

Additional abnormalities you are aware of _____

_____

From "Guidelines for Graded Exercise Testing and Exercise Prescription," ACSM, Philadelphia: Lea & Febiger, 1975.

Medical Examination—Informed Consent. Individuals over thirty-five years of age, and any individuals of any age who: (1) have any question about their health status, (2) develop symptoms during testing, or (3) have not had a medical examination in two years, must be cleared for testing by a personal physician (3). The medical referral form recommended by the American College of Sports Medicine (ACSM) is shown in table 12.1.

For all subjects undergoing maximal exercise testing, these procedures must be carefully explained. Above all, participants must know exactly what is expected of them and must be allowed to ask questions with respect to the procedures. After all questions are answered to their satisfaction, participants must sign an informed consent form such as the one recommended by ACSM and shown in table 12.2.

Date of last complete physical examination _____

Present medication _____

Please fill in the information below if it is available:

1. Urine, sp.gr._____ Alb._____ Glucose_____ Micro._____

2. Complete blood count: Hbg._____ Hct._____ WBC_____ Diff._____

3. ECG, 12 lead (enclose copy) _____

4. Blood pressure, syst._____ diast._____

5. Glucose_____mg%

6. 2-hr post-Dexicola_____mg%

7. Cholesterol_____mg% Lipoprotein electrophoresis_____

   Triglyceride_____mg%

8. Graded exercise test results (If available, enclose.)

Impression of above information _____

The above listed person is capable of participating in an exercise program as well as periodic laboratory evaluations, under the guidance and supervision of a

(   ) Physician
(   ) Exercise leader (_____) Check appropriate supervision (_____).

Signed: _____M.D.

Type or Print
Name of Physician _____

Personnel. For testing healthy school age subjects and adults under thirty-five with no known primary coronary heart disease (CHD) risk factors* or symptoms, testing may be conducted by a trained exercise technician without the presence of a physician. The exercise technicians must have had training in cardiopulmonary resuscitation (CPR) and mouth-to-mouth breathing techniques.

*Primary CHD risk factors are hypertension, hyperlipidemia, and cigarette smoking. Secondary risk factors are family history, obesity, physical inactivity, diabetes mellitus, and asymptomatic hyperglycemia. The participant's status in the last two risk factors need not be determined in young (less than thirty-five) asymptomatic participants with no risk factors. In participants thirty-five and over and in participants with risk factors or symptoms, blood lipids and blood glucose levels should be measured (3).

Table 12.2
Informed Consent for Graded Exercise Test*

1. *Explanation of the Graded Exercise Test*
   You will perform a graded exercise test on a bicycle ergometer and/or a motor-driven treadmill. The work levels will begin at a level you can easily accomplish and will be advanced in stages, depending on your work capacity. We may stop the test at any time because of signs of fatigue, or you may stop when you wish to because of personal feelings of fatigue or discomfort. We do not wish you to exercise at a level which is abnormally uncomfortable for you.

2. *Risks and Discomforts*
   There exists the possibility of certain changes occurring during the test. They include abnormal blood pressure, fainting, disorders of heartbeat, and very rare instances of heart attack. Every effort will be made to minimize them by the preliminary examination and by observations during testing. Emergency equipment and trained personnel are available to deal with unusual situations which may arise.

3. *Benefits to be Expected*
   The results obtained from the exercise test may assist in the diagnosis of your illness or in evaluating what types of activities you might carry out with no or low hazards.

4. *Inquiries*
   Any questions about the procedures used in the graded exercise test or in the estimation of functional capacity are welcome. If you have any doubts or questions, please ask us for further explanations.

5. *Freedom of Consent*
   Permission for you to perform this graded exercise test is voluntary. You are free to deny consent if you so desire.

   I have read this form and I understand the test procedures that I will perform and I consent to participate in this test.

_____

Signature of Patient

_____   _____
Date   Witness

*Where test is for a purpose other than prescription, e.g., experimental interest, this should be indicated on the Informed Consent Form.

From "Guidelines for Graded Exercise Testing and Exercise Prescription," ACSM, Philadelphia: Lea & Febiger, 1975.

A physician's presence is required when testing a participant of any age who has shown symptoms of CHD, either suspected or documented. A physician should be available at least in the general testing area if the subject is over thirty-five or if the subject exhibits major risk factors or has documented CHD (asymptomatic).

Emergency Equipment. Minimal emergency equipment should include (1) a defibrillator, (2) an oxygenator with intermittent positive pressure capability, (3) oral and endotrachial airways, and (4) a bag-valve-mask respirator. When a physician is required or available for testing, many medical instruments and pharmaceuticals would also be provided at the physician's discretion and preferably set up as an emergency cart that would be available at all times.

Exercise Protocol. As discussed in chapter 10, at least three exercise modalities can serve for testing $\dot{V}O_2$ max. The bicycle ergometer is probably the most popular because it enjoys the advantages of objectively and accurately measured power output and relative lack of movement of the upper body, which greatly lessens the problems of instrumentation, producing a "cleaner" electrocardiogram, fewer leaks around the mouthpiece, and other advantages. The treadmill is almost as popular because it allows the use of the familiar movements of running and walking. It also involves a slightly larger muscle mass, and consequently $\dot{V}O_2$ max is usually found to be 5%-8% higher on the treadmill. The third technique, bench-stepping, is the least desirable choice because of its great amount of body movement and its relatively small potential range of power outputs.

With any of the three exercise modalities, the protocol may be as follows.

1. *Intermittent incremental loading.* Exercise loads are usually started at a low level, and each load is applied for three to six minutes to allow an approximate steady state to develop. Rest intervals are allowed between exercise loads to prevent undue fatigue. Exercise loads are raised in some systematic fashion until two consecutive loads result in either a downturn in $\dot{V}O_2$, or a leveling to a plateau, or at least an insignificant increase (less than 150 ml O_2 is commonly used as a criterion). This protocol often requires participants to make two visits to the laboratory for test completion.

2. *Continuous step-incremental loading.* This procedure is typified on the bicycle ergometer by the Luft protocol in which the power output required for the first three minutes is 50 watts. Each minute thereafter the load is increased by 12.5 watts (approximately 75 kg m/min) until the subject is unable to continue. This much simpler procedure results in values for VO_2 max which are still within 5% of values gotten by the more cumbersome method described above and can be completed in most cases in twenty to twenty-two minutes (14). This protocol can also be modified to use step increments in loading such as 10 w/min, 20 w/2 min, 25 w/2 min, or 30 w/3 min without substantial differences in result (7). On the treadmill the Balke protocol is commonly used in which the subject walks at 3.3 mph for the first

two minutes on the horizontal treadmill. Every minute thereafter the incline is increased by 1% until either the heart rate reaches 180 bpm or until exhaustion. Balke considered a heart rate of 180 as the aerobic crest load, with work beyond this largely accomplished via anaerobic metabolism. However, results are more comparable with other protocols if continued until exhaustion.

3. *Continuously incremented loading (ramp loading).* Figure 12.1 shows the difference in these exercise protocols with respect to the way in which exercise intensity is increased with time. The continuously incremented or ramp-loading protocol, although not widely used yet, is becoming more popular as interest in such phenomena as anaerobic threshold (AT) and other points in the time history of the workout become more important. Breath-by-breath analyses of the gas parameters also greatly enhance the values to be derived from ramp loading. This method obviously has the advantage of providing an infinitely gentle progression of work load (no sudden increments to disturb the subject's equilibrium), but it also requires an electronic ergometer which must be modified in most cases to provide a ramp program.

Environmental Considerations. Environmental conditions such as ambient temperature, relative humidity, and air movement have a considerable effect on how the available cardiac output is divided between the active muscles and the cutaneous vessels for cooling. Thus, for example, lengthy heat exposure may lead to reduction of the central blood volume, concomitant loss of cardiac output, and spuriously low values of $\dot{V}O_2$ max. To make either intraindividual or interindividual comparisons of PWC, environmental conditions should be controlled. The World Health Organization recommends that the testing environment be maintained in the range 18°–22° C (64°–72° F), with the relative humidity below 60%, and in still air. The upper limit can be increased by about 2° C if the effective temperature is reduced by the use of a large fan (4). Normally, exercise tests are not conducted in a cold environment, and testing should be discouraged if the room temperature is below 10° C (50° F).

Time of Day, Diet, and Other Variables. Since many of the physiological functions entering into the determination of $\dot{V}O_2$ max are affected by time of day (circadian rhythm), this factor should be recorded and maintained constant in test-retest evaluations.

Ingestion of food results in a rise in both heart rate and ventilation for an hour or more, while a complete fast may result in low blood sugar during testing. Therefore a compromise is necessary, and the subject is instructed to eat only a light meal at least an hour before testing.

Unusually strenuous exertion should be avoided on the day prior to testing, and on the day of testing no other strenuous activity should precede the test. A rest period in the laboratory of at least one hour prior to testing is highly desirable. Anxiety concerning the test procedure can be a significant problem in submaximal tests, which depend on heart rate, but there is little if any

Physiology Applied to Health and Fitness

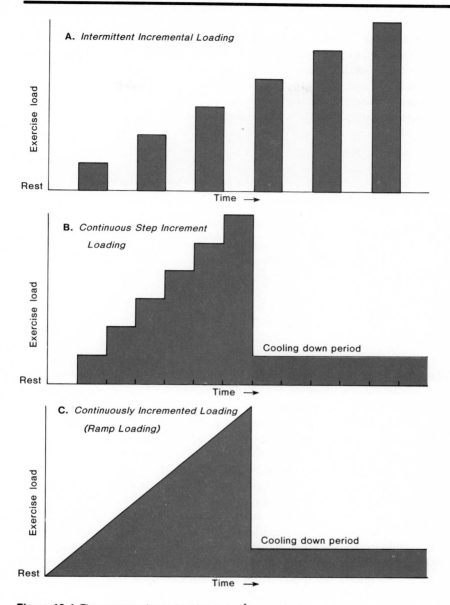

Figure 12.1 Three commonly used protocols for $\dot{V}O_2$ max testing.

effect on $\dot{V}O_2$ max. Indeed, recent work suggests it may not be a consideration even in submaximal tests (18).

Parameters To Be Measured. Since this text is directed to physical education students, the parameters to be measured will be limited to noninvasive (bloodless) techniques. In spite of this limitation, a great deal of important information can be extracted with respect to cardiovascular and respiratory function. Ideally, the following primary parameters, from which many secondary parameters can be derived, should be recorded at each exercise load:

A. *Primary Data*
1. FO_{2_E}—% O_2 in expired gas ⎫
2. F_{CO2_E}—% CO_2 in expired gas ⎬ for calculation of $\dot{V}O_2$
3. V_E—minute ventilation ⎭
4. HR—preferably from cardiotachometer, but can be taken later from recorded ECG
5. BP—systolic is gotten clearly; diastolic may be difficult
6. ECG—lead CM_5 if only one lead can be taken

B. *Derived Data* (calculated from primary data)
1. $R = \dot{V}_{CO_2}/\dot{V}_{O_2}$
2. O_2 pulse $= \dot{V}_{O_2}$ per heart beat
3. Double product $=$ HR $\times$ BPsyst/100
4. Ventilation equivalent for $O_2 = \dot{V}_E/\dot{V}_{O_2}$
5. Anaerobic threshold (see chap. 10)

The significance of the primary data is probably obvious. Of the derived data, the respiratory exchange ratio (R) is of greatest interest in helping define the point when anaerobic metabolism becomes prominent. As such, it is also useful as one criterion that $\dot{V}O_2$ max has indeed been achieved. Various authorities use different criteria, but in the absence of an R $<$ 1.05, it would be unlikely that a true $\dot{V}O_2$ max had been achieved. Before the onset of the AT, R is of course a reflection of the energy substrate utilized.

O_2 pulse provides some insight into the behavior of stroke volume and arteriovenous O_2 difference (see chap. 10).

The double product of heart rate and systolic blood pressure (HR $\times$ BPsyst/100) has been found to be well correlated with myocardial O_2 consumption and therefore reflects the work of the heart at each level. This will be important in our subsequent discussion of cardiac rehabilitation (chap. 14).

The ventilation equivalent for O_2 tells us how many liters of lung ventilation are required for one liter of O_2 uptake by the tissues. Thus this is an important reflection of the efficiency of O_2 transport mechanisms.

Physiology Applied to Health and Fitness

AT appears to be a very important determinant of the level of endurance performance that can be sustained in distance running events. Recent evidence suggests that it has important implications independent of the $\dot{V}O_2$ max (see chap. 10).

Criteria for Ending the Test.   According to ACSM recommendations (3), when an exercise test is being conducted by a nonphysician, it should be stopped for the following reasons. If a physician is conducting the test, he may decide to use other criteria.

1. Symptoms of significant exertional intolerance
    a. dizziness or near-syncope
    b. angina
    c. unusual or intolerable fatigue
    d. intolerable claudication or pain
2. Signs of intolerance
    a. staggering or unsteadiness
    b. mental confusion
    c. facial expression signifying disorders (strained or blank facies)
    d. cyanosis or pallor (facial or elsewhere)
    e. rapid, distressful breathing
    f. nausea or vomiting
    g. a definite fall in systolic blood pressure with increasing work load
3. Electrocardiographic changes
    a. S-T segment displacement of 0.2 mV. below the base line
    b. supraventricular or ventricular dysrhythmias or ectopic ventricular activity occurring before the end of a T-wave (R-on-T phenomenon). *It is recommended that a test be terminated in the presence of three or more successive ectopic ventricular complexes or with a significant increase in their occurrence—about ten per minute, depending on clinical judgment.*
    c. major left intraventricular conduction disturbances
4. Inappropriate blood pressure responses, such as a decrease in systolic blood pressure with an increase in work load

Extrapolation of Submaximal $\dot{V}O_2$ Data to Estimate $\dot{V}O_2$ Max.   If heart rate and metabolic rate ($\dot{V}O_2$) are both measured at steady state, the relationship is approximately linear, as shown in figure 12.2. Thus, if two or more paired $\dot{V}O_2$-HR values are plotted, as in figure 12.2, the resulting straight line can be extrapolated to the predicted maximum heart rate, and the $\dot{V}O_2$ max can then be read off the graph.

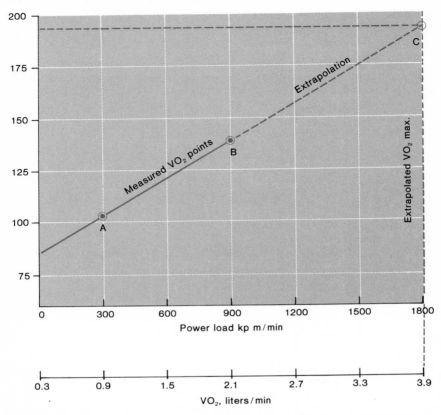

Figure 12.2 Extrapolation of submaximal $\dot{V}O_2$ data to estimate $\dot{V}O_2$ max. Linear extrapolation of the two measured $\dot{V}O_2$–HR points (A and B) to point C, the intersection with the estimated maximum heart rate line, provides an estimate of $\dot{V}O_2$ max = 3.9 liters/min.

Estimation of PWC from Heart Rate at Submaximal Loads

It has long been known that heart rate rises linearly with increasing work loads (within limits). Furthermore, the rate of rise in heart rate for the same increments of work has been used in several different methods for evaluating PWC. Figure 12.3 illustrates this principle for two middle-aged men, one in good athletic condition, the other in the untrained condition.

PWC-170 Test. Sjostrand (21) and Wahlund (24) used this principle to evaluate physical working capacity. It was modified by Adams (1) for use on elementary and junior high school boys and girls to compare the physical fitness of California and Swedish children. This test consists of two consecutive six-minute bicycle ergometer rides in which the work loads are selected to

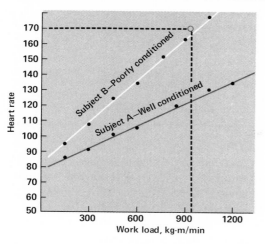

Figure 12.3 The rate of rise in heart rate with increasing exercise load as a function of physical condition.

produce heart rates of approximately 140 and 170 per minute. The working capacity is calculated by plotting (on graph paper) the heart rate against the work load at the end of each trial. A straight line is drawn through the two points to intersect the line of 170 bpm. The estimated amount of work that corresponds to a heart rate of 170 is then recorded as the individual's PWC-170. The heart rate of 170 is used rather arbitrarily as a point beyond which little increase in aerobic metabolism is expected. Use of this principle for the unconditioned subject in figure 12.3 would thus give an estimated PWC-170 of 975 kg-m/min.

This PWC-170 test has been found to correlate rather well with the measured maximal O_2 consumption of college men in the author's laboratory: $r = 0.88$. The standard error of prediction of maximal O_2 consumption from the PWC-170 test was found to be $\pm 9.4\%$, which seems to be an entirely acceptable value for this type of test (10). The test eliminates all of the laboratory technique of maximal O_2 consumption tests, but is still unsuitable for use with large groups.

Astrand-Ryhming Nomogram. This test utilizes the same basic principle as the PWC-170. Astrand and Ryhming (6) found, when working at a load that required 50% of maximal O_2 consumption, that the heart rate for a group of healthy male subjects averaged 128 after six minutes of work. The corresponding heart rate for female subjects was 138. When their subjects worked with a heavier load, thus demanding oxygen consumption of 70% of their aerobic capacity, the average heart rate was 154 for males and 164 for females. The standard deviation was eight or nine beats per minute.

Physical Fitness Testing   257

Figure 12.4 The adjusted nomogram for calculation of aerobic work capacity from submaximal pulse rate and O_2 uptake values (cycling, running or walking, and step test). In tests without direct O_2 uptake measurement, it can be estimated by reading horizontally from the body weight scale (step test) or work load scale (cycle test) to the O_2 uptake scale. The point on the O_2 uptake scale ($\dot{V}O_2$, 1) shall be connected with the corresponding point on the pulse rate scale, and the predicted maximal O_2 uptake should be read on the middle scale. A female subject (61 kg) reaches a heart rate of 156 at step test; predicted maximal $\dot{V}O_2$ = 2.41. A male subject reaches a heart rate of 166 at cycling test on a work load of 1,200 kg-m/min; predicted maximal $\dot{V}O_2$ = 3.61 (exemplified by dotted lines). (From Astrand, I. *Acta Physiol. Scand.,* 49 [suppl. 169], 1960.)

Astrand and Ryhming used these data to develop a nomogram (fig. 12.4) for predicting maximal O_2 consumption from heart rate for one six-minute submaximal work load. They found that the accuracy of prediction varied with the level of the work load selected. On the bicycle ergometer at 900 kg-m/min, the standard error of prediction for men was ± 10.4%, and at 1,200 kg-m/min it was ± 6.7%. The author found a correlation of 0.74 between predicted maximal O_2 consumption by the Astrand-Ryhming method and maximal O_2 consumption as measured in his laboratory. These data yielded an error of prediction of ± 9.3%, which agrees with their figures. Tables 12.3 and 12.4 provide the nomogram data for young men and young women in more easily used form.

The Astrand-Ryhming nomogram has proved a very usable method for small groups, and it requires about ten minutes per subject. Norms, which are given in table 12.5, have been provided (5). The test can be performed with no equipment other than a stopwatch (for taking heart rate), since the nomogram includes step-test data as well as data for bicycle ergometer use.

It should be pointed out that errors of prediction in this method are larger than the reported errors with unconditioned, sedentary groups, and also that ambient temperatures, which impose a heat stress upon individuals being tested, will obviously invalidate the procedure. For subjects over twenty-five years of age, an age correction factor must be applied (23).

One problem that arises in the use of this test is, What is a suitable test exercise load for any given subject? The errors of prediction are least when a working heart rate of 160–165 is achieved. To aid in selecting an appropriate exercise load, a nomogram has been developed by Terry and others (22) and is provided in figure 12.5. The subject is tested for one minute at 600 kg=m/min. The heart rate achieved and the body weight are then used to calculate the appropriate exercise load for the test.

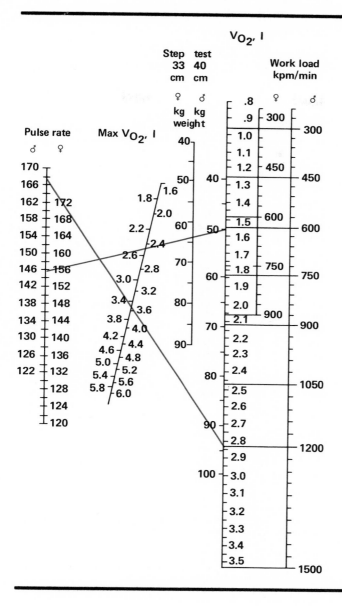

V_{O_2}, l

Step test
33   40
cm   cm

Work load
kpm/min

Pulse rate

Max V_{O_2}, l

Table 12.3
Calculation of Maximal Oxygen Uptake from Pulse Rate and Exercise Load on a Bicycle Ergometer (Men)

Working Pulse	Maximal Oxygen Uptake Liters/min					Working Pulse	Maximal Oxygen Uptake Liters/min				
	300 kpm/min	600 kpm/min	900 kpm/min	1200 kpm/min	1500 kpm/min		300 kpm/min	600 kpm/min	900 kpm/min	1200 kpm/min	1500 kpm/min
120	2.2	3.5	4.8			148		2.4	3.2	4.3	5.4
121	2.2	3.4	4.7			149		2.3	3.2	4.3	5.4
122	2.2	3.4	4.6			150		2.3	3.2	4.2	5.3
123	2.1	3.4	4.6			151		2.3	3.1	4.2	5.2
124	2.1	3.3	4.5	6.0		152		2.3	3.1	4.1	5.2
125	2.0	3.2	4.4	5.9		153		2.2	3.0	4.1	5.1
126	2.0	3.2	4.4	5.8		154		2.2	3.0	4.0	5.1
127	2.0	3.1	4.3	5.7		155		2.2	3.0	4.0	5.0
128	2.0	3.1	4.2	5.6		156		2.2	2.9	4.0	5.0
129	1.9	3.0	4.2	5.6		157		2.1	2.9	3.9	4.9
130	1.9	3.0	4.1	5.5		158		2.1	2.9	3.9	4.9
131	1.9	2.9	4.0	5.4		159		2.1	2.8	3.8	4.8
132	1.8	2.9	4.0	5.3		160		2.1	2.8	3.8	4.8
133	1.8	2.8	3.9	5.3		161		2.0	2.8	3.7	4.7
134	1.8	2.8	3.9	5.2		162		2.0	2.8	3.7	4.6
135	1.7	2.8	3.8	5.1		163		2.0	2.8	3.7	4.6
136	1.7	2.7	3.8	5.0		164		2.0	2.7	3.6	4.5
137	1.7	2.7	3.7	5.0		165		2.0	2.7	3.6	4.5
138	1.6	2.7	3.7	4.9		166		1.9	2.7	3.6	4.5
139	1.6	2.6	3.6	4.8		167		1.9	2.6	3.5	4.4
140	1.6	2.6	3.6	4.8	6.0	168		1.9	2.6	3.5	4.4
141		2.6	3.5	4.7	5.9	169		1.9	2.6	3.5	4.3
142		2.5	3.5	4.6	5.8	170		1.8	2.6	3.4	4.3
143		2.5	3.4	4.6	5.7						
144		2.5	3.4	4.5	5.7						
145		2.4	3.4	4.5	5.6						
146		2.4	3.3	4.4	5.6						
147		2.4	3.3	4.4	5.5						

Modified from I. Astrand's *Acta Physiol. Scand.* 49 (suppl. 169), 1960 by P-O. Astrand in *Work Test with the Bicycle Ergometer.* Varberg, Sweden: Monark, 1965.

Table 12.4

Calculation of Maximal Oxygen Uptake from Pulse Rate and Exercise Load on a Bicycle Ergometer (Women)

Working Pulse	Maximal Oxygen Uptake Liters/min					Working Pulse	Maximal Oxygen Uptake Liters/min				
	300 kpm/min	450 kpm/min	600 kpm/min	750 kpm/min	900 kpm/min		300 kpm/min	450 kpm/min	600 kpm/min	750 kpm/min	900 kpm/min
120	2.6	3.4	4.1	4.8		148	1.6	2.1	2.6	3.1	3.6
121	2.5	3.3	4.0	4.8		149		2.1	2.6	3.0	3.5
122	2.5	3.2	3.9	4.7		150		2.0	2.5	3.0	3.5
123	2.4	3.1	3.9	4.6		151		2.0	2.5	3.0	3.4
124	2.4	3.1	3.8	4.5		152		2.0	2.5	2.9	3.4
125	2.3	3.0	3.7	4.4		153		2.0	2.4	2.9	3.3
126	2.3	3.0	3.6	4.3		154		2.0	2.4	2.8	3.3
127	2.2	2.9	3.5	4.2		155		1.9	2.4	2.8	3.2
128	2.2	2.8	3.5	4.2	4.8	156		1.9	2.3	2.8	3.2
129	2.2	2.8	3.4	4.1	4.8	157		1.9	2.3	2.7	3.2
130	2.1	2.7	3.4	4.0	4.7	158		1.8	2.3	2.7	3.1
131	2.1	2.7	3.4	4.0	4.6	159		1.8	2.2	2.7	3.1
132	2.0	2.7	3.3	3.9	4.5	160		1.8	2.2	2.6	3.0
133	2.0	2.6	3.2	3.8	4.4	161		1.8	2.2	2.6	3.0
134	2.0	2.6	3.2	3.8	4.4	162		1.8	2.2	2.6	3.0
135	2.0	2.6	3.1	3.7	4.3	163		1.7	2.2	2.6	2.9
136	1.9	2.5	3.1	3.6	4.2	164		1.7	2.1	2.5	2.9
137	1.9	2.5	3.0	3.6	4.2	165		1.7	2.1	2.5	2.9
138	1.8	2.4	3.0	3.5	4.1	166		1.7	2.1	2.5	2.8
139	1.8	2.4	2.9	3.5	4.0	167		1.6	2.1	2.4	2.8
140	1.8	2.4	2.8	3.4	4.0	168		1.6	2.0	2.4	2.8
141	1.8	2.3	2.8	3.4	3.9	169		1.6	2.0	2.4	2.8
142	1.7	2.3	2.8	3.3	3.9	170		1.6	2.0	2.4	2.7
143	1.7	2.2	2.7	3.3	3.8						
144	1.7	2.2	2.7	3.2	3.8						
145	1.6	2.2	2.7	3.2	3.7						
146	1.6	2.2	2.6	3.2	3.7						
147	1.6	2.1	2.6	3.1	3.6						

Modified from I. Astrand's *Acta Physiol. Scand.* 49 (suppl. 169), 1960 by P-O. Astrand in *Work Test with the Bicycle Ergometer.* Varberg, Sweden: Monark, 1965.

Table 12.5
Norms for Maximal O_2 Consumption (Aerobic Working Capacity)

Women

Age	Low	Fair	Average	Good	High
20-29	1.69	1.70-1.99	2.00-2.49	2.50-2.79	2.80+
	28	29-34	35-43	44-48	49+
30-39	1.59	1.60-1.89	1.90-2.39	2.40-2.69	2.70+
	27	28-33	34-41	42-47	48+
40-49	1.49	1.50-1.79	1.80-2.29	2.30-2.59	2.60+
	25	26-31	32-40	41-45	46+
50-65	1.29	1.30-1.59	1.60-2.09	2.10-2.39	2.40+
	21	22-28	29-36	37-41	42+

Men

Age	Low	Fair	Average	Good	High
20-29	2.79	2.80-3.09	3.10-3.69	3.70-3.99	4.00+
	38	39-43	44-51	52-56	57+
30-39	2.49	2.50-2.79	2.80-3.39	3.40-3.69	3.70+
	34	35-39	40-47	48-51	52+
40-49	2.19	2.20-2.49	2.50-3.09	3.10-3.39	3.40+
	30	31-35	36-43	44-47	48+
50-59	1.89	1.90-2.19	2.20-2.79	2.80-3.09	3.10+
	25	26-31	32-39	40-43	44+
60-69	1.59	1.60-1.89	1.90-2.49	2.50-2.79	2.80+
	21	22-26	27-35	36-39	40+

Lower figure = milliliters of O_2 per kilogram body weight.
From I. Astrand, *Acta Physiol. Scand.* 49 (suppl. 169), 1960.

Zuti and Corbin (25) have used the Astrand test on over 3,000 freshman men and women at the University of Kansas, and their data provide norms that better reflect the fitness of American student populations than Astrand's norms do (see table 12.6). Comparison of the Zuti-Corbin norms with those of Astrand in table 12.5 suggests that Scandinavian students are somewhat more fit, at least for bicycle riding.

Harvard Step Test. The two tests we have described utilize heart rate *during* exercise as a criterion of PWC. Exercise on a bicycle ergometer is best suited for this approach since the subject's upper body is relatively stationary. To eliminate the need for a bicycle ergometer, tests have been devised that use bench-stepping as exercise and measure heart rate *after* exercise (during recovery). The principle is that the better the PWC of the individual, the greater the proportion of the cardiac cost that is paid during exercise, the smaller the recovery cardiac cost, and the lower the rate during recovery.

Physiology Applied to Health and Fitness

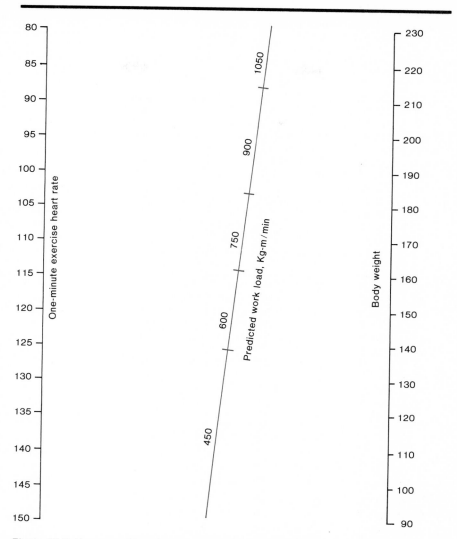

Figure 12.5 Nomogram for workload selection. (From Terry, J.W.; Tolson, H.; Johnson, D.J.; and Jessup, G.T. *J. Sports Med. Phys. Fitness* 12:361–66, 1977.)

Table 12.6
Physical Fitness Norms for Females and Males

Female

Percentiles	Height		Weight		Right Grip		Left Grip	
	cm	in.	kg	lb	kg	lb	kg	lb
100	188.0	74.0	92.3	203.0	58.2	128.0	53.6	118.0
90	173.2	68.2	69.1	152.0	34.8	76.6	31.8	70.0
80	170.2	67.0	65.0	143.0	31.4	69.0	28.9	63.6
70	167.9	66.1	62.0	136.5	29.7	65.3	26.7	58.7
60	166.4	65.5	59.7	131.3	28.0	61.7	25.0	55.0
50	164.6	64.8	58.0	127.5	26.5	58.2	24.0	52.9
40	162.6	64.0	56.2	123.7	25.0	54.9	22.4	49.2
30	160.8	63.3	54.2	119.3	23.9	52.5	28.8	45.8
20	158.8	62.5	52.1	114.7	21.4	47.0	19.9	43.8
10	156.0	61.4	49.5	109.0	19.9	43.8	17.3	38.0
0	132.0	52.0	37.7	83.0	10.0	22.0	8.6	19.0

Male

Percentiles	Height		Weight		Right Grip		Left Grip	
100	200.7	79.0	110.9	244.0	74.1	163.0	70.0	154.0
90	186.4	73.4	84.5	186.0	60.0	132.0	56.4	124.0
80	183.1	72.1	79.5	175.0	56.5	124.3	52.7	116.0
70	181.4	71.4	76.9	169.2	53.5	117.8	50.0	110.0
60	179.3	70.6	74.1	163.0	51.5	113.2	48.0	105.6
50	177.5	69.9	71.7	157.7	49.6	109.2	45.6	100.4
40	176.0	69.3	69.5	153.0	46.8	102.9	43.6	96.0
30	174.2	68.6	67.7	149.0	44.5	98.0	41.1	90.4
20	172.0	67.7	65.0	143.0	41.4	91.0	39.1	86.0
10	168.4	66.3	61.4	135.0	36.8	81.0	34.8	76.6
0	147.3	58.0	51.4	113.0	21.4	47.0	19.1	42.0

From W.B. Zuti and C.B. Corbin, *Res. Q.* 48:499–503, 1977.

The Harvard step test is probably the most widely used test, and it was devised for use with large groups. It is simple and easily administered; however, its error of prediction of maximal oxygen consumption in the author's laboratory was ± 12.5%. The test requires only a stepping bench or benches (twenty inches high and eighteen inches deep) adequate for the number of subjects (who are to step simultaneously), a stopwatch for each observer, and a metronome.

Subjects are lined up in front of the stepping bench (thirty inches of width are allowed for each), and there is one observer for each subject. The person in charge counts cadence to a metronome set at 120 counts per minute: "up—two—three—four," etc. On *up*, subjects place one foot on the bench; on *two*, they bring the other foot up and *straighten their back and legs;* on *three*, they step down with the foot that was placed on the bench first; and on *four*, they return to the starting position. Thus subjects complete one step every two

Leg Strength		Back Strength		Trunk Flexibility		EVO$_2$		% Fat	Triceps S.F.
kg	lb	kg	lb	cm	in.	l/min	ml/kg/min		mm
209	460	200	440	63.5	25.0	3.80	65.0	11.0	5.0
118	260	114	250	56.4	22.2	2.90	49.5	18.3	11.0
110	243	101	223	53.1	20.9	2.60	45.0	19.8	12.7
100	220	91	200	50.5	19.9	2.41	41.8	20.9	14.3
91	200	86	190	48.3	19.0	2.35	40.3	22.0	15.7
90	198	80	175	46.5	18.3	2.16	37.2	23.2	17.1
79	173	73	160	44.4	17.5	2.05	35.0	24.1	18.9
73	160	67	148	41.4	16.2	1.93	32.9	25.3	20.5
66	146	59	130	37.6	14.8	1.70	30.5	26.8	22.4
55	121	50	109	30.5	12.0	1.60	27.7	29.3	25.0
36	80	36	80	10.2	4.0	1.10	18.0	40.0	35.0
277	610	295	650	61.0	24.0	4.80	65.0	3.0	3.0
215	473	214	470	54.9	21.6	3.60	51.5	6.5	5.8
200	440	195	429	51.8	20.4	3.33	46.7	7.7	7.0
182	400	181	399	49.8	19.6	3.13	43.6	8.6	8.1
178	391	170	375	47.5	18.7	2.95	41.0	9.6	9.1
168	370	159	350	45.2	17.8	2.79	38.8	10.8	10.3
156	343	150	330	43.4	17.1	2.65	36.7	12.0	11.5
146	322	136	300	40.1	15.8	2.45	34.4	13.5	12.9
132	291	126	277	36.3	14.3	2.33	32.3	15.3	15.2
110	243	107	235	29.7	11.7	2.13	29.3	19.0	18.5
50	110	52	115	15.2	6.0	1.50	21.0	35.0	33.0

seconds, or thirty steps per minute. Subjects lead off with the same foot each time, although one or two changes may be made in the course of the five-minute stepping period.

If subjects fall behind the cadence because of exhaustion, they are stopped twenty seconds after falling behind the pace. When they stop (either at completion at five minutes or due to exhaustion), they sit down quietly, and observers restart their stopwatches, having recorded the duration of the stepping. Observers then take the pulse rate at the carotid artery in the neck from sixty to ninety seconds after exercise. On the basis of the duration and the recovery pulse rate, the score (in arbitrary units) is taken from table 12.7. Interpretation of the score is as follows: below 50, poor; 50 to 80, average; above 80, good.

Table 12.7
Scoring for the Harvard Step Test

Duration of Effort (Minutes)	Total Heart Beats 1 to 1½ Minutes in Recovery											
	40-44	45-49	50-54	55-59	60-64	65-69	70-74	75-79	80-84	85-89	90-94	95-99
	Score (Arbitrary Units)											
0-½	6	6	5	5	4	4	4	4	3	3	3	3
½-1	19	17	16	14	13	12	11	11	10	9	9	8
1-1½	32	29	26	24	22	20	19	18	17	16	15	14
1½-2	45	41	38	34	31	29	27	25	23	22	21	20
2-2½	58	52	47	43	40	36	34	32	30	28	27	25
2½-3	71	64	58	53	48	45	42	39	37	34	33	31
3-3½	84	75	68	62	57	53	49	46	43	41	39	37
3½-4	97	87	79	72	66	61	57	53	50	47	45	42
4-4½	110	98	89	82	75	70	65	61	57	54	51	48
4½-5	123	110	100	91	84	77	72	68	63	60	57	54
5	129	116	105	96	88	82	76	71	67	63	60	56

From C.F. Conzolazio, et al. *Physiological Measurements of Metabolic Function in Man*. Copyright © 1963. McGraw-Hill Book Company. Used by permission.

Progressive Pulse Rate Test. A test devised by Cureton (9) has been used in the author's laboratory for several years. In this test the stepping bench is seventeen inches high, and the stepping is done as in the Harvard step test, except that there are five one-minute bouts with increasing rates: 12, 18, 24, 30, and 36 steps per minute. After each one-minute stepping bout (and within ten seconds), the recovery heart rate is taken for two minutes. Subjects then rest until their pulse stabilizes within eight to twelve beats of their standing, normal rate before they start the next higher load. The pulse rates are plotted on a chart, as in tables 12.8, 12.9, and 12.10.

This test was found to correlate with measured maximal O_2 consumption, $r = 0.71$, and the standard error of prediction was $\pm 13.7\%(10)$. The progressive pulse rate test has several advantages over the Harvard step test:

1. It does not stress the subject as severely, and consequently it results in less muscle soreness.
2. It is more easily motivated because of its method of scoring.
3. Most important, it starts with low work loads, and it can be terminated at any work load under which performance drops sharply.

Cureton (9) has furnished the norms for men twenty-six to sixty, which are provided in table 12.8. The author has developed norms for college age men and women, which are provided in tables 12.9 and 12.10. The author has also developed norms for the older men and women (over sixty) in the course of his work in gerontology (11). Norms for boys and girls of elementary and secondary school age would be highly desirable since the test offers significant advantages.

One source of error in both of these step tests, and indeed in all bench-stepping procedures, is that $\dot{V}O_2$ is significantly related to limb length in this exercise modality. It was found that best muscle efficiency is achieved when bench height is near 50% of the subject's leg length (19).

Cooper Twelve-Minute Run-Walk Test. Cooper (8) developed a twelve-minute modification of the original Balke fifteen-minute run-walk field test, which he validated on 115 young air force men (mean age, twenty-two). He found that his test results on the distance covered in twelve minutes correlated .897 with measured maximal O_2 consumption. To achieve good results with this test, motivation must be high; as Cooper pointed out, "this study indicates that in young, well motivated subjects, field testing can provide a good assessment of maximum O_2 consumption; but the accuracy of the estimate is related directly to the motivation of the subjects."

Doolittle and Bigbee (12) used the test on 153 ninth grade boys and found the validity to be equally good for them ($r = 0.90$), and the test-retest data correlated $r = 0.94$. They also found the twelve-minute run-walk to be a more valid test for this age group than a 600-yard run-walk test ($r =. 0.62$).

Maksud and Coutts (15) found the test equally reproducible with boys eleven to fourteen, but the validity correlation with maximum O_2 consumption was lower ($r = 0.65$).

Shaver measured both $\dot{V}O_2$ max and performance at various distances in running for thirty untrained college men. The results show clearly that distances over one-half mile are required to produce significant correlations with VO_2 max, while distances below that are better related to anaerobic work capacity (20). See table 12.11.

Table 12.8
Scoring the Progressive Pulse Rate Test • Adult Men (26-60 Years) • Rating
Scale for Progressive Pulse Rates

Classification	Total 2-Min Pulse Count after 12 Steps/Min	Total 2-Min Pulse Count after 18 Steps/Min	Total 2-Min Pulse Count after 24 Steps/Min
Excellent	71 80 89 97	77 86 95 105	84 94 104 114
Very good	106 115 123	114 123 132	124 134 144
Above average	132 141 149	142 151 160	154 164 175
Average	158 167 175	169 178 188	185 195 205
Below average	184 193 201	197 206 216	215 225 235
Poor	210 219 227	225 234 243	245 255 265
Very poor	236 245	252 262	276 286
Mean	157.9	169.3	184.6
Sigma	28.9	30.8	33.6
Number	115	114	113
Range	100–226	100–238	120–306

From Cureton, T. K. "The Nature of Cardiovascular Condition in Normal
Humans." J. Assoc. Phys. Ment. Rehabil. 12:41–49, 1958.

Total 2-Min Pulse Count after 30 Steps/Min	Total 2-Min Pulse Count after 36 Steps/Min	Standard Score	Percentile
98	105	100	99.9
109	118	95	99.7
119	130	90	99.2
130	143	85	98.2
141	156	80	96.7
152	168	75	93.3
162	181	70	88.4
173	193	65	81.6
184	209	60	72.6
195	218	55	61.8
206	231	50	50.0
217	244	45	38.2
227	256	40	27.4
238	269	35	18.4
249	281	30	11.5
260	294	25	6.7
271	306	20	3.6
281	319	15	1.8
292	332	10	.82
303	344	5	.35
314	357	0	.14
205.8	231.0		
36.03	41.9		
112	96		
120–309	120–370		

Table 12.9
Scoring Table for Progressive Pulse Rate Test (College Age Men)

Classification	Total 2-Min Pulse Count after 12 Steps/Min	Total 2-Min Pulse Count after 18 Steps/Min	Total 2-Min Pulse Count after 24 Steps/Min	Total 2-Min Pulse Count after 30 Steps/Min	Total 2-Min Pulse Count after 36 Steps/Min	Percentile
	98	113	125	130	148	99
	104	119	131	137	155	
Excellent	110	125	137	144	162	
	116	131	143	151	169	95
	122	137	149	158	176	
	128	143	155	165	183	
	134	149	161	172	190	90
Very Good	140	155	167	179	197	
	146	161	173	186	204	80
	152	167	179	193	211	70
	158	173	185	200	218	60
	164	179	191	207	225	
Average	170	185	197	214	232	50
	176	191	203	221	239	
	182	197	209	228	246	40
	188	203	215	235	253	30
	194	209	223	242	260	20
Below	200	215	229	249	267	
Average	206	221	235	256	274	10
	212	227	241	263	281	
	218	233	247	270	288	
	224	239	253	277	295	5
Poor	230	245	259	284	302	
	236	251	269	291	309	
	242	257	271	298	316	1
Mean	170.4	184.9	196.8	213.6	232.0	
Sigma	31.2	30.3	31.4	34.4	34.4	
Number	135	135	136	139	132	
Range	99 — 237	120 — 253	125 — 260	132 — 311	147 — 320	

H.A. deVries, unpublished data.

Table 12.10
Scoring Table for Progressive Pulse Rate Test (College Age Women)

Classification	Total 2-Min Pulse Count after 12 Steps/Min	Total 2-Min Pulse Count after 18 Steps/Min	Total 2-Min Pulse Count after 24 Steps/Min	Total 2-Min Pulse Count after 30 Steps/Min	Total 2-Min Pulse Count after 36 Steps/Min
Excellent	118	133	145	150	168
	124	139	151	157	175
	130	145	157	164	182
	136	151	163	171	189
	142	157	169	178	196
Very Good	148	163	175	185	203
	154	169	181	192	210
	160	175	187	199	217
	166	181	193	206	224
	172	187	199	213	231
Average	178	193	205	220	238
	184	199	211	227	245
	190	205	217	234	252
	196	211	223	241	259
	202	217	229	248	266
Below Average	208	223	235	255	273
	214	229	243	262	280
	220	235	249	269	287
	226	241	255	276	294
	232	247	261	283	301
Poor	238	253	267	290	308
	244	259	273	297	315
	250	265	279	304	322
	256	271	289	311	329
	262	277	291	318	336

H.A. deVries, unpublished data.

Table 12.11
Correlations between Various Running Performances and $\dot{V}O_2$ Max.

Running Distance	VO_2 Max. (r)	Anaerobic Power (r)
100 yards	−.08	−.85
220 yards	−.25	−.82
440 yards	−.29	−.79
880 yards	−.35	−.32
1 mile	−.43	−.28
2 miles	−.76	−.15
3 miles	−.82	−.05

Compiled from the data of L.G. Shaver, *J. Sports Med. Phys. Fitness* 15:147–50, 1975.

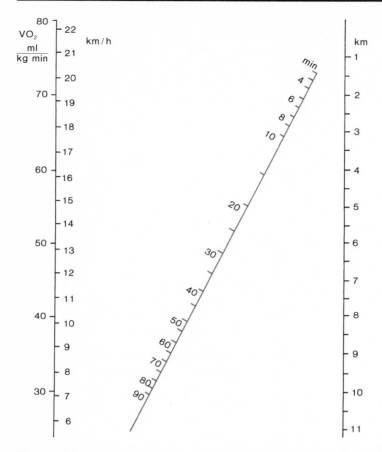

Figure 12.6 Nomogram to relate the maximal aerobic power of the subject $\dot{V}O_2$ max with the minimal time necessary in minutes, when running at maximal speed to cover the distance in kilometers. On the $\dot{V}O_2$ max line the corresponding maintenance speed of running as due to aerobic energy is also indicated. (From Margaria, R.; Aghemo, P.; and Limas, F.P. *J. Appl. Physiol.* 38:351–52, 1975.)

Since the net metabolic cost of running when referred to body weight and distance covered is essentially a constant, independent of speed, Margaria, Aghemo, and Limas (16) have provided a nomogram from which one can estimate $\dot{V}O_2$ max from best run time over distances long enough to allow reaching steady state (see fig. 12.6). For example, it can be seen from the nomogram that a subject who can run 6 km in approximately 21 minutes would have to have an aerobic capacity ($\dot{V}O_2$ max) of 60 ml/kg/min. This nomogram can be quite useful because it applies not only to athletes, but to the whole population, including women, the elderly, and children, since the

272   *Physiology Applied to Health and Fitness*

energy cost of running when referred to distance covered and to the body weight is essentially the same for all reasonably fit people. It cannot, however, be applied to those who are not fit enough to run continuously for at least 2 to 3 km, because the oxygen costs of walking and running are not the same.

Motor Fitness Tests

Many excellent tests of motor fitness have been devised. The elements of motor fitness are so many, however, and some are so difficult to define, that each motor performance test battery must be considered a compromise between the ideal of measuring all identifiable elements and the practical need to choose a number of representative elements, which allows measurement in reasonable amounts of time. Obviously, testing programs that intrude unnecessarily on instructional time cannot be tolerated.

For this reason, only examples of some of the best compromises will be offered here. For illustrative purposes, an example of the approach used by the armed forces during World War II and one example of test batteries for school age children will be offered.

Army Air Force Physical Fitness Test. The items for this test were selected to evaluate the motor fitness elements of muscle strength and endurance, cardiorespiratory endurance, speed, coordination, and power. The test items selected were sit-ups, pull-ups, and a 300-yard shuttle run (five lengths of 60 yards each). It was the author's experience during World War II that several hundred men could be tested per hour by four experienced physical training instructors. This test is one of the best for handling very large groups of adult men.

AAHPER Youth Fitness Test Battery. After the much publicized results of the Kraus-Weber test (13) and other research findings pointed out the need for a national concern about fitness, a committee of members of the national research council of the American Association for Health, Physical Education and Recreation was set up under the direction of Paul A. Hunsicker. As a result of this group's work on the Youth Fitness Project, the Youth Fitness Test was developed, designed for boys and girls from the fifth through the twelfth grade. National norms are available for these age groups (and also for college men and women and young adults from eighteen to thirty (2). This test battery consists of the following items (all for grades five through twelve).

1. Pull-up for boys; modified pull-up for girls
2. Sit-up for boys and girls
3. Shuttle run for boys and girls
4. Standing broad jump for boys and girls
5. Fifty-yard dash for boys and girls
6. Softball throw for distance for boys and girls
7. 600-yard run and walk for boys and girls

Additional tests in aquatics are recommended where facilities permit.

Interestingly, Olree and associates (17) showed that three items correlated 0.925 with the results of the entire test. Thus a great saving in time can be effected by using only pull-ups, sit-ups, and the fifty-yard run, with a loss in intelligence of less than 15%.

Physical Fitness Evaluation as a Function of Age Groups

It is obvious that the various criteria of physical fitness do not have equal importance for all age groups, and several criteria that seem likely to be very important for middle-aged and elderly groups have not yet been mentioned. Because very little experimental work has been done to identify the most important elements of physical fitness for the older age groups, much of what follows is based upon the author's survey of medical and physical education opinion.

There seems little doubt that the motor fitness elements discussed above are important in elementary and secondary school age children. PWC is at least as important for secondary school age groups as it ·is for elementary groups, and possibly somewhat less important for those on the elementary level, since elementary school age children participate in vigorous physical activity by nature. It would, however, be difficult to justify many of the elements of motor fitness as necessary or essential for middle-aged and elderly populations. For example, it is unlikely that a matron or a businessman needs high levels of speed, strength, or agility.

For the older age brackets a much better case can be made for the importance of (1) normal body weight, (2) cardiovascular fitness, (3) respiratory fitness, (4) neuromuscular relaxation, and (5) flexibility. The value of the first four elements is probably self-evident; the last, flexibility, becomes more important as aging proceeds because connective tissues tend to lose their elasticity with age, and this in turn seems to be related to many of the aches and pains of old age. Considerable evidence exists that maintenance of range of motion exerts a beneficial effect in this regard. Table 12.12 lists the factors of greatest importance in physical fitness by age groups.

Table 12.12
Suggested Values of the Components of Physical Fitness by Age Groups (in Order of Importance)

Prepuberty	Adolescence	Young Adult	Older Adult
Motor fitness	Motor fitness	PWC	PWC
PWC	PWC	Body weight	Body weight
	Body weight	Relaxation	Flexibility
		Flexibility	Relaxation

SUMMARY

1. Physical fitness can be conceptualized in two ways: (a) the *motor fitness* concept in which the elements of performance are measured, and (b) the *physical working capacity* (PWC) concept in which the capacity for O_2 transport is evaluated.

2. The following elements constitute motor fitness: strength, speed, agility, endurance, power, coordination, balance, flexibility, and body control.

3. PWC is determined by the following physiological components: cardio-vascular function, respiratory function, muscular efficiency, strength, muscular endurance, and maintenance of proper body weight.

4. A screening medical examination is necessary prior to testing $\dot{V}O_2$ max for (a) any individual over thirty-five, (b) any individual who has questions about personal health status, or develops symptoms during testing, or (c) has not had a medical exam in two years.

5. All subjects for $\dot{V}O_2$ max testing must have the testing procedures carefully explained to them, and after all questions are answered, they must sign an informed consent form.

6. Physical education personnel, trained as exercise technicians, may conduct maximal tests on healthy school age subjects and adults under thirty-five who have no known primary CHD risk factors.

7. $\dot{V}O_2$ max testing can be accomplished on a bicycle ergometer, a treadmill, or a step bench. Using any of these three exercise modalities, the protocol may be: (a) intermittent incremental loading, (b) continuous step-incremental loading, or (c) continuously incremented loading (ramp loading).

8. Environmental conditions, such as time of day, diet, and prior activity, must all be carefully controlled if either intraindividual or interindividual comparisons are to be made.

9. The primary physiological parameters to be measured and recorded are (a) percent of O_2 in expired gas; (b) percent of CO_2 in expired gas; (c) minute ventilation; (d) heart rate; (e) blood pressure; and (f) electrical activity of the heart as shown on the electrocardiogram.

10. From the primary physiological parameters one can calculate (a) respiratory exchange ratio (R); (b) oxygen pulse; (c) heart rate-blood pressure product; (d) ventilation equivalent for O_2; (e) the anaerobic threshold (AT).

11. Evaluation of PWC is best accomplished by *measuring* maximum O_2 consumption. PWC can also be *estimated* from submaximal tests if errors of measurement ranging from 10% to 15% are acceptable as a trade-off for the savings of time and effort by both subject and investigator.

12. The relative importance of the elements of physical fitness changes with increasing age. Motor fitness, which is important to children, is no longer of great importance to adults. For middle-aged and older adults the maintenance of high levels of PWC, appropriate body weight, good flexibility, and a relaxed musculature are much more important, since these factors contribute to good health.

REFERENCES

1. Adams, F.H.; Bengtsson, E.; Berven, H.; and Wegelius, C. The physical working capacity of normal school children, *Pediatrics* 28:243–57, 1961.
2. American Association for Health, Physical Education and Recreation. *Youth Fitness Test Manual.* Washington, D.C.: The Association, 1958.
3. American College of Sports Medicine. *Guidelines for Graded Exercise Testing and Exercise Prescription.* Philadelphia: Lea & Febiger, 1975.
4. Andersen, K.L.; Shephard, R.F.; Denolin, H.; Varnauskas, E.; and Masironi, R. Fundamentals of exercise testing. Geneva: World Health Organization, 1971.
5. Astrand, I. Aerobic work capacity in men and women with special reference to age. *Acta Physiol. Scand.* 49, suppl. 169, 1960.
6. Astrand, P.O., and Ryhming, I. A nomogram for calculation of aerobic capacity (physical fitness) from pulse rate during submaximal work. *J. Appl. Physiol.* 7:218–21, 1954.
7. Cherchi, A. A synthetic triangular exercise test. In *Ergometry in Cardiology,* eds. H. Denolin, K. Konig, R. Messin, and S. Degré, pp. 65–86. Mannheim, Germany: Boehringer Mannheim, 1968.
8. Cooper, K.H. A means of assessing maximal O_2 intake. *J. Am. Med. Assoc.* 203:201–4, 1968.
9. Cureton, T.K. The nature of cardiovascular condition in normal humans (part 3). *J. Assoc. Phys. Ment. Rehabil.* 12:41–49, 1958.
10. deVries, H.A., and Klafs, C.E. Prediction of maximal O_2 intake from submaximal tests. *J. Sports Med. Phys. Fitness* 5:207–14, 1965.
11. deVries, H.A. *Vigor Regained.* Englewood Cliffs, N.J.: Prentice-Hall, 1974.
12. Doolittle, T.L., and Bigbee, R. The twelve-minute run-walk: a test of cardiorespiratory fitness of adolescent boys. *Res. Q.* 39:491–95, 1968.
13. Kraus, H., and Hirschland, R.P. Minimum muscular fitness in schoolchildren. *Res. Q.* 25:178–88, 1954.
14. Luft, U.C.; Cardus, D.; Lim, T.P.K.; Howarth, J.L.; and Anderson, E.C. Physical performance in relation to body size and composition. *Ann. N.Y. Acad. Sci.* 110:795–808, 1963.
15. Maksud, M.G., and Coutts, K.D. Application of the Cooper twelve-minute run-walk test to young males. *Res. Q.* 42:54–59, 1971.
16. Margaria, R.; Aghemo, P.; and Limas, F.P. A simple relation between performance in running and maximal aerobic power. *J. Appl. Physiol.* 38:351–52, 1975.
17. Olree, H.; Stevens, C.; Nelson, T.; Agnerik, G.; and Clark, R.T. Evaluation of the AAHPER youth fitness test. *J. Sports Med. Phys. Fitness* 5:67–71, 1965.

18. Purvis, J.W., and Morgan, J.P. Influence of repeated maximal testing on anxiety and work capacity in college women. *Res. Q.* 49:512–19, 1978.

19. Shahnawaz, H. Influence of limb length on a stepping exercise. *J. Appl. Physiol.* 44:346–49, 1978.

20. Shaver, L.G. Maximal aerobic power and anaerobic work capacity prediction from various running performances of untrained college men. *J. Sports Med. Phys. Fitness* 15:147–50, 1975.

21. Sjostrand, T. Changes in the respiratory organs of workmen at an ore-smelting works. *Acta Med. Scand.,* suppl., 196:687–99, 1947.

22. Terry, J.W.; Tolson, H.; Johnson, D.J.; and Jessup, G.T. A workload selection procedure for the Astrand-Ryhming test. *J. Sports Med. Phys. Fitness* 17:361–66, 1977.

23. Von Dobeln, W.; Astrand, I.; and Bergstrom, A. An analysis of age and other factors related to maximal oxygen uptake. *J. Appl. Physiol.* 22:934–38, 1967.

24. Wahlund, H. Determination of the physical working capacity. *Acta Med. Scand.,* suppl. 215, 1948.

25. Zuti, W.B., and Corbin, C.B. Physical fitness norms for college freshmen. *Res. Q.* 48:499–503, 1977.

13

Physical Conditioning for Health and Fitness (Prescription of Exercise)

In this chapter we will deal with how to use exercise to improve health and fitness. In the light of the evidence cited in chapter 11 and that to come in chapter 14, it seems entirely likely that the physical education profession will be called upon to expand its interests beyond that of teaching skills and games to children and adolescents to include service to the community at large. As advisors in the use of exercise to improve health and fitness, we can serve not only schoolchildren, but also young, middle-aged, and elderly adults, and females as well as males.

In a letter from the President's Council on Physical Fitness and Sports (PCPFS) to physical fitness leaders in the U.S. (7), the following health and fitness problems were pointed out:

1. Fifty million of the 110 million adults in the U.S. never engage in physical activity for exercise.
2. Old, poor, and less educated Americans frequently do not understand the contributions that exercise can make to health, performance, and the quality of life.
3. Millions of Americans are willing victims of "get-fit-quick" schemes that promise fitness in a few minutes a day (thirty minutes a week) without sweat or strain.
4. The nation has a debit of well over 100,000 tons of fat.
5. The American Medical Association estimates that one-third of all American children are overweight.
6. The performances of the youth on the National Physical fitness tests have not improved since 1965, and in the opinion of the Council, the fitness levels at that time were very low!

This information from the PCPFS shows the great need for greater motivational efforts and better instruction on the part of the physical education profession in helping people apply the basics of exercise physiology to the improvement of health and fitness.

While we do not yet understand all there is to know about the prescription of exercise, we do have considerable information based on laboratory studies that can help us, as the professionals in this field, guide the layman to better use of exercise as a replacement for the physical work that has been removed by our modern life style. That at least some of the lay public recognizes the need for activity is borne out by the tremendous growth in popularity of jogging and such sports as tennis and racquetball.

Unfortunately, it is still common for a physician to prescribe to a patient, "You must get some exercise." This, of course, is roughly analogous to prescribing, "You must get some drugs," without specifying which pharmaceutical, and how much and how often to take it. Just as the medicine prescribed for a headache is quite different from that for diabetes, the exercise prescribed for developing maximal strength and muscle bulk is quite different from that needed for optimal cardiorespiratory endurance. Thus there is exercise, and there is exercise, just as there is medicine, and medicine. With exercise, there are also the questions, How much is enough? How much is too much? How much is optimal? How often? For how long?

Principles Involved in Scientific Prescription of Exercise

The following aspects of a proposed exercise program must be considered and where possible defined on the basis of scientifically derived data.

1. *Objective of the exercise program.* There is good evidence that we can bring about very desirable adaptations in human functional capacities and health-related parameters such as (a) muscular strength, (b) muscular endurance, (c) cardiorespiratory endurance, (d) muscular efficiency, (e) speed of movement, and (f) flexibility. However, each of these goals would require a different and specific exercise program.

2. *Exercise modality.* For strength gain one would prescribe progressive resistance exercise (PRE), whereas for enhancement of cardiorespiratory function one would prescribe one of many endurance type exercise programs such as jogging.

3. *Exercise intensity.* Here we are concerned with the dose-response relationship, or the level of exercise work load (power output) to the amount of adaptation brought about in the human organism.

4. *Exercise duration.* How long must the exercise be continued to bring about the desired result? Is more always better, or is there a practical limit or desirable amount that optimizes the gain for time spent?

5. *Exercise frequency.* How many times a week should one work out for best training effect?

6. *Intensity threshold for training effect.* Is there some minimal value of exercise training intensity below which no training adaptations occur?

7. *Rate of training adaptation as a function of pretraining fitness level.* Do all individuals progress at the same rate in a conditioning program, or is progress dictated by the level of fitness at entry into the program?

Our discussion in this chapter will be limited to the development of cardiorespiratory fitness since the other elements of human performance are treated in great detail in part 3 of this text, which is devoted to the training and conditioning of athletes. Furthermore we may consider cardiorespiratory fitness of threefold importance to health and fitness because the type of exercise program (endurance exercise) used for its development also contributes significantly to weight control and to relief from neuromuscular tension (relaxation). Thus we achieve three important health benefits for the price of one workout. This is not to deny the importance of muscular strength, muscular endurance, or flexibility in the overall health and fitness picture. It is merely to stress the overriding importance of cardiorespiratory fitness to our lifelong good health.

Need for Medical Evaluation prior to Participation in Endurance Exercise

We now have available to us guidelines developed by the American College of Sports Medicine (ACSM) (1) for admitting adults into exercise programs. These are based on arbitrary, but eminently sensible, separations with respect to age, symptomatology, and risk factors.

Asymptomatic Individuals under Age Thirty-five. For individuals under thirty-five with no previous history of cardiovascular disease or known primary coronary heart disease (CHD) risk factors, the risk of an increase in habitual physical activity is thought to be sufficiently low for them to proceed without special medical clearance. However, if there is any question about health status, or if they develop symptoms during pretesting, or if they have not had a medical examination in two years, then they should be referred to their personal physician before participation.

High Risk or Symptomatic Individuals and Those Thirty-five and Older. The ACSM recommends medical evaluation prior to entry into a conditioning program for all adults over thirty-five, regardless of health status. For those under thirty-five with a history or any evidence of CHD or with significant combinations of CHD risk factors such as family history, elevated blood pressure, hyperlipidemia, diabetes, cigarette smoking, or obesity, a medical evaluation is also recommended (1). Table 13.1 shows the medical findings that would contraindicate involvement in a conditioning program.

Physiological Pretest and Monitoring of Progress The type of screening test used at entry into a conditioning program will vary in level of sophistication and test parameters required with the type of population to be involved in the program. For healthy school age groups with which there is virtually no concern regarding CHD risk factors and in which large numbers must be tested in a short time, the step tests described in chapter 12 will serve. On the other hand, in adult conditioning programs, time must be taken to provide better physiological data, including $\dot{V}O_2$ max and as many of the derived parameters as the sophistication of the laboratory allows. This is necessary not only for safety reasons, but also to enhance the level of motivation of the participants. For the participants in physical fitness programs, nothing is more gratifying than the feedback from repeated testing which shows them the rewards for their efforts in terms of objectively derived health-related measurements. The discussion of the test results with the participants both individually and in groups offers excellent opportunities for education of lay people to the physiology of exercise with special emphasis on the health benefits derived from well-conceived scientific exercise programs.

Table 13.1

Contraindications to Exercise and Exercise Testing

I. Absolute contraindications
    1. Manifest circulatory insufficiency ("congestive heart failure")
    2. Acute myocardial infarction
    3. Active myocarditis
    4. Rapidly increasing angina pectoris with effort
    5. Recent embolism, either systemic or pulmonary
    6. Dissecting aneurysm
    7. Acute infectious disease
    8. Thrombophlebitis
    9. Ventricular tachycardia and other dangerous dysrhythmias (multifocal ventricular activity)
  10. Severe aortic stenosis

II. Relative contraindications*
    1. Uncontrolled or high-rate supraventricular dysrhythmia
    2. Repetitive or frequent ventricular ectopic activity
    3. Untreated severe systemic or pulmonary hypertension
    4. Ventricular aneurysm
    5. Moderate aortic stenosis
    6. Uncontrolled metabolic disease (diabetes, thyrotoxicosis, myxedema)
    7. Severe myocardial obstructive syndromes (subaortic stenosis)
    8. Marked cardiac enlargement
    9. Toxemia of pregnancy

III. Conditions requiring special consideration and/or precautions
    1. Conduction disturbance
       a) Complete atrioventricular block
       b) Left bundle branch block
       c) Wolff-Parkinson-White syndrome
    2. Fixed rate pacemaker
    3. Controlled dysrhythmia
    4. Electrolyte disturbance
    5. Certain medication
       a) Digitalis
       b) β-blocking and drugs of related action
    6. Clinically severe hypertension (diastolic over 110, grade III retinopathy)
    7. Angina pectoris and other manifestations of coronary insufficiency
    8. Cyanotic heart disease
    9. Intermittent or fixed right-to-left shunt
  10. Severe anemia
  11. Marked obesity
  12. Renal, hepatic, or other metabolic insufficiency
  13. Overt psychoneurotic disturbance requiring therapy
  14. Neuromuscular, musculoskeletal, or arthritic disorders that would prevent activity

*In the practice of medicine the value of testing often exceeds the risk for patients with these relative contraindications.

From American College of Sports Medicine, *Guidelines for Graded Exercise Testing and Exercise Prescription.* Philadelphia: Lea & Febiger, 1975.

Training Curves

The points in time and the frequency of retesting can be defined from study of figures 13.1 and 13.2. Figure 13.1 shows the training curve for a young male previously deconditioned by bed rest (34). The authors' analysis shows that these data fit an exponential curve very well ($r^2 = .997$). This type of curve has practical meaning for us because the mathematical nature of an exponential is such that the rate of change (training effect) is inversely proportional to the level of achievement at any given point in the training program. In other words the progress is most rapid when the fitness is poorest (at the beginning of training from a deconditioned state). Progress grows slower and slower as the fitness level is improved by conditioning.

Figure 13.2 shows the training curve for five older men who were monitored over forty-two weeks in the author's laboratory (10). This group shows a curve of the same exponential type ($r^2 = .969$), but interestingly, the rate of improvement is much slower. The half-time ($t_{1/2}$), or time to achieve one-half the potential improvement, is 1.3 weeks for the young and 12 weeks for the old men. Whether this is a true age difference or simply a result of even

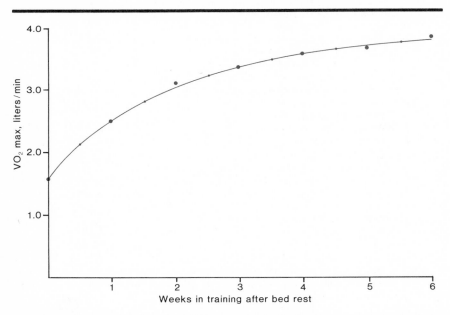

Figure 13.1 Training curve for young male after bed rest. Subject made 31% improvement over prebed-rest control value. (Drawn from the data of Saltin, B., et al. *Circulation* [suppl. VII], vols. 37–38, November 1968.

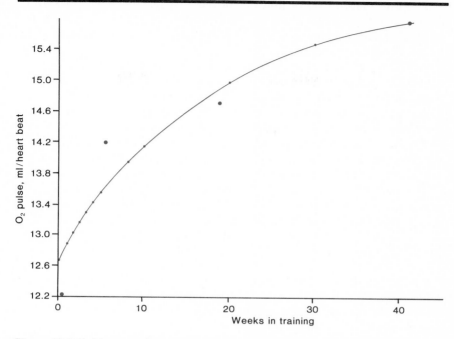

Figure 13.2 Training curve for old men ($N = 5$) who made 29% improvement in O_2 transport ($P<.05$). (From deVries, H.A. *J. Gerontol.* 25:325–36, 1970.)

greater deconditioning in the old men after a lifetime of sedentary living is not known. In any event, it is obvious from the two figures that the first retest should be at about six weeks to show the dramatic early improvement. After that, retesting at six-month intervals for healthy normals is probably sufficient.

Interval Training versus Continuous Exercise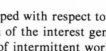

In recent years considerable enthusiasm has developed with respect to use of interval training for the training of athletes. Much of the interest generated probably resulted from the very interesting study of intermittent work done by Scandinavian investigators and discussed in chapter 10. Their work showed that subjects could handle very heavy exercise loads with surprisingly low accumulations of O_2 debt and lactic acid when work and rest intervals were interspersed.

However, Saltin (33), who is one of the Scandinavian authorities in this area of investigation, reviewed the evidence and came to the conclusion that interval training does not appear to have an advantage over continuous training in enhancing endurance capacity.

Roskamm (32), a German pioneer in this area, performed a carefully controlled experiment comparing the training effect of continuous and interval training. He found very small differences when testing the training response at maximal exercise. When heart rate at moderate exercise loads was used as the criterion for testing, continuous exercise produced far better results.

Perhaps even more important for our interests in adult fitness, Pollock (29, 30) has found that the dropout rate in a high intensity interval training program for adults was double that of a continuous jogging program.

It should be pointed out that interval training for athletes may have an advantage over continuous training, in that the faster pace of interval training may come closer to game conditions and therefore may favor the involvement of the same muscles, fiber types, and muscle recruitment patterns utilized in the competitive situation.

Until more scientific evidence is available, we must conclude that although interval training may have some small advantages (as yet unproven) over continuous training for the competitive athlete, continuous exercise at lower intensity levels is both safer and better received for purposes of health and fitness in both young and older adults.

Exercise Modality

It must be recognized from the outset that many different exercise modalities (types of exercise) can be used to enhance cardiorespiratory fitness. The exercise modality to be used for this purpose should

1. involve a large proportion of total muscle mass.
2. maximize the use of large muscles.
3. minimize the use of small muscles.
4. maximize dynamic muscle contraction.
5. minimize static muscle contraction.
6. be rhythmic, allowing relaxation phases alternating with contraction phases.
7. minimize the work of the heart per unit training effect.
8. be quantifiable with respect to intensity.

The types of exercise commonly used are jogging, walking, swimming, and cycling. Many other activities of an endurance nature can be used. Even dancing, when developed as an aerobic endurance exercise through control of rhythm and cadence as done by Jacki Sorenson, can elicit a $\dot{V}O_2$ as high as 40 ml/kg/min (15). Among the modalities commonly used—jogging, walking, swimming, and cycling—there is probably little difference in training effect if equal levels of total work are used (28, 31). The inclusion of walking may be surprising, but Pollock and associates (26) found large and very significant improvement in $\dot{V}O_2$ max in healthy, middle-aged, sedentary men from walking. It would, of course, be an insufficient challenge for a young person of average fitness. This will be discussed in the next section.

Of the eight criteria listed above for choice of exercise modality, seven relate to the importance of getting the most exercise for the least heart strain (work of the heart). The author (13) has provided data (fig. 13.3) that show clearly the differing effects of walking, cycling, and a crawling type of exercise on the relationship of heart strain to total body work in older men (mean age 69). At all levels of total body work, crawling required more work of the heart than cycling, and the work of the heart rose with increasing total work more rapidly in both crawling and cycling than in walking. Obviously, when working with sedentary, middle-aged and older people, it is desirable to minimize the ratio of cardiac effort to total body effort for maximum conditioning at minimum risk. The most important determinant of the work of the heart is the amount of rise in blood pressure caused by the exercise. This is least when large muscles are used rhythmically in dynamic contractions (12). It is greatest when small muscles are used at high fractions of their capacity or when muscles are held in static contraction. Thus the crawling exercise is worst

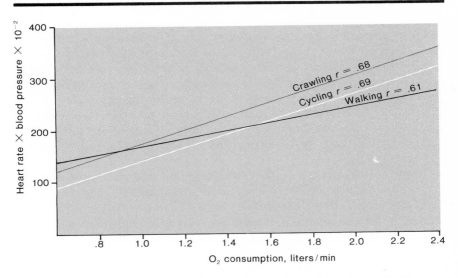

Figure 13.3 The relationship of cardiac effort to total body effort for the three different types of exercise (crawling, cycling, and walking). Each regression line represents the data on twelve subjects at five levels of O_2 consumption. (From deVries, H.A., and Adams, G.M. *J. Sports Med. Phys. Fitness* 17:41, 1977.)

Physical Conditioning for Health and Fitness

because it involves heavy use of the small muscles in the shoulder girdle and upper limbs, and it also involves static contractions of the trunk muscles to maintain the crawling posture. Cycling also caused greater blood pressure effects than walking because, as we showed by EMG techniques (12), there is considerable static contraction in the upper limb muscles. Recent work with young men supports our findings: they showed greater blood pressure responses to the bicycle ergometer than to the treadmill at equal, heavy exercise loads (3).

Exercise Intensity

When the exercise modality has been chosen, the next questions are, How much? How long? and How often? The first deals with *intensity,* the second with *duration,* and the third with *frequency.* These are the three independent variables we can manipulate (dose) to bring about the desired training effect (response). Thus we can think about exercise in dose-response terms in a way completely analogous to the methods of prescription of medicine. The most important determinant of response is *intensity,* and we shall consider it first.

Nomenclature. The first decision we need to make is whether to set up communications on the basis of dose or response. Thus we can define the intensity of the exercise in the case of jogging, for example, by spelling out the distance to be accomplished and the rate of running (time for each mile). This is prescription by dose. On the other hand we can also spell out the intensity of the exercise in terms of physiological responses such as the heart rate achieved. Obviously the latter is the safer and more effective method because the physiological strain or challenge of any given workout defined as distance and rate can vary greatly with such factors as weather, changing physical fitness, or incipient illness. However, to use response as the prescription basis requires that we train our participants to take their own heart rates quickly and accurately. Our discussion will be based on the heart rate response concept, and methods for taking heart rate are discussed in a later section of the chapter.

In the use of heart rate response we are again faced with two choices. We can express heart rate (HR) as a percent of maximum HR or as a percent of heart rate range (HRR). Because we start with a resting HR, which is a large fraction of the maximum HR, there is not a very good proportionality between percent maximum HR and the exercise $\dot{V}O_2$, which is really the measurement we are concerned with (9). Figure 13.4 shows the clear advantage in expressing the exercise target HR in percent HRR. It can be seen that percent HRR relates quite accurately to the actual $\dot{V}O_2$. That is to say, 50% HRR is about the same as 50% of max $\dot{V}O_2$. But it would require about 68% of max HR in a young subject to produce an actual $\dot{V}O_2$ of 50%, and in the elderly the percent max HR would be about 75 to produce the 50% $\dot{V}O_2$ max.

Physiology Applied to Health and Fitness

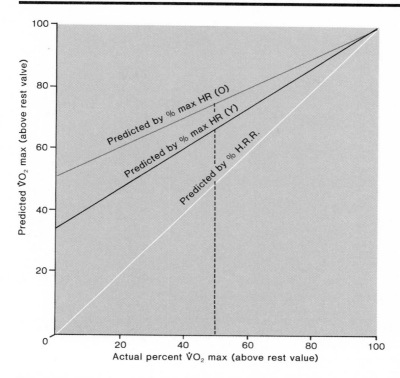

Figure 13.4 Illustration of the errors in using percent maximum heart rate for exercise prescription compared with the use of percent heart rate range. Y = college ages, O = 80 year old.

Intensity Threshold. Since the classic study of Karvonnen and colleagues (20) in 1957, we have been aware that some threshold or certain minimal level of exercise intensity must be reached before measurable training effects are achieved. Karvonnen showed that to achieve a training effect required 60% of HRR. However, his findings were based on the study of only six young male subjects, and no consideration could be given to the possible effects of age, sex, and physical fitness differences. In general, later work has supported the threshold concept, and the 60% HRR seems to be well supported with respect to young subjects of average fitness (4, 6, 35, 37). However, the author (11) has found considerable difference with respect to age and fitness. The threshold value for men in their sixties and seventies is only 40% HRR, and there is a well-defined effect of level of fitness at the beginning of the training program. On the basis of the available evidence, figure 13.5 would seem to

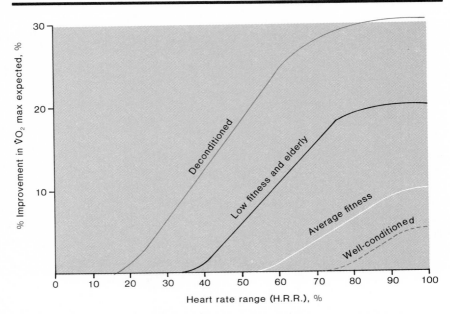

Figure 13.5 Family of curves showing exercise intensity threshold, dose-response relationship, and probable maximum results as related to pretraining $\dot{V}O_2$ max. (Conceptualization based on data of deVries, H.A. *Geriatrics* 26:94–101, 1971, and Saltin, B., et al., AHA Monograph #23, 1968.)

provide the best conceptualization of the dose-response relationship. The important points to be noted are (1) there is a point in the HRR below which no training effect is achieved (intensity threshold); (2) the intensity threshold grows higher with higher levels of fitness; (3) once the threshold is reached, the response is relatively proportional to the dose, but with somewhat less response per unit dose as fitness improves; and (4) the percentage improvement potential grows smaller with increasing fitness at the beginning of training, as would be expected from the nature of the training curves shown in figures 13.1 and 13.2.

Exercise Duration

The dose-response data presently available do not yet allow a precise graphic presentation of the relationship between exercise duration and fitness improvement, but figure 13.6 shows the author's conceptualization based on a review of the literature, with heavy emphasis on the work of Hartung and associates (18) and of Pollock (30), both of whose data points are shown. In general, it

Physiology Applied to Health and Fitness

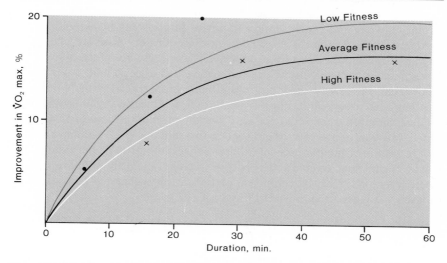

Figure 13.6 Relationship of percent improvement in V̇O$_2$ max to duration of exercise (Curves based on data of Hartung, G.H., et al., *J. Human Ergol.* 6:61, 1977 (o), and Pollock, M.L., *The Physician and Sports Med.* 6, no. 6, June 1978 (x). Hartung data at intensity of 65% HRR, Pollock data at intensity of 85%–90% HRR.

appears that a minimum duration of fifteen minutes is required at an optimal intensity before significant training changes are brought about (18). Best results probably require thirty to sixty minutes at optimal intensity (30).

It must be emphasized, however, that with previously sedentary or deconditioned subjects, one may not achieve even the fifteen-minute duration in continuous exercise. Progressive development to even that low level of duration may be required.

There is also some reason to believe that there is an interaction between intensity and duration. Pollock (30) has shown that for middle-aged men (forty to fifty-seven) walking for forty minutes per day, four days a week, produced a training effect equal to that gained from jogging thirty minutes a day, three days a week, where the weekly energy cost of the two programs was equal. We have seen similar results in our work with older men (10).

In a well-controlled study of prison inmates, Pollock and associates (29) provided some interesting data showing that the injury rate was more than double in inmates who trained for forty-five minutes as compared with those who trained for thirty minutes (54% and 24%, respectively). The injuries were largely shin splints and knee problems. In our experience shin splints can be controlled and virtually eliminated by use of static stretching (see chap. 22).

In any event the data suggest that the duration should be held to thirty minutes or less for beginning joggers to hold injury rates to a minimum. As aerobic capacity and general fitness improve, duration can be increased.

Although this chapter is directed to cardiorespiratory fitness, it should be noted that the endurance exercise prescribed here is also one of the best means of weight control (to be discussed in chap. 15), and of course, all other things being equal, the caloric expenditure is directly proportional to the duration.

It has been observed that in a typical physical education class lasting fifty minutes, children were active for only five to ten minutes, and of that short period, HR was above 150 for only one to two minutes (17). Similar findings with respect to elementary school children suggest that they do not voluntarily engage in sufficient aerobic activity during recess to be likely to improve their fitness level (19). Clearly, competent and dedicated leadership is required at all levels if we are to improve the fitness of our youth.

Frequency of Workout

Figure 13.7 shows the dose-response curve for workout frequency as conceptualized from some of the recent data (4, 16, 25, 28). In general, the available data suggest little if any improvement from one workout per week. The improvement accelerates rapidly when workouts are increased to four or five per week, with smaller payoffs for increases to six or seven per week. It seems almost certain that seven heavy workouts per week would be counterproductive, since there would be no opportunity for the muscle glycogen overshoot phenomenon to occur (see chap. 3). There is also reason to believe that adrenocortical depletion could occur on the basis of animal studies.

It appears that the optimal payoff for time spent occurs with three to five workouts per week. What effect does the spacing of the workouts have? Moffatt and associates (24) have shown that there is no difference in training effect in young males when the training is conducted on Monday, Tuesday, and Wednesday as compared with Monday, Wednesday, and Friday.

Again, for the novice jogger, injury rate is three times greater for five workouts per week than it is for three workouts per week (29). But there are significant advantages in longer and more frequent workouts for weight reduction. Thus on the basis of available evidence it would seem prudent to prescribe no more than three workouts of thirty minutes' duration per week of jogging until good fitness has been realized, at which time dosage could be increased, if desired, for weight reduction purposes to five workouts of forty-five to sixty minutes each. As the workouts became longer and more frequent, intensity would need to be reduced commensurately to prevent overtraining (staleness).

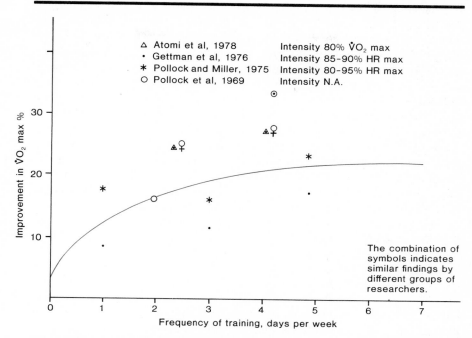

Figure 13.7 Relationship of percent improvement in VO_2 max to frequency of workout. (Curves based on data from references shown above.)

Exercise Prescription

From the preceding discussion it is obvious that, to the greatest extent possible, exercise prescription (Ex Rx) should be on an individual basis. In our experience, even though the Ex Rx is individual, large numbers of normals can be separated into four or five smaller groups by fitness category and can then exercise together as a homogeneous unit.

We shall base this discussion on the use of jog-walk and jogging as the exercise modality, although the methods and principles applied will be equally effective if cycling, swimming, walking, or other endurance exercises are chosen.

Daily Workout Plan. Every workout must include three components: (1) the warm-up period, (2) the main part of the workout, the endurance activity, and (3) the cool-down.

Warming up and cooling down are so important to the health and safety of participants that a separate chapter is devoted to these topics in part 3 of this text. Suffice to say, both the cardiovascular-respiratory and the neuromuscular systems must be brought to a state of readiness for vigorous activity by very gradual increases in intensity until myocardial blood flow and deep muscle temperatures are suitable for the challenge of the endurance workout. However, endurance activity such as jogging can be one of the safest and most effective warm-ups in itself if simply started at a rate that is slow and easy for the participant. It has been the author's experience that the calisthenics so often used as a warm-up can be more effective at the end of the jogging workout as a means of improving muscle tone and flexibility. This is especially true when the author's static stretching routines are used for prevention of muscle problems such as shin splints and soreness in the calf muscles.

In the case of jogging workouts, the cool-down period is typically a period of brisk walking until HR returns to about 120 for young and 110 and 100 for middle-aged and older adults, respectively. Thus the daily workout plan would start with five to ten minutes of slow jogging leading into the jogging workout designed to achieve target HR, followed by ten to fifteen minutes of appropriate calisthenics designed to improve muscle strength, endurance, and flexibility of the shoulder girdle and the chest as well as the abdominal wall and upper and lower back. Finally, the workout is completed with static stretching (chaps. 20 and 21).

Elements of the Ex Rx. As we have discussed above, there are four elements to the Ex Rx that must be defined. For purposes of our discussion, we will define the *exercise modality* as jogging. We shall recommend a *frequency* of three to five times a week (see fig. 13.7) and a *duration* of twenty to thirty minutes (see fig. 13.6). However, previously sedentary individuals will require from six to ten weeks of gradual increases to achieve this duration. The remainder of our discussion will be directed to the most important factor of *intensity* for which we will need the following definitions:

$$RHR = \text{resting HR}$$
$$EHR = \text{exercise HR}$$
$$MHR = \text{maximum HR (from table 13.2)}$$
$$HRR = \text{HR range} = MHR - RHR$$
$$\%HRR = \frac{EHR - RHR}{MHR - RHR} \times 100$$

Using these definitions, we now need to define three HRs for the participant:

Min HR—the intensity threshold value of HR, below which improvement is unlikely.

Target HR—the HR to which the participant should work to assure optimal training progress with minimum hazard.

Do-not-exceed HR—the HR above which intensity may be too high for optimal results and that may be counterproductive for some.

Since the author has found significant negative correlations between training response and initial fitness level (11), it is necessary to adjust the Ex Rx to the fitness level of the participant as shown in table 13.3.

Let us now calculate the intensity for a twenty-one-year-old participant who is to start in our conditioning program having been tested and found to have a $\dot{V}O_2$ max in the average category for his age. First we enter table 13.2 and find that the maximum HR would be predicted at 200. The resting HR is 70.

$$HRR = MHR - RHR = 200 - 70 = 130$$

Table 13.2

Age-fitness Adjusted Predicted MHR for Three Levels of Fitness Based on Balke Treadmill Stress Testing

Age	Predicted MHR, bpm			Age	Predicted MHR, bpm		
	Below Average	Average	Above Average		Below Average	Average	Above Average
20	201	201	196	45	174	183	183
21	199	200	196	46	173	182	183
22	198	199	195	47	172	181	182
23	197	198	195	48	171	181	182
24	196	198	194	49	170	180	181
25	195	197	194	50	168	179	180
26	194	196	193	51	167	179	180
27	193	196	193	52	166	178	179
28	192	195	192	53	165	177	179
29	191	193	192	54	164	176	178
30	190	193	191	55	163	176	178
31	189	193	191	56	162	175	177
32	188	192	190	57	161	174	177
33	187	191	189	58	160	174	176
34	186	191	189	59	159	173	176
35	184	190	188	60	158	172	175
36	183	189	188	61	157	172	175
37	182	189	187	62	156	171	174
38	181	188	187	63	155	170	174
39	180	187	186	64	154	169	173
40	179	186	186	65	152	169	173
41	178	186	185	66	151	168	172
42	177	185	185	67	150	167	171
43	176	184	184	68	149	167	171
44	175	184	184	69	148	166	170
				70	147	165	170

From Cooper, K.H., et al., *Med. Sports* eds. D. Brunner and E. Jokl, vol. 10:78–88, Basel: Karger, 1977.

Table 13.3
Exercise Rx by Fitness Level

	Low Fitness	Average Fitness	High Fitness
Minimum HR	40% HRR	60% HRR	70% HRR
Target HR	60% HRR	75% HRR	80% HRR
Do-not-exceed HR	75% HRR	85% HRR	90% HRR

For training purposes

Min HR $= 60\%$ HRR $+$ Resting HR $= 78 + 70 = 148$
Target HR $= 75\%$ HRR $+$ RHR $= 97.5 + 70 \simeq 168$
Do-not-exceed HR $= 85\%$ HRR $+$ RHR $= 110.5 + 70 \simeq 181$

Note that these values are quite different for an individual of low fitness at age sixty with a resting HR of 70. At age sixty, MHR from table 13.2 would be 158.

Target HR $= .60 (158–70) + 70 = 123$

For an individual, age twenty-one, already of better than average fitness with a resting HR of 70,

Target HR $= .80 (196–70) + 70 = 171$

Heart rate can be taken most accurately by monitoring the apical heartbeat on the lower left side of the chest wall by stethoscope and timing ten heartbeats. Stopwatches are available that read directly in HR when stopped at time for ten beats. It is important that HR be taken within five to ten seconds of the end of an exercise bout if it is to reflect the exercise HR with any degree of accuracy because HR declines very rapidly at the end of exercise. For the same reason the period of HR observation is held to as short a period as possible. The time for ten beats at a target HR of 168 is only 3.6 seconds. Thus the entire time from exercise cessation to the stopping of the stopwatch is only 8.6 to 13.6 seconds.

If such a watch is unavailable, one may elect to time ten beats with an ordinary stopwatch immediately at cessation of exercise (five to ten seconds), using table 13.4 or photocopy of it to perform the conversion to heartbeats per minute. This is considerably more accurate than counting beats over a similar time period because one can not count a fraction of a heartbeat, while the stopwatch can read a fifth or tenth of a second, thus providing far better resolution and accuracy. Even counting heartbeats for a ten-second period has been shown to be a valid method when used within ten to fifteen seconds of cessation of exercise, providing HR within 2% of the true value taken during the exercise by radio telemetry (27).

Physiology Applied to Health and Fitness

Table 13.4
Conversion of Time for Ten Beats to Heart Rate in Beats per Minute

Time	HR	Time	HR	Time	HR
10.0	60	7.4	81	4.8	125
9.9	61	7.3	82	4.7	128
9.8	61	7.2	83	4.6	130
9.7	62	7.1	85	4.5	133
9.6	63	7.0	86	4.4	136
9.5	63	6.9	87	4.3	140
9.4	64	6.8	88	4.2	143
9.3	65	6.7	89	4.1	146
9.2	65	6.6	91	4.0	150
9.1	66	6.5	92	3.9	154
9.0	67	6.4	94	3.8	158
8.9	67	6.3	95	3.7	162
8.8	68	6.2	97	3.6	167
8.7	69	6.1	98	3.5	171
8.6	70	6.0	100	3.4	176
8.5	71	5.9	102	3.3	182
8.4	71	5.8	103	3.2	188
8.3	72	5.7	105	3.1	194
8.2	73	5.6	107	3.0	200
8.1	74	5.5	109	2.9	207
8.0	75	5.4	111	2.8	214
7.9	76	5.3	113	2.7	222
7.8	77	5.2	115	2.6	231
7.7	78	5.1	117	2.5	240
7.6	79	5.0	120	2.4	250
7.5	80	4.9	122	2.3	261

Palpation of heart rate at the carotid artery is probably not a desirable procedure, since it has been shown that at rest and at slow heart rates pressure on the carotid may result in reflex activity, slowing the heart rate. Thus errors of 11 and 15 bpm were found at rest and during recovery (38).

It should be noted that using the HR methods described above results in very significant advantages over the prescription of a prescribed jogging rate and distance.

1. In hot, humid weather, working to a target HR automatically protects participants from overdoing, because the stress of the heat load will be reflected in the observed exercise HR. To avoid exceeding the THR, the individual will need to slow down, thus maintaining training stress as a constant rather than rate and distance.

2. Physiological changes due to incipient illness will similarly be corrected for.

3. As fitness improves there is no need to change the Ex Rx because, again, the exercise HR, which grows less for any given exercise bout, results in the individual's needing to increase rate to maintain the target HR.

Physical Conditioning for Health and Fitness

When working with health and fitness classes of twenty-thirty heterogeneous college age men and women, the author has found the following organizational plan to be effective (using a small track where 12 laps = 1 mile) based on fitness level as measured by the Astrand test (chap. 12). Groups A, B, C, and D are those that fall in the high, good, average, and below average categories, respectively, on the norms of table 12.5.

Group	Ex Rx for Starting Course
A	Jog ten-minute mile pace to mild fatigue or for maximum of twenty minutes (time available for this part of class)
B	Jog twelve minute-mile pace to mild fatigue or for maximum of twenty minutes
C	1. Jog three slow laps; take HR and record 2. Repeat taking HR every third lap 3. Continue to mild fatigue or twenty minutes a. If HR > THR, walk one lap after every three laps b. If HR < THR, walk one lap after every six laps
D	1. Jog two laps, take HR, walk one lap 2. Repeat for ten minutes 3. Walk remaining ten minutes at brisk rate 4. If HR > THR, jog one lap, walk one lap If HR < THR, jog three laps, walk one lap

In this fashion, the well-motivated individual in groups C or D will develop sufficient fitness to jog continuously for the whole twenty-minute period at rates dependent upon innate abilities. For the people in groups A and B, the following pace chart is provided.

The pace chart is used in a double progressive manner. Individuals progress first in endurance at an appropriate rate until they successfully complete two miles at a given mile rate without exceeding target HR. After two successful completions at a given rate, they then start the next faster pace plan—the second progression of increasing rate. Thus we work for distance (duration) first and then for rate (intensity), but always within the limits of the target HR. The same procedure is appropriate also for middle-aged healthy adults, but the rate of progress is ordinarily slower. Many other methods can be built around the basic concepts described in this chapter. The method described is offered only as one training approach that has been found to be feasible and well received by participants.

Lap	\multicolumn{7}{c}{Elapsed time for each lap of the twelve-lap mile}
	12 min	11 min	10 min	9 min	8 min	7 min	6 min
1.	1:00	:55	:50	:45	:40	:35	:30
2.	2:00	1:50	1:40	1:30	1:20	1:10	1:00
3.	3:00	2:45	2:30	2:15	2:00	1:45	1:30
4.	4:00	3:40	3:20	3:00	2:40	2:20	2:00
5.	5:00	4:35	4:10	3:45	3:20	2:55	2:30
6.	6:00	5:30	5:00	4:30	4:00	3:30	3:00
7.	7:00	6:25	5:50	5:15	4:40	4:05	3:30
8.	8:00	7:20	6:40	6:00	5:20	4:40	4:00
9.	9:00	8:15	7:30	6:45	6:00	5:15	4:30
10.	10:00	9:10	8:20	7:30	6:40	5:50	5:00
11.	11:00	10:05	9:10	8:15	7:20	6:25	5:30
12.	12:00	11:00	10:00	9:00	8:00	7:00	6:00

Effect of Sex and Age on Training Adaptations

Until recently, virtually all of the research dealing with the training parameters of intensity, duration, and frequency had been conducted on the male of the species. Recently, however, studies have been conducted to better elucidate the training adaptations of the female. While the data are still sparse, there appears to be no great sex difference in training responses with respect to intensity threshold (21), duration (41), or the magnitude of response (2, 5, 36). Thus the principles discussed above would appear to apply equally well to both sexes.

With respect to age, work from the author's laboratory has shown that the trainability of older men (10) and women (2) is equally as good as that of the young if looked at on a relative basis. Although the elderly start at a much lower fitness level, the percentage by which fitness can be improved is not significantly different in the old than it is in the young (10), nor is there apparently any great difference in trainability between the sexes in old age (2).

Specificity of Training

As was discussed in chapter 3, training results in adaptations that are specific to exercise type, intensity, and duration. Indeed, even the use of energy substrate was seen to be specific to these factors. A good review of the subcellular bases for the specificity of training is available (23). It is not surprising that when the training effect of jogging is measured ($\dot{V}O_2$ max) on the treadmill, it may be almost threefold greater than that measured during a maximal swim test (22). However, there are also central circulatory adaptations to training such that training in one modality such as jogging can result in highly significant and relatively similar reductions in heart rate during other modes of *submaximal* exercise. For example, it has been shown that ten weeks of training at jogging resulted in approximately equal reductions in submaximal HR at such dissimilar activities as treadmill walking, leg cycling, arm cycling, and load carrying (36).

We may conclude that the training effects for high level performance are highly specific, but that the health benefits with respect to reduced heart stress at submaximal work are quite general.

Potential Physiological Changes Resulting from Training

It has been found in many well-controlled experiments that very significant physiological changes are brought about by conditioning previously sedentary subjects. One of the better experiments (14) showed, for example, that sixteen weeks of training (cross-country running and interval training) three times a week produced the following benefits:

1. 52% increase of total work output at exhaustion.
2. Decrease in heart rate at a submaximal task from 170 to 144 bpm.
3. Increase of 16.2% in maximal O_2 uptake.
4. Maximal cardiac output increase of almost 2 liters/min.
5. Stroke volume increase of 13.4%.
6. Significant arteriovenous oxygen difference increase.
7. Lower blood lactate levels at a given submaximal load.
8. Significant improvements in mechanical efficiency at the higher sub-maximal work loads.

Pursuit of Excellence. We do not begin life with an equal, innate capacity for mental or physical achievement. Consequently, we cannot pursue excellence in physical fitness on an absolute scale; everyone should, however, strive for the highest level of physical fitness possible within the limits of individual physical potential. Then, and only then, can one begin to live a full life.

Training as a Stressor

As with all other good things in life, exercise and training can be carried to extremes, and the resulting stress may be detrimental instead of beneficial. The time course of normal training challenges shows an adaptive response. In the case of the endocrine system, the sympathoadrenal response to the training workout may show a gradual lowering in the level of blood catecholamines. It has been shown, for example, that the adaptation to a standard exercise challenge of thirty to fifty minutes per day for six days per week can result in plasma epinephrine dropping to one-third and norepinephrine to one-half their pretraining values. This must be considered a beneficial response in that the organism appears to be under less stress after the training period of seven weeks (40).

Physiology Applied to Health and Fitness

On the other hand, if the exercise is overdone the opposite may be the case. Williams and Ward (39) performed hematological studies before and after a relay marathon race in which each team member ran approximately one mile all out every hour for twenty-four hours. They found significant rises in white blood cell counts and percentage of polymorphonuclear leucocytes, and significant decreases in lymphocytes and eosinophils after the twenty-four hourly runs. These changes are typical of the response to high levels of stress. In addition they found highly significant increases in bilirubin, which suggest red blood cell damage, and certain muscle enzymes, which suggest muscle tissue damage.

Thus we may conclude that while exercise or training in appropriate amounts can result in better responses to a stressful environment, immoderate indulgence in exercise of too heavy intensity and duration may result in very high levels of stress with possibilities for tissue damage.

SUMMARY

1. There is a felt need for physical education professionals to assist the lay public in learning how to use exercise for health and fitness purposes.
2. To use exercise for improvement of fitness requires consideration of objectives, type of exercise, intensity, duration, frequency, intensity threshold for training effect, and rate of training adaptation as a function of pretraining fitness level.
3. The need for medical evaluation prior to participation in endurance exercise depends upon age, symptomatology, and risk factors. Medical screening examinations should be required for all persons over thirty-five, and also for those younger, if they are symptomatic or at high risk for cardiovascular disease.
4. The type of physiological screening and monitoring performed must vary in sophistication with the type of population to be involved.
5. The change in cardiovascular fitness ($\dot{V}O_2$ max) during the course of a training program appears to be well described as an exponential relationship. This means that at any point in time the rate of gain varies inversely with the fitness status at that time.
6. Interval training may have some small advantage over continuous training (as yet unproven) for the competitive athlete. However, continuous exercise at lower intensity levels is equally effective, safer, and better received for purposes of health and fitness in both young and older adults.
7. Among the exercise modalities commonly used, jogging, walking, swimming, and cycling, there is probably little difference in training effect if equal levels of total work are used.
8. When working with previously sedentary, middle-aged, and older people, it is desirable to minimize the ratio of cardiac effort to total body effort

to produce maximum conditioning at minimum risk. This is done by using exercise modalities that use large muscles rhythmically in dynamic contractions.

9. Using dose-response relationships with respect to prescribing exercise requires consideration of the factors of *intensity, duration,* and *frequency.*

10. The intensity threshold below which no cardiorespiratory training effect is likely is about 60% of aerobic capacity for the young individual of average fitness. The threshold is lower for the unfit and elderly and higher for those at better levels of fitness.

11. A minimal exercise *duration* of about fifteen minutes at optimal intensity is required before significant training changes are realized. Best results probably require thirty to sixty minutes at optimal intensity.

12. There is likely to be little or no improvement from one workout per week. The rate of gain grows rapidly with increasing frequency to four or five per week with smaller payoff for time spent beyond that.

13. The daily workout plan must include three components: (a) warm-up, (b) the body of the workout (endurance activity in the case of cardiorespiratory conditioning), and (c) the cool-down period.

14. Procedures based on the monitoring of heart rate are described for the prescription of exercise in terms of minimum HR, target HR, and do-not-exceed HR.

15. Exercise can be prescribed either on the basis of dose (rate and distance of run, for example) or on the basis of physiological response (heart rate). The latter is preferable because the use of heart rate, for example, automatically will adjust the workload dosage for such variables as weather, personal well-being, and changes in fitness.

16. There appear to be no great sex differences in training responses with respect to intensity threshold, duration, or magnitude of response.

17. The trainability of older men and women is equally as good as that of the young if looked at on a relative basis.

18. The training effects for high level performance are highly specific, but the health benefits with respect to reduced heart rate at submaximal work are quite general.

19. Physiological benefits derived from a conditioning program that improves cardiorespiratory fitness include at least the following: (a) large increases in physical working capacity, (b) significant gains in aerobic capacity, (c) increased capacity for cardiac output, and (d) more efficient achievement of submaximal work loads at lower heart rates.

20. While exercise in appropriate amounts can result in better responses to a stressful environment, immoderate indulgence in exercise of too heavy intensity and duration may result in very high levels of stress with possibilities for tissue damage.

REFERENCES

1. ACSM Position Statement, The recommended quantity and quality of exercise for developing and maintaining fitness in healthy adults. *Med. Sci. Sports* 10:7–10, 1978.

2. Adams, G.M., and deVries, H.A. Physiological effects of an exercise training regimen upon women aged 52–79. *J. Gerontol.* 28:50–55, 1973.

3. Adams, G.E.; Bonner, E.A.; Ribisl, P.M.; and Miller, H.S. Blood pressure during heavy work on the treadmill and bicycle ergometer. *Med. Sci. Sports* 10:50, 1978.

4. Atomi, Y.; Ito, K.; Iwasaki, H.; and Miyashita, M. Effects of intensity and frequency of training on aerobic work capacity of young females. *J. Sports Med. Phys. Fitness* 18:3–9, 1978.

5. Burke, E.J. Physiological effects of similar training programs in males and females. *Res. Q.* 48:510–17, 1977.

6. Burke, E.J., and Franks, B.D. Changes in $\dot{V}O_2$ max resulting from bicycle training at different intensities holding total mechanical work constant. *Res. Q.* 46:31–37, 1975.

7. Conrad, C.C. Progress report of the President's Council on Physical Fitness and Sports. December 31, 1975.

8. Cooper, K.H.; Purdy, J.G.; White, S.R.; Pollock, M.L.; and Linnerud, A.C. Age-fitness adjusted maximal heart rates. *Med. Sports* 6:1–11, 1976.

9. Davis, J.A., and Convertino, V.A. A comparison of heart rate methods for predicting endurance training intensity. *Med. Sci. Sports* 7:295–98, 1975.

10. deVries, H.A. Physiological effects of an exercise training regimen upon men aged 52–88. *J. Gerontol.* 25:325–36, 1970.

11. ———. Exercise intensity threshold for improvement of cardiovascular-respiratory function in older men. *Geriatrics* 26:94–101, 1971.

12. deVries, H.A., and Adams, G.M. Total muscle mass activation vs relative loading of individual muscles as determinants of exercise response in older men. *Med. Sci. Sports* 4:146–54, 1972.

13. ———. Effect of the type of exercise upon the work of the heart in older men. *J. Sports Med. Phys. Fitness* 17:41–48, 1977.

14. Ekblom, B.; Astrand, P.O.; Saltin, B.; Stenberg, J.; and Wallstrom, B. Effect of training on circulatory response to exercise. *J. Appl. Physiol.* 24:518–28, 1968.

15. Foster, C. Physiological requirements of aerobic dancing. *Res. Q.* 46:120–22, 1975.

16. Gettman, L.R.; Pollock, M.L.; Durstine, J.L.; Ward, A.; Ayres, J.; and Linnerud, A.C. Physiological responses of men to 1, 3, and 5 day per week training programs. *Res. Q.* 47:638–46, 1976.

17. Goode, R.C.; Virgin, A.; Romet, T.T.; Crawford, P.; Duffin, J.; Pallandi, T.; and Woch, Z. Effects of a short period of physical activity in adolescent boys and girls. *Can. J. Appl. Sports Sci.* 1:241–50, 1976.

18. Hartung, G.H.; Smolensky, M.H.; Harrist, R.B.; Rangel, R.; and Skrovan, C. Effects of varied durations of training on improvement in cardiorespiratory endurance. *J. Human Ergol.* 6:61–68, 1977.

19. Hovell, M.F.; Bursick, J.H.; Sharkey, R.; and McClure, J. An evaluation of elementary students voluntary physical activity during recess. *Res. Q.* 49:460–474, 1978.

20. Karvonen, M.J.; Kentala, E.; and Mustala, O. The effects of training on heart rate. *Ann. Med. Exper. Fenn* 35:307–15, 1957.

21. Kearney, J.T.; Stull, G.A.; Ewing, J.L.; and Strein, J.W. Cardiorespiratory responses of sedentary college women as a function of training intensity. *J. Appl. Physiol.* 41:822–25, 1976.

22. McArdle, W.D.; Magel, J.R.; Delio, D.J.; Toner, M.; and Chase, J.M. Specificity of run training on $\dot{V}O_2$ max and heart rate changes during running and swimming. *Med. Sci. Sports* 10:16–20, 1978.

23. McCafferty, W.B., and Horvath, S.M. Specificity of exercise and specificity of training: a subcellular review. *Res. Q.* 48:358–71, 1977.

24. Moffatt, R.J.; Stamford, B.A.; and Neill, R.D. Placement of tri-weekly training sessions: importance regarding enhancement of aerobic capacity. *Res. Q.* 48:583–91, 1977.

25. Pollock, M.L.; Cureton, T.K.; and Greninger, L. Effects of frequency of training on working capacity cardiovascular function and body composition of adult men. *Med. Sci. Sports* 1:70–74, 1969.

26. Pollock, M.L.; Miller, H.S.; Janeway, R.; Linnerud, A.C.; Robertson, B.; and Valentino, R. Effects of walking on body composition and cardiovascular function of middle aged men. *J. Appl. Physiol.* 30:126–30, 1971.

27. Pollock, M.L.; Broida, J.; and Kendrick, Z. Validity of the palpation technique of heart rate determination and its estimation of training heart rate. *Res. Q.* 43:77–81, 1972.

28. Pollock, M.L., and Miller, H.S. Frequency of training as a determinant for improvement in cardiovascular function and body composition of middle aged men. *Arch. Phys. Med. Rehabil.* 56:141–45, 1975.

29. Pollock, M.L.; Gettman, L.R.; Milesis, C.A.; Bah, M.D.; Durstine, L.; and Johnson, R.B. Effects of frequency and duration of training on attrition and incidence of injury. *Med. Sci. Sports* 9:31–36, 1977.

30. Pollock, M.L. How much exercise is enough? *The Physician and Sports Med.* 6: June, 1978.

31. Roberts, J.A., and Morgan, W.P. Effect of type and frequency of participation in physical activity upon physical working capacity. *Am. Corr. J.* 25:99–104, 1971.

32. Roskamm, H. Optimum patterns of exercise for healthy adults. *Can. Med. Assoc. J.* 96:895, 1967.

33. Saltin, B. Intermittent exercise: its physiology and practical application. John R. Emens Lecture, Ball State University, Muncie, Ind., February 20, 1975.

34. Saltin, B.; Blomquist, G.; Mitchell, J.H.; Johnson, R.L.; Wildenthal, K.; and Chapman, C.B. Response to exercise after bed rest and after training. American Heart Association, Monograph 23. New York: The Association, 1968.

35. Shephard, R.J. The development of cardiorespiratory fitness. *Med. Services J. Canada,* 21:533–44, 1965.

36. Van Handel, P.J.; Costill, D.L.; and Getchell, L.H. Central circulatory adaptations to physical training. *Res. Q.* 47:815–23, 1976.

37. Wenger, H.A., and MacNab, R.B.J. Endurance training: the effects of intensity, total work, duration and initial fitness. *J. Sports Med. Phys. Fitness* 15:199–211, 1975.

38. White, J.R. EKG changes using carotid artery for heart rate monitoring. *Med. Sci. Sports* 9:88–94, 1977.

39. Williams, M.H., and Ward, A.J. Hematological changes elicited by prolonged intermittent aerobic exercise. *Res. Q.* 48:606–16, 1977.

40. Winder, W.W.; Hagberg, J.M.; Hickson, R.C.; Ehsani, A.A.; and McLane, J.A. Time course of sympathoadrenal adaptation to endurance exercise training in man. *J. Appl. Physiol.* 45:370–74, 1978.

41. Yeager, S.A., and Brynteson, P. Effects of varying training periods on the development of cardiovascular efficiency of college women. *Res. Q.* 41:589–92, 1970.

14

Exercise Physiology in the Prevention and Rehabilitation of Cardiovascular Disease

To those of us who have been involved with exercise physiology for the last several decades, it seems like only yesterday that we were defending participation in competitive athletics and heavy physical training against the charges that such activity would result in dire consequences such as "athlete's heart." We are now likely to be more concerned with moderating the claims for exercise as a panacea for all ills by some of our lay enthusiasts in the pseudoscientific health areas. The truth, of course, lies at neither extreme, but somewhere in between.

This change in attitude toward the effects of exercise on the cardiovascular system, and on the heart in particular, probably came about largely as a result of stimulation of cardiologists by the work of Eckstein (14). He simulated the effect of coronary disease in dogs by surgically restricting blood flow in the coronary arteries. His experiments showed that while this arterial narrowing by itself failed to initiate collateral vessel growth to take over as a result of the partially occluded vessels, the addition of exercise did in fact promote an effective collateral circulation. Although this work was published in 1957, we do not yet have any clear evidence of this physiological adaptation in humans. Much evidence has been presented pro and con, and the issue is still in question. However, there are other physiological bases upon which one can construct a case for the use of exercise in the prevention of and rehabilitation from coronary heart disease (CHD). We have already presented in chapter 11 much of the epidemiological evidence for the importance of exercise in prevention of CHD. In this chapter we will consider the contribution that exercise physiology can make to (1) the prevention of CHD and (2) to the rehabilitation of CHD patients. This discussion will involve the theoretical bases as well as the how-to concepts in the realization that the future will probably bring demands for ever greater numbers of physical educators who can work with cardiologists and other physicians in the scientific application of exercise to the needs of CHD patients and, even more important, the preventive aspects of this serious health problem. It has been reported (16) that 77.3% of the autopsies on Korean battle casualties showed gross evidence of coronary arteriosclerosis in spite of the fact that mean age was only twenty-two. It has also been pointed out that the most common mode of death in persons with either symptomatic or presymptomatic CHD is sudden death which accounts for over half of all coronary fatalities under age sixty-five. Furthermore, of these fatalities, over 65% have shown no prior symptoms, and death occurs totally unexpectedly outside the hospital before cardiac resuscitation teams can reach the victims (25). These facts underscore the need for preventive measures on a population basis since the physician may never have the opportunity to work his art in 65% of the cases.

It is the author's belief that health and physical education professionals can help improve the life style of our American people at all ages, particularly through teaching the appropriate use of physical conditioning, the optimization of body weight and nutrition, and relaxation skills.

Although the evidence for the benefits to be derived from exercise in CHD remains somewhat controversial, we would be well advised to consider the rules under which we accept or reject scientific evidence, as pointed out so well by Dr. Sam Fox, the former director of the National Heart Disease and Stroke Control Program (20). Levels of acceptance of evidence beyond proven versus unproven do exist: (1) proven beyond reasonable doubt; (2) of probable benefit, but not above question; (3) a prudent action, on the basis of a good chance of benefit and acceptably low hazard; and (4) promising, but more good data needed. In 1969, Dr. Fox concluded that the evidence is such that it is prudent to include increased habitual physical activity in a program to prevent or manage nonacute CHD. If anything, the evidence appears to the present author to have strengthened in the decade since Dr. Fox's conservative estimate.

In any event, it is important to realize from the outset that there is absolutely no evidence to suggest a *curative* effect from physical conditioning after the CHD process has become established. At that point the best that can be expected is a *palliative* effect—one that serves to relieve or alleviate the symptoms without reversing the underlying disease processes. Nor does this chapter propose to train physical educators in cardiology. But hopefully, we can instill the interest and sufficient knowledge in physical educators to allow them to communicate and interact successfully with the physicians who must always bear the ultimate responsibility for the welfare of their patients.

Anatomy and Physiology of the Coronary Arteries

Figure 14.1 shows the larger coronary arteries of the anterior wall of the heart. These arteries arise directly from the aorta as the right and left coronary arteries, which with their subdivisions supply the blood necessary to support myocardial metabolism and thus the muscular work of the heart.

With respect to CHD, the blood supply to the left ventricle is most important since this chamber is responsible for the total systemic circulation, and therefore any major interruption of blood flow in the left coronary artery must have serious functional consequences.

As has been noted in the earlier chapters, the skeletal muscles can function for short periods of time in the absence of O_2, that is, anaerobically. The myocardium, however, is capable of very little anaerobic metabolism. In addition, the heart already extracts the largest part of the O_2 supply available to it while at rest, in contradistinction to the skeletal muscles whose extraction at rest is very low. Thus virtually the only reserve in function for the heart muscle depends on increasing the O_2 supply through increasing coronary blood flow when the body's demands for O_2 are increased as in exercise. In the normal heart, any increase in myocardial O_2 demand (MVO_2) is precisely balanced by increases in coronary blood flow, which is affected by mechanical, neural, humoral, and metabolic factors that together reduce coronary vascular resistance and increase coronary blood flow as a consequence.

Because of its almost complete dependence upon a pay-as-you-go O_2 supply, the heart muscle is vulnerable to any imbalance between supply and demand. Thus any tendency toward hypoxia, whether because of too much demand or too little supply, is associated with rapid deterioration of function such as losses in contractility and cardiac output, which thus comprise the heart's ability to meet either the pressure or volume requirements of the general circulation. These losses in function are accompanied by abnormal electrophysiology which the physician sees as an abnormal electrocardiogram. If the cellular hypoxia is sufficiently severe, within minutes irreversible structural and functional damage can occur (heart attack).

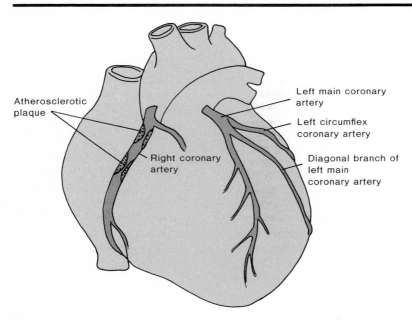

Figure 14.1 Illustration of the larger coronary arteries showing the partial occlusion of the right coronary artery by atherosclerotic plaque formation.

Physiology Applied to Health and Fitness

Nature of Coronary Heart Disease (CHD)

When the coronary arteries become rigid and narrow through atherosclerosis (the process of deposition of lipids in the intima of the arteries), or if the arteries in the inner layers of the heart muscle are being compressed from inside the left ventricle by high pressure, they can no longer adequately compensate for increased amounts of O_2 consumed by the myocardium. A critical ischemia will result leading to various functional and ultimately structural disturbances discussed below.

Angina Pectoris.   A severe constricting pain in the chest, often radiating from the precordium (the anterior wall of the chest overlying the heart and great vessels) to the left shoulder and down the left arm is angina pectoris. It is due to ischemia of the heart muscle usually, but not always caused by coronary disease. This condition is typically transient and reversible in terms of symptoms and associated functional alteration.

The recently developed medical diagnostic technique of angiography provides X-ray-type evidence of the closing down of the coronary arteries by the lipid deposits of the atherosclerotic process. Such evidence has shown that clinically manifest myocardial ischemia is usually the result of occlusion of more than 70% of the lumen of one or more arteries (2). A majority of patients with diagnosed CHD such as angina pectoris have major stenoses in two or three of the major coronary arteries. Less than 30% have only one vessel involved.

The process of atherosclerosis is long term, as the autopsies data of battle casualties from the Korean war have shown (16). It is obvious that we must become concerned with life-style modifications in high school and college students if we are to have optimal chances for success in the prevention of CHD.

Myocardial Infarction.   When the myocardial ischemia is of sufficient severity and duration to cause cell death (necrosis), the individual is said to have suffered a myocardial infarction (MI), which the layman refers to as a heart attack. The area of necrosis has, of course, suffered irreversible damage. The necrotic part of the heart muscle is replaced by noncontractile, fibrous tissue during the healing process. If this area of scar tissue is small, appropriate rehabilitative procedures can so improve the contractile function of the remaining muscle tissue that normal function can be regained. Indeed, one case has been reported in which a well-motivated man not only reversed his loss of function post-MI, but succeeded in achieving a work capacity some 44% greater than his pre-MI level (40). This must, of course, be considered an unusual case in which the myocardial damage could not have been extensive, combined with great motivation on the part of the patient. This improvement was maintained over the following fourteen-year period. This case would seem to define the outside limits for rehabilitation of CHD patients.

Sudden Death. As was pointed out earlier in this chapter, sudden death accounts for the majority of deaths from CHD. This catastrophic event seems to be the result of ischemic changes that affect the rhythm of the heart, causing such potentially lethal effects as ventricular tachycardia and fibrillation. That such cases of sudden death are, in fact, caused by severe CHD has been shown by postmortem studies in which 75% of the cases were found to have had multivessel involvement (43).

Congestive Heart Failure. Congestive heart failure is a clinical condition resulting from failure of the heart to maintain adequate circulation of the blood. It can happen in the later stages of CHD when tissue loss has resulted in mechanical or pumping inadequacies that lead to venous congestion and edema in the tissues affected by the venous congestion.

Theories Regarding Causation of CHD

Thus far we have exemplified the CHD processes as occurring by virtue of the accumulation of lipid deposits in the walls of the coronary arteries, the process we call atherosclerosis. While this process probably accounts for much, if not most of the prevalent CHD, it must be recognized that there is a considerable fraction of infarction cases in which the coronary arteries are found to be only moderately or even quite insignificantly involved at autopsy.

Wilhelm Raab, a pioneer in preventive cardiology, spent much of his professional lifetime in calling our attention to the importance of neurogenic and metabolic factors in the etiology of CHD ("ischemic heart disease" in his terms, since he objected to the term "coronary heart disease" for reasons that will become clear in this discussion). Figure 14.2 illustrates his conception of the pathophysiology involved in CHD. While he recognized the importance of atherosclerosis as the most common predisposing factor in the origin of degenerative heart disease, he deplored the lack of attention directed to the heart muscle's own nerve and hormone regulated metabolic processes. Thus in figure 14.2 one sees the possibilities for interaction among (1) the vascular mechanical factor, (2) the neurogenic-metabolic factors, and (3) the hormonal metabolic factor (46). Note that in this conceptualization, the effect of the decreased O_2 supply brought about by the atherosclerotic process is *potentiated* by the increased O_2 demand due to the sympathetic adrenergic preponderance, thereby exacerbating the myocardial hypoxia. In addition, the increased secretion of cortisol due to emotional stress is synergistic with the myocardial hypoxia in bringing about the electrolyte imbalance that ultimately causes the clinical events seen as angina, myocardial infarction, or sudden death. Raab has presented abundant evidence for the reasonableness of each step, but the total theory remains to be fully tested. Evidence has been presented from animal studies that (1) stimulation of certain brain areas can induce cardiac ischemic changes and arrythmias (39); (2) prolonged electrical

Physiology Applied to Health and Fitness

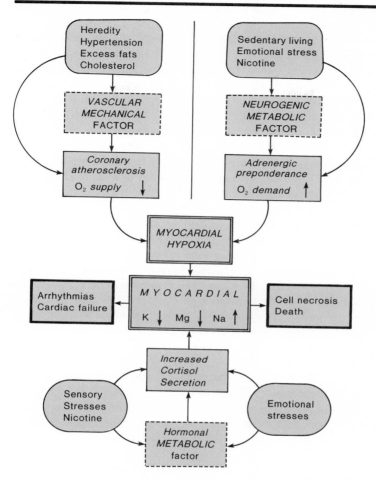

Figure 14.2 Structural formula of pluricausal degenerative, so-called coronary, heart disease. Coordination and integration of today's available clinical and experimental data reveal three inter-related main categories of potentially pathogenic interferences in the heart muscle's electrolyte-dependent functions and structure. Vascular, neurogenic, and hormonal factors appear to be jointly, and in various combinations and degrees, responsible for a typical pattern of electrolyte imbalance throughout the ventricular myocardium, namely: low potassium, magnesium, and K^+/Na^+ ratio; high sodium. Qualitatively uniform, but widely varying in degree, this pattern is characteristically present at autopsy in the following forms of cardiac pathology: subclinical and overt coronary insufficiency, ventricular hypertrophy, congestive heart failure, and myocardial infarction (including the noninfarcted tissue). (From Raab, W. *Ann. N.Y. Acad. Sci.* 147:666, 1969.)

stimulation of the stellate ganglia produces systolic hypertension and subendocardial hemorrhages in dogs (35); (3) certain types of acute cardiac necroses can be prevented by pretreatment with such stressors as physical exercise and cold baths.

In the human, it has been shown that the first day of imprisonment (serious stress) resulted in consistent findings of ECG changes suggestive of ischemia (36), and there seems to be a consistent relationship between an individual's emotional arousal, whatever its quality, and catecholamine output (37).

Friedman and his collaborators at the Harold Brunn Institute in San Francisco have built a case for the importance of personality type A in the predisposition to the development of CHD. It is their position that the major cause of coronary artery and heart disease is a complex of emotional reactions that they designate Type A behavior pattern, which is observed in an "individual who is *aggressively* involved in a *chronic incessant* struggle to achieve more and more in less and less time, and if required to do so, against the opposing efforts of other things or other persons" (23). Since the inception of this concept in the midfifties, they have done voluminous research that supports the importance of individual emotional patterns as strong contributing factors to the etiology of CHD. Friedman has summarized their findings as follows: "Extraordinarily efficient prevention could be achieved if *each of these four* measures was adopted: (1) *drastic* elimination of dietary cholesterol and animal fat, (2) avoidance of behavior pattern A, (3) lifetime participation in an *extensive* degree of physical activity, and (4) exclusion of cigarette smoking" (23).

There is basic agreement between Raab and Friedman in that both accept the importance of the atherosclerotic factor but feel that this is an incomplete picture of what should be considered a pluricausal disease entity in that neural, hormonal, and metabolic factors (and their emotional antecedents) are important contributors to its cause.

The Risk Factor Concept in CHD

The Framingham study (24) and other epidemiological studies have shown that the likelihood of developing CHD can be predicted for *groups of persons* well in advance of the appearance of symptoms. Among the risk factors are male sex, age over thirty-five, cigarette-smoking habit, elevated blood pressure, high level of serum cholesterol, glucose intolerance, and ECG abnormalities. These are not all-inclusive, as later discussion will show, but they are a set of proven merit widely used by physicians because they are readily measured by an office nurse or technician without hazard or trauma to the patient (25).

In general, in the use of these risk factor tables, the more risk factors present, or the greater the degree of abnormality of any one factor, the greater the risk. As an example, a forty-year-old man who does not smoke cigarettes, does not have glucose intolerance or ECG abnormalities, but who has a systolic blood pressure of 165 and cholesterol of 285 mg/100 ml, would have a probability of 4.9 in 100 of developing CHD within six years. On the other hand a man of the same age who smokes cigarettes, is glucose intolerant, with ECG abnormalities, blood pressure of 195, and cholesterol of 335 mg/100 ml would have a probability of 37.1 in 100 of developing CHD in the next six years. These figures compare with the average Framingham male of this age with average values for the risk factors who has a probability of only 2.3 in 100. Even this value is probably much higher than need be were we all to live optimal life styles.

It will be noted in the light of previous discussion that emotional stress and personality factors were not included in the Framingham study. This was apparently because "investigation of a possible relation between 'emotional stress' and CHD has been severely hampered by methodological problems" (32). However, Kannel who was a principal in the Framingham study, and his colleagues go on to say, "In spite of these basic difficulties, the concept of emotional stress as a potent force of morbidity and mortality in cardiovascular disease has gained increased prominence and acceptance" (32).

As a followup to the Framingham study, the Western Collaborative Group Study (WCGS) considered the same risk factors, but also evaluated the effect of personality type A (5). This was a prospective, epidemiological study of 3,154 initially well men. Using the Framingham risk prediction equation, their data over an 8½-year followup for CHD correlated well with those taken at Framingham. In addition, they estimated that removal of the excess risk associated with type A behavior would correspond to a 31% reduction of CHD incidence in their WCGS study.

Physiological Bases for Use of Exercise in CHD Prevention

The question may well be asked; On what physiological bases, or through what physiological mechanisms, is physical conditioning likely to be beneficial in the prevention of CHD? Table 14.1 shows the factors that are thought to operate in the direction of such benefit. Recognizing the importance of maximizing the ratio of myocardial O_2 supply to O_2 demand, the table is broken down into those factors that could increase the former and those that may decrease the latter. It should be pointed out that the evidence supporting these various factors ranges from that which is well proven to that which is still largely controversial. For example, the decreased $HR \times BP$ product (work of the heart) at any given submaximal work load as a result of physical conditioning is well accepted, while such factors as increased arterial O_2 saturation and improved blood distribution are still controversial.

We must also consider the potential benefits to be derived from the postulated effects of exercise on the CHD risk factors described above. There

Table 14.1

Physiological Mechanisms that May Have Potential Benefit in the Prevention (or Postponement of Symptomatology) of CHD.

Factors That May Increase Myocardial O_2 Supply		Factors That May Reduce Myocardial O_2 Demand	
1.	size of the coronary tree	1.	HR X BP product
2.	collateral circulation	2.	ability to relax tension in skeletal muscles
3.	arterial O_2 saturation	3.	emotional responses
4.	myocardial contractility	4.	obesity
5.	blood volume	5.	resting O_2 consumption by skeletal muscles
6.	blood viscosity	6.	blood distribution
7.	red blood cells		
8.	ventricular pressure		
9.	resting period of cardiac cycle		

Factors Thought to Act in Favor of CHD Risk Factor Reduction	
1.	serum triglycerides
2.	serum cholesterol (in the presence of weight loss)
3.	blood coagulability
4.	relaxation and resistance to emotional stress
5.	blood pressure
6.	glucose tolerance
7.	ECG abnormalities

seems little doubt that vigorous exercise regimens can have significant effect upon some of the physiological mechanisms that are at least statistically related to CHD incidence.

Since the research literature in this area is voluminous, documentation for the above discussion would require hundreds of references. However, we are fortunate that recent reviews of this literature are available for the interested reader. For the general area of physical activity and cardiovascular health, the reader is referred to Fox, Naughton, and Gorman (21), Hartung (27), and Scheuer and Tipton (49). For further insight into risk factor modification by chronic physical exercise see Bonnano (4). The control and modification of stress emotions through chronic exercise are well reviewed by Folkins and Amsterdam (19).

Exercise Physiology in Cardiac Rehabilitation

The term *cardiac rehabilitation* rather than the more limiting term *coronary rehabilitation* is used here because the principles and practices to be discussed are equally appropriate for patients who have undergone heart surgery for correction of congenital defects or for coronary bypasses, as they are for patients with coronary disease, whether they suffer angina pectoris or are postinfarction. Indeed, it is possible that the greatest good might come from the inclusion of individuals without overt symptoms of CHD who are at high risk for its development.

Naughton (40) defines cardiac rehabilitation as "a form of longitudinal comprehensive care through which selected patients are restored to and maintained at their optimal medical, physiological, psychological, social, vocational, and recreational status." Since one of the more important sequelae of heart disease is usually impairment of physical working capacity (PWC), it is quite natural to see the developing interest of exercise physiologists in this domain. It has been estimated that 80% of the survivors of myocardial infarction should be able to return to work and normal levels of activity (51). However, we probably fall considerably short of this goal, and it is to be hoped that the assistance of skilled exercise management by trained physical educators might contribute to the ultimate rehabilitation of the greatest possible numbers.

We have discussed the physiological mechanisms that might be effective in the *prevention* of CHD. Let us now turn our attention to the potential for improvement in PWC and its attendant health benefits in bringing about improvement in general body function when the myocardium has already sustained some measure of insult and is compromised to greater or lesser extent in its functional capacity. We must consider two possible but not mutually exclusive routes to this end: (1) reversal of the losses in myocardial functional capacity, and (2) compensatory improvement in other tissues such as skeletal muscle and organ systems such as the pulmonary system.

In the light of reports that document the capability of CHD patients to run the full marathon (42.2 km) after appropriate training (13) and of the case alluded to earlier of the patient who was able to increase his $\dot{V}O_2$ max by 44% over its preinfarction value (40), it is tempting to believe that at least a partial reversal of myocardial functional losses has occurred. While many investigators have shown very significant improvement in PWC (8, 17, 18, 41, 50), there is as yet no clear evidence for improvement of coronary collateral circulation. Angiographic techniques have not as yet shown any improvement in vascularization, but only the resting state has been investigated. Nor does there appear to be an improvement in myocardial O_2 supply (8, 17, 18, 50). Recently Ferguson and associates (18) have provided evidence that strongly suggests that the major effect of exercise training on angina pectoris patients is related to a reduction in O_2 demand (therefore lower coronary flow requirement). Symptom-limited exercise capacity was found to improve by 43% with training in their patients, and this appeared to be due to a decreased sympathetic drive to the heart and systemic arteries as reflected by lower HR and BP, and lower levels of catecholamines in arterial and coronary sinus blood for a given work load. It was postulated that peripheral adaptive changes in the trained skeletal muscles may be responsible for the decreased sympathetic drive. This is supported by earlier work which showed training to improve blood flow distribution in CHD patients.

Although we may not yet fully understand the physiological mechanisms, there is fairly good agreement with respect to the benefits to be derived from improved functional capacity and reduction in clinical symptoms. To begin with, experience from many different locations around the world has shown that typically only 40% of the patients with MI who survive the acute attack recover sufficiently to resume their original work. On the other hand, it has been repeatedly reported that in groups of patients with MI where a rehabilitation program was applied, 80% or more were able to resume their original work or start a new job (44).

Even more important, Gottheiner (26), one of the pioneers in Europe in the use of vigorous exercise conditioning for CHD, showed that 1,103 trainees in his program had a mortality rate of only 3.6% compared with 12% in a similar series of physically inactive patients in Israel.

Data have now been reported from five other longitudinal studies with very similar results (28). However, Haskell (28) points out that there are serious questions as to the comparability of the active and nonactive groups in each study with respect to the severity of illness and the likelihood of reinfarction.

Virtually all investigators have agreed in finding that physical training does increase the exercise angina threshold of most patients who adhere to the program (6, 22, 29, 50). This is of great importance to patients because it may enable participation in everyday life activities that were formerly barred to them. Hellerstein (29), who was one of the American pioneers in this field, showed that ECG responses to exercise also improved in 79% of the rehabilitation patients at the Cleveland clinic who succeeded in improving their physical fitness. This, of course, is related to, and supports the findings of, the increased anginal threshold.

 In a very interesting study by Adams, McHenry, and Bernauer (1) it was shown that most subjects trained after MI can achieve the performance levels of normal sedentary subjects, although some do not exhibit a classic training effect, probably because of residual myocardial dysfunction. Successful coronary artery bypass surgery on the other hand did not entirely normalize work performance, metabolic or hemodynamic function, but physical training after the bypass surgery resulted in further improvement.

Principles of Exercise Testing

It is to be noted that the purposes of exercise testing in cardiac rehabilitation are somewhat different from those of exercise testing of normal, healthy individuals, which was discussed in chapter 12. The purposes here are twofold: (1) to *diagnose* ischemic heart disease and investigate physiologic mechanisms underlying cardiac symptoms (angina, arrhythmias, inordinate blood pressure

Physiology Applied to Health and Fitness

rise, functional valve incompetence), and (2) to *measure functional capacity* for work, sport, or participation in a rehabilitation program or to estimate response to medical or surgical treatment (7).

Physicians often find it necessary to estimate the level of physical activity that is appropriate for a cardiac patient and to prescribe exercise on the basis of type, intensity, duration, and frequency for those patients who may benefit from such a program. While this sort of exercise testing and prescription lies entirely within the domain of the physician, many doctors have availed themselves of the assistance of physical educators in the implementation of such programs, just as they were able to improve their effectiveness by employing medical technicians to provide other highly specialized services.

In this area, as with normals, exercise tests can be designed to use steps, treadmill, bicycle, or arm-cranking ergometers. But regardless of the instrumentation, the following principles (7) apply:

1. Continuous step-incremental loading should be used and load intensity should be measurable.
2. The test should start at an exercise load considerably below the estimated level of impairment.
3. Achievement of steady state at each exercise load level is desirable.
4. The minimal parameters to be monitored are blood pressure, heart rate, and the ECG. These must be monitored at rest, at each exercise load, and during recovery.
5. Informed consent should be obtained prior to testing.
6. Commonly used criteria for stopping a test include:
    a. attainment of true maximum as described in chapter 12.
    b. attainment of an end point based on emergence of signs or symptoms of a disease process.
    c. attainment of a predetermined end point, such as 85% of age-related MHR, arbitrary HR as in PWC 150, or diagnostic ECG change.

The normal responses to exercise testing have been discussed in chapter 12. The abnormal responses are, of course, the primary responsibility of the physician who supervises the test, but the trained physical educator may function as the first line of defense and should have basic knowledge in this regard.

The American Heart Association lists the following responses to exercise testing as abnormal (7):

1. A decrease in, or failure to increase, systolic blood pressure in response to increasing load is indicative of inadequate pump function of the heart.
2. Generally, individuals with greater impairments of function respond with greater increases in HR to increasing load (even at very low loads). But, occasionally, bradycardia due to the onset of complete heart block or other abnormalities of sinus regulation may occur.

3. Various ECG abnormalities that are diagnosed by the physician.
4. Symptoms such as chest discomfort or pain, severe dyspnea or faintness, or claudication (pain in exercising muscles due to ischemia).
5. Signs such as pallor, cyanosis, or cold sweat.

Whereas PWC tests on normals typically go to true maximal capacity as described in chapter 12, most clinical investigators use tests that terminate short of maximum. In fact, heart patients often reach a level of discomforting symptoms far below their physiologic maximum. Thus symptom-limited peak performance may and often does differ considerably from physiologic maximum.

While in exercise physiology, we generally express exercise loads in terms of power output such as watts or kg-m/min, in cardiac rehabilitation it is more common to use the O_2 cost associated with a specific task. The use of the MET as a measure has become popular because it allows better communication with the layman. The resting $\dot{V}O_2$ is approximately 3.5 ml/kg/min and is defined as one MET. Any level of physical activity can then be expressed as a multiple of the O_2 cost of rest, that is, 1 MET, 2 METs, 3 METs, and so on. Tables 14.2, 14.3, and 14.4 show the energy costs in METs for bicycle ergometer, bench-stepping, and treadmill, respectively.

According to the New York Heart Association functional class system (31, p. 468) for evaluating performance of heart disease patients, a patient who becomes limited at 2.0 METs or less is class IV; 3 or 4 METs, class III; 5 or 6 METs, class II; and a patient who achieves 7.0 METs or more is considered

Table 14.2
Energy Expenditure in METs during Bicycle Ergometry

Body Weight		Work Rate on Bicycle Ergometer (kg—m/min and watts)												
		75	150	300	450	600	750	900	1050	1200	1350	1500	1650	1800	(kg-m min⁻¹)
(kg)	*(lb)*	12	25	50	75	100	125	150	175	200	225	250	275	300	(watts)
20	44	4.0	6.0	10.0	14.0	18.0	22.0								
30	66	3.4	4.7	7.3	10.0	12.7	15.3	17.9	20.7	23.3					
40	88	3.0	4.0	6.0	8.0	10.0	12.0	14.0	16.0	18.0	20.0	22.0			
50	110	2.8	3.6	5.2	6.8	8.4	10.0	11.5	13.2	14.8	16.3	18.0	19.6	21.1	
60	132	2.7	3.3	4.7	6.0	7.3	8.7	10.0	11.3	12.7	14.0	15.3	16.7	18.0	
70	154	2.6	3.1	4.3	5.4	6.6	7.7	8.8	10.0	11.1	12.2	13.4	14.0	15.7	
80	176	2.5	3.0	4.0	5.0	6.0	7.0	8.0	9.0	10.0	11.0	12.0	13.0	14.0	
90	198	2.4	2.9	3.8	4.7	5.6	6.4	7.3	8.2	9.1	10.0	10.9	11.8	12.6	
100	220	2.4	2.8	3.6	4.4	5.2	6.0	6.8	7.6	8.4	9.2	10.0	10.8	11.6	
110	242	2.4	2.7	3.4	4.2	4.9	5.6	6.3	7.1	7.8	8.5	9.3	10.0	10.7	
120	264	2.3	2.7	3.3	4.0	4.7	5.3	6.0	6.7	7.3	8.0	8.7	9.3	10.0	

From American College of Sports Medicine. *Guidelines for Graded Exercise Testing and Exercise Prescription.* Philadelphia: Lea & Febiger, 1975.

Physiology Applied to Health and Fitness

Table 14.3
Energy Expenditure in METs during Stepping at Different Rates on Steps of Different Heights

Step Height (cm)	Step Height (in.)	Steps per min 12	Steps per min 18	Steps per min 24	Steps per min 30
0	0.0	1.2	1.8	2.0	2.4
4	1.6	2.1	2.5	2.9	3.7
8	3.2	2.4	3.0	3.5	4.5
12	4.7	2.8	3.5	4.1	5.3
16	6.3	3.1	4.0	4.7	6.1
20	7.9	3.4	4.5	5.4	7.0
24	9.4	3.8	5.0	6.0	7.8
28	11.0	4.1	5.5	6.7	8.6
32	12.6	4.4	6.0	7.3	9.4
36	14.2	4.8	6.5	8.0	10.3
40	15.8	5.1	7.0	8.7	11.7

From American College of Sports Medicine. *Guidelines for Graded Exercise Testing and Exercise Prescription.* Philadelphia: Lea & Febiger, 1975.

Table 14.4
Approximate Relative Energy Expenditure in METs during Walking or Running Treadmill Tests

Grade %	km/h	Speed of Walking 2.7	3.2	4.0	4.8	5.5	5.6	6.4	6.8
	mph	1.7	2.0	2.5	3.0	3.4	3.5	4.0	4.2
0.0		1.7	2.0	2.5	3.0	3.4	3.5	4.6	5.0
2.5		2.3	2.7	3.3	4.0	4.5	4.7	6.0	6.5
5.0		2.9	3.4	4.2	5.0	5.7	5.9	7.3	7.9
7.5		3.4	4.0	5.0	6.0	6.9	7.1	8.7	9.3
10.0		4.0	4.7	5.9	7.0	8.0	8.3	10.0	10.8
12.0		4.5	5.3	6.6	7.9	9.0	9.2	11.1	11.9
12.5		4.6	5.4	6.8	8.0	9.2	9.5	11.4	12.2
14.0		4.9	5.8	7.3	8.7	10.0	10.2	12.2	13.0
15.0		5.2	6.1	7.6	9.0	10.3	10.7	12.8	13.6
16.0		5.4	6.4	8.0	9.5	10.8	11.1	13.3	14.2
17.5		5.8	6.8	8.5	10.0	11.5	11.8	14.1	15.0
20.0		6.3	7.5	9.3	11.0	12.7	13.0	15.5	16.5

Grade %	km/h mph	Speed of Running 9.7 / 6.0	11.3 / 7.0	12.9 / 8.0	14.5 / 9.0
0.0		10.0	11.5	12.8	14.2
2.5		11.4	12.7	14.1	15.4
5.0		12.7	14.0	15.4	16.7
7.5		13.9	15.3	16.6	18.0
10.0		15.2	16.5	17.9	19.3
12.5		16.5	17.8	19.2	20.5

A commonly used test is to have the subject walk at 3 mph with 2.5% grade increments utilized each 2 minutes. To convert from mph to km/h, multiply by 1.6093.
From American College of Sports Medicine. *Guidelines for Graded Exercise Testing and Exercise Prescription.* Philadelphia: Lea & Febiger, 1975.

class I. Thus class IV shows symptoms at rest, class III shows symptoms with less than ordinary activity, class II shows symptoms with ordinary activity, and class I is not symptom-limited, although clinical evidence such as an abnormal ECG is present.

Parameters to Be Measured. The single most important measurement recorded is, of course, the ECG. Recording and interpretation of the ECG lies in the domain of the physician (usually a cardiologist) who decides the lead system to be used and all pertinent methods and procedures. However, exercise program directors, leaders, and technicians should avail themselves of every opportunity to learn the basics of ECG analysis. Many cardiologists have found it helpful to provide courses for paramedical personnel dealing with ECG interpretation.

The HR $\times$ BP product, sometimes called the double product or rate-pressure product, is a very important measurement because it shows a consistent relationship with angina under various kinds of work load with anginal pain appearing time and again at the same value of double product (47). The reproducibility of this measurement is excellent (3), and it has now also been validated against measured myocardial O_2 consumption in humans with a correlation of $r = .90$ (34). Thus HR $\times$ BP is thought to provide a good estimate of the work of the heart, which is quite different from the total work of the whole body, as was discussed in chapter 13 (see fig. 13.3). However, it has also been pointed out by Ellestad (15) that under certain conditions (some patients with hypertension and valvular disease), HR alone is a better estimate of the work of the heart than the double product.

It is becoming more common to make some estimate of $\dot{V}O_2$ max when assessing the responses of the postcoronary patient to an exercise rehabilitation program. Kavanagh and Shephard (33) performed tests to $\dot{V}O_2$ max on thirty-six postmyocardial infarction patients and concluded that in patients who have recovered sufficiently to enter an exercise rehabilitation program, predictions of $\dot{V}O_2$ max (as in chap. 12, Astrand test) have about the same accuracy ($\pm$ 10%) as in healthy subjects. They also concluded that direct measurements of $\dot{V}O_2$ max can often be pursued to an O_2 plateau without undue risk. Such a decision will, of course, rest with the physician responsible for each patient. Direct knowledge of the patient's aerobic capacity can be very helpful in formulating exercise prescription and in monitoring progress.

It should be pointed out that age-adjusted maximum heart rates as used in chapter 13, while valid for healthy subjects, do not hold for cardiac patients (40). Only about 15% of the cardiac population can achieve their age-adjusted MHR without other abnormalities intervening. The remaining 85% will be limited by some abnormality at a HR level 85% or less than age-predicted MHR (40). There are some cardiac patients whose HR response is lower than predicted at each work load (in the absence of training effect). Ellestad (15) uses the term chronotropic incompetence to describe such responses and has found this to have a bad prognostic implication.

Safety and Litigation Experience. The techniques, safety, and litigation experience of seventy-three medical centers have been surveyed and reported (48). The mortality rate was 16 deaths in 170,000 stress tests, or about 0.01%. The combined incidence of mortality and morbidity was about 0.04%. Successful litigation with an out-of-court settlement was reported in one instance.

The Exercise Prescription

In general the principles discussed with respect to prescription of exercise for healthy normals in chapter 13 are still applicable here, but with a shifting of responsibility from the physical educator to the physician. Also, in the cardiac rehabilitation program, exercise is no longer just a means for improving health and fitness, but now becomes a definite therapeutic agent designed to promote a beneficial *clinical* effect, and as such, has specific indications and contraindications as well as potential toxic or adverse effects. Therefore the design of the exercise prescription is usually the responsibility of the physician (often with the help of a trained physical educator), and the implementation is carried out by the physical educator under the supervision of the physician.

Types of Exercise. All of the principles governing type of exercise discussed in chapter 13 are appropriate here, with added emphasis on the precautions necessary to hold to a minimum the work of the heart by utilizing exercise modalities that minimize the use of small muscles and isometric contraction and maximize the use of large muscles in rhythmic contractions with ample relaxation phases (10, 11, 12, 42).

Additionally, patient preferences and motivation should be considered. An interview can define the individual's skills in various exercises and sports activities and preference for group or individual program.

Exercise Intensity. The concept of prescription of intensity (as in chap. 13) on the basis of minimum HR, target HR, and do-not-exceed HR is equally appropriate here, but the implementation is quite different. The three HR levels will ordinarily be defined by the physician on the basis of limiting symptoms rather than on performance parameters. The target HR to be maintained during the active training period should be one which does not incur more than 0.3 mv of ST displacement, excessive blood pressure rise, or symptoms, but is 70%–85% of the maximum HR (MHR) attained on the exercise tolerance test (7, 30). Thus a person whose test was terminated because of angina at HR = 120 is told to train at a HR between 84 and 102 (target HR and do-not-exceed HR, respectively) (30).

Exercise Duration. As has been discussed in chapter 13, the training period should be preceded by a warm-up of five to ten minutes and followed by a cool-down period of five to ten minutes. The duration of the actual training period should be from twenty to sixty minutes but must be adjusted to the individual's physical capacity (7). Those who are poorly conditioned or

Table 14.5
Interval Training Prescription for Calibrated Bicycle

Exercise tolerance test information: The patient was stopped by symptoms after two minutes of bicycling at 600 kpm work load.

First Exercise Prescription
Interval Training Prescription for Calibrated Bicycle
  Type: Bicycle exercise at 50 rpm three times a week

Period	Elapsed Time (min)	Intensity (kpm)	Duration Work: Rest (min)	Frequency (repetitions)
Warm-up	0–3	150	2:1	1
	3–6	300	2:1	1
Training	6–27	450*	2:1	6
Cool-down	24–27	300	2:1	1
	27–30	150	2:1	1

*The training work load is 75% of maximum work load attained on the stress test. This prescription recommends twelve minutes of effort at the training level for the first week. During the second week, seven repetitions (fourteen minutes at training level) should be done; during the third week, eight repetitions, etc.

Subsequent Exercise Prescriptions

Prescription Number	Starting Date	Work Load (kpm)	Duration Work: Rest Ratio (min)	Frequency (repetitions)
1	January 1	450	2:1	6–9
2	February 1	450	2:1	7–10
3	March 1	450	3:2	6–9
4	April 1	450	3:2	7–10
5	May 1	450	4:2	6–9
6	June 1	450	4:2	7–10
7	July 1	600	2:1	6–9

• During the training phase, the work load should be 150 kpm below that which induced clinical symptoms or 2 mm ST alteration.
• On alternate prescriptions the number of repetitions is increased, while on the intervening prescription the length of the work periods may be increased.
• When the patient reaches 4:2 work:rest ratio (in minutes) at seven to ten repetitions, the patient may increase the intensity of the effort.
• Patients may be monitored on every other prescription. Progress is thus evaluated at least every eight to twelve weeks. In the recreational setting ST change should be less than 2 mm (0.2 mv).
From *Exercise Testing and Training of Individuals with Heart Disease or at High Risk for Its Development: A Handbook for Physicians.* American Heart Association, 1975.

very ill require longer durations at lower intensities than those who are better trained or less severely ill, who may train at higher intensities for shorter periods (7).

Frequency of Workout. The principles established in chapter 13 for frequency of workout for the normal, healthy individual apply equally well here.

Format for the Exercise Prescription. Examples of exercise prescription based on tolerance tests as recommended by the American Heart Association (7) are shown in tables 14.5 and 14.6. Patients may also be trained on a jog-walk program, using the methods of chapter 13 but substituting the HR at the symptomatic threshold for the MHR used for normals. Then the target HR and do-not-exceed HR become 70% and 85% of those values respectively

Table 14.6
Unsupervised Walking Program

Exercise tolerance test information: A patient is stopped by 3+ angina (heart rate 130) after 2.5 minutes at 3 mph on a 10% upgrade on the treadmill. Angina and ST segment depression began at 2.5 mph at a heart rate of 120.

First Exercise Prescription
    2.5 mph at 10% grade = 6 METs = 21 ml O_2/kg·min.
This is the angina threshold.
    Train at 75% of 6 METs = 4.5 METs = walking at 3.0–3.5 mph on level ground, daily.

Period	Intensity (METs)	Intensity (ml O_2/kg·min)	Equivalent Exercise
Warm-up	2–3	7–11	Walk ¼ mile in 7.5 min (approx. 2 mph)
Training	4.5	14–18	Walk 1 mile in 20 min (3 mph)
Cool-down	2–3	7–11	Walk ¼ mile in 7.5 min

Subsequent Exercise Prescriptions
    Using the same warm-up and cool-down patterns, alter the training period as follows:
    1. Walk two miles in forty minutes daily for three weeks (4.5 METs for twice the duration).
    2. Walk two miles in thirty-five minutes daily for three weeks (approx. 7 METs for nearly the same duration).
    3. Retest. If the patient completes the 4 mph stage at 10% grade on the treadmill with 3 mm ST depression (28 ml O_2/kg·min, 8 METs) and develops 1+ angina at the 3.5 mph stage, the patient should then:
    4. Walk two miles in thirty-five minutes, increasing to three miles in fifty-one minutes within three weeks.
    5. Increase to three miles in forty-five minutes for three weeks.

From *Exercise Testing and Training of Individuals with Heart Disease or at High Risk for Its Development: A Handbook for Physicians.* American Heart Association, 1975.

(7). The patients are then advised to monitor their HR after two-minute and then five-minute stints in any organized program of supervised exercise of an endurance type such as jog-walk, cycling, or swimming. If HR approaches target HR and remains between target and do-not-exceed HR for the necessary training period when monitored at five-minute intervals, the supervised program is probably acceptable. Here the trained physical educator can provide instruction in lessening the intensity of the training, when necessary, by decreasing cadence or slowing speed of running or swimming.

Program Development

The entire program, including the underlying principles and the methods to be used, must be explained to the patient at the beginning of the program. The collaborative efforts of physician and physical educator probably result in the best communication.

1. *Physiological Basis of the Disease.* The patient needs to be aware of the rudimentary anatomy and physiology underlying the cardiac disease processes. Well-prepared audiovisual aids can help the patient understand the limitations imposed by heart disease and also the potential for rehabilitation.

2. *Physiological Basis for Rehabilitation.* The principles of training and conditioning as presented in chapter 13 can be used to help the patient better understand what the rehabilitation program is all about.

3. *Monitoring Heart Rate.* Since the best single parameter for monitoring the response of the patient to the combined stresses of the exercise plus environment is the exercise heart rate, the patient must be taught how to take his own pulse rapidly and accurately. This ability must be checked and can be done most easily by use of a cardiotachometer, but can also be done by the methods described in chapter 13 with a stopwatch and stethoscope.

4. *Precautions to Be Observed.* The patient should be made aware of the various symptoms and their meaning. One's own sensing of special types of discomfort or distress may signal the presence of an adverse response. The patient should be clearly informed that these are signals to stop or reduce the intensity of effort. Above all, the patient must be impressed with the fact that overdoing is counterproductive, even for normal individuals, and is hazardous as well for the cardiac patient.

It must also be pointed out in the clearest terms that there is to be no interindividual competition. This may require frequent admonition for those who have led competitive lives. This point is important not only because of potential overdoing, but also because of the doubly hazardous likelihood of undesirable catecholamine responses.

5. *Prevention of Muscle Soreness.* Previous investigators have reported rather large dropout rates due to muscle and joint soreness (38, 45). However, the author (9) has found it possible to minimize this problem, even when working with men in their seventh, eighth, and ninth decades in jog-walk

training programs. The important points here are (1) start at a level well within the individual's capacity; (2) use very gentle progression in increasing the intensity and duration; (3) minimize accelerated and decelerated (jerky) movements; and most important, (4) apply the static stretching principles of chapter 21 following each workout.

6. *Retesting Schedule.* According to AHA recommendations, the patient should report back to the physician after the first week to discuss heart rate responses, symptomatic responses, and exercise pattern. Within six weeks, the patient should be reevaluated clinically and probably also by tolerance test or simple ECG monitoring to revise the exercise prescription (7).

7. *Is Athletic Competition Possible?* The discussion thus far has dealt only with the program in the early (and probably most hazardous) period in the rehabilitation program. How far can the individual go with respect to picking up old interests in competitive athletics and such physically demanding activities as marathon running? The answer to this question depends to the greatest extent on the severity of the disease processes, and these can only be evaluated by the skilled cardiologist. However, the outer boundaries have been explored, and while the results of investigation in this field cannot be termed conclusive, they do permit some cautious optimism for those with lesser levels of coronary disease.

As early as 1968, Hellerstein, one of our American pioneers in cardiac rehabilitation, reported on 254 CHD patients whose training results were very impressive. Outstanding among his patients was a forty-two-year-old businessman who had suffered a documented myocardial infarct and who trained to the point where he could run five miles in thirty-nine minutes without ischemic changes or discomfort (29). This compares favorably with the outstanding results found by Naughton (40) on one patient and reported on earlier in this chapter. Perhaps the most startling evidence of high level performance by CHD patients was furnished by Gottheiner of Tel Aviv, Israel. This author had the good fortune to listen to Gottheiner's presentation in Rome at the sports medicine meeting held in conjunction with the 1960 Olympics where he presented movie evidence of his patients performing in competition in sprint running, distance running, and even weight lifting. It was not until some years later that his data were published. They showed that great care in selection of patients and training over periods of several years preceded such performances (26).

Gottheiner used a progressive classification system of seven levels of physical activity groups. Three were preparatory and the upper four were sports classes. Class 1 started with breathing exercises and slow walking for twenty minutes per day. In accordance with their cardiac status, the class 1 participants added warming-up and strength-building exercises to enter classes 2 and 3. Admission to the lowest level sports class (class 4) was usually possible after nine months of preparation. Further progression was slower until qualification for participation in competitive sports teams in class 7 was attained.

Gottheiner's results over the years, involving some 3,000 subjects to the date of his report, showed 55% graduating through all seven classes, 25% reaching classes 5 and 6, and 20% remaining in the lower classes. These data appear overly optimistic, however, in that 910 of the 3,000 were precardiacs with no definite cardiovascular diagnosis. The philosophy is worthy of consideration in that the program for rehabilitation by means of outdoor sports activities, including certain competitive sports, offers more variety and stimulation than most exercise programs, which may result in superior emotional, environmental, and physical advantages. Whether such a program is feasible in the medicolegal environment of the U.S. is another question.

More recent evidence regarding high level performance by CHD patients showed that five patients with recent histories of disease ranging from asymptomatic to severe coronary artery disease were capable, after a training program of long distance running, to compete in the 1974 Honolulu marathon (standard course of over twenty-six miles) (13). Thus it is possible for at least some carefully selected CHD patients to undergo and benefit by a rigorous aerobic conditioning program, but this requires close medical supervision and careful coaching.

SUMMARY

1. Evidence from autopsies performed on battle casualties of the Korean War suggest that the majority of young people have the beginnings of coronary artery disease.

2. It seems likely that health and physical education professionals can make important contributions to improving the life style of the American population at all ages through the appropriate use of physical conditioning, weight control, improvement of nutrition, and instruction in relaxation skills.

3. The myocardium is capable of very little anaerobic metabolism, and since O_2 extraction is already close to maximum at rest, the heart is almost completely dependent upon increased coronary blood flow to meet the increased O_2 demand of exercise.

4. Angina pectoris is a severe constricting pain in the chest that results from ischemia in the heart muscle usually caused by coronary artery disease. It is typically transient and reversible when myocardial O_2 demand is decreased to the level of the O_2 supply made available by the coronary arteries.

5. When myocardial ischemia is of sufficient severity and duration to cause cell death, the individual is said to have suffered a myocardial infarction, which the layman calls a heart attack.

6. Sudden death accounts for the majority of deaths from CHD and is due to ischemic changes that affect the rhythm of the heart, causing such

potentially lethal effects as ventricular tachycardia and fibrillation with ultimate cardiac standstill and death.

7. Although the major cause of CHD is undoubtedly the occlusion of the coronary arteries by atherosclerotic plaque formation, some cardiologists believe there is also a contributing effect brought about by neurogenic-metabolic factors and hormonal effects.

8. It is also believed by some cardiologists that personality pattern type A, which is typified by aggressive involvement in a chronic, incessant struggle for achievement, is an important determinant of CHD.

9. Epidemological studies have identified the most important risk factors for CHD as male sex, age over thirty-five, smoking, high blood pressure, high serum cholesterol, glucose intolerance, and ECG abnormalities.

10. While physiological evidence for the therapeutic and prophylactic effects of exercise with respect to CHD is still inconclusive, many physiological changes known to be effected by conditioning can be mustered to support this hypothesis. Important literature reviews are available and cited.

11. It has been estimated that 80% of the survivors of myocardial infarction should be able to return to work and normal levels of physical activity.

12. While many investigators have demonstrated significant improvement in PWC after cardiac rehabilitation, there is as yet no clear evidence for improvement of coronary collateral circulation in humans.

13. Recent evidence suggests that the major effect of exercise training on angina patients is related to a reduction in myocardial O_2 demand.

14. Available evidence suggests that exercise training in CHD results in a considerable reduction in mortality rate.

15. Virtually all investigators seem to find that exercise training results in an increased angina threshold.

16. Most subjects trained after a heart attack can achieve the performance levels of normal, untrained sedentary subjects.

17. Normal and abnormal responses of the CHD patient to exercise testing are defined and discussed.

18. The minimal parameters to be measured and recorded in CHD patient exercise testing are ECG, and HR and BP from which the double product is calculated. Also desirable is the measurement of VO_2 with all possible derived parameters.

19. In a survey of seventy-three medical centers that had conducted a total of 170,000 exercise stress tests, the mortality rate among cardiac patients was 0.01%, and the combined mortality and morbidity rate was 0.04%. Successful litigation with an out-of-court settlement was reported in one case.

20. For CHD patients, exercise intensity must be prescribed by the physician on the basis of maximal limits determined by *symptoms* instead of performance as used with normals.

21. The CHD patient should be educated before entry into the rehabilitation program about the nature of the limitations imposed by the disease and the potential for improvement, as well as methods of HR monitoring and important precautions to be observed.

REFERENCES

1. Adams, W.C.; McHenry, M.M.; and Bernauer, E.M. Long-term physiologic adaptations to exercise with special reference to performance and cardiorespiratory function in health and disease. Chap. 24 in *Exercise in Cardiovascular Health and Disease,* eds. E.A. Amsterdam, J.H. Wilmore, and DeMaria. New York: Yorke Medical Books, 1977.
2. Amsterdam, E.A., and Mason, D.T. Coronary artery disease: pathophysiology and clinical correlations. Chap. 2 in *Exercise in Cardiovascular Health and Disease,* eds. E.A. Amsterdam, J.H. Wilmore, and DeMaria. New York: Yorke Medical Books, 1977.
3. Blomqvist, G., and Atkins, J.M. Repeated exercise testing in patients with angina pectoris: reproducibility and follow-up results. Abstract of paper to American Heart Association, Anaheim, Calif., 1971.
4. Bonanno, J.A. Coronary risk factor modification by chronic physical exercise. Chap. 19 in *Exercise in Cardiovascular Health and Disease,* eds. E.A. Amsterdam, J.H. Wilmore, and DeMaria. New York: Yorke Medical Books, 1977.
5. Brand, R.J.; Rosenman, R.H.; Shultz, R.I.; and Friedman, M. Multivariate prediction of coronary heart disease in the western collaborative group study compared to the findings of the Framingham Study. *Circulation* 53:348–55, 1976.
6. Clausen, J.P.; Larsen, O.A.; and Trap-Jensen, J. Physical training in the management of coronary artery disease. *Circulation* 40:143–54, 1969.
7. Committee on Exercise, American Heart Association. *Exercise Testing and Training of Individuals with Heart Disease or at High Risk for Its Development: A Handbook for Physicians.* New York: The Association, 1975.
8. Costill, D.L.; Branam, G.E.; Moore, J.C.; Sparks, K.; and Turner, C. Effects of physical training in men with coronary heart disease. *Med. Sci. Sports* 6:95–100, 1974.
9. deVries, H.A. Physiological effects of an exercise training regimin upon men aged 52–88. *J. Gerontol.* 25:325–36, 1970.
10. deVries, H.A., and Adams, G.M. Comparison of exercise responses in old and young men: 1. The cardiac effort/total body effort relationship. *J. Gerontol.* 27:344–48, 1972.
11. ———. Muscle mass vs. relative loading of individual muscles as determinants of exercise response in older men. *Med. Sci. Sports* 4:146–54, 1972.
12. ———. Effect of the type of exercise upon the work of the heart in older men. *J. Sports Med. Phys. Fitness* 17:41–47, 1977.
13. Dressendorfer, R.H.; Scaff, J.H.; Wagner, J.O.; and Gallup, J.D. Metabolic adjustments to marathon running in coronary patients. *Ann. N.Y. Acad. Sci.* 301:466–83, 1977.
14. Eckstein, R.W. Effect of exercise and coronary artery narrowing on coronary collateral circulation. *Circ. Res.* 5:230–35, 1957.

15. Ellestad, M.H. Stress testing, principles, and practice. Philadelphia: F.A. Davis Company, 1975.

16. Enos, W.F.; Holmes, R.H.; and Beyer, J. Coronary disease among United States soldiers killed in action in Korea. *J.A.M.A.* 152:1090–93, 1953.

17. Ferguson, R.J.; Petitclerc, R.; Choquette, G.; Chaniotis, G.; Gauthier, P.; Huot, R.; Allard, C.; Jankowski, L.; and Campeau, L. Effect of physical training on treadmill exercise capacity, collateral circulation, and progression of coronary disease. *Am. J. Cardiol.* 34:764–69, 1974.

18. Ferguson, R.J.; Cote, P.; Gauthier, P.; and Bourassa, M.G. Changes in exercise coronary sinus blood flow with training in patients with angina pectoris. *Circulation* 58:41–47, 1978.

19. Folkins, C.H., and Amsterdam, E.A. Control and modification of stress emotions through chronic exercise. Chap. 20 in *Exercise in Cardiovascular Health and Disease,* eds. E.A. Amsterdam, J.H. Wilmore, and DeMaria. New York: Yorke Medical Books, 1977.

20. Fox, S.M., and Paul, O. Physical activity and coronary heart disease. *Am. J. Cardiol.* 23:298–306, 1969.

21. Fox, S.M.; Naughton, J.P.; and Gorman, P.A. Physical activity and cardiovascular health. *Mod. Concepts Cardiovasc. Dis.* 41:17–20, 1972.

22. Frick, M.H., and Katila, M. Haemodynamic consequences of physical training after myocardial infarction. *Circulation* 37:192–202, 1968.

23. Friedman, Meyer. Pathogenesis of coronary artery disease. New York: McGraw-Hill Book Co., 1969.

24. Gordon, T.; Sorlie, P.; and Kannel, W.B. *Coronary heart disease, atherothrombotic brain infarction, intermittent claudication—a multivariate analysis of some factors related to their incidence: Framingham Study, 16-year followup,* Section 27. Washington, D.C.: U.S. Government Printing Office, 1971.

25. Gordon, T., and Kannel, W.B. *Coronary risk handbook.* New York: American Heart Association, 1973.

26. Gottheiner, V. Long-range strenuous sports training for reconditioning and rehabilitation. *Amer. J. Cardiol.* 22:426–35, 1968.

27. Hartung, G.H. Physical activity and coronary heart disease risk—a review. *Am. Correct. Ther. J.* 31:110–15, 1977.

28. Haskell, W.L. Physical activity after myocardial infarction. *Am. J. Cardiol.* 33:776–83, 1974.

29. Hellerstein, H.K. Exercise therapy in coronary disease. *Bull. N.Y. Acad. Med.* 44:1028–47, 1968.

30. Hellerstein, H.K.; Hirsch, E.L.; Ader, R.; Greenblot, N.; and Siegel, M. Principles of exercise prescription for normals and cardiac subjects. In *Exercise Testing and Exercise Training in Coronary Heart Disease,* eds. J. Naughton, and H.K. Hellerstein. New York: Academic Press, 1973.

31. Hurst, J.W., and Logue, B., eds. *The Heart.* New York: McGraw-Hill, 1974.

32. Kannel, W.B.; Castelli, W.P.; Verter, J.; and McNamara, P.M. Relative importance of factors of risk in the pathogenesis of coronary heart disease: the Framingham Study. In *Coronary Heart Disease,* eds. H.I. Russek, and B.L. Zohman. Philadelphia: J.B. Lippincott Co., 1971.

33. Kavanagh, T., and Shephard, R.J. Maximum exercise tests on "postcoronary patients." *J. Appl. Physiol.* 40:611–18, 1976.

34. Kitamura, K., et al. Hemodynamic correlates of myocardial oxygen consumption during upright exercise. *J. Appl. Physiol.* 32:516–22, 1972.

35. Klouda, M.A., and Randall, W.C. Subendocardial hemorrhages during stimulation of the sympathetic cardiac nerves. Chap. VI in *Prevention of Ischemic Heart Disease,* ed. W. Raab, pp. 49–56. Springfield: Charles C Thomas, 1966.

36. Lapiccirella, V. Emotion-induced cardiac disturbances and possible benefits from tranquil living. Chap. XXVI in *Prevention of Ischemic Heart Disease,* ed. W. Raab, pp. 212–16. Springfield: Charles C Thomas, 1966.

37. Levi, L. Life stress and urinary excretion of adrenaline and noradrenaline. Chap. X in *Prevention of Ischemic Heart Disease,* ed. W. Raab, pp. 85–95. Springfield: Charles C Thomas, 1966.

38. Mann, G.V.; Garrett, H.L.; Farhi, A.; Murray, H.; and Billings, F.T. Exercise to prevent coronary heart disease. *Am. J. Med.* 46:12–27, 1969.

39. Melville, K.I. Cardiac ischemic changes induced by central nervous system stimulation. Chap. IV in *Prevention of Ischemic Heart Disease,* ed. W. Raab, pp. 31–38. Springfield: Charles C Thomas, 1966.

40. Naughton, J. Cardiac rehabilitation: principles, techniques, applications. Chap. 26 in *Exercise in Cardiovascular Health and Disease,* eds. E.A. Amsterdam, J.H. Wilmore, and DeMaria. New York: Yorke Medical Books, 1977.

41. Neill, W.A. Coronary and systemic circulatory adaptations to exercise training and their effects on angina pectoris. Chap. 9 in *Exercise in Cardiovascular Health and Disease,* eds. E.A. Amsterdam, J.H. Wilmore, and DeMaria. New York: Yorke Medical Books, 1977.

42. Nutter, D.O.; Schlant, R.C.; and Hurst, J.W. Isometric exercise and the cardiovascular system. *Mod. Concepts Cardiovasc. Dis.* 41:11–15, 1972.

43. Perper, J.A.; Kuller, L.H.; and Cooper, M. Arteriosclerosis of coronary arteries in sudden, unexpected deaths. *Circulation,* suppl. 3, 52:27–33, 1975.

44. Pisa, Z. Programme of the European office of WHO in rehabilitation of cardiac patients. *Acta Cardiol.,* suppl. 14, 1970.

45. Pollock, M.L.; Gettman, L.R.; Milesis, C.A.; Bah, M.D.; Durstine, L.; and Johnson, R.B. Effects of frequency and duration of training on attrition and incidence of injury. *Med. Sci. Sports* 9:31–36, 1977.

46. Raab, W. Myocardial electrolyte derangement: crucial feature of pluricausal so-called coronary heart disease. *Ann. N.Y. Acad. Sci.* 147:627–86, 1969.

47. Robinson, B.F. Relation of heart rate and systolic blood pressure to the onset of pain in angina pectoris. *Circulation* 35:1073–83, 1967.

48. Rochmis, P., and Blackburn, H. Exercise tests—a survey of procedures, safety, and litigation experience in approximately 170,000 tests. *J.A.M.A.* 217:1061–66, 1971.

49. Scheuer, J., and Tipton, C.M. Cardiovascular adaptations to physical training. *Ann. Rev. Physiol.* 39:221–51, 1977.

50. Sim, D.N., and Neill, W.A. Investigation of the physiological basis for increased exercise threshold for angina pectoris after physical conditioning. *J. Clin. Invest.* 54:763–70, 1974.

51. Wenger, N.K.; Hellerstein, H.K.; Blackburn, H.; et al. Uncomplicated myocardial infarction; current physician practice in patient management. *J.A.M.A.* 224:511–14, 1973.

15

Metabolism and Weight Control

Weight control is a major component of physical fitness for adults; it is a less important factor in the physical fitness of children, but only because children are usually more active. We immediately recognize the desirability of normal weight in respect to appearance, and in physical performance obesity is a distinct disadvantage because a large proportion of the body weight, which does not contribute to performance, must nevertheless be moved at a definite cost in terms of energy. Thus an athlete who carries an excess twenty pounds of fat would compete on equal terms with athletes of normal weight only if they were forced to carry twenty-pound weights about their middle, and the rules of athletic participation have not yet provided such an equalizer for the fat man.

Most important, obesity has been shown to be associated with increased incidence of diabetes, gallstones, high blood pressure, and heart disease. Man has learned only very recently to produce food in superabundance, and only in the industrially advanced cultures; consequently, obesity as an endemic problem is also relatively new. As with most emerging health problems, passage of time is required before the facts can be sifted from the misinformation and the old wives' tales. It is the purpose of this chapter to synthesize the experimentally established facts into a practical approach to the problem of weight control.

Physiology of Weight Gain and Weight Loss

First, we must recognize the fact that the human organism is a heat exchange engine, and, although "wondrously and fearfully" constructed, must obey all the physical laws that govern energy exchange. The net energy exchange that expresses the process of metabolism most simply can be written:

Caloric balance = kilocalories from food − (kilocalories of basal metabolism + kilocalories of work metabolism + kilocalories lost in excreta)

It can be seen that if the energy intake exceeds the energy outgo, an individual is in positive energy balance. Since the law of the conservation of energy tells us that energy can neither be gained nor lost but only changed in form, we must look for this energy to be deposited in the form of body fat, which is indeed what happens. One gram of fat produces (or can be considered equal to) approximately 9.3 kcal. Allowing for the water content and connective tissue in fat tissue, then one pound of fat will be deposited in the body when an excess of approximately 3,500 kcal has been consumed.

Conversely, if the energy spent is greater than the energy consumed, there is a negative caloric balance. For a negative balance of 3,500 kcal, a pound of fatty tissue would have been lost.

Physiology Applied to Health and Fitness

A most important point here is that the metabolism equation does not dictate the *rate* at which weight can be gained or lost. Obviously a pound of weight is lost if we go into a negative caloric balance of 3,500 kcals at a rate of 100 kcals per day for thirty-five days or 350 kcals for ten days. In the first case we lose one pound in thirty-five days, and in the second case one pound in ten days. Many people have been discouraged from using exercise to reduce weight because of misleading salesmanship which stated the need for thirty-six hours of walking or other ridiculously heavy work loads to lose one pound of weight. It is indeed undesirable, as well as impossible, to lose one pound per day in this fashion. However, by applying only a very low level of salesmanship and *sound physiology,* we might say that walking an extra half hour per day would result in a weight loss of five pounds per year; and it is the *long haul* that counts.

Metabolism of Carbohydrate, Fat, and Protein. It may be asked how it comes about that fatty tissue is deposited—in keeping with the energy balance equation—if a person eats a balanced diet that consists of all three basic foodstuffs, or even a pure carbohydrate or protein diet. The discussion in chapter 3 and figure 3.2 illustrate the fact that the three different foodstuffs have a common path in the final stages of their metabolic breakdown.

In the case of a negative caloric balance, as exists during dieting, it is easy to understand how stored fat may be utilized as a source of energy. In fact, the fatty tissues that are found beneath the skin between the muscles and padding the viscera are in a constant state of flux. Neutral fat from the blood constantly replenishes the fat stores of the various fat cells, which release them when they are needed for energy purposes.

When a positive energy balance exists, synthesis of fat tissue from the excess carbohydrate or protein occurs in the liver, from whence it is transported to the fat cells. Some synthesis of fat from glycerol and fatty acids also occurs in the fat tissue cells themselves. In this fashion, weight gain (of fatty tissues) occurs when the food intake (energy) is greater than the energy output.

Although the basic laws of energy balance are always applicable, evidence is accumulating that people vary in the methods by which they metabolize food (14), and because of this there are differences in the efficiency with which food is converted into energy. Differences in the efficiency of food utilization probably account for the fact that some people can "eat like a horse" and remain thin, while others "eat like a bird" and become obese.

What Is Normal Weight?

Overweight and *underweight* are widely used terms, and this use implies that we know what constitutes normal weight for a given individual. Ordinarily, normal weight is predicted from tables that have been developed by insurance actuaries and that provide minimum, average, and maximum weights for any given age, height, and sex. But does such a table really tell us what one's proper weight may be? We could answer this question in the affirmative only if the data from which the tables were calculated were taken from a population of people whose weights were normal, and obviously this situation does not exist.

By the use of such age-height-weight tables, gross errors are not uncommon in assessing normal weight. For example, a man six feet tall, with a very light skeletal framework, might be 30 to 40 pounds overweight at 200 pounds, whereas an extreme mesomorph (heavy skeleton and musculature) might be at his best weight for athletic competition at 200 pounds.

Furthermore, age-height-weight tables commonly allow small increments in body weight with increasing age, and this concept also is erroneous. Air Force standards have been recommended that do not allow weight increases with age; these tables essentially retain the current recommended weight for ages twenty-six through thirty as applicable to all ages. This is a more logical approach to the prediction of normal weight because of evidence that during each decade after age twenty-five the body loses about 3% of its metabolically active cells. If this loss of tissue is replaced, it is probably replaced by fat tissue, so that even if an individual maintains constant weight while growing older, that person probably carries an increasing proportion of fat tissue. Obviously, it is the proportion of fat tissue in the body's composition rather than the reading on a scale that is of paramount importance.

According to U.S. Air Force standards, 115% of the standard weight is defined as overweight. For young males, *obesity* is frequently defined as the condition in which more than 20% of the body weight is composed of fat tissue. Thirty percent is the cut-off point for females.

When fifty-one male USAF personnel were compared by these two standards (37), it was found that fifteen males who were not 15% over the standard weight were nevertheless obese (more than 20% body fat). Furthermore, six men who would have been considered overweight by the tables were found to have less than 20% body fat, and consequently were not really obese. Thus twenty-one of the fifty-one cases would have been incorrectly classified by use of the age-height-weight tables alone. This clearly illustrates the need for estimation of body composition rather than a complete reliance upon tables of averages. It should be obvious that overweight due to a preponderance of bone and muscle does not have the same significance as overweight due to fatty tissue. Fortunately, methods have been devised for this discrimination.

Methods for Estimation of Body Composition

Underwater Weighing. It is common knowledge that fat people float better than thin people, and this is because fat tissue is less dense than other tissue (except lung tissue). Consequently, underwater weighing, which provides measures of body density and specific gravity, can also provide reasonably accurate estimates of the proportions of *lean body weight* and *body fat tissue* (21).

In this procedure the subject is completely submerged. Then, by Archimedes' principle, an individual's specific gravity is calculated:

$$\text{Specific gravity} = \frac{\text{Dry weight}}{\text{Loss of weight in water}}$$

This value must be corrected for residual lung volume, which is determined by a nitrogen washout of the lungs. With the corrected specific gravity, one may enter tables to arrive at the percent of body fat. The normal body fat percentage for young men has been estimated at 10% to 15% by various investigators; the normal value for young women is slightly higher (15%-20%).

Measurement of Body Volume. Specific gravity of the human body can also be calculated if its volume is known.

$$\text{Specific gravity} = \frac{\text{Weight of body (dry)}}{\text{Weight of equivalent volume of water}}$$

This technique also involves complete submersion of the subject, with measurement of the water displacement by introduction of the subject's body into a small tank whose shape is such that small volume changes make large changes in water level. The tank is called a *volumeter,* and the measurement must be corrected for residual lung volume.

Hydrometric Method. This method depends upon the principle that the proportion of fat-free body weight that is water can be assumed to be constant, at approximately 72%; therefore, any of the chemical methods by which the dilution of a solute by the body water can be calculated can yield data on the total body water and, indirectly, the fat-free body weight. This is usually done by having a subject drink a measured amount of *heavy water,* deuterium oxide. Since this heavy water is handled by the human body in exactly the same way as regular water, the amount of deuterium oxide excreted in the urine can be used as the basis for calculation of total body water. The calculation for fat-free body weight is simply:

$$\text{Fat-free weight} = \frac{\text{Total body water}}{0.72}$$

The methods discussed thus far require considerable time and laboratory facilities. Simpler methods have been proposed that depend upon skin-fold measures to estimate subcutaneous fat tissue or upon a combination of anthropometric measures to estimate the size of the bony framework. These methods are of a lower order of accuracy, but are nevertheless far better criteria of the degree of obesity or normality of body weight than age-height-weight tables.

Estimation of Body Fat from Skin-Fold Measures. Over the past several decades many studies have dealt with the use of skin-fold measurements to estimate body fatness. Different investigators have used different body sites at which skin-folds were measured and have developed various regression equations relating the thickness of the various skin-folds to the percentage of fat measured by underwater weighing as the criterion variable. To add further confusion, until very recently no one had developed a procedure that could be applied to both sexes and all ages.

Recently, however, Durnin and Womersley (10) have provided regression equations for both sexes and ages from seventeen to seventy-two for estimating percent body fat from the sum of four standard skin-folds. Table 15.1 provides the data divided by sex and age. The skin-folds are taken to the nearest millimeter, except for low values (5 mm or less), which are taken to the closest 0.5 mm. Several different skin-fold caliper instruments are available that provide accurate measurements under constant skin-fold pressure. The pressure between the caliper jaws should be 10 grams/mm² regardless of the width of the jaws. The skin should be lifted by grasping a fold firmly between the thumb and forefinger about 1 cm from the site at which the skin-fold is to be measured. The skin-folds are ordinarily taken on the right side of the body in standing position for standardization, but there is no significant difference between measurements of the right and left side (38). Measurements are taken with the skin-fold vertical except where natural folding is in opposition, as is usually the case over the subscapular site where a 45° angle usually works best.

The specific sites used to apply the method of Durnin and Womersley are as follows:

Biceps—This skin fold is measured at a point halfway between the anterior axillary fold and the antecubital fossa on the anterior midline with the arm hanging freely in extension. The placement of the midpoint is best made with the arm in 90° flexion.

Triceps—This skin fold is measured at a point halfway between the tip of the acromion process and the tip of the olecranon process on the posterior midline with the arm in relaxed extension. The placement of the midpoint is best made with the arm in 90° flexion.

Subscapular—This skin fold is taken at the tip of the scapula on an oblique angle along the natural line of folding of the skin.

Suprailiac—This skin fold is taken just above the iliac crest in the midaxillary line.

Having gotten these four skin-fold values, one takes the sum and enters the appropriate column in table 15.1 to arrive at the estimated value of fat content as a percentage of body weight (with a standard error of ± 3.5% for women and ± 5% for men). Morbidity and mortality statistics are needed to define the optimal value of percent body fat for good health. In the absence

Table 15.1

The Equivalent Fat Content as a Percentage of Body Weight for a Range of Values for the Sum of Four Skinfolds (Biceps, Triceps, Subscapular, and Suprailiac) of Males and Females of Different Ages.

Skinfolds (mm)	Males (age in years)				Females (age in years)			
	17–29	30–39	40–49	50+	16–29	30–39	40–49	50+
15	4.8	—	—	—	10.5	—	—	—
20	8.1	12.2	12.2	12.6	14.1	17.0	19.8	21.4
25	10.5	14.2	15.0	15.6	16.8	19.4	22.2	24.0
30	12.9	16.2	17.7	18.6	19.5	21.8	24.5	26.6
35	14.7	17.7	19.6	20.8	21.5	23.7	26.4	28.5
40	16.4	19.2	21.4	22.9	23.4	25.5	28.2	30.3
45	17.7	20.4	23.0	24.7	25.0	26.9	29.6	31.9
50	19.0	21.5	24.6	26.5	26.5	28.2	31.0	33.4
55	20.1	22.5	25.9	27.9	27.8	29.4	32.1	34.6
60	21.2	23.5	27.1	29.2	29.1	30.6	33.2	35.7
65	22.2	24.3	28.2	30.4	30.2	31.6	34.1	36.7
70	23.1	25.1	29.3	31.6	31.2	32.5	35.0	37.7
75	24.0	25.9	30.3	32.7	32.2	33.4	35.9	38.7
80	24.8	26.6	31.2	33.8	33.1	34.3	36.7	39.6
85	25.5	27.2	32.1	34.8	34.0	35.1	37.5	40.4
90	26.2	27.8	33.0	35.8	34.8	35.8	38.3	41.2
95	26.9	28.4	33.7	36.6	35.6	36.5	39.0	41.9
100	27.6	29.0	34.4	37.4	36.4	37.2	39.7	42.6
105	28.2	29.6	35.1	38.2	37.1	37.9	40.4	43.3
110	28.8	30.1	35.8	39.0	37.8	38.6	41.0	43.9
115	29.4	30.6	36.4	39.7	38.4	39.1	41.5	44.5
120	30.0	31.1	37.0	40.4	39.0	39.6	42.0	45.1
125	30.5	31.5	37.6	41.1	39.6	40.1	42.5	45.7
130	31.0	31.9	38.2	41.8	40.2	40.6	43.0	46.2
135	31.5	32.3	38.7	42.4	40.8	41.1	43.5	46.7
140	32.0	32.7	39.2	43.0	41.3	41.6	44.0	47.2
145	32.5	33.1	39.7	43.6	41.8	42.1	44.5	47.7
150	32.9	33.5	40.2	44.1	42.3	42.6	45.0	48.2
155	33.3	33.9	40.7	44.6	42.8	43.1	45.4	48.7
160	33.7	34.3	41.2	45.1	43.3	43.6	45.8	49.2
165	34.1	34.6	41.6	45.6	43.7	44.0	46.2	49.6
170	34.5	34.8	42.0	46.1	44.1	44.4	46.6	50.0
175	34.9	—	—	—	—	44.8	47.0	50.4
180	35.3	—	—	—	—	45.2	47.4	50.8
185	35.6	—	—	—	—	45.6	47.8	51.2
190	35.9	—	—	—	—	45.9	48.2	51.6
195	—	—	—	—	—	46.2	48.5	52.0
200	—	—	—	—	—	46.5	48.8	52.4
205	—	—	—	—	—	—	49.1	52.7
210	—	—	—	—	—	—	49.4	53.0

In two-thirds of the instances, the error was within ±3.5% of body weight as fat for the women and ±5% for the men.

From J.V.G.A. Durnin and J. Womersley, *Br. J. Nutr.* 32:77–97, 1974.

of such data, reasonable values are between 10%-15% body fat for the male and 15%-20% body fat for the female. When the male reaches 20% and the female 30%, the beginning of obesity is at hand.

Gaining Weight

The purposeful gaining of weight is a problem for a relatively small segment of our population; however, some persons desire to pad a slender or frail frame for purposes of appearance, and sometimes to provide a greater mass for body contact sports, such as football. It is obvious from the foregoing that this can be accomplished simply by ingesting more calories (in the form of food) than are spent in energy every day. For some individuals this constitutes a problem because their energy output may be prodigious due to a restless (nervous) temperament. Furthermore, this approach to the problem results in deposition of fat tissue that—beyond the normal values mentioned earlier—is not desirable.

The best method for gaining weight for any normal, healthy young individual consists of an exercise program designed to provide muscular hypertrophy with a minimum of energy expenditure. This type of exercise is best provided by professionally directed weight training, and almost invariably results in weight gains—of muscle tissue, not fat. This approach, if undertaken in moderation, is suitable for girls as well as boys.

Girls frequently express concern over the possibility of becoming muscular, but sex differences prevent this from happening in all but the most extreme exercise programs. Even in extremely heavy resistance exercise programs, there is little evidence to support the belief that girls may develop large muscles. It is more likely that excessively muscular girls are seen in association with heavy resistance sports because of a selective process: they choose to participate in sports in which they will excel.

Reducing Weight

Although the theory underlying weight reduction is beautifully simple, the practice for many millions of Americans is definitely not a simple process after obesity has set in. This is probably due to the interaction of psychological (emotional) and social problems with the physiology underlying the obesity (36).

From the standpoint of energy metabolism, obesity is necessarily the end result of a *positive energy balance*. Although this is a great oversimplification, it will aid temporarily in understanding the problem. From this point of view, only three alternative methods are available for the reduction of weight:

1. Increased energy expenditure and constant food intake
2. Decreased food intake and constant energy expenditure
3. A combination of methods 1 and 2

The first method can be accomplished by exercise programs, the second by diet.

Since the easiest *cure* for obesity is prevention, let us consider the etiology of this health problem. A few years back it was fashionable to place the blame for obesity upon endocrine malfunction; this was a popular theory in that an obese person could absolve himself of blame. Medical research, however, has not substantiated this theory; on the contrary, evidence is accumulating that indicts our sedentary way of life as the real culprit.

Greene (15), who studied 350 cases of obesity, found inactivity was associated with the onset of obesity in 67.5% of the cases, and that a history of increased food intake was found in only 3.2%. Pariskova (32), who analyzed the body composition of 1,460 individuals of all ages, concluded: "One of the most important factors influencing body composition is the intensity of physical activity, and this is true in youth, adulthood, and old age." In a study by Corbin and Pletcher (6), using 16 mm movie films to evaluate the activity level of elementary school children, it was shown that body fatness was significantly correlated ($r = -.520$) with lack of physical activity but not with caloric intake ($r = .155$). Many other investigations, too numerous to cite, provide indirect support for the belief that lack of physical activity is the most common cause of obesity. Thus a clear-cut case can be made for the importance of habitual, lifelong, vigorous physical activity as a preventive measure against obesity.

The Fat Cell Theory.   In the past decade much interesting research has been done to elucidate the relationships of fat cell size and number to obesity. Hirsch and Knittle (19) developed a painless method of needle aspiration of adipose tissue from subcutaneous fat depots such as the arm, abdomen, or buttock. From these tissue samples, it is possible to estimate both cell size and number. The results of such research suggest the following:

1. The number of fat cells in the human body is established during periods of rapid growth: (a) the latter part of gestation, (b) the first year of infancy, and (c) adolescence (18).
2. Once established, the number of fat cells appears to be fixed in spite of experimental weight gain or weight loss (18,5,35).
3. Nonobese human adults have about 27 billion fat cells, while the obese have from 42 to 106 billion (17).
4. Fat cell size is labile and varies with experimental weight gain or loss.
5. Comparing obese and nonobese human adults, cell size is only some 40% greater, while cell number is about 190% higher (17) in the obese.
6. Comparing individuals with varying percentages of ideal weight shows that there is a high correlation between cell number and percent ideal weight ($r = .8117$), but no significant correlation with cell size (17). Others have found moderate relationships between cell size and percent ideal weight, but only up through the range of relatively normal fat levels beyond which cell size no longer relates to total body fat (3,4).

The sum total of this very interesting research permits a cautious conclusion that gross obesity (typically of childhood-onset) is the result of an abnormally large number of fat cells laid down during growth. Lesser levels of obesity (possibly up to 150% ideal weight and typically adult-onset obesity) would appear to be due to enhanced cell size. The latter conclusion is supported by the work of Sims (35) who induced normally nonobese volunteers to greatly increase their body fat by overeating. The size of the adipose cells correlated closely with the increase in body fat, thus suggesting no change in number. Furthermore, Sims saw considerable variability in the ease with which these volunteers could gain weight. There appeared to be a correlation between the ease of weight gain and the fat cell number, thus validating to some extent the complaint of the obese individual that is so commonly heard, "My friend eats like a horse and stays thin, while I eat like a canary and gain weight."

The important question remaining is, To what extent can gross obesity (the increase in cell *number*) be prevented by exercise and or food restriction? Although we do not yet have data on human subjects, well-controlled studies have been performed by Oscai and his colleagues on rats (30,31). Their data show quite clearly that significant differences in future weight gain can be brought about by both exercise and dietary restriction during the growth period of rats. To the extent that these data may be extrapolated to the human, it would seem possible to prevent the laying down of excess numbers of adipose cells by providing adequate physical activity for young children and adolescents. Physical activity appears to be even more important than eating habits.

Feasibility of Weight Loss through Exercise. Planned exercise such as jog-walk combinations as a feasible method of weight reduction in the obese, even in the absence of any dietary restriction, has been well demonstrated by Moody, Kollias, and Buskirk (26). Obese college age women lost on the average 5.3 pounds over an eight-week period in which they participated in about an hour of jogging-walking for an average of four times per week. Skinfold measurements suggested the weight loss was the result of a much larger loss of fatty tissue with a concomitant gain of solid tissue (fat-free weight). This latter observation is very important because the converse is true when weight is lost by fasting. Fasting has been shown to result in weight losses that are, to a large extent, losses in lean body tissue, which is undesirable. Under total fasting conditions, the weight loss due to losses in lean body mass appears to be from 8% (2) to between 30% and 45% (9,12) of the total weight loss. Zuti and Golding (40) have shown clearly the advantages of including exercise in weight reduction programs. Twenty-five women were randomly assigned to a diet group, an exercise group, or a combination of the two. In each case, a negative caloric balance of 500 cal/day was established. All the women lost similar amounts of weight, but those who exercised lost more fat and gained lean tissue, while those on diet alone lost lean body mass and lesser amounts of fat.

Physiology Applied to Health and Fitness

Can we infer from the above that exercise is also the best means for treating the severely obese? Not necessarily. Although exercise can certainly make a contribution, medical supervision is necessary to protect the severely obese person from overstraining the cardiovascular system, the connective tissues, and so on. For the moderately obese (10% to 30% above predicted normal weight), a combination of diet and exercise is probably the optimal procedure.

Misconceptions in Exercise and Weight Control. Despite the evidence that has been cited in favor of exercise as a means of weight control, in recent years it has been popular to ridicule this practice. Data are presented that illustrate the need for thirty-five miles of running, or thirty-six hours of walking—or some other ridiculous amount of physical activity—to lose one pound of weight. Two outstanding authorities in the area of nutrition, Mayer and Stare (23), have given the lie to such statements, and they have provided experimental evidence to rectify other misconceptions. They have pointed out that it is neither necessary nor desirable to expend the energy required to lose one pound in one exercise bout; further, that "a half hour of handball or squash a day would be equivalent to 19 pounds per year."

Another general misconception is that exercise is not effective in weight reduction because appetite is automatically increased in direct proportion to the increased activity. Mayer and colleagues (24) have shown that while appetite follows activity in the range of normal activity in animals, this is not so in the low levels of activity. Figure 15.1 illustrates their work, and shows that sedentary animals (those most apt to be obese) actually display a decrease in appetite with an increase of up to one hour of daily exercise. Mayer and his collaborators have also shown that this principle applies to humans.

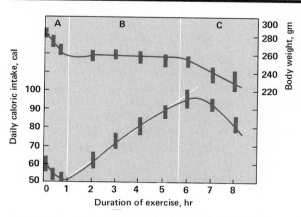

Figure 15.1 The relationship between activity and appetite and the effects upon body weight in animals. (From Mayer, J., et al. *Am. J. Physiol.* 177:544, 1954.)

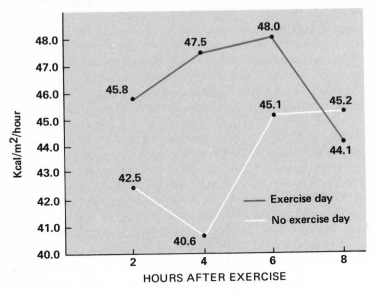

Figure 15.2 Metabolic aftereffects of a vigorous workout compared with a similar day in which no workout was taken. Each experimental point represents the mean of at least six observations for each of two middle-aged male subjects.

Metabolic Aftereffects of Exercise. The increase in metabolic rate incurred during physical activity is the main cause of energy loss. As long ago as 1933, the work of Margaria, Edwards, and Dill (22) mentioned an increased *resting* metabolic rate that lasted for several hours after completion of exercise and that could not be attributed to repayment of oxygen debt.

This increased metabolic rate was further investigated in the author's laboratory (8). In a controlled experiment it was found that the resting metabolic rate was from 7.5% to 28% higher four hours after a vigorous workout than it was at the same time of day on *nonexercise* control days. This higher metabolic rate was shown to persist for at least six hours after exercise, and this effect of exercise—over and above the energy cost of the exercise itself—would have resulted in a weight loss of four or five pounds per year if the individuals tested had exercised daily. Figure 15.2 illustrates the results of the experiment.

What Kind of Exercise Is Best? To be effective in reducing weight, exercise must be of the vigorous, endurance type so that energy expenditure may be maximized. In designing such an exercise program, the following seven factors must be considered.

1. The exercise must allow *gradual progression* from low levels to higher levels of energy expenditure.

Physiology Applied to Health and Fitness

2. Participants must be protected from injury to bony and connective tissues in the early stages.

3. The exercise must be vigorous enough to result in increased body heat, as evidenced by sweating.

4. Intensity of the exercise should be as high as possible, consistent with a duration of thirty to sixty minutes. Maximal total energy output cannot be obtained if the musculature is quickly exhausted by a few quick maximal repetitions, as in weight lifting or sprint workouts.

5. Exercise involving large components of anaerobic work do not appear to be as effective as aerobic exercise (13).

6. Soreness should be prevented, or relieved, by the procedures described in chapter 21.

7. After a minimum level of fitness is achieved, the program should be built around activities that are enjoyable and thus self-motivating.

Passmore and Durnin (33) have provided a rather complete survey of the energy requirements of various activities. Some of their data are provided in table 15.2.

Table 15.2
Energy Requirements of Various Activities

No. of Subjects	Age	Sex	Body Weight (kg)	Activity	kcal/min
12	32	M	66	Lying at ease	1.4
16	34	M	66	Sitting at ease	1.6
7	38	M	64	Standing at ease	1.9
112	young	M	..	Sitting, playing cards	2.4
3	19	M	64	Driving car	2.8
1	..		68	Volleyball	3.5
1	29	M	63	Golf	5.0
2	20	M	69	Archery	5.2
1	23	M	69	Dancing (rhumba)	7.0
4	..	M	68	Canoeing (4 mph)	7.0
7	19	M	70	Tennis	7.1
3	..	M	73	Horseback riding (trot)	8.0
3	..	M	73	Horseback riding (gallop)	10.0
4	25	M	65	Cross-country running	10.6
1	..	M	71	Cycling (13.1 mph)	11.1
1	21	M	68	Swimming (backcrawl)	11.5
1	21	M	90	Swimming (crawl, 45 yd/min)	11.5
1	..	F	57	Skiing (level, hard snow; moderate speed)	15.9
1	..	F	68	Skiing (uphill, hard snow; maximum speed)	18.6

Based on data from R. Passmore and J.V.G.A. Durnin, "Human Energy Expenditure," *Physiol. Rev.* 35:801–35, 1955.

Table 15.3
Energy Cost of Walking in Calories per Hour

Speed mph	Weight in Pounds						
	80	100	120	140	160	180	200
2.0	114	132	155	174	192	210	228
2.5	138	162	186	210	228	252	270
3.0	162	186	216	240	264	288	318
3.5	186	216	252	276	300	324	366
4.0	210	246	282	312	348	384	420

Recalculated from R. Passmore and J.V.G.A. Durnin, "Human Energy Expenditure," *Physiol. Rev.* 35:801, 1955.

For jogging or running, the energy expenditure can be estimated as 1 kcal per kilogram of body weight per kilometer covered (25). Table 15.3 shows the energy costs for various rates of walking. The data for running and walking can be applied to both male and female (11).

Dieting to Lose Weight. Severe dietary restriction is a procedure that requires medical supervision; it cannot be properly treated in this text. Moderate dietary restriction, which results in weight losses of one or two pounds per week, can be accomplished by estimation of the daily energy expenditure and maintenance of a daily food intake of 500 to 1,000 kcals per day below the expenditure level. These figures are readily available in various texts on health and nutrition. (7)

Some interesting new concepts about the physiology of weight reduction deserve comment here. It has been shown (20) that rats trained to eat their entire daily food ration in one to two hours gain more weight than animals eating ad libitum. It was further demonstrated that the trained rats increased the rate at which their adipose tissue incorporated food breakdown products into lipids (fats) by twenty-five times.

This work was extended and applied to the medical treatment of obesity in humans by Gordon, Goldberg, and Chosy (14). They initiated treatment in obese patients with a forty-eight-hour fast (to break the metabolic pattern of augmented lipogenesis), then instituted a 1,320-kcal diet that consisted of 400 kcals of protein, 720 kcals of fat, and 200 kcals of carbohydrate. This diet was given in six feedings daily, corresponding to breakfast, midmorning, lunch, midafternoon, supper, and bedtime; all feedings were approximately similar in size. They report that results have been encouraging, and they have been surprised that no patient has complained of hunger at any time, although some have lost as much as 100 pounds.

If these results can be generalized, it would seem that concentrating a large part of the daily food intake into one large meal has unfavorable metabolic

Physiology Applied to Health and Fitness

consequences (increased lipogenesis, or fat deposition). Furthermore, skipping breakfast seems undesirable. Indeed, if these data are substantiated by other investigators, a change in American eating patterns is indicated.

Water Retention in Weight Reduction Programs

Obese people are very frequently discouraged by the results of their dieting; after one, two, or even three weeks of semistarvation the scale still reads the same. If they have honestly adhered to a negative caloric balance, the reason for this phenomenon is probably water retention. It has been shown that even though body tissues are being oxidized and the end products excreted, weight may not demonstrate this loss because sufficient water is retained by the tissues to offset the weight of the oxidized tissues. This water retention cannot continue indefinitely, however, so that the predicted weight change eventually occurs although it may not follow the day-to-day caloric deficit (27). Figure 15.3 illustrates the phenomenon.

The physiology of this water retention seems to be explained by the fact that the water formed as a by-product of the metabolism of the body's fat stores is not excreted immediately via the kidney in obese subjects because of an increased level of antidiuretic hormones.

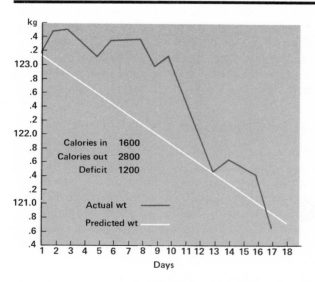

Figure 15.3 Water retention in weight reduction programs. (From Newburgh, L.H. *Physiol. Rev.* 24:18, 1944.)

Gordon, Goldberg, and Chosy (14) have also demonstrated the *water-binding effect* of consumption of an appreciable quantity of concentrated carbohydrate food. A severely obese man, who had been losing weight successfully, was given an 800-gm carbohydrate, 4,000-kcals per day diet for two days. He promptly gained eighteen pounds, which required three weeks to lose. The weight gained was shown to be water.

It is extremely important that anyone embarking upon a diet to lose weight be aware of these facts so as not to be discouraged when results are delayed by water retention.

Spot Reducing

Women are frequently encouraged to use localized exercises to reduce fatty stores in the areas of greatest fat deposition, usually buttocks, hips, and thighs. Schade and associates (34) conducted an experiment on twenty-two overweight college women in which one group used *spot reducing exercises* and the other general exercise. They concluded there was no significant difference in the effect of spot and generalized exercise on fat distribution in their subjects. Conflicting evidence was reported by Olsen and Edelstein (29). The most convincing evidence against the spot reducing concept is that tennis players whose playing arm showed a mean difference of 2.25 cm in girth (hypertrophy) compared with the nonplaying arm nevertheless showed no significant difference in skin-fold dimensions (16). A recent study by Noland and Kearney (28) supports this position and provides an excellent review of the literature on this subject.

The best evidence available seems to indicate that in negative caloric balance situations the fat comes off the area of greatest concentration, regardless of how the exercise is performed.

The Long-Haul Concept of Weight Control

It should be emphasized that weight control is best performed as a matter of childhood habit formation, which then automatically takes care of the *long haul*. This habit formation should include the development of physical skills that will permit, indeed encourage and motivate, regular participation in a vigorous activity. Such sports as tennis, badminton, handball, skiing, and horseback riding are ideally suited to lifetime needs since they are vigorous, enjoyable, and can be performed with a minimum of cooperation by other persons. Individuals who develop these skills and participate in them regularly—at least twice, and preferably three or four times a week—will seldom have to be unduly concerned about their diet. With normal moderation in eating, including a gradual decrease in caloric intake with age above twenty-five years, they will in all likelihood have no problem with obesity.

When obesity gets a start due to inactivity or overeating, it is wise to get started with corrective efforts as early as possible. The corrective measures, exercise and diet, should be set up with habit formation in mind. Thus one cannot set up a habit pattern that involves any of the many popular *crash diets;* when normal weight is regained, a change is necessary, and all too often the change is a reversion to the pre-diet obesity-causing regimen. A series of cycles of weight gain and loss is then set up, and discouragement ensues, with complete loss of control.

On the other hand, a sensible approach would involve a dietary restriction of only 300 to 500 kcals per day, with a progressive buildup of exercise such as an hour of tennis or horseback riding (another loss of 400 to 600 kcals). In this fashion, pleasurable habits can be formed that require no changes and that do not cause the distress of semistarvation on a crash diet. The process may take a little longer, but it will be infinitely more successful.

If the effectiveness of exercise for losing weight is questioned, the layman can most readily see the answer not in research but in the empirical wisdom of the experience of mankind. Obesity is practically unknown among vigorous, healthy athletes, ditchdiggers, heavy laborers, many of whom daily consume prodigious amounts of food.

Making Weight in Athletics

In many sports, such as wrestling and boxing, competition is organized into divisions by body weight; also, some states set up high school and junior high school athletics on the basis of a classification system that depends—at least in part—on the weight of the athletes. It does not require a great deal of sophistication, in coach or athlete, to see the advantages of competing in the lowest possible weight class.

For mature athletes, *making weight* is not a large health problem because experience has taught them what their normal weight should be; however, the author has seen many extreme cases of weight loss by high school athletes— on occasion as much as 15 pounds in a boy whose normal weight was 150 or 160 pounds. This practice should be condemned in the strongest terms, for these short-term weight losses can be obtained only by drastic changes in water metabolism (sweating it off, plus water restriction), and there are attendant changes in kidney and cardiovascular function (whose consequences are still difficult to evaluate). In secondary school and college athletes, a 5% weight loss is certainly the outside limit of prudence, and it is quite likely that even this amount (without medical supervision) is too great for some of the leaner athletes.

Recent evidence has shown that not only high school wrestlers, but NCAA championship team members as well, enter competition in a dehydrated state (39). In an effort to correct such practices, the American College of Sports Medicine (1) issued the following instructions as part of its position stand with respect to weight loss in wrestlers:

1. Assess the body composition of each wrestler several weeks in advance of the competitive season. Individuals with a fat content less than 5% of their certified body weight should receive medical clearance before being allowed to compete.

2. Emphasize the fact that the daily caloric requirements of wrestlers should be obtained from a balanced diet and determined on the basis of age, body surface area, growth, and physical activity levels. The minimal caloric needs of wrestlers in high schools and colleges will range from 1,200 to 2,400 kcals/day; therefore, it is the responsibility of coaches, school officials, physicians, and parents to discourage wrestlers from securing less than their minimal needs without prior medical approval.

3. Discourage the practice of fluid deprivation and dehydration. This can be accomplished by:
   a. Educating the coaches and wrestlers on the physiological consequences and medical complications that can occur as a result of these practices.
   b. Prohibiting the single or combined use of rubber suits, steam rooms, hotboxes, saunas, laxatives, and diuretics to "make weight."
   c. Scheduling weigh-ins just prior to competition.
   d. Scheduling more official weigh-ins between team matches.

4. Permit more participants per team to compete in those weight classes (119–145 pounds) which have the highest percentages of wrestlers certified for competition.

5. Standardize regulations concerning the eligibility rules at championship tournaments so that individuals can only participate in those weight classes in which they had the highest frequencies of matches throughout the season.

6. Encourage local and county organizations to systematically collect data on the hydration state of wrestlers and its relationship to growth and development.

SUMMARY

1. Besides the aesthetic disadvantages, obesity has been shown to be associated with many degenerative diseases.

2. Weight gain and loss follow the laws of thermodynamics. A positive energy balance results in a gain in weight, and a negative energy balance results in a loss.

3. Carbohydrate, fat, and protein follow the same final pathway in their metabolism, and thus fat and protein can substitute for carbohydrate in furnishing energy. All three foodstuffs, if eaten in excess, can result in deposition of fat tissue.

4. Estimation of what constitutes normal or proper body weight is subject to great inaccuracy unless measurements of body composition are made. Scale readings are relatively meaningless unless we know the proportion of the reading that is due to fatty tissues.

5. A gain in weight is best accomplished by heavy resistance, low repetition type exercise, such as weight training, in which weight gain is brought about through muscular hypertrophy.

6. Weight loss can be accomplished by increased activity (exercise) or dietary restriction. If the weight is not grossly abnormal, a progressive exercise program, although slower, is the sounder approach.

7. Recent biopsy data have shown that the *number* of fat cells is established in early life and does not vary significantly in later life even with gross changes in body fat.

8. The number of fat cells in the nonobese is about 25–30 billion. Grossly obese individuals may have three times as many fat cells as nonobese individuals.

9. Fat cell *size* varies proportionately with body fat changes in a given individual.

10. In rats, the laying down of fat cells and the level of fatness in later life can be influenced by both exercise and diet in the adolescent growth period.

11. Moderate increases in activity level, contrary to popular opinion, do not result in increased food intake if the individual has been sedentary previously.

12. Vigorous exercise not only creates an immediate increase in metabolism, it also brings about a longer-lasting (six- to eight-hour) rise in resting metabolism, which further contributes to weight loss.

13. Considerable evidence indicates that concentrating the daily food consumption into one or two large meals results in a greater tendency toward obesity.

14. In dieting to lose weight, a temporary water retention (up to three weeks) may obscure the true fat-tissue loss that is occurring.

15. *Spot reducing* rests on no sound physiological basis; rather, the best available evidence indicates that regardless of the part of the anatomy exercised, weight loss first occurs in the largest fat deposits.

16. Maintaining normal weight should be a long-term process that involves hygienic habit formation. *Crash diets* are usually foredoomed to failure because they do not accomplish habit formation.

17. *Making weight* in athletics must be considered potentially hazardous for adolescents. Weight losses for this purpose should never exceed 5% (short term), and even these losses should be effected under competent supervision.

REFERENCES

1. American College Sports Medicine. Position stand on weight loss in wrestlers. *Med. Sci. Sports* 8:xi, 1976.
2. Ashley, B.C.E. Drastic dietary reduction of obesity. *Med. J. Aust.* 1:593–96, 1964.
3. Bjorntorp, P., and Sjostrom, L. Fat cell size and number in adipose tissue in relation to metabolism. *Isr. J. Med. Sci.* 8:320–24, 1972.
4. Bonnet, F.; Gosselin, L; Chantraine, J.; and Senterre, J. Adipose cell number and size in normal and obese children. *Rev. Eur. Etudes Clin. Biol.* 15:1101–4, 1970.
5. Booth, M.A.; Booth, M.J.; and Taylor, A.W. Rat fat cell size and number with exercise training, detraining and weight loss. *Fed. Proc.* 33:1959–63, 1974.
6. Corbin, C.B., and Pletcher, P. Diet and physical activity patterns of obese and nonobese elementary school children. *Res. Q.* 39:922–28, 1968.
7. deVries, H.A. Health Science: The Positive Approach. Santa Monica, Calif.: Goodyear Publishing Co. Inc., 1979.
8. deVries, H.A., and Gray, D.E. Aftereffects of exercise upon resting metabolic rate. *Res. Q.* 34:314–21, 1963.
9. Drenick, E.J. The effects of acute and prolonged fasting and refeeding on water, electrolyte and acid-base metabolism. Chap. 32 in *Clinical Disorders of Fluid and Electrolyte Metabolism,* eds. M.H. Maxwell and C.R. Kleeman. New York: McGraw-Hill, 1972.
10. Durnin, J.V.G.A. and Womersly, J. Body fat assessed from total body density and its estimation from skinfold thickness; measurements on 481 men and women from 16–72 years. *Br. J. Nutri.* 32:77–97, 1974.
11. Falls, H.B., and Humphrey, L.D. Energy cost of running and walking in young women. *Med. Sci. Sports* 8:9–13, 1976.
12. Gilder, H., et al. Components of weight loss in obese patients subjected to prolonged starvation. *J. Appl. Physiol.* 23:304–10, 1967.
13. Girandola, R.N. Body composition changes in women: effects of high and low exercise intensity. *Arch. Phys. Med. Rehabil.* 57:297–300, 1976.
14. Gordon, E.S.; Goldberg, M.; and Chosy, G.J. A new concept in the treatment of obesity. J.A.M.A. 186:50–60, 1963.
15. Greene, J.A. A clinical study of the etiology of obesity. *Ann. Intern. Med.* 12:1797–1803, 1939.
16. Gwinup, G.; Chelvam, R.; and Steinberg, T. Thickness of subcutaneous fat and activity of underlying muscles. *Ann. Intern. Med.* 74:408–11, 1971.
17. Hirsch, J. Adipose cellularity in relation to human obesity. *Adv. Intern. Med.* 17:289–300, 1971.
18. Hirsch, J. Can we modify the number of adipose cells? *Postgrad. Med.* 51:83–86, 1972.
19. Hirsch, J., and Knittle, J. Cellularity of obese and nonobese human adipose tissue. *Fed. Proc.* 29:1516–21, 1970.

20. Hollifield, G., and Parson, W. Metabolic adaptations to a "stuff and starve" feeding program. *J. Clin. Invest.* 41:250–53, 1962.

21. Keys, A., and Brozek, J. Body fat in adult man. *Physiol. Rev.* 33:245–325, 1953.

22. Margaria, R.; Edwards, H.T.; and Dill, D.B. The possible mechanisms of contracting and paying the O_2 debt, and the role of lactic acid in muscular contraction. *Am. J. Physiol.* 106:689–715, 1933.

23. Mayer, J., and Stare, F.J. Exercise and weight control: frequent misconceptions. *J. Am. Diet. Assoc.* 29:340–43, 1953.

24. Mayer, J.; Marshall, N.B.; Vitale, J.J.; Christensen, J.H.; Mashayek, M.B.; and Stare, F.J. Exercise, food intake and body weight in normal rats and genetically obese adult mice. *Am. J. Physiol.* 177:544, 1954.

25. McMiken, D.F., and Daniels, J.T. Aerobic requirements and maximum aerobic power in treadmill and track running. *Med. Sci. Sports* 8:14–17, 1976.

26. Moody, D.L.; Kollias, J.; and Buskirk, E.R. The effect of a moderate exercise program on body weight and skinfold thickness in overweight college women. *Med. Sci. Sports* 17:75–80, 1969.

27. Newburgh, L.H. Obesity and energy metabolism. *Physiol. Rev.* 24:18, 1944.

28. Noland, M., and Kearney, J.T. Anthropometric and densitometric responses of women to specific and general exercise. *Res. Q.* 49:322–28, 1978.

29. Olson, A.L., and Edelstein, E. Spot reduction of subcutaneous adipose tissue. *Res. Q.* 39:647–52, 1968.

30. Oscai, L.B.; Babirak, S.P.; Dubach, F.B.; McGarr, J.A.; and Spirakis, C.N. Exercise or food restriction: effect on adipose tissue cellularity. *Am. J. Physiol.* 227:901–4, 1974.

31. Oscai, L.B.; Babirak, S.P.; McGarr, J.A.; and Spirakis, C.N. Effect of exercise on adipose tissue cellularity. *Fed. Proc.* 33:1956–58, 1974.

32. Pariskova, J. Impact of age, diet and exercise on man's body composition. In *International Research in Sport and Physical Education,* eds. E. Jokl and E. Simon. Springfield, Ill.: Charles C Thomas, 1964.

33. Passmore, R., and Durnin, J.V.G.A. Human energy expenditure. *Physiol. Rev.* 35:801–35, 1955.

34. Schade, M.; Hellebrandt, F.A.; Waterland, J.C.; and Carns, M.L. Spot reducing in overweight college women. *Res. Q.* 33:461–71, 1962.

35. Sims, E.A.H. Studies in human hyperphagia. In *Treatment and Management of Obesity* eds. G.A. Bray and J.E. Bethune. New York: Harper & Row, 1974.

36. Stunkard, A.J. Obesity and the social environment: current status, future prospects. *Ann. N.Y. Acad. Sci.* 300:298–319, 1976.

37. Wamsley, J.R., and Roberts, J.E. Body composition of USAF flying personnel. *Aerospace Med.* 34:403–5, 1963.

38. Womersley, J., and Durnin, J.V.G.A. An experimental study on variability of measurements of skinfold thickness on young adults. *Hum. Biol.* 45:281–92, 1973.

39. Zambraski, E.J.; Foster, D.T.; Gross, P.M.; and Tipton, C.M. Iowa wrestling study: weight loss and urinary profiles of collegiate wrestlers. *Med. Sci. Sports* 8:105–108, 1976.

40. Zuti, W.B., and Golding, L.A. Comparing diet and exercise as weight reduction tools. *Physi. Sports Med.* 4:49–53, 1976.

16

Age
and Exercise

If we consider the human life span in terms of the biblical concept of three score and ten, we have a lifetime of seventy years; however, if we judge the interest of the physical education profession by the nature of the curricula offered in teacher training institutions, it would seem that virtually all of our efforts are directed to ten years of that life span, the years involved in secondary and college education. Should physical education start in junior high school and end after two or four years of college? This seeming preoccupation of physical education with only 14% of the total life span is certainly undesirable.

That the need for physical education exists at all ages is amply demonstrated by the success of various athletic programs for children (Little League, Pop Warner League, and age-group swimming) and by weight training and conditioning gyms for adults. It seems inevitable that the scope of physical education must somehow grow to include programs that are organized and administered by professional people for *all ages* and not just for high school and college students who need the attention least. It behooves us, then, to consider the physiological changes that occur as a function of the aging process.

With respect to the entire age range of human life, physical performance measures in general improve rapidly from early childhood to a maximum somewhere between the late teens and about thirty years of age. In most cases a slow decline occurs during maturity and becomes more rapid with increasing age. The decline in physical performance with age deserves a great deal more emphasis by scientific investigators than it has been accorded in the past.

Indeed the entire body of knowledge regarding the loss of function with increasing age must be viewed with caution since in very few cases has the effect of habitual physical activity been controlled or ruled out. Wessel and Van Huss (62) have shown that physical activity decreases significantly with increasing age. This is not surprising news but does provide scientific validation of the need for consideration of this variable in all investigations directed toward aging changes in performance. To support this contention further they showed that agewise losses in physiological variables important to human performance were more highly related to the *decreased habitual activity* level than they were to *age itself.*

Statistics on population trends for the United States indicate that we are rapidly becoming a nation of older people. The absolute number, as well as the proportion of our older population segments, is increasing rapidly. In evaluation of the effects of the aging process on human performance, several problems arise. First, it is difficult to separate the effects of aging per se from those of concomitant disease processes (particularly cardiovascular problems) that become more numerous as age progresses. Second, the sedentary nature of adult life in the United States makes it very difficult to find *old* populations for comparisons with *young* populations at equal activity levels. Third, very little work has been done on longitudinal studies of the same population over a period of time. Conclusions drawn from cross-sectional studies in which

Physiology Applied to Health and Fitness

various age groups are compared must be accepted with reservations because the weaker biological specimens are not likely to be represented in as great numbers in the older populations tested as in the younger (due to a higher mortality rate).

Just as various individuals age at different rates, various physiological functions seem to have their own rates of decline with increasing age (see fig. 16.1). Indeed, some functions do not seem to degenerate with age (51); under

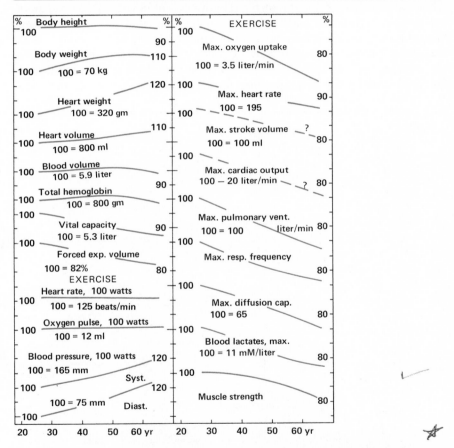

Figure 16.1 Functional variables with age. Data have been collected from various subjects, including healthy men. For data on the same function, only one study has been consulted. The values for the twenty-five-year-old subjects = 100%; for the older ages the mean values are expressed in percentage of the twenty-five-year-old individuals' values. The values should not be considered normal values, but values that illustrate the effect of aging. Note that heart rate and oxygen pulse at a given work load (100 w or 600 kpm/min, oxygen uptake about 1.5 liters/min) are identical throughout the age range covered, but the maximal oxygen uptake, heart rate, cardiac output, etc., decline with age. The data on cardiac output and stroke volume are based on few observations and are therefore uncertain. (From Astrand, P-O., and Rodahl, K. *Textbook of Work Physiology.* New York: McGraw-Hill Book Company, 1977.)

resting conditions, there seem to be no changes in blood sugar, blood pH, or total blood volume. In general, the functions that involve the coordinated activity of more than one organ system decline most with age, and, as might be expected, changes due to the aging process are most readily observed when the organism is stressed. Homeostatic readjustment is considerably slower with increasing age.

Age Changes in Muscle Function

All investigators have found that rapid improvement in strength accompanies the growth of children, and maximal strength is found to occur for most muscle groups between the ages of twenty-five and thirty (26). Rodahl (19) has shown that this increase in strength is almost entirely accounted for by the increased size of the muscle. Even sex differences in muscle quality are not very large. When strength is expressed per unit of cross-sectional area

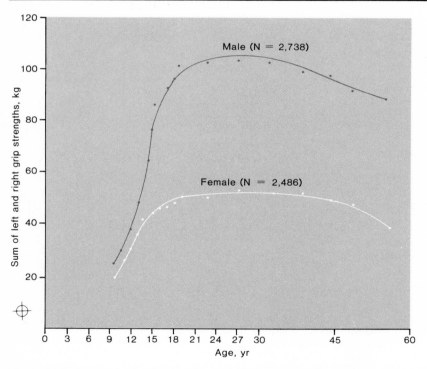

Figure 16.2 Changes in grip strength with age. (Data from Tecumseh study. Montoye, H.J., and Lamphiear, D.E. *Res. Q.* 48:109, 1977.)

Physiology Applied to Health and Fitness

(kilograms per square centimeters), differences due to age and sex are very small.

Strength decreases very slowly during maturity. After the fifth decade, strength decreases at a greater rate, but even at age sixty the loss does not usually exceed 10%-20% of the maximum, with women's losses being somewhat greater than those of men.

Figures 16.2 and 16.3 show the age changes in arm strength and grip strength found by Montoye and Lamphiear (35) in the Tecumseh, Michigan, study in which the entire community was studied. These data probably represent the best controlled study to date. Interestingly, in another study where maximal grip strength was investigated in 100 men who all did similar work in a machine shop, no change in either grip strength or endurance was found from age twenty-two to sixty-two (44). These data suggest that in this age bracket the more typical finding of small losses with age may be due largely to disuse phenomena rather than a true age effect. However, in old age there is little question that sizable decrements in strength do occur.

Changes at the Cellular Level. Animal studies have shown that important age changes occur at the cellular level. First there is a loss of contractile elements, which accounts for the decrement in strength. While this loss could be the result of losses in motor nerve fibers, this explanation has been ruled

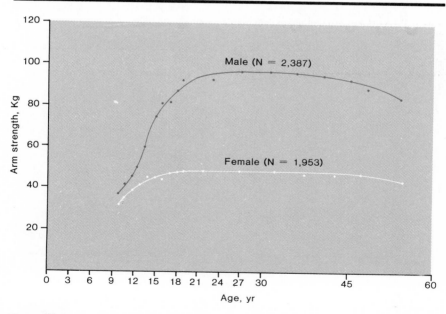

Figure 16.3 Changes in arm strength with age. (Data from Tecumseh study. Montoye, H.J., and Lamphiear, D.E. *Res. Q.* 48:109, 1977.)

Age and Exercise

out by studies on rats, which have shown that while muscle fiber numbers may be down by about 25% in old rats, no change occurs in nerve fibers (28). The second important change at the cellular level is a reduction in respiratory capacity, and this accounts for losses in muscle endurance and capacity for recovery (24).

Recently it has also been shown that the loss in human muscle tissue with age can entirely account for the downward trend in basal metabolism, which has been an accepted fact in metabolic studies for nearly a century (59).

Age and Capacity for Hypertrophy. Goldspink and Howells (27) taught hamsters to lift weights to evaluate cellular hypertrophy. After weight training for five weeks, the mean fiber area of the biceps in the young animals increased by a significant 35.6%. Fiber area in the old animals increased by 17.7%,

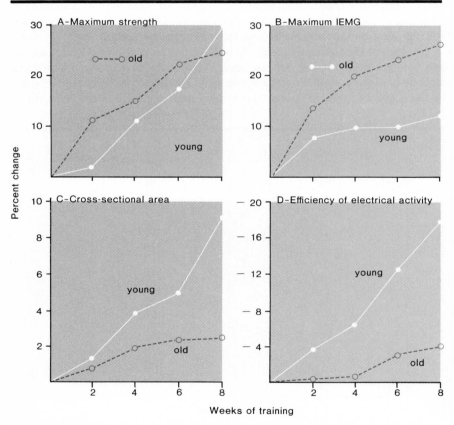

Figure 16.4 Changes occurring in the elbow flexor muscles of young and old men during weight training. (From Moritani, T., and deVries, H.A., in press.)

Physiology Applied to Health and Fitness

which was of marginal significance. All signs of hypertrophy were lost in fifteen weeks.

With respect to human strength gain, Moritani and deVries (36) investigated the time course of strength gain through weight training in old and young men to define the contribution of hypertrophy and such neural factors as disinhibition to the total change in strength over a period of eight weeks. Young and old men showed similar and significant percentage increases in strength, although the young made greater absolute gains, of course. However, the physiological adaptations of the two groups were quite different, as shown in figure 16.4. While young subjects showed highly significant hypertrophy, the strength gained by the old men was almost entirely due to learning to achieve higher activation levels as measured by EMG methods.

Damon (14) has shown that these age decrements exist whether measured in isometric, concentric, or eccentric muscle contraction and also whether measured as maximal instantaneous force achieved, or as a mean value over a finite time period. However, his work showed isotonic strength to be affected to a greater extent than isometric. The maximum velocity produced against any given mass is less for the old than the young, although the shape of the force velocity curve is similar (see fig. 16.5). Thus loss of strength with age

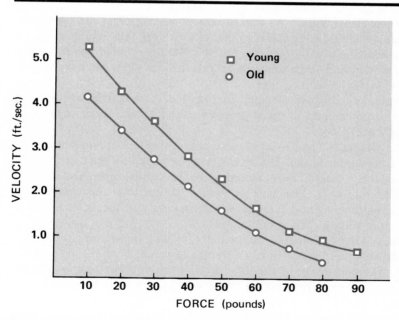

Figure 16.5 Comparison of young and old groups on force-velocity relationship. (From Damon, E.L. *An Experimental Investigation of the Relationship of Age to Various Parameters of Muscle Strength,* Doctoral dissertation, Physical Education, USC, 1971.)

Age and Exercise

consists of two components: (1) a decrease in ability to maintain maximum force statically, and (2) a decrease in ability to accelerate mass.

With respect to muscular endurance, or fatigue rate, Evans (25) has shown with the EMG fatigue curve technique that fatigue rate is significantly greater in the old than the young when isometric contractions are held to 20%, 25%, 30%, 35%, 40%, or 45% of MVC.

Age and the Cardiovascular System

The effects of a lifetime of vigorous exercise upon the cardiovascular system have not yet been investigated extensively by scientific methods. Evidence, however, has been presented from observations on isolated individuals who have trained very hard into old age. Clarence DeMar, the famous marathon runner, made it a habit to run twelve miles every day, and this level of training was maintained throughout his lifetime. He was still competing in twenty-five- and twenty-six-mile marathons at age sixty-five, and he ran his last fifteen-kilometer race at sixty-eight—two years before his death (from cancer). At the autopsy it was found that this unusually strenuous exercise had not only *not* hurt his heart, but had left him with a myocardium that was unusually well developed, valves that were normal, and coronary arteries estimated to be two or three times normal size (8).

Maximum Heart Rate. As was discussed in chapter 6, the maximum heart rate attainable during exercise decreases with age. Maximum heart rate for young adults is usually between 190 and 200 bpm; in old age this value decreases gradually. The maximum heart rate in older adults can be estimated as follows: MHR = 220 − age.

Data from animal studies suggest that the reduction in MHR with age is due to intrinsic changes in the myocardium itself, rather than changes in neural influences (13).

Cardiac Output. The at-rest cardiac output declines approximately 1% per year after maturity (8). This evidence is supported by the fact that the strength of the myocardium measured by ballistocardiography also declines at a similar rate (57). The most important parameter is the capacity for cardiac output at maximal exercise, which appears to decrease at a similar rate to that of resting cardiac output (31).

Coronary Artery Changes. Simonson (54) has shown that in normal hearts, the cross-sectional area of the lumen of coronary arteries is reduced with age. The percentage of the total arterial cross section that is open to blood flow is 29% less in age group forty to fifty-nine than it is in the age group ten to twenty-nine.

Circulatory Changes. One method for assessing vasomotor responsiveness to stress is strapping a subject to a tiltboard so that orientation in space can be changed quickly from supine to vertical posture, as well as to intermediate postures, thus bringing about quick changes in hydrostatic pressures within

Physiology Applied to Health and Fitness

the circulatory system. After a tilt to forty-five degrees, older subjects showed larger decreases in and slower recovery of systolic blood pressure than the younger group. When tilted to standing, the older group had larger decreases in and slower recovery of diastolic blood pressure. In both cases, the younger group had a greater increase in heart rate (a desirable response), which increased the blood pressure available to compensate for the increased hydrostatic pressure (37).

There appear to be no significant differences in the blood flow to the extremities at rest or in the vasomotor reflex responses to warming and cooling between healthy young and old adults. However, the response of blood flow to the stimulus of exercise is markedly less in the old than in the young subjects (45), although blood pressure shows a greater response in the old than in the young (40). Lower flow with greater pressure would be the logical result of increased peripheral resistance (see chap. 7).

Circulation time from arm to thigh was measured in 237 normal subjects and found to be slowed by 30%-40% in older subjects (over sixty) compared with the values for the young (7).

Capillary density does not appear to change with age, but the ratio of capillary to muscle fibers decreases because of the greater number of fibers per unit cross section due to atrophic processes (41).

A more hopeful note is sounded by Russian workers who have reported that the hardening of the arteries associated with aging may be reversible through systematic physical conditioning. They found a 14% slowing of pulse wave velocity after six to seven months of training in a group whose age averaged fifty-four years (60). A slower pulse wave propagation is associated with better elasticity in the arterial wall.

Changes in Pulmonary Function

Lung Volumes and Capacities. It has been firmly established that vital capacity declines with age (38, 39, 48). There appears to be no very good evidence for any change in total lung capacity and consequently residual volume increases with age (38, 39). Aging increases the ratio of residual volume (RV) to total lung capacity (TLC), and anatomic dead space also increases with age (12).

Thoracic Wall Compliance. Some tissues of the lungs and chest wall have the property of elasticity. Thus, in inspiration, the muscles must work against this elasticity, which then aids the expiration phase through elastic recoil. This relationship between force required (elastic force) per unit stretch of the thorax is called *compliance*. It is measured by the size of the ratio of volume change per unit pressure change. It may be thought of as the elastic resistance to breathing. That is to say, the less compliant the tissues, the more elastic force must be overcome in breathing. There are two tissues which offer elastic resistance to breathing, the lung tissue itself and the wall of the thoracic cage.

Age and Exercise 363

The evidence suggests that lung compliance increases with age (58), but more important, thoracic wall compliance decreases (34, 47, 58). Thus the older individual may do as much as 20% more elastic work at a given level of ventilation than the young, and most of the additional work would be performed in moving the chest wall (58).

It seems entirely likely that the age differences in lung volumes and capacities noted above can be explained largely on the basis of this lessening mobility of the chest wall with age.

Pulmonary Diffusion. A significant decrease in the capacity for pulmonary diffusion both at rest and at any given work load accompanies the aging process (22).

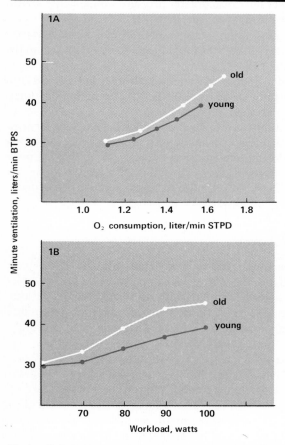

Figure 16.6 *Top:* expiratory minute volume (liters/min BTPS) as a function O_2 consumption (liters/min STPD). *Bottom:* expiratory minute volume as a function of work load in watts. (From deVries, H.A., and Adams, G.M. *J. Gerontol.* 27:350, 1972.)

Ventilatory Mechanics in Exercise. In view of the changes in pulmonary function already cited, it is not surprising to find that the process of breathing becomes less efficient with age. Figure 16.6 presents data from the author's laboratory which show clearly the need for greater ventilation in old men compared with young men at any given level of work or O_2 consumption (20). Interestingly, there is a difference in the mechanics by which the old subjects met the increased ventilatory demand. While the young first increased breathing frequency, the older men increased their tidal volume (the more efficient mechanism), thus reaching their maximal tidal volume (TV) early at work loads where the young still had large reserves of TV for work at higher loads.

Age and Physical Working Capacity (PWC)

Maximal O_2 Consumption. As was mentioned earlier, the best single measure of physical working capacity (PWC) is *maximal oxygen consumption,* and two excellent studies have related this variable to age. Robinson (48) tested a total of seventy-nine male subjects, ranging in age from six to seventy-five; the results are shown in table 16.1. Astrand (2) tested forty-four women, ranging in age from twenty to sixty-five, and these results are tabulated in table 12.5.

For boys, it is seen (table 16.1) that, in terms of absolute quantities, maximal O_2 consumption increases rapidly with age. When the results are stated in terms of O_2 consumption per unit of body weight, however, there are

Table 16.1
Highest Oxygen Intake Attained in Maximal Work as Related to Body Weight and Age

Age Group	No. of Subjects	Age in Years (Mean)	Weight in kg (Mean)	Maximal O_2 Intake[*]			
				Liters/min		Ml/kg/min	
				(Mean)	(Extremes)	(Mean)	(Extremes)
I	4	6.1	21.0	0.98	0.80-1.30	46.7	42.8-49.5
II	9	10.4	30.0	1.56	1.24-2.00	52.1	49.0-56.1
III	9	14.1	55.8	2.63	1.89-3.41	47.1	36.4-55.4
IV	11	17.4	68.5	3.61	2.96-4.20	52.8	44.6-62.5
V	11	24.5	72.5	3.53	2.56-4.50	48.7	41.9-55.6
VI	10	35.1	79.3	3.42	2.76-3.97	43.1	37.6-52.8
VII	9	44.3	74.1	2.92	2.30-3.62	39.5	33.7-46.5
VIII	7	51.0	68.7	2.63	2.24-3.35	38.4	33.7-43.2
IX	8	63.1	67.4	2.35	1.64-3.15	34.5	30.2-41.7
X	3	75.0	67.4	1.71	1.43-1.90	25.5	21.8-29.6

[*]Dry gas at 0° C and 760 mm Hg
After S. Robinson, *Arbeitsphysiologie* vol. 10, p. 279 (1938). Verlag Julius Springer, Berlin.

Age and Exercise

only very small changes with age, which may not be significant. Thus the mean maximal O_2 consumption for boys of 6.1 years is 0.98 liters/min, compared to 2.63 liters/min for boys of mean age 14.1. But when growth (or body size) is canceled out, the six-year-olds can achieve 46.7 ml per kilogram of body weight, which is virtually as good as the 47.1 ml per kilogram for fourteen-year-olds and not very far from the best value for all ages, which is achieved at mean age 17.4 years. Thus the working capacity of young boys is probably not limited by ability to transport and consume oxygen. (Equivalent data for girls are lacking.)

For adults, there is a gradual decline in maximal O_2 consumption with age for both sexes. For men, the maximal values were found at mean age 17.4 years, and they declined to less than half those values at mean age 75. For women, the maximal values were found in the age group twenty to twenty-nine, and they fell off by 29% in the age group fifty to sixty-five.

More recent data suggest that for both men (15) and women (23) the rate of decline may be slower in those who are physically active. Indeed, Kasch and Wallace (32) have provided longitudinal data suggesting that the usual 9%-15% decline in $\dot{V}O_2$ max from age forty-five to fifty-five can be forestalled by regular endurance exercise (32). Hodgson and Buskirk (30) have summarized the data from many cross-sectional studies (fig. 16.7), which suggest that active athletic subjects start at higher values and at age sixty are still at or above the level of the sedentary twenty-year-olds, although the rate of decline is similar.

It is of interest to consider the physiological functions whose decline with increasing age might contribute to this loss of ability to transport and utilize O_2. The following functions are probably the most important in achieving maximal O_2 consumption: (1) lung ventilation, (2) lung diffusion capacity for O_2, (3) heart rate, (4) stroke volume, and (5) O_2 utilization by the tissues. The implication of direct or indirect evidence is that all of these functions decline with age.

Muscular Efficiency. According to the data of Robinson (48), adults tended to be more economical in their adjustment to work than boys. He found no clear-cut differences with increasing age during maturity. Astrand (2) found significant decreases in efficiency with age in women. These differences were very small, however: 21.9% for the twenty to twenty-nine age group, and 19.6% for the fifty to sixty-five age group.

The only longitudinal data available on muscular efficiency were gathered from D.B. Dill. Dill and his associates (21) found that his efficiency in running on a treadmill declined greatly between ages forty-one and sixty-six; however, two confounding factors operated in this comparison. First, Dill had gained sixteen and a half pounds in the interim; second, although the same work load was used in both instances, it represented his maximum effort at age sixty-six, but only some 66% of his maximal effort at forty-one. Consequently, considerably more energy utilized at age sixty-six came from anaerobic

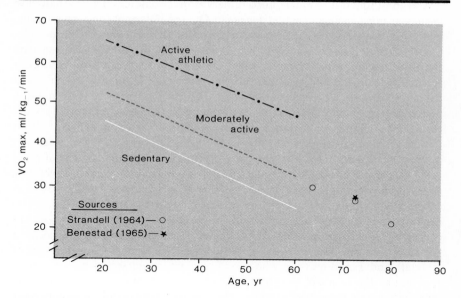

Figure 16.7 The decrease of maximum oxygen consumption with age in men twenty to sixty from a review of cross-sectional studies.

Sedentary men	$VO_2max = 54.1 - 0.41$ (age)
Moderately active men	$VO_2max = 61.6 - 0.44$ (age)
Active athletic men	$VO_2max = 76.1 - 0.48$ (age)

(From Hodgson, J.L., and Buskirk, E.R. *J. Am. Geriatr. Soc.* 25:386, 1977.)

sources, which are less efficient. In summarizing these data, it would appear that muscular efficiency decreases with age, but probably to a very slight degree.

Age and the Nervous System

Age changes (slowing) in reaction time and speed of movement have been verified. Birren and his co-workers (5,6), who have done extensive investigation in this area, have reached the conclusion that this psychomotor slowing is probably an effect of the aging of the central nervous system because the slowing is common to several sensory modalities and to several motor pathways. The decreases in conduction time, both afferent and efferent, are insufficient to account for the total slowing.

Recent work suggests that a life style of vigorous physical activity has a beneficial effect in lessening the decline in reaction and movement times (11, 56).

Cerebral function is much more vulnerable to circulatory deficits than most tissues; it must have a constant source of O_2, and it cannot function anaerobically (as can muscle tissue, for instance). For this reason it is tempting to associate the effects of aging with a decreased cerebral blood flow and with the resulting hypoxia; however, when the effects of aging per se have been separated from the effects of arteriosclerosis, which frequently accompanies the aging process, it appears that it is the arteriosclerosis that is at fault. Aging per se, in the absence of arteriosclerotic changes, probably does not result in circulatory or metabolic changes in cerebral function (5,6).

The German neurophysiologists, C. and O. Vogt, have made the extremely interesting observation that the degree of activity of a particular type of nerve cell has a great effect on its aging process. They have found that involution (part of the aging process) "is delayed not only by normal but also by such excessive activity of nerve cells as results in their hypertrophy (61)." This suggests that physical activity involving overuse of neural pathways of the central nervous system may have beneficial effects, such as we know occur in the case of muscle tissue. Their work has received support from the work of Retzlaff and Fontaine (46) who also found improved spinal motor neuron function as the result of conditioning in rats. Much more scientific research into the possible benefits of vigorous exercise for aging populations is needed.

Age and Body Composition

As has been discussed in earlier chapters, it is typical for aging humans to increase their weight, and Brozek (9) has provided interesting data on the composition of the human body as it ages (table 16.2). Clearly, this weight gain represents a mean increase in Brozek's sample of 12.24 kg (27 lb) of fat, while the fat-free body weight has actually decreased from age twenty to age fifty-five. It is obvious that, to maintain a constant proportion of body fat as one ages, weight must not merely be maintained at a constant level, it must be decreased. It is conceivable, however, that the loss in fat-free weight represents disuse atrophy of muscle tissue and may not be a necessary component of the aging changes if vigorous exercise is maintained.

Shock and his co-workers (52) have furnished compelling evidence that this loss of active tissue is the cause of the well-known decline in the basal metabolic rate (BMR) with age. When they computed BMR on the basis of body water (which reflects the amount of active tissue cells in the body), no significant changes were observed in relation to age. The customary method of calculating BMR (per unit of surface area) does not differentiate between fat and active tissue; thus we find that the aging body contains fewer and fewer cells, although the activity of individual cells probably does not change significantly. Again, we should like to know the effects vigorous exercise programs might have on this process.

Physiology Applied to Health and Fitness

Table 16.2
Average Estimated Changes in Body Composition during Maturity (20-55 Years) (Standard Weights Refer to Men 176 Cm Tall)

Age	Standard Weight (kg)	Standard Fat (%)	Fat (kg)	Fat-free Weight (kg)
20	67.6	10.30	6.96	60.6
25	69.9	13.42	9.38	60.5
30	71.3	16.20	11.55	59.8
35	72.9	18.64	13.59	59.3
40	74.1	20.74	15.37	58.7
45	75.5	22.50	16.99	58.5
50	76.0	23.92	18.18	57.8
55	76.8	25.01	19.20	57.6

From J. Brozek, "Changes of Body Composition in Man during Maturity and Their Nutritional Implications," *Fed. Proc.* 11:787, 1952.

Stature. It has been shown that, on the average, we also grow shorter as we grow older and fatter. DeQueker, Baeyens, and Claessens showed the loss rate to average about one-half inch per decade after age thirty (16).

Effects of Physical Conditioning on Agewise Losses in Functional Capacities

All of the agewise changes described thus far can only be said to *accompany* the aging process. *Causal relationships* have not been established. We may infer that changes in these various functional capacities observed in different groups of subjects at increasing age levels may result from a combination of at least three factors: (1) true aging phenomena, (2) unrecognized disease processes whose incidence and severity increase with age, and (3) *disuse phenomena* or the increasing sedentariness of our life style as we grow older. Since we can do little to modify the first two factors, and since the third factor offers potential for modification by the methods of conditioning and training already well known to our profession, the author and other investigators have addressed themselves to the question of how trainable the older human organism is.

The capacity for improvement of performance by training in children and young adults has long been established. For middle-aged adults the results of investigations over the last two decades that demonstrate the beneficial effects of physical conditioning are also too numerous to cite. However, only recently have we turned our attention to the problem of maintaining and improving physical fitness and associated functional capacities in the elderly male and female, arbitrarily defined here as those persons over 60. Table 16.3 shows the results of recent research in this direction.

Table 16.3
Effects of Physical Conditioning on the Functional Capacities
of Older Men and Women (Mean Age over 60)

Measurement	Sex	N	% Improvement	Source*
Cardiovascular system				
1. Decrease in HR at submaximal work	M F	5 } 3 }	19	Barry et al. (3)
	M	8	3–5	Niinimaa and Shephard†
	F	7	3	Niinimaa and Shephard†
2. O₂ pulse	M	13	11	Benestad (4)
	M	48	4 (6 weeks)	deVries (17)
	M	5	29 (42 weeks)	deVries (17)
	F	17	7 (12 weeks)	Adams and deVries (1)
	M	8	1–5 (11 weeks)	Niinimaa and Shephard†
	F	7	0–3 (11 weeks)	Niinimaa and Shephard†
3. Increased blood volume	M	13	9	Benestad (4)
4. Increased total Hb.	M	13	7	Benestad (4)
5. Cardiac output at submaximal work	M	34	0	deVries (17)
	M	8	9	Niinimaa and Shephard†
	F	7	0	Niinimaa and Shephard†
6. Stroke volume at submaximal work	M	34	6	deVries (17)
	M	8	0	Niinimaa and Shephard†
	F	7	0	Niinimaa and Shephard†
7. Resting systolic BP	M F	5 } 3 }	13	Barry et al. (3)
	M	66	2	deVries (17)
	F	17	0	Adams and deVries (1)
8. Resting diastolic BP	M F	5 } 3 }	6	Barry et al. (3)
	M	66	4	deVries (17)
	F	17	0	Adams and deVries (1)
9. Regression of ECG abnormalities	M F	5 3	50% of abnormalities definitely improved	Barry et al. (3)

Table 16.3 — *Continued*

Respiratory system				
1. Vital capacity	M	5 }	0	Barry et al. (3)
	F	3 }		
	M	66	5 (6 weeks)	deVries (17)
	M	8	20 (42 weeks)	deVries (17)
	F	17	0	Adams and deVries (1)
2. Maximum ventilation during exercise	M	47	12 (6 weeks)	deVries (17)
	M	7	35 (42 weeks)	deVries (17)
	M	5 }	50	Barry et al. (3)
	F	3 }		
	M	13	0	Benestad (4)
	F	17	0	Adams and deVries (1)

Physical work capacity, VO_2 max				
	M	61	9 (6 weeks)	deVries (17)
	M	8	16 (42 weeks)	deVries (17)
	F	17	37	Adams and deVries (1)
	M	5 }	76	Barry et al. (3)
	F	3 }		
	M	14	29	
				Sidney and Shephard (53)
	F	28	29	Sidney and Shephard (53)
	M	14	11	Suominen et al.‡
	F	12	12	Suominen et al.‡

Muscular strength				
	M	68	6 (6 weeks)	deVries (17)
	M	8	12 (42 weeks)	deVries (17)
	M	5 }	50	Perkins and Kaiser (43)
	F	15 }		

*Unless otherwise noted, complete source data are given in the references section at the end of the chapter.

†Niinimaa, V., and Shephard, R.J. *J. Geront.* 33:362–367, 1978.

‡Suominen, H., et al., *Eur. J. Appl. Physiol.* 37:173–180, 1977.

There seems little doubt that the PWC of the older individual can be improved by significant increments. While improved PWC may not add years to our life, it most certainly does add "life to our years." The improvement of PWC is tantamount to increasing the *vigor* of the older individual, and this can make a very important contribution to the later years of life, certainly in terms of life style and possibly even in terms of *health*.

The physiological basis for the improvement in PWC is still very much in question. All of the data reported seem to agree that the older organism is trainable and that the capacity for improvement percentagewise is probably not greatly different from that of the young. The capacity for maximum achievement is, of course, severely compromised since the older subject starts from a lower level. As to the mechanism by which improved PWC is brought about, the results reported by Saltin and colleagues (50) and Hartley and co-workers (29) suggest a difference in the mechanisms of adaptation from young adulthood to middle age (thirty-four to fifty-five years) in that the young respond with increases of (1) cardiovascular dimensions (heart size), (2) better redistribution of blood flow to the active tissues, and (3) increased cardiac output and other functional improvements. In middle-aged men the third factor alone seems to account for the improvement of aerobic capacity. Whether this is also true for the older population (sixty and over) remains unanswered at this time. The data of table 16.3 suggest that improvement of respiratory function may be another important factor in improvement of PWC in the older male.

Important effects of physical conditioning on bone and connective tissues have also been reported. One of the very serious problems for older people, especially older women, is the loss of bony tissue (osteoporosis). Smith and Reddan (55) showed that twenty women with a mean age of 82 *gained* 4.2% in bone mineral composition as the result of exercise compared with controls who lost 2.5% over the thirty-six-month experimental period.

Losses of joint mobility constitute another serious problem for the elderly. Here again, appropriate exercise can result in significant improvement (10).

Principles for Conduct of Conditioning Programs for Older Men and Women (over Sixty)

The principles set forth in this section have been developed specifically around programs for the elderly and constitute an extension of the discussion in chapter 13.

Medical Examination. It is absolutely essential that every individual over sixty be examined and have the approval of a physician before entering any physical conditioning program. In recent years, many cardiologists have added stress tests to their examinations in which the individual's ECG is monitored *during* the stress of progressively increasing exercise work loads. Such data are invaluable in the conduct of conditioning programs for middle-aged and older people.

Physiological Monitoring (in the Laboratory). Ideally, the individual's initial condition and progress in the program would be evaluated in depth, including responses with respect to O_2 consumption, cardiovascular function (blood pressure, cardiac output, ECG), respiratory function, muscular status, and anthropometric measurements. This is seldom feasible, and it is fortunate that simple measures can provide considerable insight into the individual's status (assuming prior medical clearance). At a minimum expenditure of money, time, and effort, at least the following parameters can and should be measured: (1) HR and BP response to submaximal exercise (Astrand test), (2) resting BP, (3) strength of selected muscle groups, (4) body weight, and (5) percentage of body fat estimated from skin folds. Such measurements form the basis for a scientific approach to the use of exercise in conditioning older people in that they (1) allow prescription of exercise on a dose-response basis, and (2) provide considerable motivation for the participants who can thus see their own progress with respect to health benefits they can understand.

Physiological Monitoring (Gym or Field). Participants should be taught to take their own heart rate, usually at the radial artery. They should be taught to find the artery quickly (in five to ten seconds) and to count pulse beats accurately over a fifteen-second period immediately following exercise. While such a count immediately after exercise in the young may involve considerable error because of the very rapid exponential decline in rate, the rate of decline in older people is much slower, and the rate thus counted is a valuable criterion of the adequacy of response to any given exercise workout in the middle-aged and elderly.

Prescription of Exercise (Dose-Response Data). Figure 16.8 shows that the threshold for a training effect in older people requires that they work above that percentage of their heart rate range (HRR) represented by their Astrand score for estimated maximal O_2 consumption in milliliters per kilogram per minute. For example, an estimated maximal O_2 of 30 ml/kg/min would require exercising at levels that would bring HR at least 30% of the

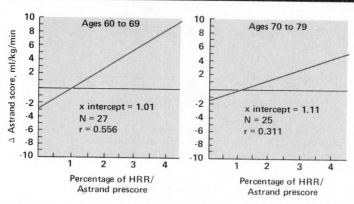

Figure 16.8 Change in Astrand test score after six weeks of training as a function of percentage of heart rate range/Astrand prescore. (From deVries, H.A. *Geriatrics* 26:94, 1971.)

Age and Exercise

Table 16.4
Maximal Heart Rates in Older Men

Age	HR	Age	HR	Age	HR	Age	HR
50	174	60	166	70	156	80	147
51	173	61	165	71	155	81	146
52	172	62	164	72	154	82	145
53	172	63	163	73	153	83	145
54	171	64	162	74	152	84	144
55	170	65	161	75	152	85	143
56	169	66	160	76	151	86	143
57	168	67	159	77	150	87	142
58	168	68	158	78	149	88	141
59	167	69	157	79	148	89	141

From *Arbeitsphysiologie* 10:251–323, 1938.

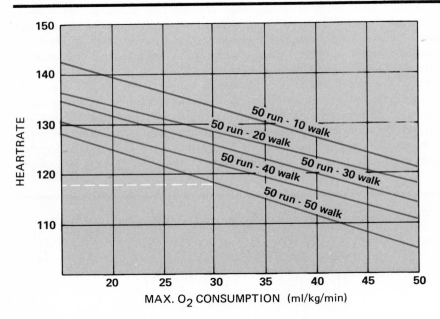

Figure 16.9 Nomogram for the estimation of heart rate response to a given dose of jogging for men sixty to seventy-nine. Example: For a man in this age bracket with a measured (or estimated from Astrand test) maximal oxygen consumption of 30 ml/kg/min, go vertically from 30 on the horizontal axis to the intersection with the 50 run-50 walk regression line. Now go horizontally to the heart rate axis to read 118 which represents the mean response to this dose. The standard error for the 5 regression lines is 8 to 10 beats. (From deVries, H.A. *Geriatrics* 26:110, 1971.)

way from resting toward maximal (18). Thus a man seventy-five years old with a resting rate of seventy and a maximal rate of 152 (taken from table 16.4) would need to work at an HR above

$$70 + .30 (152-70) = 70 + 25 = 95.$$

Since ninety-five represents the *threshold* for a training effect and since there are errors in our calculations one might raise that value by 15%-20% and estimate the desirable working HR as approximately 109 to 114 which would be a safe load for the healthy normotensive older individual. Indeed, Sidney and Shephard (53) have reported the use of HRs as high as 130–140.

Now, using those figures as a target HR, we may enter the nomogram of figure 16.9 (19) to find what combination of jog-walk would furnish the appropriate challenge. For a maximum O_2 consumption of 30 ml/kg/min even the fifty steps jog-fifty steps walk would raise HR to 118 after five sets of 50–50, and therefore the subject would be instructed to start with only two to three sets of fifty jog-fifty walk and to monitor HR carefully.

Progression. The system of gradual progression of exercise work load used in the author's laboratory and geriatric exercise program is shown in table 16.5. The warm-up is accomplished by calisthenics, the cardiovascular-respiratory challenge is provided by the run-walk program, and static stretching is used to improve joint mobility and to prevent muscle problems (17).

Table 16.5
Exercise Regimen (Three Times a Week)

A. Calisthenics (15-20 min)
   1. 5BX
   2. President's Council & Administration on Aging Series (1968)
   3. Others
B. Run-walk program (15-20 min)
   1. 50 steps run, 50 steps walk
      a. 5 sets the first day
      b. Each day increase the number of sets by one until 10 sets have been completed
      c. Use the same set procedure for each new series of run-walk
   2. 50 steps run, 40 steps walk
   3. 50 steps run, 30 steps walk
   4. 50 steps run, 20 steps walk
   5. 50 steps run, 10 steps walk
   6. 75 steps run, 10 steps walk
   7. 100 steps run, 10 steps walk
   8. 125 steps run, 10 steps walk
   9. 150 steps run, 10 steps walk
  10. 175 steps run, 10 steps walk
  11. 200 steps run, 10 steps walk
  12. Individual program
C. Static stretching to prevent soreness and to improve joint mobility (15-20 min)

The run or jog phase of the run-walk is done at the cadence and stride length normal and comfortable to the individual with no attempt at regulation for time, etc., at this age level. This program has been shown to be both safe and effective (on a three times per week basis) for a normal population of older men (17) and women (1) in the presence of medical and physiological monitoring.

Type of Exercise as a Determinant of Heart Stress in Older People. DeVries and Adams conducted an experiment on twelve older men (mean age, sixty-nine) in which the work of the heart relationship to total body work was explored for exercises involving (1) heavy but rhythmic arm and leg work (crawling), (2) heavy rhythmic leg work with moderate static contraction of the upper limbs (cycling), and (3) heavy rhythmic leg work without any static muscular activity (walking). The data indicated that the cardiac effort rises more slowly in walking (rhythmic) exercise with increasing loads of total body work than it does in either cycling or crawling effort due to the static muscular contractions (see fig. 13.3). It is important in exercising older people to reduce the sympathetic adrenergic vasoconstrictor reflex to a minimum by maximizing the rhythmic activity of large muscle masses and by minimizing (1) high activation levels of small muscles and (2) static muscle contraction of any kind. The natural activities of walking and running (jogging) are well suited to this purpose.

Implications for Physical Education and Athletics

After the foregoing review of the effects of aging upon physical performance and related physiological functions, the future for the college-age student using this text must seem unattractive indeed. However, two important mitigating factors must be considered.

First, each performance measure discussed has wide variability in its measurement. Since all of the discussion has centered about comparisons of *means* for various age groups, it is extremely important to recognize that, in all of the measurements, it is possible for a superior individual in an older age bracket to surpass the performance of an inferior young individual (superior or inferior in respect to his age group).

Second, for many of the measurements discussed in this chapter, we have no way of knowing how much of the age decrement can be attributed to aging per se and how much to increasingly sedentary habits or increasing degrees of unrecognized disease processes such as arteriosclerosis.

That some decline in physical performance must occur with advanced age is certainly undeniable, but that the ravages of old age can be slowed down by a sensibly vigorous regimen of physical exercise is supported by the available scientific evidence. Let us now consider the implications of the foregoing information for the philosophy and curricula of physical education in regard to four arbitrarily defined age groups: (1) childhood, six to twelve years; (2) adolescence, thirteen to twenty years; (3) maturity, twenty-one to sixty years; and (4) old age, sixty-one and over.

Childhood. There seems to be no physiological disadvantage for this age group with respect to energy supply. In relation to their mass, children can transport and utilize oxygen at rates comparable to young adults.

In regard to strength, it is well known through observation of norms for motor performance tests that strength per unit of weight increases with age. Or one might say that the younger the child, the less able the child is to handle its own body weight.

In regard to the nervous system and coordination, all evidence indicates that reaction time and speed of movement are at a low level in early childhood and improve rapidly to their maximal values in young adulthood.

According to the evidence cited, it would seem that childhood is a time for games and activities that are vigorous enough to bring about maximal physical development. Such activities as gymnastics (at least apparatus events), in which strength per unit of body weight is at a premium, would be most successfully accomplished during the adolescent years. Activities that test, and possibly develop, speed of reactions and movements are challenging and are well received in childhood. Baseball, basketball, and soccer are ideal activities.

Adolescence. During these years an individual achieves or closely approaches the maximum in all performance and physiological measurements. Kobayashi and co-workers (33) have shown that during the adolescent growth spurt, training increased aerobic capacity above the normal increase attributable to age and growth. No great increases resulted from training at earlier ages. There are two considerations for physical education for this age range: (1) supplying the immediate needs for exercises that will provide the greatest physical development, and (2) forming the foundation for exercise habits in later years.

The first consideration requires the inclusion of vigorous strength and endurance activities, and there are no limitations for a healthy individual in this age bracket. Every individual should be exposed to and challenged by heavy-resistance activities for developing strength and activities of sufficient intensity and duration of effort to build muscular and cardiorespiratory endurance. Examples of heavy-resistance activities are gymnastics, wrestling, and weight training. Endurance-building activities include swimming and distance running (against time, at least occasionally), and all of the vigorous sports activities—football, soccer, handball, tennis, badminton.

The second consideration, forming the foundation for exercise in later years, requires a good instructional program that is geared to the needs of adults and that is readily provided by the physical recreational opportunities of a particular geographic area. Primary consideration must be given to the fact that adults cannot always count on even one available partner or opponent much less seventeen for baseball or nine for basketball. Individual and dual sports must be introduced and skills must be taught so that future *participation* is encouraged rather than *spectatoritis*.

Maturity. It is the author's conviction that limitations on exercise in the adult years are self-imposed by a lack of consistency and continuity in an individual's exercise habits. Probably every activity that is mastered during adolescence or young adulthood can be continued by a healthy individual well into old age. This is true, however, only if participation occurs a minimum of three times weekly. If occupational obligations prevent this degree of participation, calisthenics or weight training programs can be designed to provide the mid-week conditioning necessary to prevent the weekend's activity from becoming a strain instead of an exhilarating experience.

For example, skiing, especially cross-country skiing, can be one of the most demanding of all sports; few adults can ski every weekend, let alone three times per week, but it is a simple matter for a trained physical educator to provide midweek exercise programs for maintaining the physical capacities needed for successful weekend skiing. Suitable calisthenic or weight training exercises can maintain the muscular strength and range of movement needed, while simple application of interval training principles to bench-stepping or stair-climbing can maintain the requisite cardiorespiratory fitness.

From the standpoint of *preventive medicine,* it is most important that every healthy adult possess skill in one or several sports that are vigorous enough to maintain optimum levels of cardiorespiratory fitness. Such activity will also provide benefits in maintaining body weight and relieving nervous tension (chap. 11). Ideally, this activity should be so enjoyable and challenging that it is self-motivating. Activities that are interpreted as drudgery are not likely to be long continued, no matter how rewarding. Excellent activities that meet the above criteria are tennis, handball, volleyball (if played correctly), badminton, skiing, surfboarding, etc. It will be noted that some excellent activities, from a fun standpoint have been omitted; golf, for instance, is not vigorous enough for young adults (nor is swimming, as ordinarily performed) to form the *only* physical recreation, if maximum benefits in physical condition are desired.

It should also be emphasized that any layoff from vigorous activities must be followed by a period of progressive rebuilding to the former level of competency. In general, the older the individual, the lower the first work load should be, and the longer the period of progressive rebuilding of condition.

Old Age. Depending upon the health of the individual, a greater or lesser reduction in work load (both intensity and duration) is indicated. Most activities allow for such adjustment without complete cessation. In tennis and badminton one plays doubles instead of singles; in skiing one skis for shorter periods of time with increasing lengths of rest intervals. Obviously, in all competitive dual sports, one modification exists simply in playing opponents of roughly equivalent age.

There is no evidence that vigorous exercise can in any way injure a *healthy* individual in the older age brackets; however, frequent physical examinations, at least yearly, are necessary to protect the individual from overstrain during an incipient illness.

Indeed, there is a growing body of experimental evidence to show that healthy old individuals improve their functional capacities through physical conditioning much as do young people. Percentagewise their improvement is comparable to that in the young, although they start at and progress to lower achievement levels and probably require less training stimulus to bring about the desired response. In general, it may be said that the effects of physical conditioning upon middle-aged and older individuals are opposite in direction to those commonly associated with the aging process.

SUMMARY

1. The entire body of knowledge with respect to the loss of function with increasing age must be viewed with caution, since in very few cases has the effect of decreasing levels of physical activity been controlled or ruled out.

2. Various physiological functions seem to have differing rates of decline with increasing age, with some functions being relatively unaffected.

3. Muscular strength decreases very slowly during maturity with an increasing rate of loss in old age.

4. The capacity for muscular hypertrophy is greatly decreased in old age although strength gains are still possible. Such gains in strength appear to result from a "learning" to innervate a larger portion of the motor neurone pool.

5. Cardiac output decreases after early maturity by a little less than 1% per year. This seems to be true both at rest and at maximal effort.

6. Peripheral blood flow in general is slowed down with typically increasing resistance to flow and concomitant increases in blood pressure.

7. Pulmonary function declines with age with the most important changes being seen with respect to vital capacity, lung diffusion, and thoracic wall compliance.

8. During adulthood, there is a gradual decline in aerobic capacity with age for both sexes. Some recent evidence suggests the decline may be slower in those who are physically active.

9. In our culture, it is typical (though not desirable) to gain weight as we grow older, although the lean body mass decreases. Therefore, the typical gain in weight is due to increasing levels of body fat.

10. As we grow older, our reaction and movement times are slower. This is probably due to age changes in the CNS.

11. The trainability of the older individual is roughly equal to that of the young adult when expressed on a relative basis.

12. Important health benefits have been reported to result from physical conditioning of the previously sedentary elderly. Such benefits include: (a) improved O_2 transport and aerobic capacity, (b) lowered blood pressure, (c) improved breathing capacity, (d) reduction in osteoporotic changes, (e) improved joint mobility, and (f) a tranquilizer effect that reduces neuromuscular tension (anxiety).

13. The principles of exercise physiology that can be brought to bear on the development of conditioning programs for the elderly are discussed in detail.

REFERENCES

1. Adams, G.M., and deVries, H.A. Physiological effects of an exercise training regimen upon women aged 52–79. *J. Gerontol.* 28:50–55, 1973.
2. Astrand, I. Aerobic work capacity in men and women with special reference to age. *Acta Physiol. Scand.* vol. 49 (supp. 169), 1960.
3. Barry, A.J.; Daly, J.W.; Pruett, E.D.R.; Steinmetz, J.R.; Page, H.F.; Birkhead, N.C.; and Rodahl, K. The effects of physical conditioning on older individuals. *J. Gerontol.* 21:182–91, 1966.
4. Benestad, A.M. Trainability of old men. *Acta Med. Scand.* 178:321–27, 1965.
5. Birren, J.E.; Imus, H.A.; and Windle, W.F., eds. *The Process of Aging in the Nervous System.* Springfield, Ill.: Charles C Thomas, 1959.
6. Birren, J.E.; Butler, R.N.; Greenhouse, S.W.; Sokoloff, L.; and Yarrow, M.R., eds. *Human Aging, a Biological Behavioral Study.* Washington, D.C.: Public Health Service Publication no. 986, 1963.
7. Boerner, W.; Moll, E.; Schroder, J.; and Rau, F.P. Abhangigkeit der kreislaufzeit von alter geschlecht und korperlange. *Archiv fur Kreislaufforschung.* 43:221–35, 1964.
8. Brandfonbrener, M.; Landowne, M.; and Shock, N.W. Changes in cardiac output with age. *Circulation* 12:557–66, 1955.
9. Brozek, J. Changes of body composition in man during maturity and their nutritional implications. *Fed. Proc.* 11:784–93, 1952.
10. Chapman, E.A.; deVries, H.A.; and Swezey, R. Joint stiffness: Effects of exercise on old and young men. *J. Gerontol.* 27:218–21, 1972.
11. Clarkson, P.M. The effect of age and activity level on fractionated response time. *Med. Sci. Sports* 10:66, 1978.
12. Comroe, J.H.; Forster, R.E.; Dubois, A.B.; Briscoe, W.A.; and Carlsen, E. *The Lung.* Chicago: Year Book Medical Publishers, Inc., 1962.
13. Corre, K.A.; Cho, H.; and Barnard, R.J. Maximum exercise heart rate reduction with maturation in the rat. *J. Appl. Physiol.* 40:741–44, 1976.
14. Damon, E.L. An experimental investigation of the relationship of age to various parameters of muscle strength. Doctoral dissertation, Physical Education, USC, 1971.
15. Dehn, M.M., and Bruce, R.A. Longitudinal variations in maximum oxygen intake with age and activity. *J. Appl. Physiol.* 33:805–7, 1972.
16. DeQueker, J.V.; Baeyens, J.P.; and Claessens, J. The significance of stature as a clinical measurement of aging. *J. Am. Geriatr. Soci.* 17:169–79, 1969.

17. deVries, H.A. Physiological effects of an exercise training regimen upon men aged 52–88. *J. Gerontol.* 25:325–36, 1970.

18. ———. Exercise intensity threshold for improvement of cardiovascular-respiratory function in older men. *Geriatrics* 26:94–101, 1971.

19. ———. Prescription of exercise for older men from telemetered exercise heart rate data. *Geriatrics* 26:102–11, 1971.

20. ———. Comparison of exercise responses in old and young men: II. Ventilatory mechanics. *J. Gerontol.* 27:349–52, 1972.

21. Dill, D.B.; Horvath, S.M.; and Craig, F.N. Responses to exercise as related to age. *J. Appl. Physiol.* 12:195–96, 1958.

22. Donevan, R.E.; Palmer, W.H.; Varvis, C.J.; and Bates, D.V. Influence of age on pulmonary diffusing capacity. *J. Appl. Physiol.* 14:483–92, 1959.

23. Drinkwater, B.L.; Horvath, S.M.; and Wells, C.L. Aerobic power of females, age 10–68. *J. Gerontol.* 30:385–94, 1975.

24. Ermini, M. Aging changes in mammalian skeletal muscle. *Gerontology* (Basel) 22:301–16, 1976.

25. Evans, S.J. An electromyographic analysis of skeletal neuromuscular fatigue with special reference to age. Doctoral dissertation, Physical Education, USC, 1971.

26. Fisher, M.B., and Birren, J.E. Age and strength. *J. Appl. Psychol.* 31:490–97, 1947.

27. Goldspink, G., and Howells, K.F. Work-induced hypertrophy in exercised normal muscles of different ages and the reversibility of hypertrophy after cessation of exercise. *J. Physiol.* 239:179–93, 1974.

28. Gutman, E.; Hanzlikova, V.; and Jakoubek, B. Changes in the neurovascular system during old age. *Exp. Gerontol.* 3:141–46, 1968.

29. Hartley, L.H.; Grimby, G.; Kilbom, A.; Nilsson, N.J.; Astrand, I.; Bjure, J.; Ekblom, B.; and Saltin, B. Physical training in sedentary middle aged and older men: III. Cardiac output and gas exchange at submaximal and maximal exercise. *Scand. J. Clin. Lab. Invest.* 24:335–49, 1969.

30. Hodgson, J.L., and Buskirk, E.R. Physical fitness and age: with emphasis on cardiovascular function in the elderly. *J. Am. Geriatr. Soc.* 25:385–92, 1977.

31. Julius, S.; Amery, A.; Whitlock, L.S.; and Conway, J. Influence of age on the hemodynamic response to exercise. *Circulation* 36:222–30, 1967.

32. Kasch, F.W., and Wallace, J.P. Physiological variables during 10 years of endurance exercise. *Med. Sci. Sports* 8:5–8, 1976.

33. Kobayashi, K.; Kitamura, K.; Miura, M.; Sudeyama, H.; Murase, Y.; Miyashita, M.; and Matsue, H. Aerobic power as related to body growth and training in Japanese boys: a longitudinal study. *J. Appl. Physiol.* 44:666–72, 1978.

34. Mittman, C.; Edelman, N.H.; Norris, A.H.; and Shock, N.W. Relationship between chest wall and pulmonary compliance and age. *J. Appl. Physiol.* 20:1211–16, 1965.

35. Montoye, H.J., and Lamphiear, D.E. Grip and arm strength in males and females, age 10 to 69. *Res. Q.* 48:109–20, 1977.

36. Moritani, T., and deVries, H.A. Neural factors versus hypertrophy in the time course of muscle strength gain in young and old men, in press.

37. Norris, A.H.; Shock, N.W.; and Yiengst, M.J. Age changes in heart rate and blood pressure responses to tilting and standardized exercise. *Circulation* 8:521–26, 1953.

38. Norris, A.H.; Shock, N.W.; Landowne, M.; and Falzone, J.A. Pulmonary function studies: age differences in lung volume and bellows function. *J. Gerontol.* 11:379–87, 1956.

39. Norris, A.H.; Shock, N.W.; and Falzone, J.A. Relation of lung volumes and maximal breathing capacity to age and socio-economic status. In *Medical and Clinical Aspects of Aging,* ed. H.T. Blumenthal, pp. 163–71. New York: Columbia University Press, 1962.

40. Palmer, G.J.; Ziegler, M.G.; and Lake, C.R. Response of norepinephrine and blood pressure to stress increases with age. *J. Gerontol.* 33:482–87, 1978.

41. Pariskova, J.; Eiselt, E.; Sprynarova, S.; and Wachtlova, M. Body composition, aerobic capacity and density of muscle capillaries in young and old men. *J. Appl. Physiol.* 31:323–25, 1971.

42. Pemberton, J., and Flanagan, E.G. Vital capacity and timed vital capacity in normal men over forty. *J. Appl. Physiol.* 9:291–96, 1956.

43. Perkins, L.C., and Kaiser, H.L. Results of short term isotonic and isometric exercise programs in persons over sixty. *Phys. Ther. Rev.* 41:633–35, 1962.

44. Petrofsky, J.S., and Lind, A.R. Aging, isometric strength and endurance, and cardiovascular responses to static effort. *J. Appl. Physiol.* 38:91–95, 1975.

45. Redisch, W.; Meckeler, K.; and Steele, J.M. Changes in peripheral blood flow and vascular responses with age. In *Medical and Clinical Aspects of Aging,* ed. H.T. Blumenthal, pp. 453–59. New York: Columbia University Press, 1962.

46. Retzlaff, E., and Fontaine, J. Functional and structural changes in motor neurons with age. In *Behavior Aging and the Nervous System,* ed. A.T. Welford and J.E. Birren. Springfield: Charles C Thomas, 1965.

47. Rizzato, G., and Marazzini, L. Thoracoabdominal mechanics in elderly men. *J. Appl. Physiol.* 28:457–60, 1970.

48. Robinson, S. Experimental studies of physical fitness in relation to age. *Arbeitsphysiologie* 10:251–323, 1938.

49. Rodahl, K. Physical work capacity. *Arch. Environ. Health* 2:499–510, 1961.

50. Saltin, B.; Hartley, H.; Kilbom, A., and Astrand, I. II. Physical training in sedentary middle aged and older men. *Scand. J. Clin. Lab. Invest.* 24:323–34, 1969.

51. Shock, N.W. Current concepts of the aging process. *J.A.M.A.* 175:654–56, 1961.

52. Shock, N.W.; Watkin, D.M.; Yiengst, M.J.; Norris, A.H.; Gaffney, G.W., Gregerman, R.I.; and Falzone, J.A. Age differences in the water content of the body as related to basal oxygen consumption in males. *J. Gerontol.* 18:1–8, 1963.

53. Sidney, K.H., and Shephard, R.J. Frequency and intensity of exercise training for elderly subjects. *Med. Sci. Sports* 10:125–31, 1978.

54. Simonson, E. Changes of physical fitness and cardiovascular functions with age. *Geriatrics* 12:28–39, 1957.

55. Smith, E.L., and Reddan, W. Physical activity—a modality for bone accretion in the aged. *Am. J. Roentgenol.* 126:1297, 1977.

56. Spirduso, W.W. Reaction and movement time as a function of age and physical activity level. *J. Gerontol.* 30:435–40, 1975.

57. Starr, I. An essay on the strength of the heart and on the effect of aging upon it. *Am. J. Cardiol.* 14:771–83, 1964.
58. Turner, J.M.; Mead, J.; and Wohl, M.E. Elasticity of human lungs in relation to age. *J. Appl. Physiol.* 25:664–71, 1968.
59. Tzankoff, S.P., and Norris, A.H. Effect of muscle mass decrease on age-related BMR changes. *J. Appl. Physiol.* 43:1001–6, 1977.
60. Vasiliyva, V.Y. The effect of physical exercise on the cardiovascular system of elderly persons. *Excerpta Medica Gerontology and Geriatrics* 5:5, no. 641. From papers presented at 2d conference on Gerontology and Geriatrics at Moscow, 1962.
61. Vogt, C., and Vogt, O. Aging of nerve cells. *Nature* 158:304, 1946.
62. Wessel, J.A., and Van Huss, W.D. The influence of physical activity and age on exercise adaptation of women aged 20–69 years. *J. Sports Med.* 9:173–80, 1969.

Part 3

Physiology of Training and Conditioning Athletes

17

Physiology of Muscle Strength

The desirability of a *minimum quantity* of strength has long been recognized in athletics; however, the advantages of maximum levels of strength for all sports in which power is a factor were not recognized by physical educators, athletes, or coaches until quite recently. This strange neglect of the strength factor in athletes was the result of an unscientific acceptance by virtually everyone concerned of an old wives' tale that claimed that the development of large amounts of strength in the musculature (through activities such as weight training) inevitably resulted in a condition known as *muscle-bound*. Being muscle-bound was supposed to limit both range and speed of movement in those who participated in weight training. Therefore it was anathema to all but the most heretical coaches. This belief persisted until shortly after World War II.

At the end of World War II the need for rehabilitation procedures to restore strength to various body segments of injured veterans was acute. This need brought about a scientific evaluation of weight training procedures, and the pioneering work of De Lorme and Watkins (13) brought acceptance of weight training for rehabilitation purposes. Acceptance of weight training by the medical profession apparently stimulated the research-oriented members of the physical education profession to put the muscle-bound hypothesis to the test by the scientific method. The results are now history, for many well-controlled investigations laid the ghost of the muscle-bound myth. It is now generally accepted that properly conceived weight training programs not only do not slow or restrict joint motion, but may even improve these factors while providing very substantial gains in strength.

The importance of strength in athletics is not always obvious. However, in an activity such as shot-putting, the need for maximum power is readily apparent. As discussed earlier, power is the rate of doing work (producing force). We may therefore think of power as the result of two factors: (1) strength to produce the force and (2) speed to increase the rate at which the force can be applied. In other words, we can improve an athlete's power in three different ways: (1) increase speed, strength remaining constant, (2) increase strength, speed remaining constant, and (3) improve speed and strength.

From practical experience in coaching, speed of movement improves rapidly in the training program to a plateau, from which it can be increased only with great difficulty. On the other hand, very few athletes have even begun to approach their maximum strength levels, and thus large gains in power are possible by improving strength while simply maintaining speed.

This chapter will discuss the physiological bases of strength and its improvement; it will culminate in a set of principles, based upon our present knowledge, for the formulation of programs designed to improve the strength (and power) factor in athletes.

I. Physiology of Strength

Hypertrophy versus Hyperplasia. It is a well-known fact that when a muscle is trained with heavy resistance exercise, it grows larger in girth. This growth could be the result of either the enlargement of each muscle fiber (hypertrophy) or an increasing number of cells (hyperplasia). The latter was ruled out in the 1800s by Morpurgo. However, in recent years several investigators have reported fiber-splitting as the result of heavy resistance training, but no one has found evidence of true cell division (mitosis). For all practical purposes we may still accept hypertrophy as the basis for the growth of a muscle in response to heavy exercise.

We may, however, ask whether the hypertrophy is selective with respect to fiber type. The best evidence available suggests that heavy resistance exercise does result in a selective hypertrophy of the FT fibers (11, 39). Although there is no gain or loss of FT or ST fiber *numbers,* the selective hypertrophy of the FT fibers has been reported to bring about marginally significant changes in the cross-sectional *areas,* with the FT increasing by 5%-12% as the ST decrease by roughly equivalent amounts (11). The increased area of FT fibers appears to be accounted for entirely by growth of the IIa (FOG) fibers, with no change in the IIb (FG) fibers.

Morphological versus Neurological Factors in Strength Gain. It has been shown that maximal electrical stimulation of human muscles can bring about approximately 30% greater expression of strength than can be elicited by maximal voluntary contraction (26). Therefore it is obvious that strength (or at least the *expression* of strength as we measure it) can also be improved by greater activation of the muscle tissue by central nervous system influences. In the young individual, neural factors account for the greatest part of the early gain in strength (first three to four weeks) after which hypertrophy accounts for virtually all of the strength gain (32). See figures 4.14 and 4.15, and the discussion in chapter 4. As we pointed out in chapter 16, the elderly depend almost entirely on improving activation level, since apparently very little capacity for hypertrophy remains beyond age sixty (33).

Disuse and Atrophy. Anyone who has had a limb in a cast can attest to the basic fact of muscular atrophy (shrinking of a muscle). Scientific evidence has been presented (14, 24) to show that immobilization of a limb (whether mechanically, as by cast, or by denervation) results in decreased fiber size and (at least over a short period of time) in no loss in number of fibers. Thus we may say that all changes in muscle tissue brought about by training programs are impermanent, and that training must be carried on systematically throughout a lifetime. Otherwise, a degree of atrophy from disuse will set in.

There is growing evidence that stretch of muscle in vivo may not only retard atrophy of a denervated muscle, but may even induce muscular hypertrophy (7, 16). Animal experiments provide a biochemical basis for this finding in that amino acid transport into the stretched muscle is improved.

Orthopedists now try to cast limbs in positions that allow the greatest resting muscle lengths.

Mechanical Factors in Strength. If strength development and muscle hypertrophy proceed together, we might expect to find no change in the ratio of strength per unit of cross section as training proceeds. Indeed, Hettinger and Mueller (22) found this to be true in their work with isometric training. Furthermore, it has been suggested that this ratio remains very constant regardless of age or sex (36).

In any event, strength per unit of cross section is a theoretical concept, and it tells us little about the forces that can be produced at the ends of the bony lever systems. It will be recalled from earlier discussions that two factors interact to determine the force available at the end of a bony lever: (1) the length at which the muscle is working, and (2) the angle of pull. Figure 17.1 illustrates this in respect to the third-class lever that works at the elbow during elbow flexion. At full extension (180 degrees) where the muscle length is greatest, the angle of pull is poor. At full flexion (40 degrees), the muscle is at twice the disadvantage because it is short and because it works at a poor angle. The best combination of the two factors seems to be at approximately 115 degrees (9).

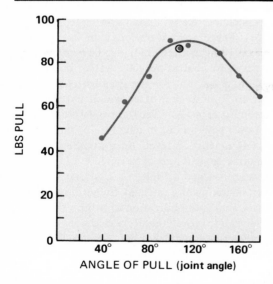

Figure 17.1 Relationship of force available in respect to joint angle for elbow flexion. (From Clarke, H.H. *Arch. Phys. Med.*, 31:81, 1950.)

Physiology of Training and Conditioning Athletes

When a muscle works directly in pulling upon a bone (without going through a mechanical lever, as in pulling a rope over a pulley), the curve approximates a length-tension diagram in which the muscle length is the dominant factor in determining the force available. In figure 17.2 it is seen that the best force of contraction results when the hip is fully flexed at 50 degrees because this position has the hip extensors fully stretched. Thus it can be seen that the strength available for doing useful work varies from joint to joint and also with the angle of each joint. This is a very important point to bear in mind when one considers the effects of an exercise. It will be discussed in greater detail in a following section.

The Necessary Stimulus for Strength Gain. Our training procedures for the development of strength could undoubtedly be greatly improved if we knew precisely what quantity and quality of stimulus were required to bring about a training effect.

In a very general way, we know that we must bring about overload conditions to elicit the hypertrophic response (or strength gain). That is to say, we must load the muscle beyond its normal everyday use to bring about an adaptive response. The original work of Hettinger and Mueller (22) defined 20% maximal voluntary contraction (MVC) for 1 sec/day as the load required to prevent atrophy and 35% MVC as the threshold value at which a training response began to appear. However, holding muscle tensions between 35% and 100%, MVC brings about many effects such as (1) increased tension in muscle and connective tissue, (2) gradual occlusion of blood flow,

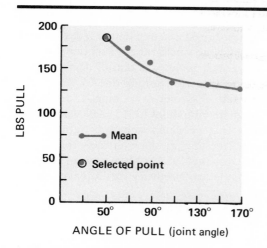

Figure 17.2 Relationship of force available in respect to joint angle for hip extension. (From Clarke, H.H. *Arch. Phys. Med.*, 31:81, 1950.)

Physiology of Muscle Strength

(3) increased local temperature, (4) hypoxia, and (5) increased levels of metabolic end products. The question is, Which of these factors provides the signal to elicit the hypertrophy response? Early thinking suggested hypoxia as the most likely signal but that was ruled out by studies that showed no relationship between experimental occlusion and training effect when equal loads of tension were applied (21). At Harvard, Goldberg and associates (16) have done extensive work on this problem. After reviewing their own work and that of others, they point out that the factor that best explains all experimental observations with respect to muscle hypertrophy, including the fact that even passive stretching can improve protein turnover and amino acid transport and can stimulate muscle O_2 consumption, is that of tension development. However, the mechanisms by which tension development is coupled with the metabolic events remain to be discovered.

Specificity of Strength Gain. Considerable evidence is accumulating that suggests that strength gained is relatively specific to certain aspects of the training method.

Joint Angle during Exercise. Logan demonstrated that the strength training effect is specific to the angle at which the greatest resistance is applied (30). Of his three experimental groups (fifteen in each), one used weight resistance and one used spring resistance to strengthen the knee extensors, while the third group was a control. Using weight resistance, the greatest resistance was encountered from about 155 degrees to full extension, and with spring resistance (on this particular device), the greatest resistance was offered at 115 degrees. The weight training group made significantly greater gains than the spring resistance group when tested at 155 degrees, and the spring resistance group showed significantly greater gains when tested at 115 degrees.

It should be noted that these differences existed despite the fact that both groups exercised isotonically throughout the whole range of motion, and the only real difference in resistance at various angles was that of degree. These results implied that even greater differences in training due to joint angle specificity would be found as the result of isometric training, and this has indeed been the case. Two independent investigations (3, 15) have shown that strength tested at angles other than that at which isometric training took place may show gains less than 50% of that at the exercised angle. This difference in gains appeared when the test angle was as little as twenty degrees removed from the exercise angle. The lesser specificity for isotonic training has also been demonstrated (12).

It has also been shown that the rate of strength gain is greater when a muscle is trained isometrically at a short length compared with a long length. This was found to be true for all muscles tested: elbow flexor and extensor, and wrist pronators and supinators (37).

Exercise Velocity. The strength gains from heavy resistance exercise appear to be limited to velocities at or below those used in training. Two studies using isokinetic methods have agreed that the significant strength benefits gained in training could not be demonstrated when the velocity of contraction was greater than that at which training occurred (29, 31). This effect was best demonstrated by Ikai (26) who plotted force-velocity curves (see chap. 4 and fig. 4.8) before and after training. He found that training at 100% MVC improved force but not velocity; training at 30%-60% MVC improved both force and velocity; training with no load and maximum velocity improved velocity but not force. This work was done with an inertia wheel and has the greatest implications for the training of athletes where the force-velocity curve changes are of far greater concern than capacity for a maximal static contraction or even a constant velocity contraction (isokinetics), neither of which is a common occurrence in athletics.

These experimental findings are entirely consistent with our basic knowledge of the neurophysiology involved. Henneman and his colleagues (18, 19, 20) have provided conclusive evidence for the *size principle* of motor unit recruitment. They showed that the smallest motoneurons had the lowest thresholds (ST fibers), the largest had the highest thresholds (FT fibers), and motor units are recruited and drop out in accordance with this principle. Consequently it is necessary to train at both the force and velocity that are to be applied in the athletic event. These findings also explain the low level of correlation found between strength measured statically and various power functions of the same muscle groups (10).

Training Curves—The Time Course of Strength Development. The most definitive work in this area was done by Mueller and Rohmert (34) at the Max Planck Institute in Dortmund, Germany. Unfortunately their important work was published only in German and consequently has remained almost unknown in the U.S.A. They showed that strength gain during isometric training approximates an exponential function with time: the rate of gain in strength at any given time is inversely proportional to the difference from the plateau value *(Endkraft)* to be achieved by that particular method of training. There is reason to believe this also applies to isotonic and isokinetic training as well. This means that the untrained individual gains at a much greater rate than the relatively trained one. Lack of appreciation of this fact has clouded most of the research directed to comparisons of different training methods.

The method used by Mueller and Rohmert depends on the following relationships:

$$S_R = 100 \; \frac{S_B}{S_E} \; \text{in percent}$$

where S_R = relative strength

S_B = strength at beginning of any training period of time

S_E = end strength, the plateau value achieved when strength gain over three to four weeks is less than standard error of measurement

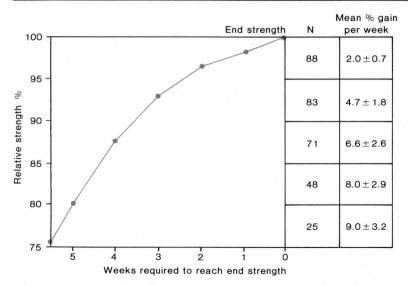

	End strength	N	Mean % gain per week
		88	2.0 ± 0.7
		83	4.7 ± 1.8
		71	6.6 ± 2.6
		48	8.0 ± 2.9
		25	9.0 ± 3.2

Figure 17.3 Gain in relative strength with isometric training of the trunk extensor muscles by one maximal contraction for one second daily. (Redrawn from the data of Mueller, E.A., and Rohmert, W. *Int. Z. Angew, Physiol. einschl. Arbeitsphysiologie* 19:403, 1963.)

Figure 17.3 illustrates this method. The important point is that beginning strength really has no physiological meaning. It depends to a very large extent on training status, which cannot be defined at the beginning of training. On the other hand, the end strength is a well-defined point, and if every subject is trained to this plateau value, the weekly values of relative strength can be plotted as in figure 17.3.

To compare the effects of various training regimens, one then plots the mean weekly rate of strength gain as a function of relative strength at the start of training. This is shown in figure 17.4, where the mean total strength gain per week (V_{TOT}) is calculated as:

$$V_{TOT} = \frac{100 - S_R}{\text{weeks in training}}$$

Subjects who enter into training experiments have greatly varying levels of training status, and this method has the obvious advantage of controlling for these levels.

Physiology of Training and Conditioning Athletes

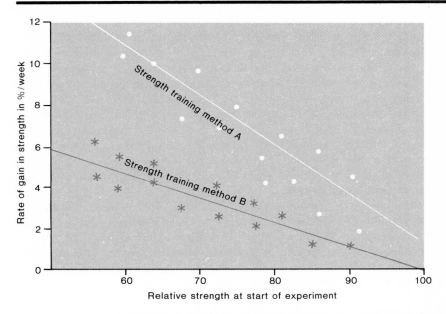

Figure 17.4 Conceptualization of the method of Mueller and Rohmert for comparison of strength training methods.

Methods for Measurement and Training of Strength

In general, there are in a physiological sense only five ways in which the contractile elements of muscle can produce force through the various bony levers available in the human body. They are (1) isometric contraction (a static contraction); (2) concentric isotonic contraction (shortening); (3) eccentric isotonic contraction (lengthening); (4) isokinetic contraction in which the angular velocity of the limb segment is constant; and (5) more or less normally accelerated movement against resistance, as applied in the use of the inertia wheel. Each of these types of muscle contraction can be used for both measurement and training purposes. Although many comparisons among methods have been attempted, none has yet been satisfactory from the standpoint of research design. For example, to achieve a meaningful comparison between isometric and isotonic methods of strength training one would need to equate (1) the training stimulus and (2) the trainability of the subjects, or at least assure randomization of trainability. To the author's knowledge, no investigation has yet satisfied these requirements, although methods have been proposed for their accomplishment. The use of the force-time integral (38)

Physiology of Muscle Strength   395 would satisfy the first requirement, and the use of Mueller and Rohmert's method would satisfy the second (34). Consequently we shall review various training methods without comment on their comparative effectiveness.

Isometric Training. Until quite recently, when man wanted to increase his strength he usually used some form of weight lifting in application of the overload principle. In 1953 Hettinger and Mueller (of Germany) published their work on isometric training (22). Their findings indicated that a maximum training effect could be obtained from one daily six-second isometric contraction against two-thirds of an individual's maximal contraction strength. Greater force, duration, or numbers of repetitions did not seem to increase the rate of strength gain, which they found averaged 5% per week when training was performed five times per week. Strength improved in various muscles from 33% to 181%.

Advantages of Isometric Methods. The chief advantages here are administrative in nature. If only one contraction is used per day (as in the original work of Hettinger and Mueller), great savings in time are possible. Furthermore, a little ingenuity can reduce the equipment needed to that already available in the gym or on the practice field. The elimination of the need for equipment like barbells or dumbbells also makes it possible to work out larger groups in shorter periods of time.

However, recent work from the same German laboratory (34) has modified the original work of Hettinger and Mueller, and it now appears that the rate of strength gain approximately doubles when maximal contraction strength is used instead of two-thirds maximum. Also, a higher end strength can be reached by increasing the number of six-second repetitions to between five and ten.

If these changes are made in the procedure, plus working each joint at three to four angles to eliminate the specificity problem, isometric results may well match isotonic results, but making these changes would nullify the advantage in time. It is also important to recall the discussion of chapter 7 in which the well-defined effect of isometric tension in raising the arterial blood pressure was pointed out. Isometric exercises must be considered potentially hazardous for people in cardiac rehabilitation or adult exercise programs because of the greater rise in blood pressure.

Methods in the Use of Isometric Contraction. The work of Mueller and Rohmert (34) seems to be definitive, so there are few choices to be made as to frequency, intensity, and the like. Best results appear to be obtained by using maximal contraction strength, held for five seconds and repeated five to ten times daily. It would be desirable to apply these contractions at varying points in the range of motion if a well-rounded workout is desired, or if the activity that is trained for demands strength or power, throughout the entire range of motion. In some cases, as in ballistic movements, analysis may indicate the need for maximal power at the beginning of the motion, and exercises should be designed accordingly.

Physiology of Training and Conditioning Athletes

Measurement of isometric tension is done easily, quickly, and is fairly precise, but does this measure the same capacity that is developed by isotonic exercise? A well-controlled study showed no significant correlation between isotonic and isometric measurements of strength gains (4). Even absolute strength, when measured by isotonic and isometric methods, yielded a correlation of only 0.622–0.800 (1, 4). The result of isotonic programs should therefore be measured isotonically and the results of isometric programs should be measured isometrically.

Cable Tension Strength Tests. Of all the methods for measuring isometric strength, probably the simplest and most widely used is the cable tension testing method of Clarke. Figure 17.5 illustrates the cable tensiometer instrument used in the testing, and figure 17.6 shows the application of the method in testing elbow flexion strength. Objectivity coefficients of 0.90 and above were obtained when the tests were administered by experienced testers. Thirty-eight such tests have been devised and validated for testing the various muscle groups of the body (8).

Isotonic Training. Probably the greatest advantage in isotonic methods is that strength gains are specific to the angle at which the resistance is encountered. Thus isotonic exercises can be designed to work the entire range of motion in one contraction, but several contractions would be needed at different angles to work the whole range of motion with isometric methods.

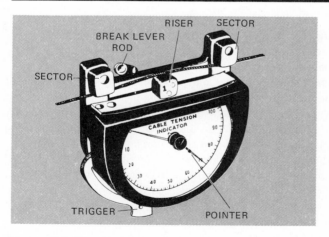

Figure 17.5 The cable tensiometer used for evaluation of muscular strength. (From Clarke, H.H., and Clarke, D.H. *Developmental and Adapted Physical Education.* Copyright © 1963, Prentice-Hall, Inc., Englewood Cliffs, N.J.)

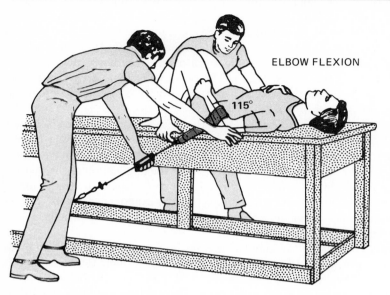

ELBOW FLEXION

115°

Figure 17.6 Use of the cable tensiometer in testing elbow flexion strength. (From Clarke, H.H., and Clarke, D.H. *Developmental and Adapted Physical Education.* Copyright © 1963, Prentice-Hall, Inc., Englewood Cliffs, N.J.)

Another advantage for isotonic methods may be that the individual sees work being done, and this appears to be a psychological advantage for those who find a static contraction boring.

Training Methods Using Isotonic Contraction. Many combinations of resistance, repetitions, and number of sets are possible here, but let us first define our terms. *Repetitions* (although used incorrectly, the term is firmly entrenched in the literature) means the total number of executions. *Execution maximum (EM)* and *repetition maximum (RM)* can be used interchangeably, and they indicate the maximum weight that can be lifted for the indicated number of repetitions; that is 10 EM (RM) is the greatest weight that can be lifted ten times. *One set* is the number of repetitions done consecutively, without resting.

The investigations performed in this area are not in close agreement, but a general picture seems to emerge. The classic work of De Lorme and Watkins (13) recommended the following program:

1 set of 10 repetitions with ½ 10 RM
1 set of 10 repetitions with ¾ 10 RM
1 set of 10 repetitions with full 10 RM

Physiology of Training and Conditioning Athletes

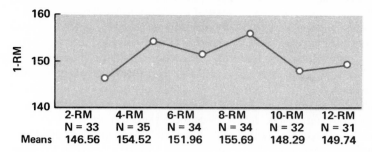

	2-RM	4-RM	6-RM	8-RM	10-RM	12-RM
	N = 33	N = 35	N = 34	N = 34	N = 32	N = 31
Means	146.56	154.52	151.96	155.69	148.29	149.74

Figure 17.7 Mean strength resulting from weight training programs involving six different methods. (From Berger, R.A. *Res. Q.*, 33:334, 1962.)

Other investigators have furnished support for the effectiveness of this method of weight training (2); however, systematic investigations of the value of varying numbers of repetitions seem to indicate that fewer repetitions may be even more effective: four, five, or six. Berger's data (5) in figure 17.7 provide rather good evidence that between four and eight repetitions provide maximal results in terms of strength gain. Another approach that has been validated under laboratory conditions as being very effective is that of doing ten repetitions in which each repetition is done against the maximum possible for that particular execution (6). Thus individuals start with their own 1 RM and use as large a weight as possible for each successive lift. In comparison with the use of one set of ten repetitions with the full 10 RM, it was shown that significantly greater gain in strength was achieved, although the total work in foot-pounds was greater in the 10 RM groups. Muscle hypertrophy seems to be best brought about by the De Lorme and Watkins procedure (2).

The optimal number of workouts per week is probably between three and five, depending on the amount of other vigorous activity a given individual may be indulging in (work or play) beyond the weight training program.

Measuring Strength by Isotonic Methods. Measuring strength by isotonic methods degenerates into a trial-and-error situation. For example, if we wish to determine the maximal weight that can be lifted by a muscle group, we may start with a weight that is estimated to be less than maximal and work up to the maximal in small increments; however, the number of trials needed to establish the maximum will vary from subject to subject (according to our "guessing" ability), and varying levels of fatigue will influence the maximum attained. This is hardly objective measurement, and consequently it is not often used for scientific purposes. Also, this method is too time-consuming for classwork.

Isokinetic Training.   In recent years, measurement of strength under conditions of constant angular velocity muscular contractions has become popular. This is called *isokinetic strength* measurement and probably got its impetus from the work of Asmussen and associates (1), who designed a device using strain gauges with which the force of contraction could be measured and recorded during movement, but also could be limited to constant angular velocity conditions. Figure 17.8 shows the results of their work in which it can be seen that, at all velocities, eccentric contraction can develop more force and concentric contraction can develop less force than isometric contraction. They also showed that the correlation between maximum isometric and dynamic contraction was 0.80 at a velocity of 15% of arm length per second.

While this approach represents a distinct improvement over isometric tests and those using repeated measures of ability to lift weight, it still does not measure the quality of muscle that is really most important to athletic ability, the ability to produce force in accelerated movement that is typical of almost all skilled muscle action in sports. Interestingly the basis for such measurement of force (strength) in accelerated movement against an inertia wheel was provided in the early 1920s by A. V. Hill (25). It hardly seems too early to rejuvenate the methods of Hill to study the production of force by muscles under game conditions, which are really not isometric or isokinetic but in fact necessitate the production of force in accelerating-decelerating movements.

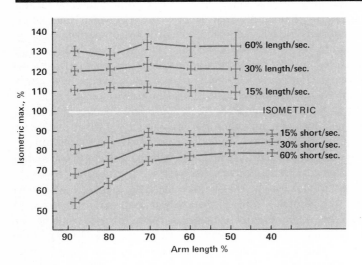

Figure 17.8 Concentric force (lower three curves) and eccentric force (upper three curves) at three different velocities of movement, expressed as percentages of isometric force in the different positions. Extended arm to the left. Horizontal and vertical bars denote ± 1 s.e. (From Asmussen, E., Hansen, O., and Lammert, 0. *Communications from Testing and Observation Institute of the Danish National Association for Infantile Paralysis,* 1965.)

Physiology of Training and Conditioning Athletes

Unfortunately, neither the isokinetic instruments nor the inertia wheel is well suited to the training of groups. The isokinetic instrumentation is very expensive, and the inertia wheel must be specially constructed. Both types of equipment can be considered research instruments.

Variability in Strength/Individual Variability. Strength may be expected to vary somewhat in an individual from day to day. The amount of this variability has been found to range from 1.5% to 11.6% for women and from 5.3% to 9.3% for men (calculated as the standard deviation from the mean) (41).

Generality versus Specificity. We often speak of a strong man or a weak man, and imply that strength is a general quality, that all muscles are strong or weak to the same degree. Obviously the relationships of strength among the various muscles of any individual are not perfect, but there is a high degree of generality. When the strength of single muscle groups was correlated against the total of twenty-two representative muscle groups, all correlations were positive and were significant at better than 0.01 level of confidence. Muscle groups that correlated highest (most representative of general strength) were leg extensors, 0.89; hip flexors, 0.72; knee extensors, 0.70; handgrip, 0.69; and elbow flexors, 0.64 (40). Thus strength tests that use several of these muscle groups can estimate the general strength quite accurately.

Quantity and Quality of Muscle Tissue. It is apparent that *quantity* and *quality* enter into the determination of the ultimate strength of a muscle. When we relate strength and size of the same muscle group in different subjects, substantial correlations are found. The author has found this relationship to be between $r = 0.80$ to 0.90 for well-trained, nonobese young men in the elbow flexor group.

The quantity (or absolute muscle force) varies from muscle to muscle within an individual, however, and the quality of muscle tissue is best measured as strength per unit of physiological cross section. Physiological cross section is the same as anatomic cross section only when the muscle fibers are parallel to the tendon of the muscle; otherwise it is obtained by dividing the volume of the muscle by the length (35). The mean strength per unit of cross section has been estimated for various muscle groups, and it varies from 4.36 kg/cm^2 for the rectus femoris (female) to 14.76 kg/cm^2 for the triceps muscle (male).

The available evidence indicates that in strength training programs the strength and size of muscles increase proportionately; consequently it is the quantity (volume, not number of fibers) that increases, not the quality (22).

Effect of Various Factors on Strength

Age.   As a child grows from infancy to adulthood, strength grows commensurately with the growth in the size of the muscles. Changes in the quality of muscle tissue seem to be small in magnitude since almost all the gain in strength can be accounted for by the increased size (36). Although experimental data are not available for the older segments of the population, it is quite probable that the decline in strength beyond age thirty can be attributed to decreased quantity of muscle tissue rather than to qualitative changes.

Sex.   Most of the difference in strength between the sexes can be accounted for by the difference in muscle size (36). It is possible that differences in motivation (culturally inspired) may be responsible for the observed qualitative differences, and that in a histological sense no difference exists.

Diurnal Variation.   A clear-cut picture of variation in grip strength with time of day has been presented. Figure 17.9 shows the diurnal variation in grip strength, and its close relationship to the diurnal variation in oral body-temperature readings (42). The closeness of the relationship does not establish cause and effect, but other evidence (in relationship to athletic warm-up) tempts one to relate the two factors causally.

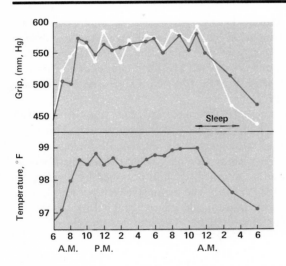

Figure 17.9 Diurnal variation of grip strength and body temperature. (From Wright, V. *Res. Q.*, 30:114, 1959.)

Another investigator, working with other muscles (including the elbow flexor and knee extensor groups), was unable to show a systematic variability in strength that was related to time of day in these muscles (40). It is possible there are differences from muscle to muscle.

Seasonal Effects on Strength Gain.   Experiments in Germany on twenty-one subjects who were observed for strength gains as a result of isometric training indicated seasonal variations. A minimum strength gain was found to occur in January and February, and a tenfold higher gain rate was observed in September and October. The investigators attributed this to changes in the diet: availability of fresh fruits and vegetables (23). This finding is interesting, but it requires verification by other investigators.

Effects of Heat and Cold on Strength.   Immersion of the arm in hot water (120° F) for eight minutes resulted in small but significant gains in grip strength in eight of twelve subjects. With the same subjects, immersion of the arm in cold water (50° F) resulted in a mean decrease in grip strength of 11%, which was highly significant (17). It would be of interest to extend these observations to varying temperatures and to other muscle groups.

Psychological Factors in Strength.   It has long been known on empirical bases that strength, as expressed by voluntary maximal contractions, is limited by psychological factors before physiological limits are reached, and many examples illustrate this point. In one instance, a young man working under his car was pinned there when the jack failed; his mother, a woman of less-than-average size, lifted a corner of the car to release him. Unquestionably, her usual inhibitory processes were themselves inhibited, allowing her to exert her full physiological strength.

Some interesting experiments support this view. It was shown that forearm flexor strength can be significantly increased by the following factors:

1. A pistol shot from two to ten seconds before strength effort, +7.4%.
2. A subject shouting at application of force, +12.2%.
3. Hypnosis that suggests a greater strength, +26.5%; hypnosis that suggested weakness brought about a significant decline in strength, 31.7% (27).

Plasticity of the Human Body.   At the age of twenty-five one of the best U.S. weight lifters in the heavyweight class decided to discontinue his weight lifting career and to take up long-distance running. Three years later he ran the marathon distance in 3 hours, 3½ minutes. Figure 17.10 shows the changes in physique and performance that had taken place within the period under review. He lost 64 lbs. body weight; his strength (total snatch, and clean and jerk) declined by 196 lbs.; but per unit of body weight he was now slightly stronger.

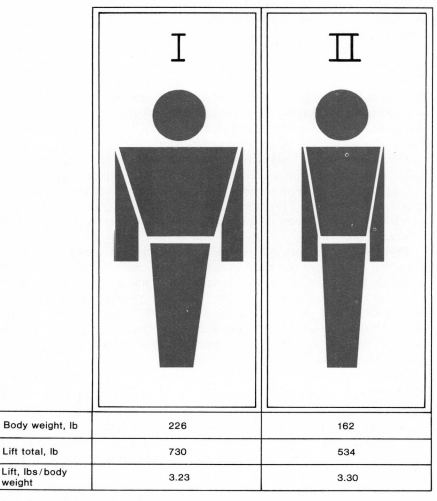

	I	II
Body weight, lb	226	162
Lift total, lb	730	534
Lift, lbs/body weight	3.23	3.30

Figure 17.10 Heavyweight weight lifting champion turned marathon runner. (From Jokl, E., and Jokl, P. *Am Corr. Ther. J.* 33:61, 1979.)

SUMMARY

Muscular Strength

1. Strength is a very important factor in any physical activity in which muscular forces move the body or extraneous sports implements that have appreciable mass.
2. Athletic power is the rate of producing force. Power can usually be improved to the greatest extent by increasing the available force (strength).
3. Increased girth of muscle in response to training is the result of hypertrophy, largely of the fast-twitch fibers.
4. Early responses to strength training are the result of better innervation, while hypertrophy becomes the dominant factor after three to four weeks. Strength training in the elderly results largely in improved activation of muscle tissue with little hypertrophy.
5. The overload principle means that gains in muscular strength and hypertrophy are brought about only when a muscle works against considerably greater resistance than that to which it is accustomed. For isometric training, a minimum resistance of one-third maximum contraction is required to furnish a training stimulus.
6. Strength gains are specific to the angle in the range of motion at which the resistance is met in training and also to the velocity of training.
7. The rate of strength gain is most rapid when a muscle has achieved only a small proportion of its possible maximal end strength. The rate of gain slows down as muscle strength approaches its maximal end strength.
8. The rate of strength loss after training ends is a very much slower process than strength gain. Retention of strength can probably be brought about by as little as one maximal contraction per week.
9. The strength of any muscle is the result of both the quantity and quality of the muscle tissue. Quality (strength per unit of cross-sectional area) varies considerably from muscle to muscle within an individual, but strength gains appear to be brought about largely by quantitative increases in fiber size.
10. To achieve maximal accuracy, strength testing should be applied under conditions that are constant in time of day, ambient temperature, and psychological factors.

Training Methodology and Isometric Training

1. Maximal contraction produces the fastest gains.
2. Duration of five seconds is optimal.
3. Higher end strength values can be attained by increasing repetitions from one to five to ten a day.
4. If a contraction strength less than the maximum is used, it should be based on a maximum that is measured weekly.
5. Workouts four or five times per week appear to be optimal.

Physiology of Muscle Strength

Training Methodology and Isotonic Training

1. All contractions should be made through the full range of motion.
2. For the development of *strength,* a program based on two or three sets of four to ten repetitions each, using maximum resistance for the number of repetitions, seems to rest on sound experimental bases.
3. For the development of *hypertrophy,* the De Lorme technique is probably most effective.

> 1st set: 10 repetitions with ½ 10 RM
> 2d set: 10 repetitions with ¾ 10 RM
> 3d set: 10 repetitions with full 10 RM

4. Workouts should be scheduled no less than three and no more than four times weekly for optimal results.
5. The total exercise program should be scheduled so that no more than one workout per week approaches exhaustion.

REFERENCES

1. Asmussen, E.; Hansen, O.; and Lammert, O. The relation between isometric and dynamic muscle strength in man. *Communications from the Testing and Observation Institute of the Danish National Association for Infantile Paralysis* 20:1–11, 1965.
2. Barney, V. S., and Bangerter, B. L. Comparison of three programs of progressive resistance exercise. *Res. Q.* 32:138–46, 1961.
3. Bender, J. A., and Kaplan, H. M. The multiple angle testing method for the evaluation of muscle strength. *J. Bone Joint Surg.* 45A:135–40, 1963.
4. Berger, R. A. Comparison of static and dynamic strength increases. *Res. Q.* 33:329–33, 1962a.
5. ———. Optimum repetitions for the development of strength. *Res. Q.* 33:334–38, 1962b.
6. Berger, R.A., and Hardage, B. Effect of maximum loads for each of ten repetitions on strength improvement. *Res. Q.* 38:715–18, 1967.
7. Booth, F.W. Time course of muscular atrophy during immobilization of hindlimbs in rats. *J. Appl. Physiol.* 43:656–61, 1977.
8. Clarke, H.H., and Clarke, D.H. *Developmental and Adapted Physical Education.* Englewood Cliffs, N.J.: Prentice-Hall, Inc. 1963.
9. Clarke, H.H.; Elkins, E.C.; Martin, G.M.; and Wakim, K.G. Relationship between body position and the application of muscle power to movements of the joints. *Arch. Phys. Medi.* 31:81–89, 1950.
10. Considine, W.J., and Sullivan, W.J. Relationship of selected tests of leg strength and leg power on college men. *Res. Q.* 44:404–16, 1973.
11. Costill, D.L.; Coyle, E.F.; Fink, W.F.; Lesmes, G.R.; and Witzmann, F.A. Adaptations in skeletal muscle following strength training. *J. Appl. Physiol.* 46:96–99, 1979.
12. Darcus, H.D., and Salter, N. The effect of repeated muscular exertion on muscle strength. *J. Physiol.* 129:325–36, 1955.

13. De Lorme, T.L., and Watkins, A.L. *Progressive Resistance Exercise.* New York: Appleton-Century-Crofts, 1951.

14. Eichelberger, L.; Roma, M.; and Moulder, P.V. Tissue studies during recovery from immobilization atrophy. *J. Appl. Physiol.* 18:623–28, 1963.

15. Gardner, G.W. Specificity of strength changes of the exercised and nonexercised limb following isometric training. *Res. Q.* 34:98–101, 1963.

16. Goldberg, A.L.; Etlinger, J.D.; Goldspink, D.F.; and Jablecki, C. Mechanism of work-induced hypertrophy of skeletal muscle. *Med. Sci. Sports* 7:248–61, 1975.

17. Grose, J.E. Depression of muscle fatigue curves by heat and cold. *Res. Q.* 29:19–31, 1958.

18. Henneman, E.; Somjen, G.; and Carpenter, D.O. Functional significance of cell size in spinal motoneruons. *J. Neurophysiol.* 28:560–80, 1965.

19. Henneman, E., and Olson, C.B. Relations between structure and function in the design of skeletal muscle. *J. Neurophysiol.* 28:581–98, 1965.

20. Henneman, E.; Somjen, G.; and Carpenter, D.O. Excitability and inhibitability of motoneurons of different sizes. *J. Neurophysiol.* 28:599–620, 1965.

21. Hettinger, T. Der einfluss der muskeldurchblutung beim muskel-training auf den training-erfolg. *Arbeitsphysiologie* 16:95–98, 1955.

22. Hettinger, T., and Mueller, E.A. Muskelleistung und muskeltraining. *Arbeitsphysiologie* 15:111–26, 1953.

23. ———. Die trainierbarkeit der muskulatur. *Arbeitsphysiologie* 16:90–94, 1955.

24. Hettinger, T., and Mueller-Wecker, H. Histologische und chemische veranderungen der skeletmuskulatur bei atrophie. *Arbeitsphysiologie* 15:459–65, 1954.

25. Hill, A.V. The maximum work and mechanical efficiency of human muscles and their most economical speed. *J. Physiol.* 56:19–41, 1922.

26. Ikai, M. Training of muscle strength and power in athletes. *Br. J. Sports Med.* 7:43–47, 1973.

27. Ikai, M., and Steinhaus, A.H. Some factors modifying the expression of human strength. *J. Appl. Physiol.* 16:157–63, 1961.

28. Jokl, P., and Jokl, E. Heavyweight weight lifting champion turned marathon runner. *Am Corr. Ther. J* 33:61, 1979.

29. Lesmes, G.R.; Costill, D.L.; Coyle, E.F.; and Fink, W.J. Muscle strength and power changes during maximal isokinetic training. *Med. Sci. Sports* 10:266–69, 1978.

30. Logan, G.A. Differential applications of resistance and resulting strength measured at varying degrees of knee flexion. Doctoral dissertation, USC, 1960.

31. Moffroid, M., and Whipple, R. Specificity of speed of exercise. *J. Amer. Phys. Ther. Assoc.* 50:1699, 1970.

32. Moritani, T., and deVries, H.A. The time course of strength gain: neural factors versus muscular hypertrophy. *Am. J. Phys. Med.,* 58:115–130, 1979.

33. ———. Neural factors versus hypertrophy in the time course of muscle strength gain in young and old men, in press.

34. Mueller, E.A., and Rohmert, W. Die geschwindigkeit der muskelkraft zunahme bei isometrischen training. *Int. Z. Angew. Physiol.* 19:403–19, 1963.

35. Ralston, H.J.; Polissar, M.J.; Inman, V.J.; Close, J.R.; and Feinstein, B. Dynamic features of human isolated voluntary muscle in isometric and free contractions. *J. Appl. Physiol.* 1:526–33, 1949.

36. Rodahl, K. Physical work capacity. *AMA Arch. Environ. Health* 2:499–510, 1961.
37. Rohmert, W., and Neuhaus, H. Der einfluss verschiedener ruhelange des muskels auf die geschwindichkeit der kraftzunahme durch isometrisches training. *Int. Z. Angew. Physiol. einschl. Arbeitsphysiologie* 20:498–514, 1965.
38. Starr, I. Units for the expression of both static and dynamic work in similar terms and their application to weight lifting experiments. *J. Appl. Physiol.* 4:21–29, 1951.
39. Thorstensson, A. Observations on strength training and detraining. *Acta Physiol. Scand.* 100:491–93, 1977.
40. Tornvall, G. Assessment of physical capabilities. *Acta Physiol. Scand.* 53, suppl. 210:1–102, 1963.
41. Wakim, K.G.; Gersten, J.W.; Elkins, E.C.; and Martin, G.M. Objective recording of muscle strength. *Arch. Phys. Med.* 31:90–100, 1950.
42. Wright, V. Factors influencing diurnal variation of strength of grip. *Res. Q.* 30:110–16, 1959.

18

Development of Muscular and Circulorespiratory Endurance

The ability to persist in physical activity, to resist muscular fatigue, is referred to as endurance. If there is any one most important factor in human performance, it is endurance. It is probably the most important component of physical fitness in that it reflects the state of some of the physiological systems that are most important to the general health of an individual. In athletics, there are few sports in which endurance is not a factor, and in many sports all of the training and conditioning programs are directed toward this end.

The concept of endurance is not altogether simple. Analysis of this factor into its component parts is apt to mislead a student into thinking that the components are discrete elements, when in fact the elements are interwoven and interrelated, and basically inseparable. Realizing that analysis of endurance into its several components is essentially artificial, we shall nevertheless make the analysis to enable better understanding of the physiology involved.

Endurance as a Factor in Human Performance

A. Psychological elements
   1. Motivation
   2. Willingness to take pain
B. Physiological elements
   1. Local endurance: involvement of only one, or several, localized muscle groups
      a. Strength of a particular muscle group
      b. Energy stores
      c. Peripheral circulatory factor
   2. General endurance: whole body activity
      a. Strength of general musculature
      b. Energy stores
      c. Systemic circulatory factor
         (1) Aerobic activity: limited by maximal O_2 consumption
            (a) Respiratory function
            (b) Cardiac output
            (c) O_2 carrying capacity of blood
            (d) Vascularization of muscle tissues
            (e) Aerobic capacity of muscle tissue
         (2) Anaerobic activity
            (a) Muscle glycogen
            (b) ATP and CP stores
            (c) Alkaline reserve: blood buffers
      d. Efficiency of heat regulatory mechanisms
      e. Effectiveness of the nervous system in maintaining high levels of skill and coordination
   3. Muscular efficiency: energy input required to bring about desired level of muscular performance

We shall dismiss the psychological elements (although they are very important) as being beyond the scope of this text, and muscular efficiency warrants a chapter by itself. Thus the remainder of this chapter will be concerned with the physiology of endurance with respect to its two major components: (1) local, or muscular, endurance, and (2) general, or systemic, endurance.

Local or Muscular Endurance

Strength and Endurance. Let us consider for a moment the simplest possible illustration of muscular endurance. Let us concern ourselves with the maintenance of an isometric contraction of the elbow flexors against a load that is at least 60% of the *maximal* voluntary contraction (MVC). A load of this magnitude results in virtually complete occlusion of the blood vessels that supply the muscle tissue because the pressure of the contracted muscle exceeds systolic arterial pressure (35).

Under these conditions the duration of the isometric contraction may be limited by a finite energy supply or by the buildup of acid metabolic end products. Thus the contracting muscle group utilizes its only available sources of direct energy, the breakdown of ATP and creatine phosphate, and subsequently the energy available from the breakdown of glycogen. Neither of these energy sources can be replenished because circulation is occluded; consequently the duration of this isometric contraction becomes a function of the amount of energy stored, the rate of energy depletion, and the concomitant drop in tissue pH that decreases the contractility of the muscle. The exact factor that actually sets the limit for duration of this isometric contraction is still in question. Recent work (1) suggests that depletion of muscle glycogen does not set the limit for isometric contraction. Depletion of high energy phosphates also seems unlikely, but at intermediate isometric tensions, lactate buildup appears to be a likely cause (27). Decreases in pH *within the muscle cell* may offer a possible explanation for rapid fatigue from very heavy work (22). Such pH decreases in the muscle cell may reduce the binding capacity for calcium ion through an inactivation of the fibrillar protein, troponin (see chapter 3).

Figure 18.1 illustrates the relationship of muscular endurance to the magnitude of the load imposed upon the muscle. It will be noted that load is designated in terms of the percentage of MVC; this is necessary because strength is such a large factor in local muscular endurance. It is obvious that if a load of fifty pounds were used for subjects of widely varying strength, a weak subject would find that holding it required close to maximal contraction and the duration of the hold would be only a few seconds. For a very strong subject, the load would be light, and the duration might be several minutes (even longer if the contraction is below that which causes occlusion). Thus

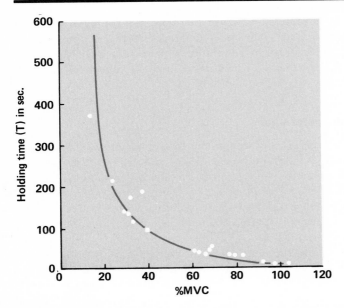

Figure 18.1 Relationship between observed holding times of contractions sustained to fatigue vs. percent MVC. (From Ahlborg, B., et al. *J. Appl. Physiol.* 33:224, 1972.)

the test results would be more closely related to strength than to muscular endurance.

To further illustrate this fact, many investigators have found the relationship between strength and *absolute endurance* to be high, from $r = 0.75$ to $r = 0.97$, while the relationship between strength and *relative endurance* is practically nil (4, 38) or even negative (23). Absolute endurance is measured by using the same load for everyone, whereas relative endurance is measured by using a given percentage of each individual's MVC (thus ruling out strength as a factor).

Measurement of Muscular Endurance

Maintenance of Isometric Tension. The simplest and most direct method has already been described, but another method that involves isometric tension is illustrated by figure 18.2. The subject holds a *maximal contraction,* and the effects of energy depletion upon maximum strength are observed over time. It is also of interest to note that the effects of the circulatory factor do not become apparent (there are no differences between open circulation and circulation occluded by a pressure cuff) until the contraction strength falls below (approximately) 60% of MVC (this value is also supported by data for

Physiology of Training and Conditioning Athletes

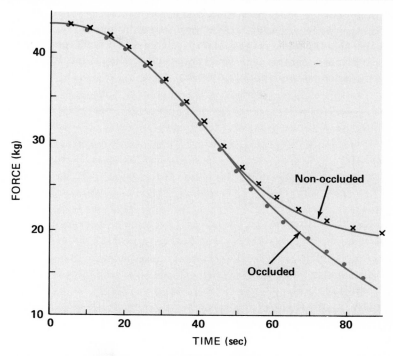

Figure 18.2 Isometric fatigue curves for forearm muscles. In subjects holding maximum voluntary contraction (MVC), the decrement in MVC is a function of fatigue. (From Royce, J. *Res. Q.*, 29:204, 1958.)

the arm musculature). However, the percentage of MVC at which occlusion occurs undoubtedly varies from muscle to muscle (it is probably very much lower in such muscles as the gastrocnemius, with bipennate fiber orientation). The plotting of these fatigue curves is a laboratory procedure however, and does not lend itself readily to practical situations.

Recovery from Fatigue. It has been shown that recovery of maximal strength after isometrically induced fatigue is remarkably fast, being complete in about ten minutes (18). Interestingly, the recovery of endurance function is related to the tension of the fatiguing contraction, whereas strength recovery is not (18).

Isotonic Testing by Ergographic Methods. F.A. Hellebrandt and her colleagues (20) have modified the classic Mosso ergograph into a very useful instrument for muscle endurance testing by isotonic methods (see fig. 18.3).

Development of Muscular and Circulorespiratory Endurance 413

The main advantage in the use of this instrument is that the applied force is constant throughout the range of motion, and consequently the range of contraction decreases with increasing fatigue, resulting in fatigue curves or *ergograms* (as in fig. 18.4). It is possible to set up structured, progressive therapeutic exercise programs with this instrument, and so it has found use in physical medicine and rehabilitation work.

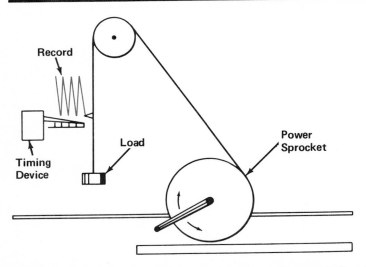

Figure 18.3 A schematic representation of Kelso-Hellebrandt ergograph. (From Hellebrandt, F.A., Skowlund, and Kelso. *Arch. Phys. Med.*, 29:21, 1948.)

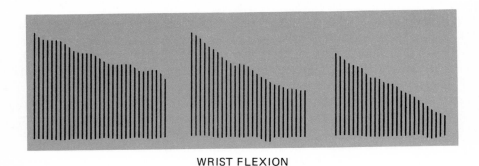

WRIST FLEXION

Figure 18.4 Ergograms from three workbouts of wrist flexion showing increasing levels of fatigue from left to right. (From Hellebrandt, F.A., Skowlund, and Kelso. *Arch. Phys. Med.*, 29:21, 1948.)

Strength-Decrement Index. H.H. Clarke (6) and his associates have demonstrated that as heavy work is performed over a sufficient period of time to bring about fatigue, decrements in the strength of the muscles involved can be observed with their cable-tension strength testing procedures. They have suggested the use of the percentage strength loss (of preexercise value) as a measure of the level of fatigue. Conversely, the *strength decrement index* (SDI) also provides information relative to muscular endurance.

Electromyographic Evaluation of Fatigue. EMG testing is mentioned at this point not because it has practical use but because it sheds additional light on the physiology of fatigue. It has been shown that as normal human muscles maintain a constant isometric tension, the electrical activity in the muscle increases as the muscle fatigues (14). The most obvious explanation for this phenomenon is that as the contraction continues and fatigue occurs, each motor unit is able to contribute less force to the contraction, and consequently more and more motor units must be recruited to maintain the same level of tension. The author has found that the rate of increase in electrical activity with time is highly correlated with isometric endurance measured on a hydraulic dynamometer. This relationship may have practical value for estimating muscular endurance in subjects who are unable or unwilling to cooperate fully in testing.

Factors Affecting Muscular Endurance

1. *Age.* Relatively little information is available concerning the effect of age. Rich has shown that there seems to be no greater fatiguability in young children than in older (high school age) children (32). Evans (15) has shown that older men fatigue more rapidly than young, although the loss of endurance is smaller than expected.

2. *Sex.* Most available data seem to agree that there is no significant difference in muscular endurance due to sex, if the strength factor is ruled out. Indeed, recent data suggest that relative endurance may be superior in the female because of a higher critical occluding tension (24). However, it is also possible that MVC of the female does not come as close to true physiological maximum as does that of the male, due to cultural differences and/or less experience in exerting MVC. Thus any percent MVC would represent a lower true fraction of capacity and thus account for the greater endurance time.

3. *Temperature.* It appears that rate of fatigue and total amount of work done (handgrip dynamometer) are adversely affected by immersion of the arm for eight minutes in a water bath at 120° F (19). The effects of cold were shown to be advantageous, until an optimum muscle temperature of 80° F was reached. This appears to be an optimal temperature, and temperatures below this produce poorer performance (7).

4. *Cross-education Effect.* It has been demonstrated that training one limb brings about changes in its untrained partner (25). When the endurance of the trained limb was increased 966%, the untrained contralateral limb improved by 275%.

5. *Circulation.* Twenty-nine weeks of isometric training have been shown to bring about a decrease in the ratio of blood flow debt per unit of exercise effort (40). This is in agreement with the work of Rohter, Rochelle, and Hyman (33), who have shown significant improvement in muscle blood flow as the result of the training regimen of college swimmers. It seems highly probable that improved peripheral circulation, by virtue of improved vascularization of active muscle tissue, is one of the most important mechanisms in the development of improved muscular endurance levels.

Improvement of Muscular Endurance. In a classic piece of work, with which every student of physical performance should become familiar, Hellebrandt and Houtz (21) provided definitive answers to some of the basic questions that must be answered to place exercise programs upon a scientific, systematized basis. Using ergographic procedures, they performed 620 experiments that tested different training procedures on wrist flexion and extension. They used thirty-second workbouts, alternated with thirty-second rest periods, and each bout consisted of twenty-five isotonic contractions.

When experimentation with successively heavier loads is conducted, *work curves* can be plotted in which the work done (kilogram-meters) is seen to rise to an *optimum load,* then fall again. Thus in figure 18.5 it can be seen that at the initial test, 2.0 kg allowed the best combination of $F \times D$ (force times distance, distance being the result of the number of repetitions and the

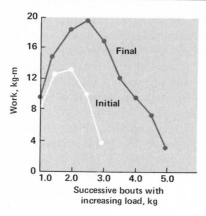

Figure 18.5 Work curve showing the total work done in successive workbouts as a function of load. (From Hellebrandt, F.A., and Houtz, S.J. *Phys. Ther. Rev.,* 36:371, 1956.)

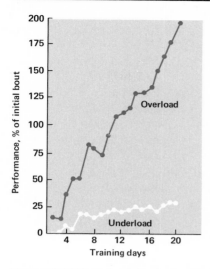

Figure 18.6 Effects of underload and overload in improving muscular endurance. (From Hellebrandt, F.A., and Houtz, S.J. *Phys. Ther. Rev.*, 36:371, 1956.)

height lifted for each repetition as recorded on the ergogram). As more weight was loaded in subsequent tests, the distance suffered by more than was gained in weight moved, and therefore the product of $F \times D$ decreased. Resistances less than the value that brings about the optimal load constitute an *underload,* and a greater-than-optimal load constitutes an *overload.* The *initial* and *final curves* of figure 18.5 show the improvement in one of their subjects in six practice periods. It can be seen that strength, power, and endurance improve simultaneously.

Figure 18.6 demonstrates the need for overload conditions in developing muscular endurance. The two groups performed the same number of contractions per day (250), three times a week for eight weeks, the only difference being that one group trained with an underload and one with an overload.

Comparison was also made of two groups that worked with underload and overload conditions when the total work done per day was held constant. In other words, the group that trained with an overload (twenty-five repetitions with the 25 repetition maximum [RM]) did fewer total bouts than the group that trained with an underload, so that work in terms of $F \times D$ was equal for all subjects. Again, figure 18.7 demonstrates the need for overload conditions.

It would be very desirable to pursue such avenues of investigation for other muscle groups, and indeed for overall body activity. Only with such systematic investigations can the training and conditioning of athletes become a science.

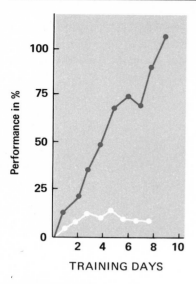

Figure 18.7 Effects of underload and overload of force overcome at each contraction when the total work done is held constant. Lower curve represents underload; upper curve represents overload. (From Hellebrandt, F.A., and Houtz, S.J. *Phys. Ther. Rev.*, 36:371, 1956.)

The results of recent work on the training effects of strength versus endurance workouts is of interest. Clarke and Stull (5, 39) conducted two series of training experiments, one of which used low resistance and high repetitions, a combination which is ordinarily considered to be *endurance type training*. In the other experiment they used the De Lorme technique of heavy resistance and low repetitions *(strength-type training)*. Surprisingly, the gains in strength were as great from endurance training as from strength training, and the gains in absolute endurance were also similar in the two training experiments, although relative endurance did not change. Two other investigations support their findings (13, 37), which lead to the conclusion that the first order of business in improving athletes' muscular endurance is to optimize their strength. Since heavy resistance training such as weight training is more efficient in the use of time this would seem to be the method of choice.

General or Circulorespiratory Endurance

When we change our frame of reference from a localized movement, such as elbow flexion or leg extension, to an activity that involves gross body movement, such as running or swimming, the problem changes from local involvement of a small percentage of the body's musculature to one that involves a

Physiology of Training and Conditioning Athletes

large percentage with many muscle groups working simultaneously. In local situations, muscle endurance is limited by a combination of energy supply and peripheral vascularization; the central systems of supply are never extended to a large degree. In gross body activity, it is the central systems of respiration, circulation, and heat dissipation, and the nervous system and the homeostatic mechanisms, in addition to peripheral muscle function, that are likely to establish the outer limits of performance.

Aerobic versus Anaerobic Work. When work begins, anaerobic energy sources are used during the transition to *steady state,* as was described in chapter 10. If the work load is greater than aerobic capacity, anaerobic mechanisms continue to contribute until maximum O_2 debt is achieved, at which point the exercise must end or slow down. The percentage contribution of aerobic and anaerobic energy depends upon the nature of the exercise as shown in figure 18.8.

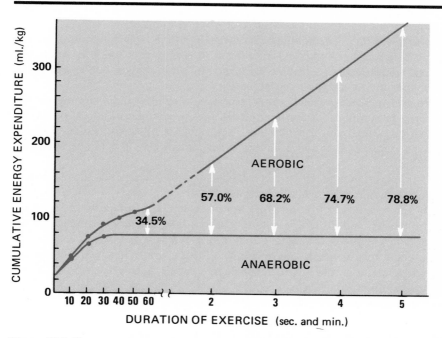

Figure 18.8 The relative importance of oxygen debt and steady oxygen intake during maximum exercise of varying duration. (From Shephard, R.J. *J. Sports Med.* 10:73, 1970.)

Table 18.1
Capacity and Power of Energetic Processes in Muscle

Processes	Total Energy (cal/kg)	Maximum Power (Kcal/kg/hr)
Aerobic	—	15.0
Alactacid O_2 debt (fast component)	95	45.0
Lactacid O_2 debt (slow component)	220	21.5

From R. Margaria, P. Cerretelli, and F. Mangili, "Balance and Kinetics of Anaerobic Energy Release during Strenuous Exercise in Man," *J. Appl. Physiol.* 19:627, 1964.

Margaria and his associates, who furnished some of the original data on O_2 debt, recently defined the limitations of each of the three processes of energy supply (30). Table 18.1 shows their results.

The alactacid debt, according to their figures, can be completely utilized in one explosive effort, in as little as a fraction of a second, and the process is completed in less demanding exercise in less than twenty seconds. When a subject exercises at a maximal rate (as in running events, up to the 440) both anaerobic processes are exhausted in about forty seconds, and exercise can continue only on a pay-as-you-go basis.

In reality, most athletic events involve both aerobic and anaerobic energy sources, the proportions depending upon the speed and duration of the event. It can be estimated, for example, that sprints are about 90% anaerobic and distance runs are about 90% aerobic; other athletic events fall in between (see fig. 18.8).

The importance of all this in our discussion of endurance is that, again, different factors are involved, even for a gross body activity, which depend upon the speed and duration of an athletic event. It can readily be seen that sprint events, and athletic events in general, that last less than one minute, depend largely upon the factors discussed in the first half of this chapter for their endurance element. The remainder of this chapter, then, will be directed toward athletic events in which the energy supply is largely aerobic (events of several minutes' duration or longer).

Determinants of Circulorespiratory Endurance

Energy Substrate. As was discussed in chapter 3, exhaustion in lengthy, severe bicycle ergometer rides seems to occur when muscle biopsy techniques show muscle glycogen to be depleted. Thus, there seems little doubt that muscle glycogen depletion is one factor which *can* set the limits of endurance. However, it must be recognized that bicycle exercise places most of the load upon relatively few muscles. Costill and co-workers (9, 10) have raised the interesting question of whether glycogen depletion is also a factor under conditions of endurance type *running events* where the load is distributed over a greater muscle mass. Since substantial amounts of muscle glycogen were found after both prolonged (ten miles) and short (maximal O_2 consumption test) exhaustive runs on the treadmill, glycogen depletion is an unlikely explanation for the fatigue at exhaustion in such efforts. Furthermore, in a subsequent study involving three consecutive days of such efforts, they found that subjects were able to initiate the ten-mile run on the third day with leg muscle glycogen concentrations lower than those measured after termination of running on the first day.

Pernow and Saltin (31) have shown the importance of *free fatty acids* (FFA) as an energy substrate. When subjects performed bicycle ergometer work to exhaustion (1 to 1.5 hours' duration), it was shown that when the glycogen stores are reduced, prolonged work can still be performed by the same muscles, but only if the intensity of the work is less than 60%–70% maximum $\dot{V}O_2$ and the supply of FFA is adequate. Elimination of both muscle glycogen and FFA seriously impairs the ability for prolonged work.

Maximum O_2 Consumption (Aerobic Capacity).   Except in a marathon run lasting several hours, or in athletic events that overload limited muscle masses (as the bicycle ergometer), the limiting factor in endurance is probably O_2 supply or tissue O_2 utilization, rather than oxidizable substrate. We may think of the possible limiting factors for O_2 supply as (1) external respiration, (2) gas transport. The available evidence suggests that the limiting factor for high level athletic performance is some combination of gas transport and tissue O_2 utilization.

Gas transport can be limited by (1) cardiac output, (2) vascular dynamics, and (3) O_2-carrying capacity of the blood (hemoglobin concentration and number of erythrocytes per unit of blood). All of these factors, operating in concert, can be evaluated by the maximal O_2 consumption test described earlier in this book. This test is undoubtedly the best single predictive measure of success in endurance-type athletic events. High correlations have been demonstrated between aerobic capacity and time on a 4.7-mile run ($r = 0.83$) (8) and endurance time on a bicycle test ($r = 0.78$) (42).

Genetic Factor.   As was discussed in chapter 10, training can bring about substantial improvement in aerobic capacity and thus markedly improve performance, but genetic factors ultimately set the ceiling. Record-breaking performance is not to be expected unless the genetic endowment for O_2 transport has been considerably better than average to begin with.

Motivation.   In an interesting experiment on the influence of motivation on physiological parameters limiting work capacity, Wilmore (41) tested the PWC of twenty-two college age males on a bicycle ergometer on three occasions, two control and one experimental. On the experimental test the subjects were motivated by competition. As might be expected, the performances were significantly better in the competitive situation, but there were *no significant differences* in the maximum *physiological responses* such as heart rate, maximum ventilation, or O_2 consumption. It may be concluded that the maximal values for the physiological variables are essentially fixed for a given individual at a given time at a given training level and that the supramaximal performances elicited by motivation were the result of increased anaerobic rather than aerobic capacity, which may be the result of reduced psychological inhibitions allowing greater tolerance to anaerobic metabolites.

Physiological Changes Resulting from Training. Continual, methodical stressing of the human organism by subjecting it to progressively increasing work loads results in responses that are seemingly directed toward making its reaction to the challenges of increased metabolic rates ever more successful. The physiological changes brought about by training can be summarized as follows:

1. Lower resting heart rate
2. Lower heart rate for any submaximal work load
3. Greater maximal cardiac output
4. Greater maximal stroke volume
5. Lower ventilation equivalent (less ventilation required per unit O_2 utilized)
6. Greater maximal O_2 consumption
7. Less utilization of anaerobic energy sources for a given work load
8. Capacity for greater O_2 debt (probably due to a combination of improved alkaline reserve and greater willingness to bear pain)
9. Less displacement of physiological function by any given level of work load, and faster recovery to base-line values after completion of exercise

Training Methods for Distance Events. Interval training consists of short periods of work alternating with short rest intervals as distinct from workbouts that are continuous. Thus an individual training for a 1,500-meter swim, instead of swimming long distances continuously, might be trained largely on 100-meter swims that are swum faster, and the endurance (or training) factor would be gained through manipulation of (1) the speed at which the 100s are negotiated and (2) the total number of 100s accomplished, plus (3) the duration of the rest interval.

From the standpoint of the exercise physiologist, interval training makes very good sense indeed. Obviously, one of the primary goals of a conditioning program is to achieve the greatest possible work load with the smallest physiological strain (fatigue), and that this can best be achieved through the methods of interval training is supported on physiological bases. Astrand and co-workers (2) found that a work load (2,160 kg-m/min) that could be tolerated for an hour when done intermittently resulted in exhaustion in nine minutes when done continuously. Thus the total work done continuously was 19,400 kg-m, while the work done intermittently was 64,800 kg-m.

The level of physiological stress can be evaluated best by the heart rates and blood lactates achieved. In the continuous work the heart rate reached 204 bpm and blood lactate rose to 150 mg percent. In the intermittent work, although more than three times as much work was done, heart rate did not exceed 150 and blood lactate was 20 mg percent, which indicates that very little of the work was done anaerobically.

More recent work has confirmed these theoretical advantages for interval training (17), and it has also been shown that by shortening the pause (rest) interval to thirty to forty-five seconds, the training stimulus to both cardiopulmonary and muscle glycolytic systems can be greatly increased (28). Thus one of the advantages of interval training lies in the *simultaneous* development of both aerobic and anaerobic capacities. Both are essential to many sports.

In spite of these obvious theoretical advantages for interval training over continuous training, when the methods are compared under careful experimental conditions for practical training effect, no clear advantage has been found (29, 34, 36) in favor of interval training. However, interval training methods are advantageous in that interval training is usually conducted at a higher pace and therefore favors the use of the same recruitment patterns and muscle fiber types as does the actual event. In addition, both aerobic and anaerobic capacities can be trained simultaneously.

The work of the Astrands and their colleagues (2) also sheds light on what constitute desirable intervals. The same work load when done in alternating three-minute intervals of work and rest required heart rates of 188 and blood lactate of 120 mg percent, but alternating intervals of thirty seconds raised these values to only 150 and 20, respectively. The value of the thirty-second work interval, which is often recommended, is thus borne out on experimental bases.

As an extension of this work, the effect of changing the length of the rest interval was also investigated (3). It was shown that the physiological stressfulness of the training is not highly related to the duration of the rest interval. Only small increases in blood lactate were observed when the rest interval was decreased from four minutes to thirty seconds, although the total work done increased fourfold.

Although maximal training effect on the circulatory system probably demands maximal loading of the O_2 transport system, this does not necessarily mean that the athlete's performance must be maximal in terms of *speed* of running or swimming. Karlsson, Astrand, and Ekblom (26) have shown that as an athlete approaches maximum performance (speeds), there is a considerable range, possibly from 80% to 100% of maximum performance capability (speed of running, etc.) at which O_2 consumption is at its maximum, although the production of lactic acid rises rapidly over this same range. Since they also found a declining O_2 pulse at the highest levels of work, the data suggest that training the O_2 transport system is best carried out at a work rate that is short of maximal performance but will still fully load the O_2 transport system. Such a small reduction in speed would imply less fatigue (lower levels of lactate) and thus permit an increase in training volume.

To summarize, interval training for events that are largely *aerobic* can be developed around work intervals ranging from thirty seconds to five minutes with alternating rest intervals of approximately the same duration. It must be realized that the shorter the interval, the greater the training effect on anaerobic capacity, and the longer the interval, the more effective for aerobic capacity, so that the selection of interval depends to a large extent on the exact nature of the event trained for. It must also be remembered that the shorter intervals allow very large total work loads to be handled. It is the author's experience with swimmers that such workouts should be reserved for bringing athletes to peak performance. Using intervals of thirty to sixty second swims can bring high school and college swimmers to a peak in four to six weeks. If such workouts are used too early in the season or over too long a period, staleness may well be the result.

Determinants of Success in Distance Events. It has recently been shown that distance running performance is most strongly related to $\dot{V}O_2$ max ($r = -.84, -.87, -.88$ for 1-, 2-, 6-mile runs, respectively). The percentage of ST muscle fibers contributes less strongly ($r = -.52, -.54, -.55$ for 1-, 2-, 6-mile runs). Muscle enzyme patterns correlated poorly but may be more important as to intraindividual changes in training state (16). At a given constant speed of running, the percentage of $\dot{V}O_2$ max required is also highly related to performance, $r = -0.94$ (12).

Analysis of Pace as an Indicator of Training Needs. In middle-distance and distance athletic events, where pace is established on a voluntary basis, considerable insight can be gained into training needs by comparing *split times* with those of championship performances in the same event. If an athlete's split times, for example, are in the same proportion as those of championship performance but slower for each split, further improvement probably depends upon increased power or better technique. In this case, off-season weight training and interval training, with rate as the variable for progression, should be utilized.

On the other hand, if an athlete meets the championship pace on the first splits but fades badly late in the race, the fault probably stems from circulorespiratory factors, and endurance work is indicated (such as increasing the number of repeats of an interval training workout).

Marathon Running. Running in a marathon (42 km, or 26.2 miles) has become the ultimate achievement for many newly converted joggers. Therefore physical educators should become familiar with some of the physiology involved in this activity. Costill has furnished an excellent review on this topic (11).

1. *Desirable physical characteristics.* The ideal marathoner should be small in stature with a small bony frame and minimal fat. Both fat and bony structure constitute dead weight to be moved at a cost in energy.

2. *Desirable physiological characteristics.* As has been discussed for distance events in general, a large $\dot{V}O_2$ max is essential. In addition, a large fraction of ST fibers in the leg muscles and the ability to utilize a large fraction of the $\dot{V}O_2$ max are important.

3. *Running efficiency.* Since oxygen (and, to a lesser extent, energy substrate) must be transported by the circulatory system, higher maximal limits can be set by the more efficient neuromuscular mechanisms, other things being equal. Differences in the mechanical efficiency of running, even among skilled distance runners, may be as large as 50% (see chap. 19).

4. *Physiological responses.* The marathon run is an extremely costly event in terms of energy consumption with an average cost of about 2,400 calories for the 26.2 miles. Energy consumption increases if the wind is blowing or if the terrain is hilly. One might think that the loss going uphill is regained on the downhill stint, but this is not the case because the gain is not so large as the loss.

Circulatory response requires near-maximal levels of heart-rate, stroke volume, and cardiac output even under normal weather conditions. Adverse weather conditions such as high heat and humidity therefore necessitate slowing the pace so that the circulatory system can meet the demand for the additional circulation required for thermoregulation. Even under normal conditions, rectal temperature rises to about 104° F and may go as high as 106° F.

Naturally, under such conditions fluid losses are extremely high, averaging about one liter per square meter body surface per hour. Despite these high fluid losses, which may range from eight to twelve pounds weight loss, even ad lib fluid ingestion cannot keep up with the loss rate because of a limited rate of gastric emptying (11). Even the partial fluid replacement that is possible is extremely important in stabilizing rectal temperature (11).

SUMMARY

Muscular Endurance

1. An optimal level of strength should be developed in the endurance training program. This allows a muscle group to work at lower percentages of its all-out capacity, and thus significantly increases endurance *(absolute endurance)*.

2. The *overload principle* applies to muscular endurance as well as to strength. Repetitions with easy work loads do not bring about optimal improvement in muscular endurance.

3. In general, suitable overload training brings about improvement in strength and muscular endurance simultaneously.

4. If either the duration of work load or the *total work* is held constant, overload training brings about far greater improvement than underload training, in strength and in endurance.
5. Overload training should not reach the point where the range of motion is curtailed.
6. The *power* (work per unit time) of muscular contraction seems to be more important than the total amount of work in bringing about a training effect.

Circulorespiratory Endurance

1. The limits of human circulorespiratory endurance are set by such psychological factors as motivation and willingness to take pain, and by many physiological factors. The most important physiological factors are oxygen transport and the metabolic capacity of the involved muscle tissue.
2. At the beginning of an endurance workbout, anaerobic energy sources are used until the transition to steady state is accomplished, at which point the energy is supplied almost entirely aerobically.
3. Alactacid energy sources can be completely used up in one explosive effort or in less demanding exercise in less than twenty seconds. Both alactacid and lactacid energy sources are completely exhausted at maximal effort in about forty seconds.
4. Events of several minutes' duration or longer are almost entirely aerobic.
5. The limiting factors in circulorespiratory endurance are primarily O_2 transport and O_2 utilization in the skeletal muscles. Energy substrate depletion can be a limiting factor if small muscle masses are used over very long periods (several hours).
6. Interval training concepts rest on sound theoretical bases. More work can be done per workout for any given physiological stressfulness if it is done intermittently rather than continuously.
7. Thirty-second workbouts are most advantageous, and the rest interval is probably best set by physiological stress as measured by heart rate. The rest interval has been adequate when the resting heart rate has returned to 120 bpm.
8. Experimental evidence is still inconclusive with respect to the practical advantages of interval training over continuous training in terms of increased PWC or $\dot{V}O_2$ max.
9. Success in distance running is strongly related to (a) $\dot{V}O_2$ max, (b) percentage of ST muscle fibers, and (c) percentage of $\dot{V}O_2$ max required at a given speed of running.
10. Training progress can be hampered by exercising to complete exhaustion. Workouts should not exceed intensity and duration levels that allow recovery from fatigue in several hours.

Physiology of Training and Conditioning Athletes

11. Athletes should not be brought along too fast early in the season for fear of hitting peak performances before championship events are scheduled. In general, the early season workout progressions should gradually increase the number of repetitions; late in the season the rate for each repetition is the important variable.

12. If split times are proportional to record performances, further improvement probably depends upon improvement of strength (power) or technique. If split times fade badly in comparison with record performances, circulorespiratory endurance needs improvement.

13. The physiology of marathon running is discussed with emphasis upon the limitations of the human organism in such activity under adverse conditions.

REFERENCES

1. Ahlborg, B.; Ekelund, L.G.; Guarnieri, G.; Harris, R.C.; Hultman, E.; and Nordesjo, L-O. Muscle metabolism during isometric exercise performed at constant force. *J. Appl. Physiol.* 33:224–28, 1972.

2. Astrand, I.; Astrand, P-O.; Christensen, E.H.; and Hedman, R. Intermittent muscular work. *Acta Physiol. Scand.* 48:448–53, 1960.

3. ———. Myohemoglobin as an oxygen store in man. *Acta Physiol. Scand.* 48:454–60, 1960.

4. Caldwell, L.S. Relative muscle loading and endurance. *J. Engineering Psychol.* 2:155–61, 1963.

5. Clarke, D.H., and Stull, G.A. Endurance training as a determinant of strength and fatiguability. *Res. Q.* 41:19–26, 1970.

6. Clarke, H.H.; Shay, C.T.; and Mathews, D.K. Strength decrement index: a new test of muscle fatigue. *Arch. Phys. Med. Rehabil.* 36:376–78, 1955.

7. Clarke, R.S.J.; Hellon, R.F.; and Lind, A.R. The duration of sustained contractions of the human forearm at different muscle temperatures. *J. Physiol.* 143:454–73, 1958.

8. Costill, D.L. The relationship between selected physiological variables and distance running performance. *J. Sports Med.* 7:61–66, 1967.

9. Costill, D.L.; Sparks, K.; Gregor, R.; and Turner, C. Muscle glycogen utilization during exhaustive running. *J. Appl. Physiol.* 31:353–56, 1971.

10. Costill, D.L.; Bowers, R.; Branam, G.; and Sparks, K. Muscle glycogen utilization during prolonged exercise on successive days. *J. Appl. Physiol.* 31:353–56, 1971.

11. Costill, D.L. Physiology of marathon running. *J.A.M.A.* 221:1024–29, 1972.

12. Costill, D.L.; Thomason, H.; and Roberts, E. Fractional utilization of the aerobic capacity during distance running. *Med. Sci. Sports* 5:248–52, 1973.

13. De Lateur, B.J.; Lehman, J.F.; and Fordyce, W.E. A test of the De Lorme axiom. *Arch. Phys. Med. Rehabil.* 49:245–48, 1968.

14. deVries, H.A. Method for evaluation of muscle fatigue and endurance from electromyographic fatigue curves. *Am. J. Phys. Med.* 47:125–35, 1968.

15. Evans, S.J. An electromyographic analysis of skeletal neuromuscular fatigue with special reference to age. Doctoral dissertation, Physical Education, USC

16. Foster, C.; Costill, D.L.; Daniels, J.T.; and Fink, W.J. Skeletal muscle enzyme activity, fiber composition and VO_2 max in relation to distance running performance. *Eur. J. Appl. Physiol.* 39:73–80, 1978.

17. Fox, E.L.; Robinson, S.; and Wiegman, D.L. Metabolic energy sources during continuous and interval running. *J. Appl. Physiol.* 27:174–78, 1969.

18. Funderburk, C.F.; Hipskind, S.G.; Welton, R.C.; and Lind, A.R. Development of and recovery from fatigue induced by static effort at various tensions. *J. Appl. Physiol.* 37:392–96, 1974.

19. Grose, J.E. Depression of muscle fatigue curves by heat and cold. *Res. Q.* 29:19–31, 1958.

20. Hellebrandt, F.A.; Skowlund, H.V.; and Kelso, L.E.A. New devices for disability evaluation. *Arch. Phys. Med. Rehabil.* 29:21–28, 1948.

21. Hellebrandt, F.A., and Houtz, S.J. Mechanisms of muscle training in man. *Phys. Ther. Rev.* 36:371–83, 1956.

22. Hermansen, L., and Osnes, J.B. Blood and muscle pH after maximal exercise in man. *J. Appl. Physiol.* 32:304–8, 1972.

23. Heyward, V. Influence of static strength and intramuscular occlusion on submaximal static muscle endurance. *Res. Q.* 46:393–402, 1975.

24. Heyward, V., and McCreary, L. Comparison of the relative endurance and critical occluding tension levels of men and women. *Res. Q.* 49:301–7, 1978.

25. Hodgkins, J. Influence of unilateral endurance training on contralateral limb. *J. Appl. Physiol.* 16:991–93, 1961.

26. Karlsson, J.; Astrand, P.-O.; and Ekblom, B. Training of the oxygen transport system in man. *J. Appl. Physiol.* 22:1061–65, 1967.

27. Karlsson, J.; Funderburk, C.F.; Essen, B.; and Lind, A.R. Constituents of human muscle in isometric fatigue. *J. Appl. Physiol.* 38:208–11, 1975.

28. Keul, J. The relationship between circulation and metabolism during exercise. *Med. Sci. Sports* 5:209–19, 1973.

29. Knuttgen, H.G.; Nordesjo, L.-O.; Ollander, B.; and Saltin, B. Physical conditioning through interval training with young male adults. *Med. Sci. Sports* 5:220–26, 1973.

30. Margaria, R.; Cerratelli, P.; and Mangili, F. Balance and kinetics of anaerobic energy release during strenuous exercise in man. *J. Appl. Physiol.* 19:623–28, 1964.

31. Pernow, B., and Saltin, B. Availability of substrates and capacity for prolonged heavy exercise in man. *J. Appl. Physiol.* 31:416–22, 1971.

32. Rich, G.Q. Muscular fatigue curves in boys and girls. *Res. Q.* 31:485–98, 1960.

33. Rohter, F.D.; Rochelle, R.H.; and Hyman, C. Exercise blood flow changes in the human forearm during physical training. *J. Appl. Physiol.* 18:789–93, 1963.

34. Roskamm, H. Optimum patterns of exercise for healthy adults. *Can. Med. Assoc. J.* 96:895–900, 1967.

35. Royce, J. Isometric fatigue curves in human muscle with normal and occluded circulation. *Res. Q.* 29:204–12, 1958.

36. Saltin, B. Intermittent exercise: its physiology and practical application. John R. Emens Lecture, Ball State University, Muncie, Indiana, Feb. 20, 1975.

37. Shaver, L.G. Effects of training on relative muscular endurance in ipsilateral and contralateral arms. *Med. Sci. Sports* 2:165–71, 1970.

38. Start, K.B., and Graham, J.S. Relationship between the relative and absolute isometric endurance of an isolated muscle group. *Res. Q.* 35:193–204, 1964.

39. Stull, G.A., and Clarke, D.H. High resistance, low repetition training as a determiner of strength and fatiguability. *Res. Q.* 41:189–93, 1970.

40. Vanderhoof, E.R.; Imig, C.J.; and Hines, H.M. Effect of muscle strength and endurance development on blood flow. *J. Appl. Physiol.* 16:873–77, 1961.

41. Wilmore, J.H. Influence of motivation on physical work capacity and performance. *J. Appl. Physiol.* 24:459–63, 1968.

42. ————. Maximal oxygen intake and its relationship to endurance capacity on a bicycle ergometer. *Res. Q.* 40:203–10, 1969.

19

Efficiency of
Muscular Activity

For the engineer and the physicist, definition of the efficiency of a machine is quite simple.

$$\text{Efficiency} = \frac{\text{Output}}{\text{Input}} = \frac{\text{Work done by machine}}{\text{Work done on the machine}}$$

The physiologist usually uses the same concept in the following terms.

$$\text{Efficiency} = \frac{\text{Work output}}{\text{Energy expended}}$$

In practice, energy expended is measured indirectly by oxygen consumption, which is converted into heat units (calories), and work output; which is measured in foot-pounds (or kilogram-meters), is also converted into heat units (3,087 ft-lb or 427 kg-m = 1 kcal) so one may work with similar units. This is a simple procedure for activities in which the work output is easily measured as force times distance, such as riding a bicycle ergometer or lifting the body weight in bench-stepping.

Problems arise, however, in defining the base line from which to measure the energy input (the denominator of our efficiency equation). Should we use the gross VO_2 during the exercise in question, or should we subtract the resting value of VO_2, in which case we would have "net efficiency"? Gross efficiency and net efficiency have been the classic methods employed, but each results in artifactual errors of computation—gross efficiency because it includes resting VO_2, which is not really attributable to the exercise, and net efficiency because the appropriate resting value is impossible to measure precisely, as will be shown below. Gaesser and Brooks (26) have shown clearly that erroneous conclusions may be derived from the classic approach. At the present time there is ongoing dialogue with respect to the best approach for the measurement of efficiency. Two new methods have been proposed, and we may define the differences as follows:

(1) Gross efficiency $= \dfrac{\text{Work output}}{\text{Energy expended}} = \dfrac{W}{E} \times 100$

(2) Net efficiency $= \dfrac{\text{Work output}}{\text{Energy expended above that at rest}} = \dfrac{W}{E\text{-}e} \times 100$

(3) Work efficiency $= \dfrac{\text{Work output}}{\text{Energy expended above that in unloaded cycling}}$

$= \dfrac{W}{E_L\text{-}E_u} \times 100$

(4) Delta efficiency $= \dfrac{\Delta\text{Work output}}{\Delta\text{Energy expended}} = \dfrac{\Delta W}{\Delta E} \times 100$

Physiology of Training and Conditioning Athletes

Where W = caloric equivalent of external work done

E = gross caloric expenditure including resting expenditure

e = resting caloric expenditure

E_L = caloric expenditure under load

E_u = caloric expenditure in unloaded pedaling

ΔW = caloric equivalent of increment in work output above previous work rate

ΔE = increment in caloric expenditure above that at previous work rate

These different methods result in quite different values of efficiency under the same experimental conditions on a bicycle ergometer, with gross efficiency showing a range of 7.5% to 20.4%, net efficiency, 9.8% to 24.1%, and delta efficiency 24.4% to 34.0%. Work efficiency proved difficult to apply because of the difficulty in obtaining a true zero work pedaling condition (26).

Since caloric expenditure goes up at least in proportion to (linearly with) and probably more rapidly than does work rate (26, 29), efficiency must either remain constant or decrease with increasing work rate. Gaesser and Brooks (26) showed that delta efficiency produced this result, while the use of gross or net values showed increasing values of efficiency with increasing work rate due to the artifacts of calculation. Thus delta efficiency seems to be the method of choice. Work efficiency would be equally satisfactory if a true zero work unloaded pedaling device were available. Unfortunately, all presently available ergometers in the author's experience provide anywhere from twenty to thirty-five watts of resistance due to such factors as friction and inertia when set at zero load.

If efficiency values are needed for application in nutritional studies where gross energy expenditure is the matter of concern, then gross efficiency is the measure to be used.

The problem becomes more complex when activities such as walking and running are considered because large proportions of the working forces are dissipated in reciprocal movements of the arms and legs.

Fenn (25) has demonstrated that even the more difficult problems are capable of solution through application of motion picture recording and subsequent analysis of the forces involved in accelerating and decelerating the various body segments. His analysis, as shown below, of the forces and energy involved in running is of interest.

13 hp Total oxygen consumption	7.8 hp Waste in recovery	Developing tension	Useful work 2.95 hp	Gravity . . . . 0.1 hp
				Velocity changes 0.5
		Maintaining tension		Acceleration of limbs . 1.68
				Deceleration of limbs . 0.67
	5.2 hp Initial energy	Shortening energy	Friction loss Waste Fixation	
		Waste		

This classic work of Fenn has recently been improved upon by Winter (43), who has further refined the calculation of the numerator of the efficiency equation by accounting for the internal work done by the limbs themselves.

The efficiency of machines varies between 10% and 20% in steam engines, 20% and 30% in gasoline engines, and 80% and 90% in electric motors. The mechanical efficiency of man varies from less than 10% to 30% or even 40% or more, depending on the activity and method of calculation.

It would be well at this point to consider in which athletic activities efficiency plays an important role. Obviously, any event that involves endurance will be very much influenced by the factor of muscle efficiency. Thus we are talking largely about running events greater than a quarter mile and swimming events beyond 100 yards. In the events where a single explosive effort is required, such as the shot put, power rather than efficiency is the critical factor. In sprint events efficiency of movement is of some importance but is probably secondary to the need for power.

An analogy from the automotive world seems appropriate. In a drag race (an acceleration contest), one does not care how many miles per gallon of gas (efficiency) the machine achieves; *power* is all-important. For an economy run, however, power is unimportant; miles per gallon determines the winner.

Thus this chapter may be considered a continuation of the last chapter in that it also is mainly concerned with endurance. The maximum speed at which humans can run distance events depends very largely on the rate at which they can supply energy (limited by maximal O_2 consumption) and their efficiency in using this energy.

Aerobic versus Anaerobic Efficiency

The efficiency of work done at the expense of incurring lactacid oxygen debt has been reported to be only about half that of work done aerobically (5, 16, 17). However, the recent work of Gladden and Welch (27) suggests that anaerobic metabolism is no less efficient than aerobic metabolism. In any event, efficiency does in fact decrease as the power output (work rate) increases (24, 26, 29). This may be the result of a true decrease in muscular efficiency, or an increased metabolic overhead (cost of heart and lung function and the like, which are not measured as work output), or both.

Effect of Speed on Efficiency

Simple Movements. In a classic experiment in 1922 of muscle physiology, A. V. Hill (30) provided evidence of an optimum speed of movement below which efficiency falls slowly and above which it falls rapidly. It must be cautioned, however, that this work was done on a simple contraction of isolated muscle groups.

Running. In another important experiment in 1926, Sargent (39) determined the oxygen consumption of a subject who ran 120 yards at varying rates of speed, up to an all-out sprint. His results indicated that O_2 consumption increased as the 3.8th power of speed. This would mean that O_2 consumption is increased almost sixteen times when speed is doubled. This would also mean a tremendous loss of efficiency as running speed increases, and many theories of pace, etc., have been built on this concept.

Sargent's subject, however, ran anaerobically, and his O_2 consumption was calculated as O_2 debt based on the O_2 consumption after the race; from this was subtracted the resting rate measured *before* the run. It has, however, been demonstrated that resting consumption after exercise remains considerably higher for six to eight hours after exercise, and this higher metabolism is not related to the O_2 debt incurred during the run. Consequently, Sargent's calculations resulted in erroneously high O_2 consumptions because he had subtracted too low a base-line value from the recovery O_2 consumption rates.

More recent work (22, 23, 36) shows that the rate of increase in energy demand is only *proportional* to increases in speed. Thus efficiency remains constant in spite of changes in speed for all distance events.

There are large differences in the efficiency of running from individual to individual. Figure 19.1 shows the differences in O_2 consumption for runners of varying skill as studied by Dill, Talbot, and Edwards in 1930. It is seen that the famous marathoner of that time, Clarence DeMar had the lowest O_2 consumption at 26 ml/kg/min while the less skilled runners went 54% higher, requiring as much as 40 ml/kg/min for the same run. This classic study demonstrates the importance of skill in running, although we still do not fully understand the methods required to optimize this factor.

Stride length and the amount of vertical movement appear to be important determinants of running efficiency. In a comparison of good and poor runners in Japan, it was found that the better performers (1) used a longer stride, (2) had a faster stride, and (3) had a greater forward lean. Most important, they found that in the 5000-meter event the poor performers did 17,968 kg-m of vertical work (wasted energy) while the better performers did only 9,407 kg-m of vertical work (37).

Two groups of investigators working with untrained subjects found that there is an optimal stride length for each individual and that the tendency is toward overstriding (inefficient) rather than understriding (9, 15).

Trained male runners appear to be more efficient than trained females, although the differences found were not large—10% or less (8, 32).

Walking. The most economical rate of walking has been found to be 4 km/hr (2.4 mph) (10) at which the energy consumption is about one-half kilocalorie per kilometer walked per kilogram of body weight. The energy consumption for aerobic running is about twice that value, regardless of speed.

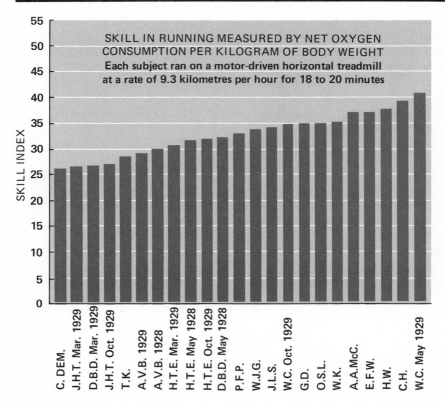

Figure 19.1 Skill in running. each man ran at the same rate on the horizontal treadmill. in contrast with the relatively uniform cost of walking at 90 m/min the cost varied widely. the most efficient was Clarence Demar, the famous marathoner of that time, 1929. (from Dill, D.B., Talbot, and Edwards. *J. Physiol.* 69:267, 1930.)

In walking more slowly than 4 km/hr, energy is apparently wasted in static components of muscle activity because weight is supported too long in respect to the useful work. In walking faster, energy consumption increases faster than the useful work done, and efficiency again falls off, presumably because of a disproportionate increase of energy wasted in accelerating and decelerating body parts.

As a result of the increasing inefficiency at increasing rates, energy consumption of walking at about 8 km/hr (4.8 mph) becomes greater than that required for running, so that running is more efficient than walking above 8 km/hr.

Cycling. Using total work loads on bicycle ergometers that were equated for energy cost, Henry (28) found that 69 rpm resulted in work output of 620 kg-m/min while 116 rpm produced only 95 kg-m/min. The difference in efficiency is obvious. Earlier workers had reported that 70 rpm is the most efficient pedaling rate. However, the most recent work shows that the most efficient pedaling rate increases with increases in power output (work rate) from 42 rpm at a light load of 40.8 watts to 62 rpm at a heavy load of 327 watts (40). These differences among investigators are probably due to the different flywheel masses involved on the different ergometers. A small flywheel mass would require higher peak force output by the leg muscles.

Storage of Elastic Energy. An explanation seems called for to rationalize the fact that many forms of physical activity have an optimum rate, above which increased speed demands disproportionately greater energy expenditure. The efficiency of running, on the other hand, seems to be unaffected by speed.

Early workers, such as Fenn (25) and Hubbard (33), discussed the possibility of the storage of mechanical energy in the muscles and tendons. The extension of a muscle and tendon that are antagonistic in one phase of reciprocating movement might store the kinetic energy of the protagonist as potential energy, which is released when the antagonist contracts.

It has been demonstrated in a laboratory preparation that such a storage of energy can and does occur (12). When a contracted muscle was forcibly stretched, a substantial amount of the work done in stretching the muscle appeared to be available in the work done in the subsequent contraction. Furthermore, the sooner the contraction followed the forced stretch the greater was the increase in work performed. It must be realized that this phenomenon could bring about considerable economy in quickly reciprocating movements, as in running, but the economy would be less as the rate slows because of the greater length of time during which the contracted muscle exerts tension (uses energy) against the stretching. It has been calculated by the same investigators that this elastic work may contribute as much as half of the total mechanical work performed in running (11). More recent work from the same laboratory has corroborated this concept and extended it into the realm of human arm and leg movement (13, 42). Using an electronic force platform, Thys, Farragiana, and Margaria (42) studied subjects in deep knee bending exercise under two conditions, (1) rebounding, where they bounced back up immediately after assuming the full squat position thus using the elastic energy stored in stretching the leg extensors in the subsequent contraction, and (2) nonrebounding exercise, identical except that the exercise stopped for a fraction of a second in the full squat position, thus allowing the elastic energy to be dissipated as heat. They found the rebounding movement to have faster maximum speed of movement (20%), more power (29%), and better efficiency (37%). This appears to be an important factor to be applied in athletics wherever pertinent. These facts would seem to offer the best explanation for the failure of the efficiency of running to decrease with increasing speed (14).

It has also been suggested that there are sex differences in the storage of elastic energy. Although the leg extensors of the male can sustain much higher stretch loads, the female may be able to utilize a greater portion of the stored elastic energy in jumping (35).

Effect of Work Rate on Efficiency

It appears from available data that if speed is held constant, the work rate (or power output) probably affects work efficiency or delta efficiency very little, at least at the light and moderate loads that have been used in investigations (26, 29).

Effect of Fatigue on Efficiency

It has been pointed out that the electrical activity of a muscle increases with time even though it maintains constant tension; figure 19.2 illustrates this fact. The best explanation for this is that, as fatigue occurs, more and more motor units are required to do the same piece of work. It is obvious that the recruitment of more units should result in greater energy consumption. Since the work *output* remains constant, the increased energy *input* must result in lowered efficiency. This will be important to our discussion of pace and efficiency.

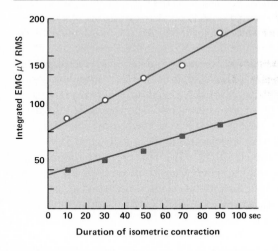

Figure 19.2 Illustration of increased electrical activity as a function of time when 40% of maximal voluntary contraction strength is maintained isometrically in the elbow flexors. open circles represent data for a subject whose maximal duration was 127 seconds; squares represent subject with maximal duration of 338 seconds.

Physiology of Training and Conditioning Athletes

Diet and Efficiency

There appears to be agreement among various investigators that the efficiency of muscular work is greatest when carbohydrate supplies the energy for muscular contraction; however, the differences due to this factor are small, and probably not more than 5%.

Effects of Environmental Temperature

It has been shown that moderate work performed at 100° F ambient temperature requires an average 13.3% higher metabolic rate than work performed at 85° F. When work was heavy, the increase was 11.7% (18). The difference between work loads is probably not significant.

This decreased efficiency is undoubtedly the result of the increased load upon the circulatory system for meeting the demand for increased peripheral circulation to transport heat from the core to the skin (increased heart rate, etc.).

Effect of Obesity on Efficiency

Dempsey et al. (19) compared the exercise responses of fourteen normal young men with fourteen obese young men. It was shown that the obese required 2.01 liters/min of O_2 at a work load of 650 kg-m/min, while the normals used only 1.54 liters/min for a slightly larger work load. Thus the obese required 30% more energy to do the job. For the moderately obese, the difference is probably not significant (44).

The Looseness Factor in Athletics

Athletes and coaches frequently speak of being *loose* or *tight* in such activities as sprint running and swimming. It has been suggested that a tight runner may be thought of as one whose movements are impeded by the resistance of antagonistic muscles and their connective tissues due to a lack of flexibility. The author has investigated this possibility by controlled experiments in which four subjects ran ten 100-yard sprints. Five of the sprints were run under normal conditions of flexibility, and five were run after flexibility of trunk flexion, ankle flexion, and ankle extension had been significantly improved by static stretching. No significant differences in speed or oxygen consumption were found (21).

In the light of much personal experience with the looseness factor by athletes and coaches, it is difficult to reject its existence; however, the author's experiments indicate that this factor is probably not a simple matter of increased or decreased static flexibility. More likely it is a matter of changes in neuromuscular coordination. There is also the possibility that it is a psychological illusion.

Acceleration-Deceleration versus Smooth Movement

In some activities (such as swimming) the smoothness of movement is important in determining efficiency. In swimming the butterfly stroke, for example, if constant velocity could be maintained throughout the various events of the single-stroke cycle (the pull, the arm recovery, and the leg drive), maximum efficiency would be achieved. This is so because at a constant speed, a constant amount of energy is required to overcome the resistance of the body's movement through the water. This resistance is called *drag*. If the stroke is coordinated so that the leg drive sustains the velocity achieved by the arm pull during the recovery phase of the arms, only the force of drag must be overcome. If the coordination is such that the swimmer comes to a virtual standstill during arm recovery, then, in addition, the body must be *accelerated* by each arm pull. The energy required for acceleration is very costly because it varies as the square of the velocity. To double the velocity, then, requires four times the energy output.

To illustrate this concept, table 19.1 was prepared from an experiment in which the author measured the velocity of outstanding butterfly swimmers for each 0.10 second in their stroke cycles (20). For the sake of comparison, arbitrary energy units were calculated for (1) an ideal swimmer, swimming

Table 19.1

Comparison of Energy Requirements for One Stroke (in Arbitrary Units)
(V = Velocity in ft/sec; V^2 is an Estimate of Energy Consumption)

Consecutive 1/10-second Periods	Ideal Swimmer		National Record holder		Inexperienced Swimmer	
	V	V^2	V	V^2	V	V^2
0–0.1	6.07	36.54	8.00	64.00	11.10	123.21
0.1–0.2	6.07	36.54	6.72	45.13	8.63	74.48
0.2–0.3	6.07	36.54	6.63	43.96	6.12	37.45
0.3–0.4	6.07	36.54	4.92	24.21	3.08	9.49
0.4–0.5	6.07	36.54	4.42	19.54	1.00	1.00
0.5–0.6	6.07	36.54	6.22	38.69	5.76	33.18
0.6–0.7	6.07	36.54	5.33	28.41	4.35	18.92
0.7–0.8	6.07	36.54	4.92	24.21	3.21	10.30
0.8–0.9	6.07	36.54	4.84	23.43	2.42	5.86
0.9–1.0	6.07	36.54	6.22	38.69	6.55	42.90
1.0–1.1	6.07	36.54	8.52	72.59	14.52	210.83
Mean	6.07		6.07		6.07	
Total energy units used for 1 stroke		405.24		422.86		567.62

Physiology of Training and Conditioning Athletes

so smoothly that no acceleration-deceleration occurs, (2) a national record holder, whose highest velocity was 1.94 times his slowest velocity, and (3) an inexperienced swimmer, whose highest velocity was 14.52 times his slowest velocity (not unusual).

Although all three swimmers achieved an *average* velocity of 6.07 feet per second for the stroke that was analyzed, even the champion used 4% more energy than was necessary, and the inexperienced swimmer used 40% more. Furthermore, it has been shown that drag increases disproportionately above five or six feet per second, so that the actual differences in efficiency would be even greater than those predicted (4).

Pace and Efficiency

The question of how best to pace an endurance event cannot be answered simply. First, we have seen that efficiency for most activities (but apparently not running) falls off beyond some optimum rate of speed. Second, the fatigue level must be considered inasmuch as efficiency falls off rapidly as fatigue brings about greater recruitment of muscle fibers to do the same job and possibly interferes with neuromuscular coordination.

In events that are largely aerobic (two-mile or more run), a constant rate is probably the most efficient. This is true even in running in which efficiency does not seem to vary with speed (aerobic condition); even though the varying rates may be equally efficient, the *changes in rate* cost additional energy.

Constant rate does not necessarily mean running or swimming equal split times. For example, it is typical in distance swimming that the number of strokes per minute remains constant throughout the 1,500-meter swim, but each successive 100 meters shows a small decrement in speed—two or three seconds, or more, depending on the swimmer. This is the result of the fatigue process; even though cadence remains the same, the force developed—and consequently the *distance per stroke*—decreases.

For middle-distance running events, some very interesting evidence *against* the constant rate has been presented by Robinson and colleagues (38), who ran two men on three different pace plans. Each man ran one trial at constant speed, one trial with a fast first minute and slower remainder, and one trial with a slow first minute and faster remainder. In both men, it was found, the slow start and faster finish resulted in slightly less O_2 consumption than the constant rate, and in considerably less than the fast start and slower remainder. The overall time for the three pace plans was kept constant on the treadmill.

They also showed in another experiment that when three men ran to exhaustion in from 2.58 to 3.37 minutes, each man's O_2 consumption increased from 60% to 143% during the last half minute, as the lactic acid rose to high levels.

Thus the explanation for the results in the pace experiment would seem to be that starting out fast results in an earlier accumulation of lactic acid, and thus more of the race is run inefficiently. On this basis, it would seem wise to run middle-distance races in which high O_2 debts are encountered on a pace plan that postpones the O_2 debt until late in the race. However, more recent studies that measured O_2 consumption to compare similar pace plans did not entirely support Robinson's data. Adams and Bernauer (3) found the steady pace to be significantly less demanding, and Kollias and co-workers (34) found no difference between steady state and slow-fast pace, but a fast-slow pace required significantly greater O_2 consumption than either of the other two conditions. Summarizing the evidence from these three metabolic studies with two radiotelemetry studies on the cardiac cost of similar pace plans (7, 41) leaves us with complete agreement only on the fact that the fast-slow pacing creates greater physiological demands for middle-distance runs. Whether the slow-fast pace is better than the steady pace will be finally resolved only by further investigation.

Efficiency of Positive and Negative Work

Positive work is done when muscle contraction provides force that works through a distance, and this is associated with the concentric contraction of muscle. Negative work results when an extrinsic force overcomes the force developed by muscle contraction, and thus the muscle lengthens during contraction (eccentric contraction). By definition, positive work lifts a weight; negative work lowers a weight to its resting place. These definitions are not altogether satisfactory, however, for if we continue increasing the velocity with which we lower the weight until we accelerate its lowering to 32 ft/sec^2, all of the work would be done by gravitational force. A very slow lowering, on the other hand, could require a large degree of effort in a physiological sense. Undoubtedly, increasing efforts by biophysicists will improve our definitions of these terms; in the meantime, some very interesting findings revolve about these definitions.

A.V. Hill and his co-workers have discovered the surprising fact that application of external mechanical work to a living muscle cell can reverse the normal biochemical processes (1, 31). Their experiments suggest that an absorption of energy by muscle tissue occurs during the forced extension of the contractile component when the muscle is in isometric contraction, and experiments on intact human muscles have supported these findings.

Asmussen (6) studied a man's energy expenditure in riding a bicycle uphill and downhill on a treadmill. He found it required from three to nine times as much energy to pedal uphill (positive work) as it did to resist the force of gravity in backpedaling downhill (negative work).

Abbott, Bigland, and Ritchie (2) found similar results. They used a pair of subjects on two bicycle ergometers, coupled in opposition (back to back) so that all of the positive work of one subject was dissipated as negative work in the other. Although the subjects used the same leg muscles in similar movements at identical speeds, at 35 rpm the O_2 consumption for positive work was 3.7 times that of negative work. This ratio increased with the speed of pedaling.

Principles for Improvement of Efficiency

1. Improve the skills involved first.
   A. Eliminate *unnecessary movements.*
   B. Eliminate *unnecessary muscle activity*—within even the necessary movement.
      (1) Maintain relaxation in antagonistic muscles.
      (2) Relax even the prime movers when possible; for example, convert tension movements into ballistic movements wherever possible; thus even the prime movers can relax during a large portion of the movement.
   C. Make all movements in the *correct directions;* for example, an arm pull that moves laterally from the body in swimming the crawl wastes a considerable component of its total force.
   D. Apply only the *necessary amount of power;* too forceful an effort is usually wasteful.
   E. Use *muscles that are best suited to the activity;* for example, use the larger leg muscles in lifting heavy weights rather than the weaker back muscles.
   F. Use the *optimum speed* if time is not a factor.
2. Improve physical condition so that a given level of work results in less fatigue. Better condition delays fatigue; fatigue increases energy cost.
3. Avoid costly acceleration. Maintain a constant cadence if possible, even though fatigue may result in progressively slower split times.
4. Use a high carbohydrate, low fat, normal protein diet.
5. Pace (for a distance event) should adhere to principle three. For middle-distance events that incur large oxygen debts, a gradually accelerating pace may have advantages.

REFERENCES

1. Abbott, B.C.; Aubert, V.M.; and Hill, A.V. Absorption of work by muscle stretched during single twitch or short tetanus. *Proc. R. Soc. Lond.* 139:86–104, 1951.
2. Abbott, B.C.; Bigland, B.; and Ritchie, J.M. Physiological cost of negative work. *J. Physiol.* 117:380–90, 1952.

3. Adams, W.C., and Bernauer, E.M. The effect of selected pace variations on the oxygen requirement of running a 4:37 mile. *Res. Q.* 39:837–46, 1968.

4. Alley, L.E. An analysis of water resistance and propulsion in swimming the crawl stroke. *Res. Q.* 23:253–70, 1952.

5. Asmussen, E. Aerobic recovery after anaerobiosis in rest and work. *Acta Physiol. Scand.* 11:197–210, 1946.

6. ———. Experiments on positive and negative work. In *Fatigue,* eds. W.F. Floyd and A.T. Welford. London: H.K. Lewis and Co., 1953.

7. Bowles, C.J., and Sigerseth, P.O. Telemetered heart rate responses to pace patterns in the one-mile run. *Res. Q.* 39:36–46, 1968.

8. Bransford, D.R., and Howley, E.T. Oxygen cost of running in trained and untrained men and women. *Med. Sci. Sports* 9:41–44, 1977.

9. Burke, E.J., and Berger, R.A. Energy cost of running at three different stride lengths. *N. Z. J. HPER* 9:96–99, 1976.

10. Cavagna, G.A.; Saibene, F.P.; and Margaria, R. External work in walking. *J. Appl. Physiol.* 18:1–9, 1963.

11. ———. Mechanical work in running. *J. Appl. Physiol.* 19:249–56, 1964.

12. ———. Effect of negative work on the amount of positive work performed by an isolated muscle. *J. Appl. Physiol.* 20:157–58, 1956.

13. Cavagna, G.A.; Dusman, B.; and Margaria, R. Positive work done by a previously stretched muscle. *J. Appl. Physiol.* 24:21–32, 1968.

14. Cavagna, G.A., and Kaneko, M. Mechanical work and efficiency in level walking and running. *J. Physiol.* 268:467–81, 1977.

15. Cavanagh, P.R.; Williams, K.R.; and Hodgson, J.L. The effect of stride length variation on O_2 uptake during distance running. *Med. Sci. Sports* 10:63, 1978.

16. Christensen, E.H., and Hogberg, P. The efficiency of anaerobical work. *Arbeitsphysiologie* 14:249–50, 1950.

17. ———. Steady state, O_2 deficit and O_2 debt at severe work. *Arbeitsphysiolgie* 14:251–54, 1950.

18. Consolazio, C.F.; Matoush, L.D.; Nelson, R.A.; Torres, J.B.; and Isaac, G.J. Environmental temperature and energy expenditure. *J. Appl. Physiol.* 18:65–68, 1963.

19. Dempsey, J.A.; Reddan, W.; Balke, B.; and Rankin, J. Work capacity determinants and physiologic cost of weight supported work in obesity. *J. Appl. Physiol.* 21:1815–20, 1966.

20. deVries, H.A. A cinematographical analysis of the dolphin swimming stroke. *Res. Q.* 30:413–22, 1959.

21. ———. The "looseness" factor in speed and O_2 consumption of an anaerobic 100-yard dash. *Res. Q.* 34:305–13, 1963.

22. Dill, D.B. Comparative physiology of oxygen transport. *J. Sports Med. Phys. Fitness* 3:191–200, 1963.

23. ———. Oxygen used in horizontal and grade walking and running on the treadmill. *J. Appl. Physiol.* 20:19–22, 1965.

24. Donovan, C.M., and Brooks, G.A. Muscular efficiency during steady-state exercise; II. Effects of walking speed and work rate. *J. Appl. Physiol.* 43:431–39, 1977.

25. Fenn, W.O. Frictional and kinetic factors in the work of sprint runners. *Am. J. Physiol.* 92:583–610, 1930.

26. Gaesser, G.A., and Brooks, G.A. Muscular efficiency during steady-rate exercise: effects of speed and work-rate. *J. Appl. Physiol.* 38:1132–39, 1975.

27. Gladden, L.B., and Welch, H.G. Efficiency of anaerobic work. *J. Appl. Physiol.* 44:564–70, 1978.

28. Henry, F.M. Individual differences in O_2 metabolism of work at two speeds of movement. *Res. Q.* 22:324–33, 1951.

29. Hesser, C.M.; Linnarsson, D.; and Bjurstedt, H. Cardiorespiratory and metabolic responses to positive and negative and minimum load dynamic leg exercise. *Resp. Physiol* 30:51–67, 1977.

30. Hill, A.V. The maximum work and mechanical efficiency of human muscles and their most economical speed. *J. Physiol.* 56:19–41, 1922.

31. ———. Production and absorption of work by muscle. *Science* 131:897–903, 1960.

32. Howley, E.T., and Glover, M.E. The caloric costs of running and walking one mile for men and women. *Med. Sci. Sports* 6:235–37, 1974.

33. Hubbard, A.W. An experimental analysis of running and a certain fundamental difference between trained and untrained runners. *Res. Q.* 10:28–38, 1939.

34. Kollias, J.; Nicholas, W.C.; Buskirk, E.R.; and Mendez, J. Oxygen requirements for running at moderate altitude. *J. Sports Med.* 10:27–35, 1970.

35. Komi, P.V., and Bosco, C. Utilization of stored elastic energy in leg extensor muscles by men and women. *Med. Sci. Sports* 10:261–65, 1978.

36. Margaria, R.; Cerretelli, P.; Aghemo, P.; and Sassi, G. Energy cost of running. *J. Appl. Physiol.* 18:367–70, 1963.

37. Miyashita, M.; Miura, M.; Murase, Y.; and Yamaji, K. Running performance from the viewpoint of aerobic power. Paper to International Symposium on Environmental Stress, Santa Barbara, Calif., September 1, 1977.

38. Robinson, S.; Robinson, D.L.; Mountjoy, R.J.; and Bullard, R.W. Fatigue and efficiency of men during exhausting runs. *J. Appl. Physiol.* 12:197–202, 1958.

39. Sargent, R.M. The relation between O_2 requirement and speed in running. *Proc. R. Soc. Lond.* 100:10–22, 1926.

40. Seabury, J.J.; Adams, W.C.; and Ramey, M.R. Influence of pedalling rate and power output on energy expenditure during bicycle ergometry. *Ergonomics* 20:491–98, 1977.

41. Sorani, R.P. The effect of three different pace plans on the cardiac cost of 1320-yard runs. Doctoral dissertation, Physical Education, USC, 1967.

42. Thys, H.; Faraggiana, T.; and Margaria, R. Utilization of muscle elasticity in exercise. *J. Appl. Physiol.* 32:491–94, 1972.

43. Winter, D. A new definition of mechanical work done in human movement. *J. Appl. Physiol.* 46:79–83, 1979.

44. Wolfe, L.A.; Hodgson, J.L.; Barlett, H.L.; Nicholas, W.C.; and Buskirk, E.R. Pulmonary function at rest and during exercise in uncomplicated obesity. *Res. Q.* 47:829–38, 1976.

20

Speed

Speed of movement is very important in athletics. It is worthy of careful analysis so that we may better understand this aspect of human performance and thus be in a better position to improve this function in athletes.

First, we must realize that, basically, speed is the result of applying force to a mass. Second, speed usually implies movement at a *constant* rate. The movement of a body (human or otherwise) at a constant rate requires sufficient driving force to balance the forces that resist movement. An airplane must have just enough force to overcome the friction of air drag to maintain a constant speed; if more than this balancing amount of force is applied, acceleration occurs (speed increases with time); if less, the aircraft decelerates.

In the human body the resisting force has several components, and we can think of a balance of positive and negative forces in respect to propulsion of the body or any of its parts (the same physical laws apply). The positive force that propels the body is provided by muscular contractions, aided in some cases by the storage of elastic energy (see chap. 19). The negative forces depend upon the nature of the activity.

In running, for example, it was shown in chapter 19 that the 2.95 hp of useful work (positive force) developed by the muscles were used to balance the negative forces, as follows: (1) gravity, 0.1 hp; (2) velocity changes, 0.5 hp; (3) acceleration of limbs, 1.68 hp; and (4) deceleration of limbs, 0.67 hp. Had this run been performed on the track instead of on a treadmill, another negative force, air resistance—possibly as much as 0.5 hp—would have had to have been overcome, and the positive or propelling force needed for maintaining a constant rate of speed would have been 3.45 hp.

From the above considerations we might hypothesize that speed may be improved by either increasing the positive or by decreasing the negative factors. In a practical sense, this suggests that improving strength would be the most important positive factor. The negative factors might be reduced through improved neuromuscular coordination (skill) and flexibility, which might conceivably decreases the values of the factors listed above. We shall examine these possibilities in this chapter.

Intrinsic Speed of Muscle Contraction

As has been pointed out, muscles differ in their ability to produce fast movement. This, of course, reflects the differences in their makeup with respect to proportions of FT and ST fibers, as discussed in chapters 2, 3, and 4. Thus there are considerable intrinsic differences between the postural extensor muscles and the faster flexor muscles within a given individual. There are also interindividual differences in speed of movement for the same muscle group or type of movement, and speed of contraction also varies greatly from one animal species to another, approximately in inverse ratio to size. This variability in speed persists even in animals deprived of innervation, so that at least in simple movements, the differences in speed of contraction must be intrinsic to the muscle tissue itself. But is this intrinsic difference the result of different lengths of sarcomeres or differences in the velocity of actin filaments sliding past myosin filaments? It seems clear that sarcomere length is relatively constant, not only among human individuals, but also among vertebrates in general (14). The sliding velocity of the filaments may vary about threefold, and this seems to account for differences in speed of contraction (3).

Comparing the intrinsic speeds of different muscles requires that we equate them by fiber length. Obviously, a muscle fiber that is ten times longer than another fiber can shorten at one end ten times as fast, although intrinsic properties are the same.

Although evidence is lacking, it seems very likely that ultimate maximal speed capacity is limited by the intrinsic speed of an individual's muscle tissue and by the nicety of that person's neuromuscular coordination patterns. Neither factor is amenable to changes as large as those that affect strength and endurance.

Force-Velocity Relationship

It has been shown with experimental muscle preparations that the force available from a muscle's shortening decreases as the rate of shortening increases. Figure 20.1 illustrates this relationship, which has important implications for athletics.

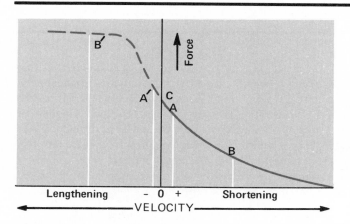

Figure 20.1 Force-velocity relation of contracting muscle: to the right, shortening; to the left, lengthening. *C*, at zero speed, represents an isometric contraction. *A* and *A'* are at the same velocity of shortening and lengthening; so are *B* and *B'*. (From Hill, A.V. *Lancet*, 261:947, 1951.)

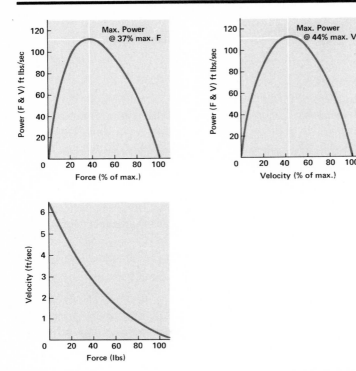

Figure 20.2 The relationships among force, velocity, and power produced in healthy young men in elbow flexion.

The shape of the curve in figure 20.1 also leads us to conclude that there must be an *optimum speed* at which a muscle can produce its greatest *power* and greatest *efficiency*. It has been found that this optimum speed is approximately one-third of the maximum speed at which it can shorten under zero load (12). Kaneko (16) has shown that maximum power is developed when force and velocity are both about 35% of their maximum values. Data from our laboratory shown in figure 20.2 are in essential agreement.

Another important implication pertains to the vulnerability to athletic injury. Hill (11) has pointed out that if a rapid flexor movement is made, and the object is suddenly wrenched in the opposite direction, the muscle or its attachments may be torn. This can happen because the forces of the muscle in the lengthening movement could be increased several times beyond what they were in shortening before a reflex inhibition (from the inverse myotatic reflex) could take place. This sudden increase in the forces, which is predicted from figure 20.1, could well exceed the elastic limits of the tissues. Figure

　　　　　Physiology of Training and Conditioning Athletes

20.1 shows clearly that a virtually instantaneous change from a shortening contraction at Point B to a lengthening contraction at point B' results in a severalfold increase in forces within the muscle involved. This is the basis for many athletic injuries!

From the standpoint of human performance in the intact individual, there are two aspects of speed. The first (really acceleration) is related to how fast an athlete can accelerate from a standstill (a football lineman's charge); this factor is an important determinant of speed for the first five or ten yards. This factor is probably determined by the shape of an individual's force-velocity curve. On the other hand, for distances more than twenty yards, the only important determinant is the maximal movement rate, which is in turn limited by intrinsic speed and neuromuscular coordination.

These two factors are not highly related. Therefore, an individual could conceivably be a slow starter with a good 100-yard speed, or a fast starter with a poor 100-yard time, or indeed be proficient in football, tennis, and other sports, where quick-starting movements are important, but be a poor 100-yard sprinter.

Specificity of Speed

It is commonplace in physical education and athletics to speak of an individual as fast or slow, but evidence is accumulating that speed has very little generality. Indeed, an individual with a fast arm movement may well have slow leg movement. In fact, this specificity extends even to the type of task and the direction of movement (2, 9, 20).

It has been shown, for example, that speed is 87% or 88% specific to the limb (20). Even within a limb, speed is 88%–90% specific to the direction of movement. This means there is practically no correlation between the speeds with which one can perform an arm movement and a leg movement and only a small relationship between speed of movement in a forward arm swing and a backward arm swing. This might lead us to describe an individual as fast in a backward swing of the right leg! Obviously, this makes no sense, but neither does it make sense to speak of a fast or slow individual. We may speak of a fast runner, but we are not justified in assuming the same individual can throw a fast ball in baseball.

Strength and Speed

It might be expected from the preceding discussions that strength and speed of movement would be highly related; however, the experimental evidence is controversial. Four different studies have agreed in findings that strength and speed are unrelated (1, 9, 10, 22). However, all of these studies were concerned with the same movement pattern, the horizontal adductive arm swing. Nelson and Fahrney (19) have shown rather strong significant and consistent correlations in three experimental subject groups of 0.74, 0.79, and 0.75 between strength and speed of elbow flexion. Thus a final decision on strength-speed relationships must await further research.

Interestingly, if movements are resisted by substantial loads on a tested limb, sizable relationships between strength and speed (up to 0.76) can be demonstrated (24). This leads us to think of speed in terms of the neuromotor specificity that has been demonstrated in so many other aspects of human performance. What we are saying, in effect, is that various speeds of movement against varying loads require different neuromotor coordination patterns. Thus, if we measure strength statically, we might expect higher correlations between strength and speed as we slow the speed (or increase the load) to more closely approximate static contraction. And this, we find, is what happens.

Carrying our discussion further requires that we visualize two strength factors: one a *static strength,* as we usually measure it by dynamometer or cable tensiometer, and the other a dynamic or *strength in action* factor, which must be measured during movement. Each may involve a separate and distinct neural coordination pattern. Dynamic strength can be described as:

$$F = \frac{2md}{t^2}$$

where F is the force of contraction, m is the mass moved, d is the distance, and t is the time. This approach has been widely used by F.M. Henry and his collaborators at the University of California. In this approach the mass is measured and the distance moved is timed electrically so that F, the force of contraction (or strength, if a maximal effort is made), can be calculated. Although the relationship between dynamic strength and speed is very close, it is hard to define; in calculating a correlation, the results are spurious because distance is usually a constant and mass varies very little. In effect, one would be correlating speed against itself.

Now for the most practical question: Can we improve speed of movement by improving strength? Even though there is no relationship between the static strength level of achievement and speed at any given time, considerable evidence (6, 12, 23) indicates a strong relationship between gain in strength and gain in speed. It has been shown that gains in strength, whether brought about by isometric or isotonic training, are associated with significant gains in speed of movement. The gain in speed has also been demonstrated to result from both strength training that used the same movement as was tested and from training that merely improved the strength of the involved muscles but avoided training in the same movement.

Interestingly, Francis and Tipton (8) found that the knee-jerk reflex time is improved by physical conditioning of the involved muscles. A significant 5% improvement in reflex time was shown after six weeks of weight training, although there appeared to be no correlation between strength and reflex time or between improvement in strength and reflex time. Thus, while the mechanism remains obscure, there seems to be good reason to include strength training in a training regimen for speed. Furthermore, it must be realized that in most speed events there is a period of acceleration to attain maximum velocity as in running the 100-yard sprint. Since acceleration of the body's mass is by definition dependent on strength (force = mass × acceleration) there can be no question of the importance of optimizing strength (force) in the practical coaching situation.

Flexibility and Speed

As we have pointed out, logic would indicate that improvement of flexibility should decrease the negative forces (resistance) involved in running and thus improve the speed. However, experiments by the author—in which speed and oxygen consumption during a 100-yard dash were measured—failed to confirm this hypothesis (5). Even though range of motion was significantly improved, the short-term effect on speed was not significant.

In another experiment, in which the long-term effects of supplementing sprint training with flexibility work and weight training were investigated, it was found that neither weight training nor flexibility added to the gains in speed of the sprint training program. When both were used, however, the gains in speed were significantly better than those by sprint training alone (6).

It must be realized that stopwatch errors in timing a 100-yard dash can be 1%–2%, and the changes brought about by an improved range of motion are not likely to be much larger than this; so the question cannot be considered closed.

Body Mechanics and Speed in Running

This area of human performance has not yet been exhaustively investigated, but some experiments have been performed and their results are interesting.

Bringing about many accelerations and decelerations of the limbs at exactly the right time, at exactly the right rate, and with precisely the appropriate amount of force to run well, obviously requires exquisite neuromuscular coordination patterns. One of the basic questions about running, Is the maximum speed limited by the maximum rate of leg alternation? was answered by a simple but clever experiment by Slater-Hammel (21), who demonstrated that the rates of leg alternation in sprinting were 3.10 to 4.85/sec. Because considerably higher rates are possible in cycling (5.5 to 7.1), he concluded that speed in running is not limited in this way.

In another interesting kinesiological analysis of running, Hubbard (13) demonstrated that improvement results from increasing the length of stride rather than the rate of movement. Applying the formula for dynamic strength,

$F = 2md/t^2$, we see that this requires more dynamic strength because d increases while t and m remain the same; thus F, the force required (strength), must be greater. Again, we see that strength is a factor in speed.

Photographic analyses have shown that efficient running is also characterized by a high knee lift, a long running stride, and placement of the feet beneath the runner's center of gravity (4).

Use of the electrogoniometer (20), which provides electrical recording of joint angle changes, has shown that experienced distance runners increase both stride length and frequency when increasing velocity from running a 440 in 2:12 to running a 440 in 60.9 seconds. Stride length is more important at the lower speeds, while frequency becomes more important at the higher speeds. The only joint function that may become limiting appears to be that of hip flexion since it is the only joint angle that increases markedly at the higher speeds. Thus the track coach would be well advised to include flexibility work in the regimen for distance runners, such as the static stretching techniques described in the following chapter.

Physiological Considerations in the Design of Running Tracks

McMahon (17) at Harvard University studied the physiological bases of running speed from the engineering viewpoint. When the university decided to build a new indoor track (completed in 1977), the Harvard track coach and the planning office sought his advice. The most important question was how much compliance to build into the track surface. On a springy track the time spent rebounding from the surface is increased so that the runner is slowed down. One might therefore suppose that the hardest surface is the fastest, but that did not turn out to be the case. They found in some interesting experiments that if they "tuned" the resiliency (compliance) of the track surface to the elastic and mechanical properties of the human runner, running speed could be increased.

How much compliance should be built into the track? They found that the most useful way to measure running speed (from the engineering viewpoint) was the ratio of step length (distance the body moves forward while one foot is on the ground) to ground contact time. On a tuned track, the ground contact time should be minimized and the step length should be maximized. At a track stiffness of approximately twice that of the runner's legs, the two factors determining running speed come together in optimal fashion; theory dictated enhanced performances of 2%–3%. In actuality, the runner's speed advantage on the new track averaged 2.91% (18).

Unfortunately, the International Amateur Athletic Federation does not recognize indoor records so that we will have to await the building of new outdoor tracks to physiological specifications before we can see the effect on world records. McMahon (18) predicts that the one-mile record could be improved by as much as seven seconds.

Physiology of Training and Conditioning Athletes

Sex Differences in Speed of Movement

In sports such as running and swimming, records for women show their speed to be 85%–90% that of men, but that this reflects true differences in speed of movement per se is questionable. First, it has been shown that speed depends to a large extent upon strength, so that the difference in speed may merely reflect the sex differences in strength. Second, the selection of women for sports represents a much smaller population of athletes, and it is possible that the fastest women athletes have not been found.

Jokl and Jokl (15) have plotted the times for running as they relate to the distance of the event for men and women, using world record times up to 1976. The relationship is simplified by using a double logarithmic plot. Figure 20.3 shows that there is an 11% difference for the 100-meter run, with gradually increasing differences up to 23% slower performance for women on the marathon. It is reasonable to assume that the women athletes are selected from a smaller fraction of their total population, so that no physiological conclusions can be drawn.

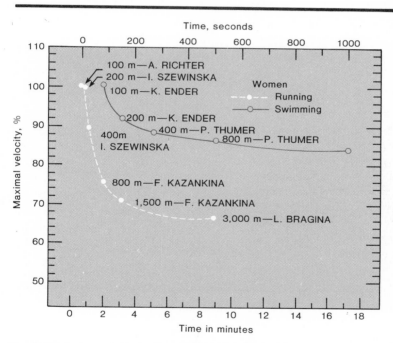

Figure 20.3 Comparison of men's and women's world records in running (100 m-42,200 m). (From Jokl, P., and Jokl, E. *J. Sports Med. Phys. Fitness* 17:213, 1977.)

In controlled experiments, arm speed was found to be 17% slower in women than in men, but when the length of the arm was removed as a factor, the sex difference was only 5%. This is probably a good evaluation of true sex differences in speed of movement.

Variation of Speed with Distance in Running and Swimming

Jokl and Jokl (15) have analyzed and compared world records through 1976 in swimming and running. It is interesting to note that their data (see fig. 20.4) show that the decline in speed with distance is greater in running than in swimming. In terms of duration, the 1,500-meter running event and the 400-meter freestyle swimming event are comparable. But the swim velocity has fallen about 10% from maximal velocity, whereas run time has fallen by 30%. The differences in rates of decline of velocity become greater with increasing distances, and on the whole, runners seem to lose speed at rates almost three times greater than swimmers (15).

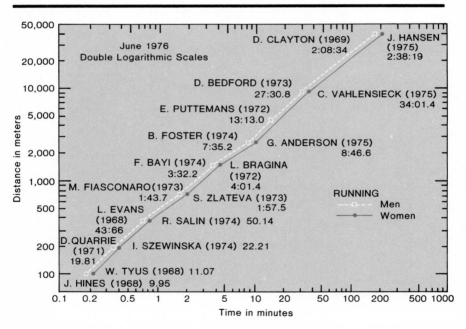

Figure 20.4 Decline of running and swimming speed with distance (women) based on world records in swimming (100 m–1,500 m) and running (100 m–3,000 m), expressed as percentage values of maximal velocity per unit of time, August 1976; 100 m world record = 100% maximal velocity. (From Jokl, P., and Jokl, E. *J. Sports Med. Phys. Fitness* 17:213, 1977.)

Jokl and Jokl explain this difference on the basis of three advantages enjoyed by the heart in swimming over running exercise: (1) weightlessness, (2) horizontal body position, and (3) cooling effect of the water.

At first, one would think that the first two items are really the same in that the horizontal body position and weightlessness both should operate through increasing venous return, but evidence is presented to show that body immersion creates a separate and additive effect upon heart size, over and above that of the horizontal position. In addition to these two factors, the cooling effect of the water should hold the vasodilatation of the skin to a minimum and thus further improve the venous return.

Limiting Factors in Speed

Speed of Single Muscle Contraction. In a simple contraction of a muscle, intrinsic speed of the muscle, which depends on physiochemical properties, is probably most important. Neuromuscular coordination patterns are a smaller, though still important, factor.

A.V. Hill (12) has pointed out the importance of muscle temperature. An animal's muscle contraction can be quickened about 20% by raising its body temperature 2° C. He suggests that such an increase might be brought about in a sprinter by diathermy, and he raises the interesting question whether an athlete might not then do 100 yards in 8.0 seconds.

Speed of Gross Motor Movements. Many important factors act and interact in determining gross motor movements. In lightly loaded and simple movements, the limitations are probably similar to those of a single muscle contraction. In lightly loaded movements of greater complexity, it is likely that ability to coordinate neuromotor patterns would set the upper limits.

In heavily loaded but simple movements, the strength factor is probably dominant. In heavily loaded and complex movements, the limits are undoubtedly set by an interaction of strength and neuromotor coordination.

New Methods for Improving Sprint Speed

In general, the methods for improving sprinting speed fall into one of two categories: (1) *sprint resisted training* in which sprint running is simulated with added resistance, the effort being to improve the dynamic strength factor, and (2) *sprint assisted training* where the effort is directed toward improving the rate of leg alternation. The first method uses devices such as uphill running and weighted clothing. The second method uses downhill running, towing behind an auto at velocities above maximum unassisted, and treadmill running at supramaximal rates (possible because of decreased air resistance). Dintiman (7) has provided an excellent review of the literature in this area for the interested reader.

SUMMARY OF PRINCIPLES FOR COACHING

1. The strength of the prime and assistant movers used in an activity should be developed to an optimum level, preferably by dynamic movements that are closely related to the skill.
2. If speed is desired, the skill should be practiced at rates at least as fast as those to be used in competition. Faster-than-competition rates can be practiced by several different methods. For sprinters, it can be accomplished by downhill running, auto towing, or treadmill running.
3. Flexibility should be improved until range of motion is such as to ensure that no resistance to movement can occur in the skill under consideration.
4. Warming-up should be long and vigorous enough to bring about increased deep-muscle temperature. Ordinarily this will require sweating.
5. A skill should be analyzed on the basis of kinesiological principles, and all improper applications of positive forces should be corrected. Any unnecessary accelerations and decelerations, or movements in the vertical dimension, should be eliminated.
6. If the speed of movement is greater than that of middle-distance running, air resistance can become an important negative factor, and it should be held whenever possible to a minimum; for example, use the crouch position in ice skating and bicycle racing.

REFERENCES

1. Clarke, D.H. Correlation between strength/mass ratio and the speed of an arm movement. *Res. Q.* 31:470–74, 1960.
2. Clarke, D.H., and Henry, F.M. Neuromotor specificity and increased speed from strength development. *Res. Q.* 32:315–25, 1961.
3. Close, R.I. Dynamic properties of mammalian skeletal muscles. *Physiol. Rev.* 52:129–97, 1972.
4. Deshon, D.E., and Nelson, R.C. A cinematographical analysis of sprint running. *Res. Q.* 35:451–55, 1964.
5. deVries, H.A. The "looseness" factor in speed and O_2 consumption of an anaerobic 100-yard dash. *Res. Q.* 34:305–13, 1963.
6. Dintiman, G.B. Effects of various training programs on running speed. *Res. Q.* 35:456–63, 1964.
7. ———. Techniques and methods of developing speed in athletic performance. In *Proceedings of the International Symposium on the Art and Science of Coaching,* eds. L. Percival and J.W. Taylor, vol. 1, pp. 97–139. Willowdale, Canada: F.I. Productions, 1971.
8. Francis, P.R., and Tipton, C.M. Influence of a weight training program on quadriceps reflex time. *Med. Sci. Sports* 1:91–94, 1969.
9. Henry, F.M. Factorial structure of speed and static strength in a lateral arm movement. *Res. Q.* 31:440–47, 1960.

10. Henry, F.M., and Whitley, J.D. Relationships between individual differences in strength, speed and mass in an arm movement. *Res. Q.* 31:24–33, 1960.
11. Hill, A.V. The mechanics of voluntary muscle. *Lancet* 261:947–51, 1951.
12. ———. The design of muscles. *Br. Med. Bull.* 12:165–66, 1956.
13. Hubbard, A.W. An experimental analysis of running and a certain fundamental difference between trained and untrained runners. *Res. Q.* 10:28–38, 1939.
14. Huxley, H.E. Factors limiting the maximum tensions and maximal speed of shortening of muscles. Chap. 7 in *Structure and Function of Muscle,* ed. G.H. Bourne. New York: Academic Press, 1972.
15. Jokl, P., and Jokl, E. Running and swimming world records. *J. Sports Med. Phys. Fitness* 17:213–29, 1977.
16. Kaneko, M. The relation between force, velocity and mechanical power in human muscle. *Res. J. Phys. Educ.* (Japan) 14:141–45, 1970.
17. McMahon, T.A. Using body size to understand the structural design of animals: quadripedal locomotion. *J. Appl. Physiol.* 39:619–27, 1975.
18. McMahon, T.A., and Greene, P.R. Fast running tracks. *Sci. Am.,* 239:148–63, 1978.
19. Nelson, R.C., and Fahrney, R.A. Relationship between strength and speed of elbow flexion. *Res. Q.* 36:455–63, 1965.
20. Sinning, W.E., and Forsyth, H.L. Lower limb actions while running at different velocities. *Med. Sci. Sports* 2:28–34, 1970.
21. Slater-Hammel, A. Possible neuromuscular mechanisms as limiting factors for leg movement in sprinting. *Res. Q.* 12:745–57, 1941.
22. Smith, L.E. Individual differences in strength, reaction latency, mass and length of limbs and their relation to maximal speed of movement. *Res. Q.* 32:208–20, 1961.
23. ———. Influence of strength training on pre-tensed and free arm speed. *Res. Q.* 35:554–61, 1964.
24. Whitley, J.D., and Smith, L.E. Velocity curves and static strength-action strength correlations in relation to the mass moved by the arm. *Res. Q.* 34:379–95, 1963.

21

Flexibility

Flexibility can be most simply defined as the range of possible movement in a joint (as in the hip joint) or series of joints (as when the spinal column is involved). The need for flexibility varies with the athletic endeavor, but in some activities it is all-important. A hurdler must have the best possible hip flexion-hip extension flexibility. In competitive swimming, shoulder and ankle flexibility can be decisive factors. And a diver who cannot execute a deep pike position will never achieve outstanding success.

Even for the armchair athlete, flexibility is important because graceful movement in walking and running are unlikely without it. More importantly, considerable evidence indicates that maintenance of good joint mobility prevents or to a large extent relieves the aches and pains that grow more common with increasing age.

It should be recognized from the outset that flexibility is specific to a given joint or combination of joints. As with speed of movement, an individual is a composite of many joints, some of which may be unusually flexible, some inflexible, and some average. Accordingly, it would be incorrect to speak of a flexible individual (9).

Physiology of Flexibility

What Sets the Limits of Flexibility? For some joints the bony structure sets a very definite limit on range of motion; for example, extension of the elbow joint and the knee joint are limited in this fashion. Also, in a very heavily muscled person it is likely that flexion of the elbow and knee joints is limited by the bulk of the intervening muscle. These are mechanical factors that cannot be greatly modified, and therefore they are only of academic interest.

In such joints as the ankle joint or hip joint, however, the limitation of range of motion is imposed by the soft tissues: (1) muscle and its fascial sheaths; (2) connective tissue, with tendons, ligaments, and joint capsules; and (3) the skin. This, then, is where our interest lies, for these factors are modifiable by physical methods, and they are important factors in human performance.

If a resting excised muscle is stretched passively (no contraction), the greater the length becomes the greater the force required to hold the stretch. It has been shown that this resistance does not lie in the contractile elements of the muscle, but is due almost entirely to the fascial sheath that covers the muscle and the sarcolemma of the muscle fiber (1, 21). Thus it is the fascial investments of muscle tissue with which we are concerned in the pursuit of flexibility.

What are the relative contributions of the various soft tissues (listed above) in limiting our movement? An ingenious experiment by Johns and Wright (10) was directed toward the problem of joint stiffness, and the author has recalculated their data to estimate the percentage contributions of the various

Physiology of Training and Conditioning Athletes

Table 21.1
Estimated Contribution of Various Tissues in Resisting Wrist
Flexion and Extension in the Cat

Tissue	Extension—48° Torque Required		Flexion + 48° Torque Required	
	gram cm	% total	gram cm	% total
1. Skin	−70	11.2	−45*	−8.7
2. Extensor muscles	−35	42.4	0	36.9
3. Flexor muscles	−230		190	
4. Tendon	−70	11.2	170	33.0
5. Joint capsule	−220	35.2	200	38.8
Total	−625	100	515	100

*°Skin aided in flexing the joint.
Calculated from the data of Johns, R.J., and Wright, V.,"Relative Importance of Various Tissues in Joint Stiffness," *J. App. Physiol.* 17:824–28, 1962.

tissues in resisting wrist flexion and extension in the cat (they showed that these functions in a cat are very similar to those of man). Table 21.1 shows the results of these calculations. It can be seen that the most important factors limiting free movement are (1) muscles (and their fascial sheaths), (2) the joint capsule, and (3) the tendons. It must be remembered that these data apply directly only to the wrist joint, but in other joints where ligaments play a more prominent role, as in the ankle joint, these structures are no doubt equally important.

Static versus Dynamic Flexibility (Stiffness). It is obvious that the ability to flex and extend a joint through a wide range of motion (which is measured virtually in a static position) is not necessarily a good criterion of the stiffness or looseness of that same joint as this applies to the ability to move the joint quickly with little resistance to the movement. Range of motion is one factor; the only one that has been widely investigated at this point. How easily the joint can be moved in the middle of the range of motion, where the speed is necessarily greatest, is quite another factor.

We should therefore consider two separate components of flexibility: (1) *static flexibility,* which is what we ordinarily measure as range of motion, and (2) *dynamic flexibility,* which has been investigated in respect to stiffness in joint disease (25), but has been neglected in physical education. It may be hypothesized that the flexibility of motion may be of much greater importance to physical performance than the ability to achieve an extreme degree of flexion or extension of a joint!

The method for such investigations was developed by Wright and Johns (25) for laboratory work, and with their methods the physical factors that contribute to joint stiffness and therefore limit dynamic flexibility have been identified and their relative contributions measured (in finger and wrist joints only). They investigated the effects of *elasticity, viscosity, inertia, plasticity,* and *friction* in normal and in diseased joints. It was found that inertia and friction were negligible; viscosity accounted for only one-tenth the torque used in moving the joint passively; and elasticity and plasticity were the major factors. These forces are wasted on the stretching of connective tissues.

Stretch Reflexes and Flexibility. The stretch reflexes as they apply to the stretching of body components for improving static flexibility were discussed at length in chapter 5, but we remind the reader that a muscle that is stretched with a jerky motion responds with a contraction whose amount and rate vary directly with the amount and rate of the movement that causes the stretch. This is the result of the myotatic reflex that originates in the muscle spindle.

On the other hand, a firm, steady, static stretch invokes the inverse myotatic reflex, which brings about inhibition not only of the muscle whose tendon organ was stretched, but also of the entire functional group of muscles involved. It has been shown, for example, that the amount of *tension increase* for a given amount of stretch is more than doubled by a quick stretch, as compared to slow stretch (22), when the degree of stretching is the same.

Some therapists have suggested the use of tension in the agonist either before or when stretching the antagonist to take advantage of the inhibition brought about by *reciprocal inhibition,* the neuromuscular function that serves to turn off one of a pair of muscles when its opponent is activated in reciprocating type movements. However, a study using EMG techniques recently reported by Moore and Hulton (19) showed that for most subjects the lowest levels of innervation during passive stretching were attained by the static stretching technique. Attempts at implementation of the reciprocal inhibition principle were not so effective in reducing activation as was the static method. Since there were no significant differences in the range of motion attained between static stretch procedures and the others, this work seems to support the use of static stretching as the method of choice.

Measuring Flexibility

Static Flexibility. In general, static flexibility can be measured in two ways: (1) by *goniometry,* the *direct measurement* of the angle of the joint in its extremes of movement, and (2) by *indirect measurement* of joint angles through measurement of how closely one body part can be brought into opposition with another body part or some other reference point.

Physiology of Training and Conditioning Athletes

Goniometry, in its classic form, uses a protractorlike device to measure the angle of the joint at both ends of its movement range, but it suffers from the serious disadvantage that body parts are not regular geometric forms, and a good deal of subjectivity is introduced during measurements in deciding where the axis of a bony lever may be. A simple but ingenious device, Leighton's *flexometer,* overcomes this disadvantage to a large extent (fig. 21.1); it is a small instrument that is strapped onto a body part and records range of motion in respect to a perpendicular established by gravity. Reliability coefficients well above 0.90 have been reported (13, 14).

An indirect method has been devised by Cureton (5) for measuring four types of flexibility: (1) trunk flexion forward, (2) trunk extension backward, (3) shoulder elevation, and (4) average ankle flexibility. The method can be exemplified by a description of the measurement of trunk flexion forward. Subjects assume a long sitting position, with hands behind the neck and feet eighteen inches apart; they then bend downward and forward to place the forehead as close as possible to the floor. The distance from forehead to floor is the score on flexibility for this measure.

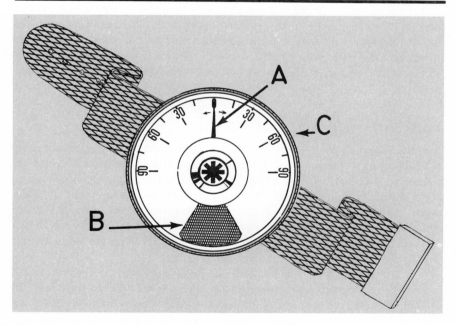

Figure 21.1 Drawing illustrating the principle of Leighton's flexometer. *A* indicates needle; *B* indicates weight that keeps needle vertical; *C* indicates housing that rotates in respect to needle with movement of body part. (From Leighton, J.R. *Arch. Phys. Med.* 36:571, 1955.)

The sit and reach test of Wells and Dillon (24) has been widely used as a test of back and leg flexibility. In the long sitting position, subjects slide their hands forward on a table that is approximately of shoulder height to the limit of their reach. The distance reached by the fingertips is the score on this test, and a reliability of 0.98 has been reported.

Dynamic Flexibility.   This method, developed by Wright and Johns (25) to measure the stiffness of normal and diseased joints, uses laboratory devices to measure the forces (torque) needed to move a joint through various ranges of motion at varying speeds. Although this method has not yet been applied to research in physical performance, it seems to have distinct possibilities for such use. It seems highly probable that this measurement can tell us more about potential performance in speed events than can static flexibility.

Effects of Anthropometric Measurements upon Measurement of Flexibility.   One of the criticisms leveled at tests that use the indirect principle for measurement of static flexibility is that the measurement depends too heavily on anthropometric measurements. For example, it can be argued that an individual with a long upper body and arms and with short legs might have little trouble with such trunk flexion tests as touching the floor with the fingertips.

Several investigators have attacked this problem with respect to men (23), women (2, 8, 16), and elementary school boys (17), but have found no meaningful relationships between static flexibility and various measurements and ratios of body parts. It appears that static flexibility can be measured indirectly, with no undue interference from varying anthropometric measurements.

Methods for Improving Range of Motion

The question of which methods are most advantageous for improving range of motion has received very little attention. The conventional calisthenic exercises used for this purpose have usually involved bobbing, bouncing, or jerky movements in which one body segment is put in movement by active contraction of a muscle group, and the momentum is then arrested by the antagonists at the end of the range of motion. Thus the antagonists are stretched by the dynamic movements of the agonists. Because momentum is involved, this may be called the *ballistic method.*

On the other hand, the methods of Yoga suggested to the author the possibility for better application of the available knowledge regarding the stretch reflexes (see chap. 5). Although the static stretch methods developed by the author (6) are in many cases derived from Hatha Yoga, and both depend on the same physiological principles (though unknown to Yoga), there are also enough differences so that the term *static stretch* was coined to separate the methods. The most important differences are (1) static stretching should be considered a generic term for the use of held stretches to apply neurophysiological principles to strictly physical and physiological uses, whereas Yoga is a system of abstract meditation and mental concentration pursued for spiritual purposes; (2) static stretching is based on neurophysiological principles with better health and performance as the goals, whereas Yoga is based on spiritual principles and pursued as a method of attaining union with the supreme spirit of the universe; and (3) Yogic stretching has been directed largely to the joints and musculature of the trunk, whereas static stretching is equally concerned with the limb muscles (7).

The static stretching method involves holding a static position for a period of time, during which specified joints are locked into a position that places the muscles and connective tissues passively at their greatest possible length. Since the neurophysiology of the stretch reflexes suggests advantages in static stretching procedures, a study was undertaken in the author's laboratory to compare the ballistic and static methods (6). The difference in the two stretching methods is illustrated by figures 21.2 and 21.3. The ballistic exercises were taken—to a large extent—from Kiphuth's (11) exercises for stretching swimmers. The static exercises were developed to best utilize the inverse myotatic reflex and were designed to parallel the ballistic exercises in the affected muscles and joints.

It was found that both methods resulted in significant gains in static flexibility (in seven thirty-minute training periods) in trunk flexion, trunk extension, and shoulder elevation; there was no significant difference between methods. We may therefore conclude that static stretching is just as effective as the conventional ballistic methods, but the former offers three distinct advantages: (1) there is less danger of exceeding the extensibility limits of the tissues involved; (2) energy requirements are lower; (3) although ballistic stretching is apt to cause muscular soreness, static stretching will not; in fact, the latter relieves soreness (see chap. 22).

It is also of interest that the changes brought about by stretching exercises persist for a considerable period of time (eight weeks or more) after stretching is discontinued (18).

Chapman has shown that dynamic as well as static flexibility can be significantly improved by exercise in the old as well as in the young (4). This finding would appear to have some importance in gerontology since declining joint mobility creates many problems for the elderly.

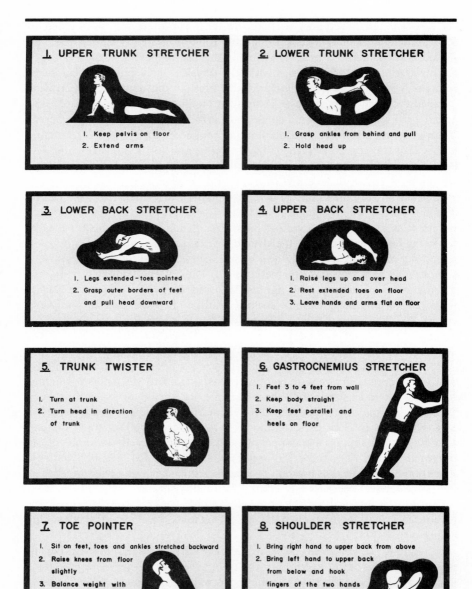

Figure 21.2 Illustration of the static stretching methods used in the author's laboratory.

1. TRUNK LIFTER

1. Hands behind neck
2. Raise head and chest vigorously
3. Partner holds feet

2. LEG LIFTER

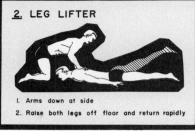

1. Arms down at side
2. Raise both legs off floor and return rapidly

3. TRUNK BENDER

1. Legs apart and straight
2. Hands behind neck
3. Bend trunk forward and down-ward in a bouncing fashion
4. Keep back straight

4. UPPER BACK STRETCHER

1. Legs crossed, sitting position
2. Try to touch head to floor
3. Use vigorous bouncing motion

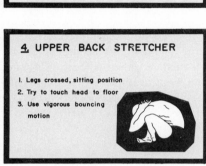

5. TRUNK ROTATOR

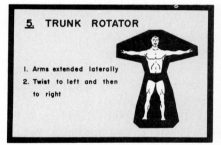

1. Arms extended laterally
2. Twist to left and then to right

6. GASTOCNEMIUS STRETCHER

1. Stand on balls of feet
2. Lower weight and return rapidly
3. Use partner to balance if necessary

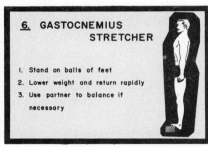

7. SINGLE LEG RAISER

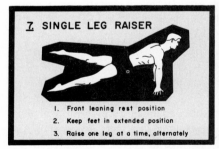

1. Front leaning rest position
2. Keep feet in extended position
3. Raise one leg at a time, alternately

8. ARM AND LEG LIFTER

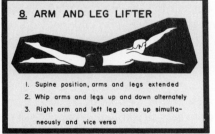

1. Supine position, arms and legs extended
2. Whip arms and legs up and down alternately
3. Right arm and left leg come up simulta-neously and vice versa

Figure 21.3 Illustration of the ballistic stretching exercises against which the static stretching method was compared experimentally.

Weight Training and Flexibility

Many investigators have shown that weight training has no harmful effects upon either speed or range of movement when properly pursued, but an interesting study by Massey and Chaudet (15), which supports these findings, indicates the need for capable guidance. In an experiment designed to evaluate the effects of weight training on range of movement, they found no appreciable effects in general, but they did find a significant decrease in ability to hyperextend the arms at the shoulder joint, a movement for which no exercise had been included. It is very probable that inclusion of an exercise for hyperextension would have prevented this decrease in mobility.

Very heavy resistance exercise can, under certain circumstances, result in a restriction of range of motion. This factor, however, is not inherent in weight training, and it can be prevented by inclusion of the proper exercises and performance thoughout the full range of motion. As Massey and Chaudet pointed out, it appears that weight lifting increases range of movement in the joints that are exercised, but may restrict range of movement in the areas not exercised. Therefore, a well-rounded workout is indicated when heavy resistance methods are used.

Factors Affecting Flexibility

Activity. It has been found that active individuals tend to be more flexible than inactive individuals (18). This is in accord with the well-known fact that connective tissues tend to shorten when they are maintained in a shortened position (as when a broken limb is placed in a plaster cast).

Sex. The results of two investigations agree that among elementary school age children girls are superior to boys in flexibility (12, 20). It is likely that this difference exists at all ages and throughout adult life.

Age. The results of many tests indicate that elementary school age children become less flexible as they grow older, reaching a low point in flexibility between ten and twelve years of age (3, 12, 20). From this age upward, flexibility seems to improve toward young adulthood, but it never again achieves the levels of early childhood. Dynamic flexibility apparently grows steadily poorer, from childhood on, with increasing age (25).

Temperature. Dynamic flexibility is improved 20% by local warming of a joint to 113° F, and it is decreased 10%–20% by cooling to 65° F (23). Experience indicates that static flexibility is probably similarly affected by temperature changes.

Ischemia. Dynamic flexibility is markedly reduced by arterial occlusion for twenty-five minutes (25). The physiology underlying this phenomenon has not been elucidated, but it would appear to have important implications for the study of joint disease.

SUMMARY

1. Two types of flexibility should be recognized: (1) *static flexibility,* a measure of range of motion, and (2) *dynamic flexibility,* a measure of the resistance to motion offered by a joint. (The following principles apply to static flexibility only because dynamic flexibility has not yet received the attention of physical educators.)
2. Flexibility can be limited by bone structure or by the soft tissues. When it is limited by soft tissues, great improvements can be brought about by the proper stretching methods.
3. After improvements have been brought about, cessation of the exercise program is not immediately accompanied by regression of flexibility. The effects of a stretching program are relatively long lasting (at least eight weeks).
4. Stretching by jerking, bobbing, or bouncing methods invokes the stretch reflexes, which actually oppose the desired stretching.
5. Stretching by static methods invokes the inverse myotatic reflex, which helps relax the muscles to be stretched.
6. Static stretching methods have been shown to be just as effective as the ballistic methods.
7. Static stretching is safer than ballistic methods because it does not impose sudden strains upon the tissues involved.
8. Ballistic stretching methods frequently cause severe soreness in muscles. Static stretching does not usually cause soreness; it may, indeed, relieve soreness when it has occurred.

REFERENCES

1. Banus, M. G., and Zetlin, A. M. The relation of isometric tension to length in skeletal muscle. *J. Cell. Comp. Physiol.* 12:403–20, 1938.
2. Broer, M. R., and Galles, N. R. G. Importance of relationship between various body measurements in performance of toe-touch test. *Res. Q.* 29:253–63, 1958.
3. Buxton, D. Extension of the Kraus-Weber test. *Res. Q.* 28:210–17, 1957.
4. Chapman, E. A.; deVries, H. A.; and Swezey, R. Joint stiffness: effects of exercise on young and old men. *J. Gerontol.* 27:218–21, 1972.
5. Cureton, T. K. Flexibility as an aspect of physical fitness. *Res. Q.* 12:381–90, 1941.
6. deVries, H. A. Evaluation of static stretching procedures for improvement of flexibility. *Res. Q.* 33:222–29, 1962.
7. ———. *Health Science: A Positive Approach.* Santa Monica, Cal.: Goodyear Publishing Co., 1979.
8. Harvey, V. P., and Scott, G. D. Reliability of a measure of forward flexibility and its relationship to physical dimensions of college women. *Res. Q.* 38:28–33, 1967.
9. Hupperich, F. L., and Sigerseth, P. O. The specificity of flexibility in girls. *Res. Q.* 21:25, 1950.

10. Johns, R. J., and Wright, V. Relative importance of various tissues in joint stiffness. *J. Appl. Physiol.* 17:824–28, 1962.

11. Kiphuth, R. J. H. *Swimming.* New York: A. G. Barnes & Co., 1942.

12. Kirchner, G., and Glines, D. Comparative analysis of Eugene, Oregon, elementary school children using the Kraus-Weber test of minimum muscular fitness. *Res. Q.* 28:16–25, 1957.

13. Leighton, J. R. A simple objective and reliable measure of flexibility. *Res. Q.* 13:205–16, 1942.

14. ———. An instrument and technic for the measurement of range of joint motion. *Arch. Phys. Med. Rehabil.* 36:571, 1955.

15. Massey, B. H., and Chaudet, N. L. Effects of systematic heavy resistance exercise on range of joint movement in young male adults. *Res. Q.* 27:41–51, 1956.

16. Mathews, D. K.; Shaw, V.; and Bohnen, M. Hip flexibility of college women as related to length of body segments. *Res. Q.* 28:352–56, 1957.

17. Mathews, D. K.; Shaw, V.; and Woods, J. B. Hip flexibility of elementary school boys as related to body segments. *Res. Q.* 30: 297–302, 1959.

18. McCue, B. F. Flexibility of college women. *Res. Q.* 24:316, 1953.

19. Moore, M. A., and Hutton, R. S. Electromyographic evaluation of muscle stretching techniques *Med. Sci. Sports* 10:66, 1978.

20. Phillips, M. Analysis of results from the Kraus-Weber test of minimum muscular fitness in children. *Res. Q.* 26:314–23, 1955.

21. Ramsey, R. W., and Street, S. The isometric length tension diagram of isolated skeletal muscle fibers of the frog. *J. Cell. Comp. Physiol.* 15:11, 1940.

22. Walker, S. M. Delay of twitch relaxation induced by stress and stress relaxation. *J. Appl. Physiol.* 16:801–6, 1961.

23. Wear, C. L. Relationships of flexibility measurements to length of body segments. *Res. Q.* 34:234–38, 1963.

24. Wells, K. F., and Dillon, E. K. Sit and reach, a test of back and leg flexibility. *Res. Q.* 23:115–18, 1952.

25. Wright, V., and Johns, R. J. Physical factors concerned with the stiffness of normal and diseased joints. *Bull. Johns Hopkins Hosp.* 106:215–31, 1960.

22

Physiology of Muscle Soreness— Cause and Relief

Immediate versus Delayed Muscle Pain
Theoretical Bases: Muscle Spasm versus Structural Damage
Attempt at Unification: Practical Aspects for Coach and Athlete
Physiology Underlying Static Stretching
Factors in the Prevention of Soreness
Relief of Muscular Soreness
Severe Muscle Problems

Immediate versus Delayed Muscle Pain

It is common experience that physical overexertion results in pain. In general, two types of pain are associated with severe muscular efforts: (1) pain during and immediately after exercise, which may persist for several hours, and (2) a localized soreness, which usually does not appear until twenty-four to forty-eight hours later.

The first type of pain is probably due to the diffusible end products of metabolism acting upon pain receptors (25, 26), but this is not a very serious problem because it is of short duration and is relieved by cessation of exercise or by short periods of rest.

The second type of pain can become chronic under certain conditions and is at least annoying enough to constitute a deterrent to further exercise. The author has seen *shin splints* develop to such a degree that an athlete has given up athletic activity rather than endure this nagging pain. This localized and delayed muscle soreness, or *lameness*, sometimes called a *myositis*, is often attributed to microscopic tears in muscle or connective tissues. It is this second type of pain that is important and that concerns us in this chapter.

Theoretical Bases: Muscle Spasm versus Structural Damage

There are two major theoretical positions with respect to the causation of delayed muscle soreness; the muscle spasm theory and the structural damage theory.

The hypothesis of torn muscle or torn connective tissues was probably first presented by Hough (13) at the turn of the century. He presented no direct evidence for this theory, nor has anyone since actually demonstrated the existence of torn tissue in relation to this type of soreness. While there is no question that violent trauma can result in the rupture of a muscle and/or connective tissue, most of the exercise that is known to result in soreness does not fall in this category. Furthermore, it seems somewhat illogical to postulate that a tissue has been structurally damaged by the very function for which it is specifically differentiated.

Over the years other investigators have furnished indirect evidence suggesting the possibility of tissue damage (2, 17). For example, Asmussen (2) tested sixteen subjects in a situation where one arm raised a weight by concentric contraction and the other arm lowered it by eccentric contraction. The muscle tension involved was the same, but the production of metabolites should have been much greater in the muscle doing positive work because it has been shown that negative work is several times more efficient (see chap. 19). Thus, if metabolic waste products were responsible for the delayed muscle soreness, as they are widely believed to be for the immediate soreness, then the arm doing positive work should have been the site for the soreness. He found, however, that the muscles doing negative work developed marked

Physiology of Training and Conditioning Athletes

soreness and palpable changes in the muscles, while the muscles doing positive work exhibited fatigue and even exhaustion, a result that hardly ever appeared in the muscles doing negative work. He concluded on this basis that muscle soreness is the result of mechanical stress and not of metabolic waste products (2).

Only very recently has there been stronger evidence presented for the possibility of structural damage. Abraham (1) provided an excellent study of urinary excretion of myoglobin and hydroxyproline (OHP) as related to delayed muscle soreness. Myoglobin was taken to represent muscle tissue damage, while OHP under certain conditions would reflect connective tissue disruption. Myoglobin excretion did not differentiate soreness from nonsoreness, and therefore it appears unlikely that muscle tissue damage is the cause of delayed soreness. On the other hand, Abraham found that OHP levels were significantly higher during maximal soreness, suggesting that this condition may be related to disruption of the connective tissue elements in the muscle. However, as he pointed out, the rise in urinary OHP levels indicated speeded-up collagen degradation, but it might also reflect simply an increased collagen synthesis. It has in fact been shown that a hypertrophying rat muscle may show a 50% greater collagen content within six days (12). In any event, it seems safe to conclude that Abraham's data show a relationship between soreness and alterations in muscle connective tissue.

Early workers had suggested the possibility of imbibition of H_2O and consequent swelling of the tissues as a possible cause for the delayed soreness, but this possibility was ruled out (3), and more recent data confirm the early findings (28).

Williams and Ward (29) studied the hematological changes elicited by prolonged intermittent running where each subject competed in a twenty-four-hour relay, running approximately one mile each hour. Analysis of the blood data suggested the possibility of skeletal muscle tissue damage. Such exercise is both prolonged and severe and may not relate to the milder exercise known to bring about delayed soreness. Furthermore, the race was run under climatic conditions conducive to heat injury, and some of the enzyme changes observed could have been related to this factor.

Thus we must accept the *possibility* of structural damage, most likely in the connective tissues within the muscle, such as the endomysium and perimysium (see chap. 2).

The spasm theory for the cause of soreness is suggested by observation of typical muscle fatigue curves in excised muscles (fig. 4.4). It is readily seen that, in addition to the decrement in amplitude of contraction with increasing fatigue, an increasing inability to achieve complete relaxation is typical, and significantly this may end in contracture. Evidence has shown the same phenomenon to occur in the intact human muscle. Petajan and Eagan (24) interpreted their findings as showing the tendency of the untrained muscle after intense exercise to remain in the contracted condition since the increased intramuscular pressure of exercise limits the availability of factors important to the recovery process. For these reasons, the author has proposed the spasm theory: the delayed localized soreness that occurs after unaccustomed exercise is caused by tonic, localized spasm of motor units.

Theoretical Basis for a Spasm Theory. A rationale based upon considerable physiological evidence can be constructed to support this hypothesis. First, it has been shown that exercise above a minimal level causes a degree of ischemia in the active muscles (11, 27). Second, ischemia can cause muscle pain, probably by transfer of *P substance* (18, 25, 26) across the muscle cell membrane into the tissue fluid where it gains access to pain nerve endings. Third, the pain brings about a reflex tonic muscle contraction, which prolongs the ischemia, and a vicious cycle is born. Evidence has been presented (23) that supports the concept of spasm caused by painful stimuli. This hypothesis agrees with the thinking of medical clinicians, who have suggested that many of the aches and pains of organic disease and anxiety states result from muscle spasm.

From the standpoint of the physical educator and athletic coach, the spasm theory becomes even more attractive in that the vicious cycle hypothesized above has a vulnerable aspect that allows application of simple corrective measures for relief. Competitive swimmers and swimming coaches know that swimmer's cramp (gastrocnemius) is promptly relieved by gently forcing the cramped muscle into its longest possible state and holding it there for a moment. This relief of cramp by stretching has also been demonstrated experimentally (23). It is very likely that the inverse myotatic reflex (see chap. 5), which originates in the Golgi tendon organs, is the basis for this relief.

The first experimental study to test the spasm theory (5) was done on seventeen college age subjects, who did a four-minute standard exercise (designed to produce soreness) that consisted of wrist hyperextension against a resistance of 9½ pounds. Both arms were exercised simultaneously; immediately after exercise, and at intervals thereafter, the wrist flexors and extensors of the nondominant arm were stretched by static methods. The dominant arm, which was not stretched, developed significantly greater levels of soreness for the group. The greatest soreness levels were found twenty-four and forty-eight hours after the exercise. The difference in soreness between stretched and unstretched arms was significant for both of these observations.

During the same period electromyographic equipment was designed to achieve very high sensitivity so that small differences in resting muscle tissue activity could be observed. Use of this instrumentation showed that static stretching markedly reduced resting EMG activity in six of seven subjects who had chronic muscular problems of the shin-splint type (fig. 22.1). Symptomatic relief seemed to parallel lowered EMG values (6). The subject who was atypical showed a marked rise in electrical activity and increased levels of pain. It was hypothesized that in this case structural damage had indeed occurred—a truly ruptured muscle.

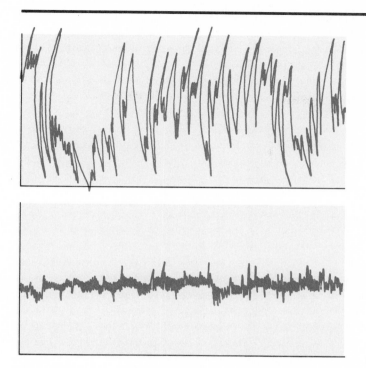

Figure 22.1 Effects of static stretching on muscle soreness of the shin-splint type. Upper trace: electrical activity before stretching. Lower trace: electrical activity after stretching. Symptomatic relief accompanied the decrease in electrical activity.

This introduced the interesting possibility that ruptured muscles could be differentiated from those that are merely in spasm by applying static stretch and observing the EMG changes. To test this hypothesis, eighteen subjects—three of whom had medically verified ruptured muscles—were tested by the same technique. Of the fifteen subjects who showed no evidence of a torn muscle, thirteen had lowered levels of electrical activity after static stretching, and the three subjects with torn muscles showed higher levels (7).

In the most recent experiment (with more sophisticated EMG equipment, described in chap. 4), it has been possible to bring about muscular soreness experimentally and to relieve it by static stretching, with the entire series of physiological changes in electrical state of the muscle under EMG observation (8). Figure 22.2 illustrates the course of events in fifteen subjects (eleven males, four females) who did arm curls (ten sets, with ten repetition maximum) with the right arm. The left arm was unexercised and thus furnished a control for comparison.

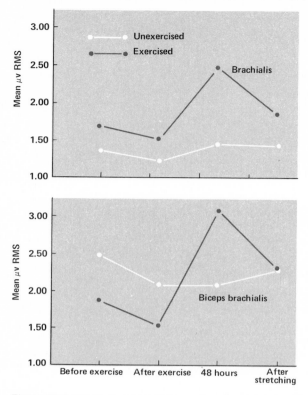

Figure 22.2 Effects of static stretching on experimentally induced soreness. Note that the electrical activity is virtually brought back to presoreness values by the static stretching. Symptomatic relief usually parallels the decreased electrical activity.

The relationship between the soreness that was present after forty-eight hours and the increased electrical activity (evidence of increased muscular activity or local spasm) is clearly shown in figure 22.2. Over the biceps, activity increased 98%, and over the brachialis (this included some biceps activity) 62%, when soreness appeared. The relief of soreness by static stretching is also shown. Immediately after the forty-eight-hour EMG observation, both arms were stretched, and EMG recordings were again taken immediately after the stretching. Again, large decreases in electrical activity paralleled the symptomatic relief.

The rise in electrical activity in the exercised muscle and the relatively unchanged activity in the paired, unexercised muscle is difficult to explain on any other basis than a tonic local muscle spasm.

Physiology of Training and Conditioning Athletes

Abraham (1) attempted to duplicate our EMG experiments, but was unable to find significant EMG changes as the result of soreness induced by methods similar to ours. However, he used bipolar electrodes in his investigation, which we have shown to have less than half the sensitivity of the unipolar lead that we used (9, 21), and furthermore his instrumentation eliminated the frequency spectrum below 30 Hz, which some of our early data suggest may be the most important area for these muscle phenomena (5).

Our EMG data also extended over a four-year period during which we examined thirty-one athletes (8) referred to our lab by coaches, trainers, and physicians, because of muscle injuries. These data are also quite conclusive in showing a very well-defined rise in EMG during the soreness phenomenon, and a return to more normal resting values after treatment as a spasm.

Thus the evidence for increased EMG activity during the delayed muscle soreness phenomenon is strong. In addition, it has been shown that when muscle pain is brought about by saline injection, EMG changes reflect this pain quite faithfully (4).

Attempt at Unification: Practical Aspects for Coach and Athlete

It would appear that the major unresolved issue is whether or not tissue damage is the primary cause in the development of delayed soreness. The best evidence (1) suggests that there are *changes* in connective tissue that do relate to the soreness phenomenon. However, these changes are at least as likely to be the result of the beginnings of hypertrophy and consequent increase in collagen turnover (12) as they are to be structural damage.

It is very difficult to explain the twenty-four to forty-eight-hour delay in appearance of the pain, if indeed there has been structural damage. Why don't we have pain at the time of the trauma? Asmussen has offered a seemingly reasonable explanation:

The injury need not necessarily be actual ruptures for these would probably have happened during actual work and so would most likely have given rise to immediate sensations. Rather, one might imagine, that the high tension has initiated a slow reaction in the connective tissue, causing the development of the actual pain producing condition. It is tentatively suggested that the soreness is caused by a swelling of or in the connective tissue, causing it to be distended at right angles to its natural direction of traction. (2)

Any distention of the tissue is, of course, likely to produce pain, but two investigators have shown that there is no tissue distention, at least in terms of a total limb volume increase, at the time delayed soreness appears (3, 28).

If we modify Asmussen's explanation to involve only localized shortening and thickening of connective tissue and thus no necessary increase in volume, then all experimentally observed facts would be reasonably well rationalized.

Thus we could explain the fact that the fatigued muscle is known to have a shorter resting length; whether we call this a spasm or connective tissue shortening and swelling becomes unimportant. The vicious cycle of the spasm theory postulated by the author, together with Asmussen's explanation, would be the best available explanation for all presently available data. Furthermore, the fact that static stretching provides relief from both traumatic and delayed soreness is also consistent with such a theoretical position. This latter fact is probably most important for physical educators, coaches, and athletes because the static stretching relief of muscle pain rests on a great deal of human experience plus experimental research findings (5, 6, 7, 8, 23). There is no question but that it works, even though the exact physiological mechanism may still be in some doubt.

Physiology Underlying Static Stretching

To test the spasm theory, several different lines of investigation have been pursued in the author's laboratory. First, it was hypothesized that, if the spasm theory had merit, the simple stretching technique that relieves a swimmer's cramp in the calf muscle should also be effective in providing prevention and relief for any sore muscle that can be put on stretch. Therefore a stretching technique was designed to take best possible advantage of the following neurophysiological concepts (see chap. 5).

1. There are two components to the spindle reflex: phasic and static (16, 22).
2. The amount and rate of the phasic response in the spindle reflex are proportional to the amount and rate of stretching (22).
3. The Golgi tendon receptor organs have a relatively high threshold, but when innervated, bring about inhibition of not only the muscle in which the receptors are situated, but also the entire functional muscle group (20).
4. Steady stretch depresses the monosynaptic response, even when the tendon organs are not active (14).
5. The amplitude of EMG in large human muscles characteristically diminishes when they are stretched (15).
6. Reduced amplitude of EMG in large muscles upon stretching is related to tendon organ activity in human muscle (19).

The author's system of *static stretching* was developed around the six concepts outlined above: a body position is held that locks the joints around the sore muscle in a position of greatest possible muscle length and with as little concomitant muscle activity as possible. (Many Yoga exercises have been found useful since they use the same principle.) This procedure results in the least possible reflex stimulation to the involved muscle. A *bouncing*

Physiology of Training and Conditioning Athletes

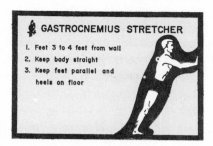

Figure 22.3 Illustration of the static stretching method as applied in the author's laboratory for relief of experimentally caused soreness (and also used for accidental soreness).

stretch, on the other hand, would invoke stretch reflexes whose end result (contraction of the sore, stretched muscle) would be undesirable. The duration used in most cases has been two sets at two minutes each with a one-minute rest intervening. Figure 22.3 illustrates the stretching principle for the gastrocnemius muscle.

Factors in the Prevention of Soreness

The theoretical considerations discussed above allow us to bring theory and practice to bear on the problem of preventing muscle pain.

Warm-up. It has long been the popular opinion of coaches and athletes that *warm-up processes* serve to prevent muscle soreness. Experimental evidence is very meager, however, probably because no one cares to set up an experiment in which the subjects may be injured. All that is known of muscle physiology tends to support the need for warm-up as a protective measure.

In an experiment in which the author investigated the effects of flexibility upon O_2 consumption during 100-yard sprints, subjects ran under conditions of no warm-up, as a control situation, compared with a static stretching flexibility warm-up. In this experiment, two of the four subjects developed sore muscles as the result of running without warm-up. This would certainly seem to support prevailing opinion and theories of muscle physiology on the necessity for warm-up.

Progression in Training Programs. In the many experiments and pilot studies conducted in the author's laboratory, one factor especially stands out in regard to muscle soreness: soreness seems to occur only when large overloads of intensity or endurance are imposed upon an individual muscle. In fact, one of the difficult problems to overcome in setting up systematic experimentation was development of standard exercises that would result in high levels of soreness in large percentages of the subjects. Thus if sore muscles are to be avoided, a *gradual increase* of work load should be planned, so that no one workout represents too great an overload for the physical condition of the musculature.

Types of Activity and Soreness.   Some types of muscular activity are more likely to result in sore muscles than others. Activities most likely to result in soreness are:

1.   Vigorous muscle contractions while a muscle is in a shortened condition; this often results in muscle cramp (23).
2.   Muscle contractions that involve jerky movements. In this case, a muscle is temporarily overloaded when a full load is placed on it before enough motor units have been recruited (see fig. 20.1).
3.   Muscle contractions that involve repetitions of the same movement over a long period of time (endurance imposed on a limited number of muscle fibers). This repetitious movement causes even more soreness if a slight rest interval is allowed between repetitions because the bout can be disproportionately increased in length and a greater total work load is demanded.
4.   Bouncing-type stretching movements. At the end of a ballistic motion, the movement is stopped by the muscle and connective tissues, which brings about reflex contraction at the time the muscle is being forcefully elongated.

Static Stretching.   On many occasions it is impossible to avoid some of the conditions that predispose toward sore muscles, but in such situations a brief (ten-minute) period of static stretching after the workout can bring about a significant degree of prevention. The author has described a well-rounded static stretching program for this purpose elsewhere (10). In any event, application of the principles of kinesiology will enable a professionally trained coach or physical educator to design the exercise for a specific situation.

For example, in a running situation where shin splints may be expected, the muscles involved are the flexors of the ankle joint. Consequently, the athletes are put into a kneeling position, with the ankles extended (plantar flexed), and the full weight of the body is brought to bear—gently—by rocking back onto the ankles. This position is held a minimum of one minute, with *no bouncing.*

Relief of Muscular Soreness

When a muscle becomes painful twenty-four to forty-eight hours (or more) after unaccustomed exercise, relief can usually be provided by the following procedure.

1.   Determine (by palpation) which muscle or muscles are involved.
2.   Determine the nature of the activity that brought about the situation.

3. Determine the muscular attachments of the involved muscle or muscles by consulting a textbook of anatomy or kinesiology.
4. Devise a simple position in which the attachments are held as far apart as possible with the least possible effort.
5. Have the subject hold this position for two-minute periods, with a one-minute rest period intervening. If the pain is severe, this should be repeated two or three times daily.

This procedure has proven effective even in chronic muscular problems (4).

Severe Muscle Problems

None of the foregoing discussion should be construed as suggesting that all painful muscles are due to muscle spasm; nor is it suggested that muscles cannot, under certain conditions, be torn (ruptured). It is obvious that a muscle can be put under such great, sudden strain that some of the tissue exceeds its elastic limits and rupture may occur, but this probably occurs much less often than athletes and coaches seem to think.

In any event, muscular pain that is severe or that persists longer than a few days, should be diagnosed by a physician, as should any muscle injury in which deformation, swelling, or inflammation occurs.

SUMMARY

1. Two types of muscle pain result from overexertion: (A) pain during and immediately after exercise, which is probably due to diffusion of metabolites into the tissue spaces, and (B) a localized, delayed soreness that appears in twenty-four to forty-eight hours.
2. There are at present two major theoretical positions with respect to the causation of delayed muscle soreness, the muscle spasm theory and the structural damage theory.
3. Until recently, only indirect evidence had been presented with respect to the structural damage theory. Recent evidence has shown a relationship between delayed muscle soreness and changes in urinary hydroxyproline. These changes can be explained as the result of either connective tissue growth or disruption.
4. The spasm theory explains the localized delayed soreness as follows: (A) exercise causes localized ischemia; (B) ischemia causes pain; (C) pain brings about greater reflex motor activity; (D) greater motor activity creates greater local muscle tension, which causes ischemia; (E) a vicious cycle is born.
5. The spasm theory is supported by the following evidence: (A) static stretching procedures, which are effective in relieving a cramp, also furnish a degree of prevention against soreness; (B) where muscle soreness exists, electromyography shows markedly higher electrical activity;

(C) when a muscle is stretched, as for relief of cramp, symptomatic relief is usually seen, along with decreased electrical activity; (D) muscular soreness has been produced and relieved under experimentally controlled conditions. EMG observations showed the predicted rise in electrical activity with soreness and its decline with relief.

6. The best explanation of all presently available experimental observation appears to be that offered by a combination of the work of Asmussen and Abraham with the author's spasm theory.

7. Several factors are important in the prevention of sore muscles: (A) proper *warm-up* is essential; (B) workouts should be designed with *progressive* increases in work load; (C) activity that involves vigorous muscle contraction with the muscle in a shortened condition, jerky movements, long-term repetition of a movement, or bouncing-type stretching is most likely to produce muscular soreness.

8. *Static stretching* has been found to be effective in providing both prevention and relief of muscular soreness.

REFERENCES

1. Abraham, W.M. Factors in delayed muscle soreness. Med. Sci. Sports 9:11–20, 1977.

2. Asmussen, E. Observations on experimental muscular soreness. *Acta Rheumatol. Scand.* 2:109–16, 1956.

3. Boyle, R.W., and Scott, F.H. Some observations on the effect of exercise on the blood, lymph, and muscle in its relation to muscle soreness. *Am. J. Physiol.* 122:569–84, 1938.

4. Cobb, C.R.; deVries, H.A.; Urban, R.T.; Leukens, C.A.; and Bagg, R.J. Electrical activity in muscle pain. *Am. J. Phys. Med.* 54:80–87, 1975.

5. deVries, H.A. Electromyographic observations of the effects of static stretching upon muscular distress. *Res. Q.* 32:468–79, 1961a.

6. ———. Prevention of muscular distress after exercise. *Res. Q.* 32:177–85, 1961b.

7. ———. Treatment of muscular distress in athletes. *Proceedings of the 65th Annual College Physical Education Association Meeting,* Kansas City, December, 1961c.

8. ———. Quantitative electromyographic investigation of the spasm theory of muscle pain. *Am. J. Phys. Med.* 45:119–34, 1966.

9. ———. Efficiency of electrical activity as a physiological measure of the functional state of muscle tissue. *Am. J. Phys. Med.* 47:10–22, 1968.

10. ———. *Health Science: A Positive Approach.* Santa Monica, Calif.: Goodyear Publishing Co., 1979.

11. Dorpat, T.L., and Holmes, T.H. Mechanisms of skeletal muscle pain and fatigue. *Arch. Neurol. Psychiatry* 74:628–40, 1955.

12. Goldberg, A.L.; Etlinger, J.D.; Goldspink, D.F.; and Jablecki, C. Mechanism of work-induced hypertrophy of skeletal muscle. *Med. Sci. Sports* 7:185–98, 1975.

13. Hough, T. Ergographic studies on muscular soreness. *Am. J. Physiol.* 7:76–81, 1902.

14. Hunt, C.C. The effect of stretch receptors from muscle on the discharge of motoneurons. *J. Physiol.* 117:359–79, 1952.

15. Inman, B.T.; Ralston, H.J.; Saunders, J.B.; Feinstein, B.; and Wright, E.W., Jr. Relation of human electromyogram to muscular tension. *Electroencephalogr. Clin. Neurophysiol.* 4:187–94, 1952.

16. Katz, B. Depolarization of sensory terminals and the initiation of impulses in the muscle spindle. *J. Physiol.* 111:261–82, 1950.

17. Komi, P.V., and Buskirk, E.R. The effect of eccentric and concentric muscle activity on tension and electrical activity of human muscle. *Ergonomics* 15:417–34, 1972.

18. Lewis, T. *Pain.* New York: Macmillan, 1942.

19. Libet, B.; Feinstein, B.; and Wright, E.W., Jr. Tendon afferents in autogenetic inhibition. *Fed. Proc.* 14:92, 1955.

20. McCouch, G.P.; Deering, I.D.; and Stewart, W.B. Inhibition of knee jerk from tendon spindles of crureus. *J. Neurophysiol.* 13:343–50, 1950.

21. Moritani, T., and deVries, H.A. Reexamination of the relationship between the surface IEMG and force of isometric contraction. *Am. J. Phys. Med.* 57:263–77, 1978.

22. Mountcastle, V.B. Reflex activity of the spinal cord. In *Medical Physiology,* ed. P. Bard, chap. 60. St. Louis: C.V. Mosby Co., 1961.

23. Norris, F.H., Jr.; Gasteiger, E.L.; and Chatfield, P.O. An electromyographic study of induced and spontaneous muscle cramps. *Electroencephalogr. Clin. Neurophysiol.* 9:139–47, 1957.

24. Petajan, J.H., and Eagan, C.J. Effect of temperature and physical fitness on the triceps surae reflex. *J. Appl. Physiol.* 25:16–20, 1968.

25. Rodbard, S., and Farbstein, M. Improved exercise tolerance during venous congestion. *J. Appl. Physiol.* 33:704–10, 1972.

26. Rodbard, S. Pain associated with muscular activity. *Am. Heart J.* 90:84–92, 1975.

27. Rohter, F.D., and Hyman, C. Blood flow in arm and finger during muscle contraction and joint position changes. *J. Appl. Physiol.* 17:819–23, 1962.

28. Talag, T.S. Residual muscular soreness as influenced by concentric, eccentric and static contraction. *Res. Q.* 44:458–69, 1973.

29. Williams, M.H., and Ward, A.J. Hematological changes elicited by prolonged intermittent aerobic exercise. *Res. Q.* 48:606–16, 1977.

23

Warming Up
and Cooling Down

Until recently, the value of warming up had not been challenged. On the basis of theoretical concepts, warming up was accepted by virtually all coaches and athletes; however, much scientific interest has lately been directed toward (1) its value in athletics, (2) elucidation of its physiological nature, and (3) comparisons of the effectiveness of various warm-up procedures.

Unfortunately the various investigators have used different methods, so that the type, intensity, and duration of the warm-ups have varied, as has the physical activity in which the level of performance was to be affected. Consequently the work of the investigators can seldom be compared, and confusion has resulted. Some investigations have been equivocal, some poorly controlled, and still others have used so little warm-up activity (in terms of intensity and duration) that no conceivable physiological changes could have been brought about. On the other hand, because some of the experiments have been properly conducted and are quite definitive in certain respects, we will try to derive some principles by which physical educators and coaches can guide their professional activities.

Practice Effect versus Physiological Warm-Up

A great source of confusion is the fact that the effects of practice in improving a skill are frequently confounded with the actual warming up in which physiological changes are brought about. Unquestionably, if skill and accuracy are important factors in a physical activity, practice can bring about improvement in performance. The question considered in this chapter has to do with the physiological aspects of warming up.

Physiology of Warming Up

On theoretical grounds it might be expected that a warming-up that resulted in increased blood and muscle temperatures should improve performance through the following mechanisms: (1) muscles relax and contract faster, (2) lower viscous resistance in the muscles increases efficiency, (3) hemoglobin gives up more oxygen at higher temperatures and also dissociates much more rapidly, (4) myoglobin shows temperature effects similar to those of hemoglobin, (5) metabolic process rates increase with increasing temperature, and (6) resistance of the vascular bed decreases with increasing temperature.

Gutin and his co-workers have suggested another rationale for the use of warm-up which is based on a mobilization hypothesis (1, 2, 19, 20). They view prior exercise (PE) as a mobilizing stimulus for the systems involved in O_2 transport, thereby allowing the subject to reach a high level of aerobic metabolism more quickly during the subsequent athletic task. This reduces the initial O_2 deficit (see chap. 10, fig. 10.5), thus leaving more of the anaerobic capacity available for later use. The results of their research are discussed later.

General versus Local Heating. Three well-controlled investigations find substantial and significant improvements in performance (1% to 8%) when the entire body is heated so that rectal and muscle temperatures are increased (3, 13, 28). This heating can be accomplished actively by vigorous exercise of various kinds, or passively by hot baths, showers, Turkish baths, or diathermy. However, local heating of only the involved limb has been shown to result in earlier fatigue and lessened work output in that limb (14, 18). It has also been shown that in local heating, the factor of major importance is probably the distribution of blood between the skin and the underlying muscles, if both are served by the same large artery (29, 30).

It seems likely, then, that the explanation for the different effects of local and general heating lies in the fact that in local heating, a large vasodilatation effect is possible in the skin to the detriment of circulation through the underlying muscle. This could well result in the magnitude of decreased performance actually observed. On the other hand, heating of the entire body must exert some, or all, of the beneficial effects stated previously, while the vasodilatation effect on the skin cannot be nearly so large and may not occur to any great extent when a large proportion of the musculature is active.

On the basis of all the available evidence, there seems little doubt that general heating of the body that results in increased core (rectal) and muscle temperatures improves performance.

Recently, theoretical evidence has shown that the effect of temperature on the force-velocity curve (chap. 4, fig. 4.8) is such that a rise in muscle temperature of 5° C should increase velocity and maximum power by about 10% (8). This is in agreement with the empirical evidence from the earlier studies.

Rectal versus Muscle Temperature. Figure 23.1 illustrates the changes in muscle and rectal temperature that occur as the result of warming up by riding the bicycle ergometer at a moderate load. It can be seen that the greatest part of the increase in muscle temperature occurs in the first five minutes, and rectal temperature increases more gradually and steadily for thirty minutes.

Figure 23.2 shows the same temperature data and relates the two temperatures to performance time for a sprint on the bicycle ergometer. Because performance has shown its greatest improvement during the time that muscle temperature has increased markedly and rectal temperature has increased very little, Asmussen and Boje (3) consider muscle temperature to be the more important factor. This contention has been supported by Carlile (13), who found no positive relationships between rectal temperature and swimming times.

More recent work has further corroborated this early work (22, 24), and we may therefore conclude that any procedure that significantly raises deep muscle temperature will improve performance.

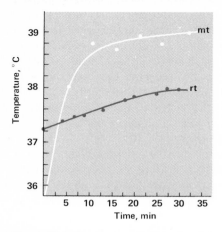

Figure 23.1 Temperature measured in lateral vastus muscle (mt) and in rectum (rt) during a work of 660 kg-m/min. (From Asmussen, E., and Boje, O. *Acta Physiol. Scand.* 10:1, 1945.)

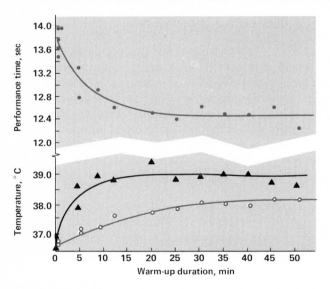

Figure 23.2 Effect of *duration* of warm-up on performance time: ●───● is performance time for sprint, △───△ is muscle temperature, ○───○ is rectal temperature. (From Asmussen, E., and Boje, O. *Acta Physiol. Scand.* 10:1, 1945.)

Physiology of Training and Conditioning Athletes

On the other hand, it has been shown that if a subject is warmed up in such a fashion as to raise rectal temperature, and muscle temperature is allowed to return to normal while rectal temperature is still elevated (rectal temperature returns to normal much more slowly), performance is still somewhat improved over control conditions (28). Thus we must say that both muscle and blood temperatures are important, but muscle temperature probably has the greater effect.

O_2 Consumption and Warm-Up.   It has been shown that maximal oxygen uptake is slightly higher after warming up, compared with cold conditions, but that the O_2 necessary for a given amount of work is reduced (3, 38). This would seem to indicate efficiency is improved as a result of warming up.

The author has attempted to separate the effects of temperature from those of increased mobility (flexibility) in warming up for 100-yard dashes. When flexibility was improved by static stretching so as to eliminate the temperature and circulatory factors, no improvement in efficiency could be demonstrated (16). Thus it seems that the improvement in efficiency is probably temperature-related.

The mobilization hypothesis of Gutin, Andzel, and co-workers has been tested in four different investigations (1, 2, 19, 20). On the basis of all available data it would appear that the hypothesis is tenable, but only with respect to athletic tasks in which the initial work loads are maximal or supramaximal. It is of questionable value in endurance tasks where aerobic capacity is important (20). Prior exercise (PE), the term used by these investigators to distinguish from true temperature warm-up effects, appears to be effective only under very specific conditions: (1) the PE must be approached in very gradual fashion; (2) the PE must be at an intensity below anaerobic threshold (Heart rate = 140 has proven successful); and (3) most important, the beneficial effect is available for only thirty to sixty seconds after the PE. The objective of the PE is to mobilize the O_2 transport function without depletion of phosphagen energy and/or stored O_2 supplies (see chap. 10). Their data suggest that a thirty- to sixty-second rest between PE and the athletic task is sufficient to replenish the O_2 stores and phosphagen energy while still retaining a partial mobilization effect. Rest intervals longer than sixty seconds are ineffective and may decrease subsequent performance.

Blood Flow through the Lungs.   An investigation of the effects of exercise on pulmonary blood flow showed that a period of moderate exercise, such as might be used for warming up, results in a decrease of total pulmonary resistance of about 13% (43). This decrease was highly significant. The reduced resistance to blood flow and its concomitant improvement of lung circulation could make an important contribution to the warm-up phenomenon.

Various Types of Warm-Up

Passive versus Active. In regard to whole-body warm-up and large-muscle activity, there seems little doubt that any procedure that increases rectal and muscle temperatures will improve subsequent athletic performance. This warm-up effect has been demonstrated for active warm-up brought about by such diverse activities as running, bicycle riding, bench-stepping, or calisthenics. Passive heating by hot baths, hot showers, Turkish baths, or diathermy has also been found effective.

Related and Unrelated Methods. *Related warm-up* is any procedure that involves the athletic activity itself, or something close to it. *Unrelated warm-up* is any procedure designed to bring about the desired physiological changes without involving the actual movement itself. Investigations in this area are somewhat inconclusive, but it can be assumed on the basis of common sense that if the desired physiological changes can be achieved by use of related warm-up procedures, these would be preferable to unrelated procedures in that a practice effect would also be gained. In many athletic events, however, the activity is not well suited for warming up (jumping), or it is too fatiguing.

Intensity and Duration of Warm-Up. Burke (12) has demonstrated that optimal combinations of intensity and duration are needed to bring about the desired warm-up effect. Too little work does not achieve optimal levels of temperature, and too much warm-up can result in impaired performance due to fatigue. The interaction of the effects of warm-up and fatigue in untrained young girls is shown clearly in the work of Richards (fig. 23.3). The girls warmed up with varying durations (one to six minutes) of bench-stepping prior to a vertical jump test. It can be seen that the warm-up effect is greater than the fatigue effect when the warm-up is carried out for one to three minutes. Longer warm-up results in more loss to fatigue than is gained in warm-up benefit (34). However, figure 23.4 shows that in a *well-conditioned* athlete, a very heavy load can be used for as long as thirty minutes, with ever-increasing muscle temperature and constantly improving performance. It should be pointed out that for the average schoolboy or poorly conditioned athlete a thirty-minute warm-up at an intensity of 1,600 kg-m/min (fig. 23.4) would result in complete exhaustion.

Bonner (11) has shown that the two-factors theory of Richards (warm-up vs. fatigue effects) also holds for increasing levels of intensity as well as increasing durations of warm-up. For a given athletic task there is a specific combination of intensity and duration that results in best performance.

Obviously, the intensity and duration of warm-up also must be adjusted to the individual athlete. As a rule of thumb, one may look for signs of development of heat from within, and in normal environment this is indicated by perspiration. For those who wish to be more scientific, an increase in rectal temperature of 1° or 2° F appears desirable.

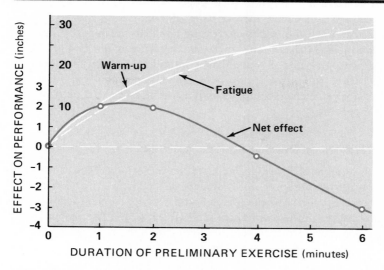

Figure 23.3 Effect of *length* of preliminary exercise on jumping performance. The inner numbers refer to the magnitude of the exponential factors. (From Richards, D.K. *Res. Q.* 39:668, 1968.)

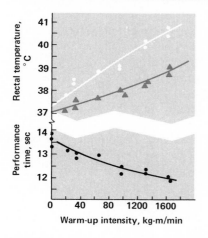

Figure 23.4 Effect of *intensity* of thirty-minute warm-up on performance time: ●———● is sprint performance time, △———△ is rectal temperature, ○———○ is muscle temperature. (From Asmussen, and Boje, O. *Acta Physiol. Scand.* 10:1, 1945.)

Overload Warm-Up. A common practice in baseball is to swing two or three bats in preparation for a turn at bat. Although this is more a practice effect than a true warm-up, it has interesting implications in both a practical and a scientific sense. It has been shown that throwing an eleven-ounce baseball for a warm-up results in significantly improved velocity in subsequent tests with a ball of regulation weight (41). The neurophysiology of this phenomenon has not yet been elucidated, and it is also possible that psychological effects are important.

Effect of Warm-Up on Various Athletic Activities

Speed. A number of investigators have shown various types of warm-up effective in improving the speed of running (9, 38), cycling (3), and arm speed (33). Other investigators, however, have found no improvement from various warm-up procedures in these activities (12, 21, 23, 25). This conflicting evidence leaves the picture rather unclear. Unfortunately, none of the investigators who found no improvement had measured muscle or rectal temperature, so we cannot be sure that a true physiological warm-up had occurred.

On the other hand, the only study that attempted psychological control over the subjects found no improvement, and it is possible that the improvements the other investigators found were caused by psychological factors. Conclusions must await further research.

Strength. An interesting picture emerges in respect to strength. The two investigators who used whole-body warm-up found significant increases in strength after warm-up (3, 12), while the three investigators who applied only local heat found no improvement after warm-up (14, 18, 37). It is therefore tempting to hypothesize that strength changes depend upon central nervous system changes that are brought about by temperature change, or circulatory change, or both.

On the basis of this evidence, plus evidence related to jumping and swimming, it seems that strength can be improved by a general body warm-up, but the explanation of the underlying physiology awaits further investigation.

Muscular Endurance. Asmussen and Boje (3), using whole-body heating, found that increased muscle temperature brought about improvement in times for riding a stationary bicycle. The work load of 9,860 kg-m would take about five minutes, and it must therefore be considered to have an element of muscular endurance.

All other investigators have used only local-muscle warm-up. Their results agree that increasing the local-muscle temperature by warming up, either actively or passively, results in no improvement (36, 37) or in a decrement in endurance (14, 18, 29, 30).

There is also agreement on the finding that local cooling that reduces skin temperature results in better endurance for the underlying muscle groups (14, 18, 29, 30). The rationale for this improvement was discussed earlier in this chapter.

Circulorespiratory Endurance. Grodjinovsky and Magel (17) found that only a vigorous warm-up consisting of five minutes of jogging, eight calisthenic exercises, plus a sprint of one-tenth mile improved time in the one-mile run. Warm-up without the sprint had no significant effect.

Power. One of the best measures of human power is the vertical jump, and complete agreement exists among the four investigations on warming up for jumping (26, 31, 32, 34). Significant improvements, ranging from 2.6% to 20.0%, were found to result from the following warm-ups: massage, running in place, isometric stretching, deep knee bends and stool-stepping. These findings, moreover, could be predicted on the basis of the strength findings, for power is really the expression of strength (force) per unit of time.

Throwing. Because throwing has a strength factor (dynamic strength), this activity would also be expected to show improvement as the result of warm-up, and the three investigations in this area support this contention. Improvement was shown to result from overload warm-up (41), related warm-up (27, 35), and unrelated warm-up (27).

Swimming. All investigations in this area show that swimming times can be improved by warming up. Hot showers of eight minutes' duration resulted in about 1.0% improvement in 40-yard times and 1.5% improvement in 220-yard times (13). Hot baths of from fifteen to eighteen minutes improved performance in the 400-meter freestyle and the 200-meter breaststroke by 2.1% to 3.9%, and in the 50-meter freestyle by as much as 2.0% (28).

Jogging and bicycle ergometer work improved subsequent swim times by 0.6% to 2.2% (28). Short-wave diathermy improved swimming times by 1.3% to 1.9%, and cold baths caused decreased performances by 3.6% to 6.3% (28). One investigator found improvement from a related warm-up (swimming), but not from an unrelated warm-up (40).

The author has attempted to compare the values of commonly used warm-up procedures for highly skilled varsity swimmers for 100-yard times in their specialty strokes. It was found that a 500-yard swim was the only warm-up that brought about significant improvement for the group as a whole (1.0%). It was also found that calisthenic warm-ups produced the best improvement (2.0%) for the butterfly and breaststroke men and that it impaired the performance of the freestylers and backstrokers (15). This phenomenon would seem to point up the need for *individualizing* warm-up procedures.

Duration of the Warm-Up Effect

In some athletic events it is not possible to warm-up after the program has begun—swimming meets in which there is only one pool. A very practical question, then, is how long a warm-up effect persists. This question cannot be answered for the practice effect, but for temperature changes in muscle tissue it has been shown that this effect persists for forty-five to eighty minutes (28, 29).

Recovery between Events

In many athletic competitions such as swimming, track, and field, an athlete competes in more than one event, and the events may be separated by various periods of time. How does the athlete best recover between events to achieve optimal performance in each event?

Physiologically the problem is that the intense muscular activity of the first event results in the production of lactic acid (LA), which inhibits the mobilization of free fatty acids and retards the rate of glycolysis by inhibiting the activity of such enzymes as lactic dehydrogenase and phosphofructokinase. Therefore the removal of LA between events becomes critical.

The breathing of 100% O_2 to enhance lactate removal has been attempted, but two groups of investigators have shown this to be ineffective (39, 42).

In recent years, it has been found that not only is LA taken up by the heart, liver, brain, and kidney, but most important it is taken up by exercising muscle. Therefore one might hypothesize that the LA level could be reduced fastest by exercising the involved muscles during recovery at a rate that would optimize LA uptake by the muscles without producing any LA production (that is, exercising below anaerobic threshold). Recent evidence from four different investigations shows that this is indeed the case (7, 10, 39, 42). Thus exercising during recovery the same muscles used in the preceding event at a rate of 30%–50% of VO_2 max would permit the highest possible skeletal muscle blood flow without producing additional LA. It has been shown that the fastest rate of reduction of LA occurs when exercise is at about 32% of VO_2 max, but even self-selected rates of exercise (about 50% VO_2 max) were almost as effective (7). Thus easy jogging after a running event or easy swimming between swimming events for a period of ten to fifteen minutes should bring about optimal recovery between events.

Warm-Up and Prevention of Muscle Injury

Although there is still some uncertainty about the value of warm-up in improving performance, warming up has been retained as standard practice on the grounds that it might prevent injury to muscles; however, there is no evidence to support this contention. The lack of evidence is understandable:

Physiology of Training and Conditioning Athletes

no investigator would intentionally subject his subjects to experiments designed to bring about injury.

Quite unintentionally, objective evidence has become available in the author's laboratory. In an unrelated study, four college-age male subjects ran 100-yard dashes (against time) to measure metabolic efficiency (16). When the subjects ran without warming up (control procedure), two of them developed muscular soreness that might have become severe in the absence of appropriate preventive measures. Thus it seems that muscle injury is indeed a real possibility when vigorous exercise is not preceded by proper warming up to bring about increased body temperatures.

Warm-Up and Heart Function

The potential for injury to skeletal muscles by vigorous exercise without warm-up has been well recognized. Until recently, however, no one has questioned the effect of strenuous exercise without warm-up on the heart—the most important muscle of all. Barnard and his colleagues at UCLA (4, 5) have conducted two important studies in this regard. In the first study (4) they ran forty-four healthy, asymptomatic individuals, ages twenty-one to fifty-two on a severe ten-second treadmill test without prior warm-up. Immediately after the run, 68% of the men had abnormal ECG changes. When two minutes of easy jogging preceded the sudden strenuous exercise, the abnormal ECG changes were eliminated or reduced in severity in almost all cases.

In a second study (5) they showed that the ECG abnormalities were the result of abnormally large increases in blood pressure that greatly increased the work of the heart (see chap. 6). When their subjects were given a fifteen- to twenty-minute warm-up prior to the sudden exercise, the ECG abnormalities were again eliminated or reduced in almost all cases. They concluded that the adaptation of coronary blood flow to a rapid increase in the work of the heart is not instantaneous and that periods of myocardial ischemia may occur even in normal hearts. These findings alone underscore the need for adequate warm-up before sudden strenuous exercise.

SUMMARY

Although all the results are not yet in for the warming-up phenomenon, intelligent coaches and athletes use the best available evidence to govern their activities, and the best available evidence justifies the following principles for warming up.

1. Whole-body warm-up that raises muscle and blood (rectal) temperatures can significantly improve athletic performance.
2. Whenever possible, a *related warm-up* that raises muscle and blood temperatures is preferred over an unrelated warm-up so that a practice effect may be simultaneously achieved.

3. Warming up is important for preventing muscle soreness or injury.
4. Warming up is most important to protect the heart from ischemic changes that otherwise occur with sudden strenuous exercise.
5. Warming-up procedures must be suited to the individual.
6. Warming-up procedures must be suited to the athletic event.
7. A combination of intensity and duration of warm-up must be achieved that produces temperature increases in the deep tissues without undue fatigue. Sweating is an indication of increased internal temperature. For high level competitive performances, the additional effort of taking the rectal temperature appears worthwhile; an increase of 1° or 2° F is desirable.
8. If active, related warm-up is impossible, passive heating can be used effectively.
9. Warming-up appears to be most important (makes the greatest contribution) in activities that directly involve strength, and indirectly in events that have a large element of power or acceleration of body weight.
10. Overload warm-up may be valuable for events in which neuromuscular coordination patterns are of major importance.
11. Tissue temperature changes brought about by warming up probably persist for forty-five to eighty minutes.

REFERENCES

1. Andzel, W.D., and Gutin, B. Prior exercise and endurance performance: a test of the mobilization hypothesis. *Res. Q.* 47:269–76, 1976.
2. Andzel, W.D. The effects of moderate prior exercise and varied rest intervals upon cardiorespiratory endurance performance. *J. Sports Med. Phys. Fitness* 18:245–52, 1978.
3. Asmussen, E., and Boje, O. Body temperature and capacity for work. *Acta Physiol. Scand.* 10:1–22, 1945.
4. Barnard, R.J.; Gardner, G.W.; Diaco, N.V.; MacAlpin, R.N.; and Kattus, A.A. Cardiovascular responses to sudden strenuous exercise—heart rate, blood pressure and ECG. *J. Appl. Physiol.* 34:833–37, 1973.
5. Barnard, R.J.; MacAlpin, R.; Kattus, A.A.; and Buckberg, G.D. Ischemic response to sudden strenuous exercise in healthy men. *Circulation* 48:936–42, 1973.
6. Barnard, R.J. Warm-up is important for the heart. *Sports Med. Bull.* (ACSM), January 1975.
7. Belcastro, A.N., and Bonen, A. Lactic acid removal rates during controlled and uncontrolled recovery exercise. *J. Appl. Physiol.* 39:932–36, 1975.
8. Binkhorst, R.A.; Hoofd, L.; and Vissers, C.A. Temperature and force-velocity relationship of human muscles. *J. Appl. Physiol.* 42:471–75, 1977.
9. Blank, L.B. Effects of warm-up on speed. *Athletic J.* 10:45–46, 1955.
10. Bonen, A., and Belcastro, A.N. Comparison of self-selected recovery methods on lactic acid removal rates. *Med. Sci. Sports.* 8:176–78, 1976.
11. Bonner, H. Preliminary exercise: a two-factor theory. *Res. Q.* 45:138–47, 1974.

Physiology of Training and Conditioning Athletes

12. Burke, R.K. Relationships between physical performance and warm-up procedures of varying intensity and duration. Doctoral dissertation, USC, 1957.

13. Carlile, F. Effect of preliminary passive warming-up on swimming performance. *Res. Q.* 27:143–51, 1956.

14. Clarke, R.S.J.; Hellon, R.F.; and Lind, A.R. The duration of sustained contractions of the human forearm at different muscle temperatures. *J. Physiol.* 143:454–73, 1958.

15. deVries, H.A. Effects of various warm-up procedures on 100-yard times of competitive swimmers. *Res. Q.* 30:11–20, 1959.

16. ———. The looseness factor in speed and O_2 consumption of an anaerobic 100-yard dash. *Res. Q.* 34:305–13, 1963.

17. Grodjinovsky, A., and Magel, J.R. Effect of warm-up on running performance. *Res. Q.* 41:116–19, 1970.

18. Grose, J.E. Depression of muscle fatigue curves by heat and cold. *Res. Q.* 29:19–31, 1958.

19. Gutin, B.; Stewart, K.; Lewis, S.; and Kruper, J. Oxygen consumption in the first stages of strenuous work as a function of prior exercise. *J. Sports Med. Phys. Fitness* 16:60–65, 1976.

20. Gutin, B.; Horvath, S.M.; and Rochelle, R.D. Physiological response to endurance work as a function of prior exercise. Abstracted in *Med. Sci. Sports* 10:50, 1978.

21. Hipple, J. Warm-up and fatigue in junior high school sprints. *Res. Q.* 26:246–47, 1955.

22. Kaijser, L. Oxygen supply as a limiting factor in physical performance. In *Limiting Factors of Human Performance,* ed. J. Keul. Stuttgart: G. Thieme, 1973.

23. Lotter, W.S. Effects of fatigue and warm-up on speed of arm movements. *Res. Q.* 30:57–65, 1959.

24. Martin, B.J.; Robinson, S.; Wiegman, D.L.; and Aulick, L.H. Effect of warm-up on metabolic responses to strenuous exercise. *Med. Sci. Sports* 7:146–49, 1975.

25. Massey, B.; Johnson, W.R.; and Kramer, G.F. Effect of warm-up exercise upon muscular performance using hypnosis to control the psychological variable. *Res. Q.* 32:63–71, 1961.

26. Merlino, L. Influence of massage on jumping performance. *Res. Q.* 30:66–74, 1959.

27. Michael, E.; Skubic, V.; and Rochelle, R. Effect of warm-up on softball throw for distance. *Res. Q.* 28:357–63, 1957.

28. Muido, L. The influence of body temperature on performances in swimming. *Acta Physiol. Scand.* 12:102–9, 1946.

29. Nukada, A. Hauttemperatur und leistungsfahigkeit in extremitaten bei statischer haltearbeit. *Arbeitsphysiologie* 16:74–80, 1955.

30. Nukada, A., and Mueller, E.A. Hauttemperatur und leistungsfahigkeit in extremitaten bei dynamischer arbeit. *Arbeitsphysiologie* 16:61–73, 1955.

31. Pacheco, B.A. Improvement in jumping performance due to preliminary exercise. *Res. Q.* 28:55–63, 1957.

32. ———. Effectiveness of warm-up on exercise in junior high school girls. *Res. Q.* 30:202–13, 1959.

33. Phillips, W.H. Influence of fatiguing warm-up exercises on speed of movement and reaction latency. *Res. Q.* 34:370–78, 1963.

34. Richards, D.K. A two-factor theory of the warm-up effect in jumping performance. *Res. Q.* 39:668–73, 1968.

35. Rochelle, R.H.; Skubic, V.; and Michael, E. Performance as affected by incentive and preliminary warm-up. *Res. Q.* 31:499–504, 1960.

36. Sedgewick, A.W. Effect of actively increased muscle temperature on local muscular endurance. *Res. Q.* 35:532–38, 1964.

37. Sedgewick, A.W., and Whalen, H.R. Effect of passive warm-up on muscular strength and endurance. *Res. Q.* 35:45–59, 1964.

38. Simonson, E.; Teslenko, N.; and Gorkin, M. Einfluss von vorubungen auf die leistung beim 100 m. lauf. *Arbeitsphysiologie* 9:152–65, 1936.

39. Stamford, B.A.; Moffatt, R.J.; Weltman, A.; Maldonado, C.; and Curtis, M. Blood lactate disappearance after supramaximal one-legged exercise. *J. Appl. Physiol.* 45:244–48, 1978.

40. Thompson, H. Effect of warm-up upon physical performance in selected activities. *Res. Q.* 29:231–46, 1958.

41. Van Huss, W.D.; Albrecht, L.; Nelson, R.; and Hagerman, R. Effect of overload warm-up on the velocity and accuracy of throwing. *Res. Q.* 33:472–75, 1962.

42. Weltman, A.; Stamford, B.A.; Moffatt, R.J.; and Katch, V.L. Exercise recovery, lactate removal and subsequent high intensity exercise performance. *Res. Q.* 48:786–96, 1977.

43. Widimsky, J.; Berglund, E.; and Malmberg, R. Effect of repeated exercise on the lesser circulation. *J. Appl. Physiol.* 18:983–86, 1963.

24

Environment and Exercise

The efficiency of the human organism in various forms of work or exercise may vary between 15% and 40%. This means that, of the energy consumed, only 15%–40% is converted into useful work, and that the remaining energy (60%–85%) is wasted as heat energy. This wasted heat energy must be dissipated; otherwise the body temperature will rise unduly. Furthermore, in a hot climate the body also absorbs heat from its environment. These two factors tend to increase the body heat stores and thus increase body temperature.

Physiology of Adaptation to Heat and Cold

There are four means by which the body can maintain thermal balance by losing heat to the environment.

1. **Conduction.** Heat exchange by conduction is accomplished through contact between one substance and another substance. The rate of exchange is determined by the temperature difference between the two substances and by their thermal conductivities. For example, the body loses heat in this manner when submerged in cold water.

2. **Convection.** In convection, heat is transferred by a moving fluid (liquid or gas). Thus in the example of a man submerged in cold water, the heat that is transferred from the body to the water by conduction is carried away from the body by convection (the water that has been warmed rises, making way for new molecules to be heated by conduction, etc.).

3. **Radiation.** The process of heat transfer by means of electromagnetic waves is radiation. These waves can pass through air without imparting much heat to it; however, when they strike a body, their energy is largely transformed into heat. This is the means by which the sun heats the earth, which also explains why one can be perfectly comfortable in air that is below the freezing point if one receives enough solar radiation. (Skiing in high mountains in subtropical latitudes is an example; the air may be cold due to the altitude, yet the sun's declination is such that it transfers considerable radiant heat.)

4. **Evaporation.** Changing a liquid into a gas is called evaporation, or vaporization, and requires large amounts of heat energy. Thus while one kilocalorie can raise the temperature of one liter of water one degree centigrade, it takes 580 kilocalories to evaporate one liter of water at body temperature. These 580 kilocalories are taken from the surroundings, and this of course is the principle that underlies the operation of a kitchen refrigerator. Human beings function much as refrigerators when they leave a swimming pool and the atmosphere absorbs the water on their skin.

Thus it seems that people's problems in adjusting to their thermal environment are twofold: (1) heat dissipation in hot climates and (2) heat conservation in cold climates. We can gain heat from two sources: environment and metabolism. We can lose heat from one, two, or more of the following factors: conduction, convection, radiation, and evaporation.

Physiology of Training and Conditioning Athletes

Under normal indoor atmospheric conditions, resting individuals maintain body temperature equilibrium within narrow limits. In this situation heat gain is entirely due to metabolism, and heat loss is estimated to occur approximately 40% by convection, 40% by radiation, and 20% by evaporation (insensible perspiration); conduction is usually negligible. Input is balanced by output, and body temperature remains constant—at, or close to, 98.6° F. In fact, within the range of about 30° F to 170° F environmental temperature, the body temperature of a nude human is maintained at a constant temperature within about 1° F of normal resting temperature. This very precise regulatory function is brought about by nervous feedback mechanisms operating through the temperature regulatory center *(thermostat)* in the hypothalamus. The temperature receptors that feed into this thermostat sense the body temperature (1) at the preoptic area of the anterior hypothalamus, (2) in the skin, and (3) probably in some of the internal organs.

When the body temperature is too high, the thermostat in the hypothalamus, having received the error signals from the temperature sensors, increases the rate of heat loss from the body in two principal ways: (1) by stimulating the sweat glands to secrete, which results in evaporative heat losses from the body, and (2) by inhibiting the sympathetic centers in the posterior hypothalamus, thus reducing the vasoconstrictor tone of the arterioles and microcirculation in the skin. This allows vasodilatation of the skin vessels and thus better transport of metabolic heat to the periphery for cooling.

When the body temperature is too low, mechanisms are brought into play to (1) produce more metabolic heat and (2) conserve the heat produced within the body. Heat production is increased by hypothalamic stimulation of (1) shivering, which can increase metabolic rate by two to fourfold, (2) catecholamine release, which increases the rate of cellular oxidation, and (3) the thyroid gland, which increases metabolic rates considerably.

Heat conservation is brought about by vasoconstriction of the skin vessels and abolition of the sweating response.

Exercise in the Cold

Some sports and athletic activities are of necessity carried on in cold environments. Skiing and ice skating depend on snow and ice, and many other sports, such as football and soccer, are occasionally played in very cold weather. A cold environment ordinarily poses few problems for athletes because increased metabolic heat due to the activity soon warms them to a normal *operating temperature,* and heat dissipation to the atmosphere occurs easily by radiation, convection, and when they start sweating, by evaporation.

The chief problem in this situation is to prevent sudden changes in temperature (chilling), and athletic dress is extremely important, especially when there are intermittent periods of activity and rest (as in football). Athletes must be dressed in attire that (1) keeps them comfortably warm while they are waiting to perform and warming up, and (2) can be removed (in part) after warm-up has been accomplished.

It is possible for metabolic rates to increase by as much as twenty-five to thirty times basal values in very vigorous activity. This means that even in the coldest weather (no wind) an athlete has large heat loads to dissipate if a sport is extremely vigorous. Many athletes sweat profusely even in cold environments, and the important consideration is that the clothing worn during actual participation (and after warm-up) be as light as possible in weight and provide as little barrier to passage of water vapor (sweat) as possible. Sweat will otherwise accumulate on the skin or in soaked jerseys, thus leading to chilling in the time between the end of exercise and showering.

Cold Acclimatization. It is well known that continued exposure to cold environments results in greater ability to withstand cold; however, the physiological adjustments are not yet well defined. The most important factor is the maintenance of *core temperature* (rectal temperature, which reflects the temperature of the central nervous system and deep viscera). Core temperature is maintained at a fairly constant 99° F even though skin temperature may fall from its normal average temperature of 92° F to as low as 60° F.

On the basis of the earlier discussion, it is seen that the body can react to cold (1) by reduction of heat loss and (2) by increased metabolism. When a resting and naked human is cooled from a comfortable environment of 85° F to approximately 72° F, no increase in metabolism occurs, and heat is conserved by vasoconstriction of cutaneous blood vessels that prevents loss of the heat carried by the blood from the core. Below 72° F, increased metabolism results from shivering; the involuntary contraction of the muscles in shivering may raise the metabolic rate from two to four times the resting rate.

For these reasons investigators have looked for changes in basal metabolic rates, peripheral circulation, and skin temperature as indicators of acclimatization. The results are controversial. An increased basal metabolic rate (BMR) of 35% in Korean women who dive for commercial purposes in winter water (temperature 50° F) has been observed (17), and Eskimos have been found to have a higher BMR than Caucasians: 46 kcal/M^2/hr compared with 37 kcal/M^2/hr (22). However, experiments on personnel during expeditions into antarctic regions have failed to find significant BMR changes.

Local adaptation to cold has been shown in the fishermen of Gaspé Bay, Canada, who gave lower pressor responses to immersion of hands and feet in ice water than did controls (21). It is interesting to note that hypnosis suppressed shivering, lowered the heart rate, and improved vigilance-task performance significantly over the controls during cold exposure at 40° F (19).

It seems likely that adaptation to cold is comprised of physiological and psychological factors, and it may well be that the interaction between the two—as well as the type of physiological changes—can vary from individual to individual.

Physiology of Training and Conditioning Athletes

Human Limitations in Cold Environments. Ability to withstand cold environments varies widely with individuals. Truly remarkable resistance to cold has been claimed by some of the adherents of religions that practice religious pilgrimages, such as Yoga. In one such pilgrimage, which was observed under scientific conditions (26), a Nepali pilgrim was uninjured by four days of exposure at 15,000 to 17,000 feet, where temperatures ranged between 5° F and 9° F at night, although he wore only light clothing and no shoes or gloves. It was found that his resistance to cold depended upon elevated metabolism.

Body build and tissue proportions are important factors in determining an individual's ability to withstand cold. Other things being equal, the more rotund *(endomorphic)* a person, the less surface area he has in relation to volume (mass of tissues); consequently, heat loss occurs at a slower rate than in a person of angular *(ectomorphic)* build. Furthermore, fat tissue is an excellent insulator against heat loss. These two factors make the round, fat person better able to withstand cold; conversely, the tall, thin person is better able to dissipate heat (and remain cool) in a hot climate.

Effect of Ice-Cold Showers. A shower of ice-cold water (32° F to 35° F) over the chest causes large increases in systolic and diastolic arterial pressures (18); it also causes increased pulse pressure, heart rate, and cardiac output. These changes are not necessarily dangerous in themselves for healthy individuals, but they would almost certainly be hazardous for anyone with impaired cardiac performance or in the presence of circulatory overload. Whether the same experimentally observed cardiovascular changes can be extrapolated to a mildly cold shower (60° F to 70° F) is not yet known.

Exercise in the Heat

Exercise in hot climates is a more serious problem than exercise in the cold. Whereas in a cold climate the increased metabolic heat production combats the increased heat loss to the environment, in a hot climate metabolism and environment combine to increase heat gain in body tissues. The problem is further complicated by the fact that when environmental temperature approaches skin temperature (approximately 92° F), heat loss through convection and radiation gradually comes to an end, so that at temperatures above skin temperature the *only* means for heat loss is *evaporation of sweat.* Radiation and convection reverse their direction and add heat to the body.

Sweating, then, is the only avenue for heat loss at temperatures above skin temperature, and it is the most important avenue at temperatures that approach skin temperature. At this point it is most important to understand that the mere process of sweating is not in itself effective in dissipating heat; *liquid sweat must be converted to a gas by evaporation before any heat loss occurs.* Sweat that merely rolls off is virtually ineffective, but large heat losses can result when the weather is so dry that the liquid evaporates from the skin rapidly. Under such conditions sweating is imperceptible. For these reasons, exercise in the heat will be discussed as two separate and distinct environmental problems: hot and dry environment, and hot and humid environment.

Hot, Dry Environment. When a person works or plays in a hot and dry environment, cooling of the skin is brought about by evaporation of sweat; there is no problem because dry air can absorb considerable moisture before becoming saturated. Cooling the skin is not the desired end result, however; it is the *internal environment* that must be cooled at all costs. To retain a normal core temperature, heat must be transported from the core to the skin, and this requires adjustments from the normal, resting circulatory state. As we discussed earlier (chap. 7), the arteriovenous anastomoses of the microcirculation open up, along with precapillary sphincters, to increase flow through the skin and subcutaneous tissues. This results in greater volumes of slow-moving blood in and close to the skin for better transfer of heat to the evaporative surfaces, and thus in better cooling.

Along with the improved cooling, however, it must be noted that the volume of the circulatory system has increased by a considerable amount. Under these conditions, venous return to the heart is somewhat impaired, and this results in a decreased stroke volume (in accord with Starling's law). To maintain a constant cardiac output for the demands of both exercising muscles and skin circulation, the heart rate must increase (28,32). Because increases in rate depress cardiac efficiency, exercise at temperatures close to or above skin temperature can impose very severe loads upon the cardiovascular system, even when the air is relatively dry.

Since the entire process of heat dissipation now depends upon elimination of water in perspiration, it is obvious that dehydration is a distinct possibility. How important this factor may be has been pointed out by Adolph (1) and his associates, who note that a man walking in the desert (temperature 100° F) will lose approximately one quart of water per hour. Furthermore, their extensive desert experimentation indicates that voluntary thirst results in adequate water replacement during rest but *not* during work or exercise.

Hot, Humid Environment. When the air surrounding an individual is not only hot but is also loaded with moisture, evaporative cooling is impaired because evaporation cannot take place unless volumes of air are available to take up the water vapor given off. To illustrate this, let us take the extreme example where the air is completely saturated (100% relative humidity) and the air temperature is higher than the skin temperature. Under these conditions no heat dissipation can occur; consequently, the metabolic heat accumulates and raises body temperature, until death ensues (108° F to 110° F).

It may therefore be concluded that the problems in a hot, dry atmosphere are related to increased cardiovascular loads and dehydration if water intake is insufficient. In a hot, humid climate the same problems exist, and are aggravated by a lessened ability to unload water vapor into an already loaded ambient atmosphere. These facts are illustrated in figure 24.1, where the hot,

Physiology of Training and Conditioning Athletes

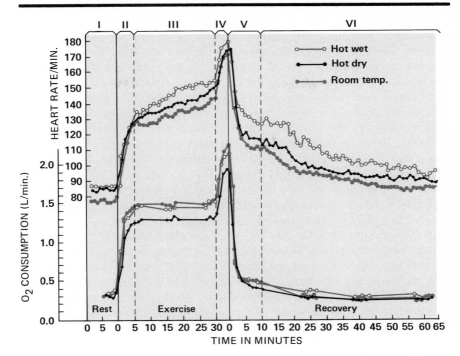

Figure 24.1 Heart rate and oxygen consumption of male subjects pedaling a bicycle ergometer at 540 kg-m/min during phases II and III, and at 720 kg-m/min during phase IV, in the different environments. (From Brouha, L.A., "Effect of Work on the Heart," chap. 21 in *Work and the Heart*, Rosenbaum, F.F., and Belknap, E.L., eds., 1959. Courtesy of Harper & Row, Publishers, New York.)

wet environment is 90° F and 85% relative humidity, the hot, dry environment is 100° F and 25% relative humidity compared with the normal or control environment (room temperature) of 72° F and 42% relative humidity. It is clear that although the temperature is lower in the hot, wet situation, it is considerably more stressful in terms of heart rate response than the hot, dry climate.

Human Limitations in the Heat

The combination of hot weather and strenuous physical activity resulted in almost 200 deaths from heatstroke in recruits at training centers in the U.S. during World War II. In just one summer (1952) there were approximately 600 heat casualties at one Marine Corps recruit training center (23). From August 1959 to October 1962, twelve heatstroke deaths were reported in football players, seven in high school and five in college (13). These statistics do not appear to be improving and show the need for greater familiarity among physical educators and coaches with the physiological effects of combinations of heat stress and physical activity.

First, we need to define the problem. Unfortunately, we cannot evaluate the heat stress of any given situation by simply reading the thermometer because, as has been pointed out earlier, the transfer of heat into or out of the body depends on the balance of heat gain from metabolism and environment against heat loss to the environment. Thus we need information regarding not only temperature, but also humidity, air movement, and heat gain from solar radiation. What is really needed is one index that is sensitive to all of the above factors to give us an *effective temperature* that tells the whole story of heat stress. Such an index was developed by Yaglou (33) in 1927 and has gained wide usage in industry and in the military. His effective temperature, or ET, was defined as that temperature with 100% relative humidity and still air that brings about an equivalent physiological response to the environment under observation. It was later corrected for radiation effect and was then called *corrected effective temperature,* or CET. Estimation of CET requires the reading of three instruments, *dry bulb, wet bulb,* and *globe* thermometers, plus the necessary calculations to arrive at CET. Botsford developed a new instrument (5), the *wet globe thermometer,* which exchanges heat with the surroundings by conduction, convection, evaporation, and radiation essentially the same way a perspiring human does (see fig 24.2); so the one reading, temperature of the globe without further calculation (WGT), provides a com-

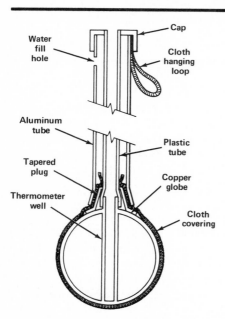

Figure 24.2 Sectional sketch showing construction of the wet globe thermometer (not to scale). (From Botsford, J.H. *Am. Ind. Hyg. Assoc. J.* 32:9, 1971.)

prehensive measure of the cooling capacity of the work environment.* This instrument provides a simple readout and consequently should be placed in every institution where heat stress can conceivably become a problem in the conduct of physical education or athletics. Figure 24.3 shows Botsford's compilation of data for maximal allowable WGT at various metabolic rates with the author's extrapolations to caloric value of various athletic activities.

One note of caution is advisable in the use of such an approach. All of the data on which figure 24.3 is based were taken on subjects who wore no artificial barriers to vapor loss during sweating. The suit of armor worn by football players creates a great load by virtue of its weight, and more important, it cuts off about 50% of the player's evaporative surface area. This was shown to increase the sweat loss by 78%. Thus the uniform both adds to the metabolic heat load and simultaneously prevents transfer of the heat away from the body (13).

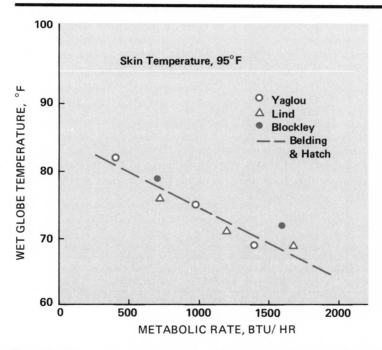

Figure 24.3 Maximum wet globe temperatures for continuous work according to various authorities. Approximate metabolic rates would be about 500 Btu or 125 kcal/hr for standing at ease, 1,000 Btu or 250 kcal for light calisthenics, 1,500 Btu or 375 kcal for playing singles tennis, and 2000 Btu or 500 kcal for heavy activity in football, basketball, or handball. (From Botsford, J.H. *Am. Ind. Hyg. Assoc. J.* 32:1, 1971.)

*This instrument is now commercially available from Howard Engineering Co., P.O. Box 3164, Bethlehem, Pa. 18017

To emphasize this point, the author has used the temperature and humidity data relating to the football heatstroke deaths reported by Fox and co-workers (13) to estimate the WGT at the time of each football fatality. Since only temperature and humidity were available, the estimates were based on an 8-mph breeze and no radiant effect.

Figure 24.3 suggests a maximal WGT of about 64° F for a metabolic rate of 2,000 Btu/hr which probably is a good approximation of the metabolic rate during football. But table 24.1 shows that only two of the nine fatalities occurred above that level of heat stress. This suggests (1) that the heat stress was considerably greater for uniformed football players than for equivalent metabolic levels of lightly clothed individuals on whom the data were taken, and (2) that the data of figure 24.3 must be used very conservatively with respect to football players.

The feasibility of reducing heat stress casualties through enlightened control of activity when heat loads are hazardous has been shown in the Marine Corps recruit training program (23). They reduced the weekly heat casualty rate from 12.4 to 4.7 per 10,000 recruits by instituting a program involving:

1. Curtailed activity when heat loads were high.
2. Gradual breaking-in period for the first week or two.
3. Increased emphasis on physical fitness.
4. Allowing water ad libitum.
5. Replacement of salt.
6. Loosening of uniform regulations to allow T-shirts, etc.

We can certainly make no less effort on behalf of our athletes. This area of concern is so important that the American College of Sports Medicine has seen fit to issue a position statement regarding distance running in the heat (2), which is reprinted as table 24.2.

Table 24.1
Estimations of the ET and WGT at the Time of Each Football Fatality

Football Fatality	Dry Bulb Temp °F	Wet Bulb Temp °F	Relative Humidity	Effective Temp (ET)	Wet Globe Temp (WGT)
1	64	64	100%	52	52
2	90	75	50	77	67
3	85	76	62	74	64
4	75	67	75	63	59
5	82	73	68	71	62
6	85	71	50	72	63
7	83	72	60	71	62
8	93	76	45	79	68
9	81	75	78	71	62

From Fox, E.L., et al. *Res. Q.* 37:333, 1966.

Table 24.2
The American College of Sports Medicine Position Statement on Prevention of Heat Injuries During Distance Running

Based on research findings and current rules governing distance running competition, it is the position of the American College of Sports Medicine that:

1. Distance races (> 16 km or 10 miles) should *not* be conducted when the wet bulb temperature–globe temperature* exceeds 28° C (82.4° F).

2. During periods of the year when the daylight dry bulb temperature often exceeds 27° C (80° F), distance races should be conducted before 9:00 A.M. or after 4:00 P.M.

3. It is the responsibility of the race sponsors to provide fluids which contain small amounts of sugar (less than 2.5 gm glucose per 100 ml of water) and electrolytes (less than 10 mEq sodium and 5 mEq potassium per liter of solution.)

4. Runners should be encouraged to frequently ingest fluids during competition and to consume 400–500 ml (13–17 oz.) of fluid 10–15 minutes before competition.

5. Rules prohibiting the administration of fluids during the first 10 kilometers (6.2 miles) of a marathon race should be amended to permit fluid ingestion at frequent intervals along the race course. In light of the high sweat rates and body temperatures during distance running in the heat, race sponsors should provide "water stations" at 3–4 kilometer (2–2.5 mile) intervals for all races of 16 kilometers (10 miles) or more.

6. Runners should be instructed in how to recognize the early warning symptoms that precede heat injury. Recognition of symptoms, cessation of running, and proper treatment can prevent heat injury. Early warning symptoms include the following: piloerection on chest and upper arms, chilling, throbbing pressure in the head, unsteadiness, nausea, and dry skin.

7. Race sponsors should make prior arrangements with medical personnel for the care of cases of heat injury. Responsible and informed personnel should supervise each "feeding station." Organizational personnel should reserve the right to stop runners who exhibit clear signs of heat stroke or heat exhaustion.

It is the position of the American College of Sports Medicine that policies established by local, national, and international sponsors of distance running events should adhere to these guidelines. Failure to adhere to these guidelines may jeopardize the health of competitors through heat injury.

*Adapted from Minard, D. Prevention of heat casualties in Marine Corps recruits. *Milit. Med.* 126:261, 1961. WB-GT = 0.7 (WBT) + 0.2 (GT) + 0.1 (DBT) From *Med. Sci. Sports* 7:vii, 1975.

Effects of Age, Sex, and Obesity. *Age.* It has been shown that prepubertal girls (11,15) and prepubertal boys (15) have less tolerance for exercise in the heat than do adults. This appears to be true even when subjects are matched for aerobic power and working at the same proportion of their maximal capacity (11). There is also evidence that elderly people do not respond as well to heat exposure as do young people (12,30). Although there seems little doubt that children and older adults are more susceptible to heat stress, the physiological mechanisms responsible for their lower tolerance are not fully understood. Drinkwater and Horvath (12) suggest that for children the greater heat risk is due to the instability of an immature cardiovascular system, while for the elderly the problem is more likely related to their decreased aerobic capacity.

Sex. The available evidence suggests that at all ages females are less heat tolerant than are males, although the difference may not be great (16,30).

Obesity. As might be expected, exercise in the heat is more stressful for the obese than the lean individual (16). This can be explained on a simple dimensional basis. Heat production is related to the *volume* of metabolizing tissue, which is a cubic function. The ability to dispose of the heat is related to the skin surface *area,* which is only a square function, and therefore the rounder (the more endomorphic) an individual becomes, the greater the difference in growth of volume in proportion to skin area and therefore the poorer the capacity for heat dissipation.

Acclimatization to Hot Environments

In these days of rapid transportation, individual athletes and whole teams frequently travel far enough for their competitions to encounter a severe climatic change. If an athlete goes from a cold to a hot climate, this will entail a considerable decrement in performance if the event involves heavy demands on the cardiovascular system.

For these reasons, coaches and exercise physiologists have explored the possibility of bringing about improved heat tolerance by physical conditioning in a normally cool environment. Considerable controversy has been generated with respect to the magnitude of the potential for such an acclimatization procedure, but there is little doubt about the fact that heat tolerance can be improved to some extent by conditioning alone. Thus appropriate conditioning can improve the function of the sweating mechanism and expand the plasma volume (29). Also the sensitivity of the sweating response is increased, so the sweating occurs at lower skin and core temperatures. Thus the trained individual stores less heat in the transient phase of starting to exercise in the heat, arrives at a thermal steady state sooner, and maintains a lower internal temperature at equilibrium (24). In the long term then, the results of physical conditioning for heat acclimatization per se are cooler skin and core temperatures, which in turn reduce the level of skin blood flow needed for regulation of body temperature (27). This results in a greater portion of the cardiac output being available to muscle blood flow, which is what improves performance in the heat.

The controversy with respect to the magnitude of the acclimatization effect through training in a cool environment has probably resulted from three factors: (1) inadequate controls of the physical characteristics of subjects by some investigators, (2) differences in the heat tolerance tests used, and (3) differences in the intensity and duration of training used (14).

When these problems are resolved, there appears to be reasonable agreement that the best improvement in heat tolerance results from intensive interval or continuous training at intensities greater than 50% $\dot{V}O_2$ max for eight to twelve weeks (14,25). Utilization of such procedures appears to produce about 50% of the total adjustment resulting from actual heat acclimatization (25).

Thus the evidence seems clear cut. When competition is scheduled for a different, hot climate, artificial acclimatization is a must for preventing serious decrements in performance.

Even after full acclimatization has been brought about, two other precautions should be followed for maintaining optimum health and performance in hot climates. First, and most important, athletes must maintain adequate intakes of water. The simplest way to check this is by recording weight records under consistent conditions. Dehydration shows up quickly as a loss in weight, and any weight loss of more than two or three pounds should be corrected by encouraging fluid consumption. Second, athletes should be encouraged to decrease the protein in their diet because the specific dynamic action of protein digestion causes more heat formation than the other foodstuffs. Inclusion of more foods with high water and mineral content, such as fruits and salads, is also advisable. Recent evidence (20) also suggests that ingestion of 250–500 mg of vitamin C daily increases heat tolerance.

Fluid and Electrolyte Replacement

We are indebted to Costill and his co-workers at Ball State University for considerable improvement in our understanding of the needs for fluid and electrolyte replacement in athletes who must train or compete in the heat.

Until very recently, it had been believed that the loss of electrolytes such as sodium, potassium, and chloride in heavy sweating had to be replaced by taking salt pills or using various "athletic drinks" that included these electrolytes in their makeup. However, Costill and colleagues (7) have shown that such practice is of minimal value for athletes who are losing water at a rate of 3% of body weight daily or less, if they are permitted to eat and drink ad lib. Indeed, their experimental subjects accumulated even more sodium ions when rehydrating with water than with electrolyte solutions. However, in some occupational and athletic situations, it is possible to lose more than 8% of body weight by heavy sweating, and it seems unlikely that such losses could be replaced without electrolyte supplementation. Losses of fluids and electrolytes up to 3% body weight are made up by normal mineral ingestion in food, together with a greatly reduced rate of excretion by the kidneys and the formation of a hypotonic sweat (much lower salt concentration).

Another important fact is that little exchange of water occurs in the stomach, and therefore the rate of movement of fluids from the stomach to the intestine is very important in fluid replacement. Many factors can effect the rate of gastric emptying, the more important being the volume, temperature, and sugar content of the drink. Costill and Saltin (6) showed that a drink volume of about 400–600 ml leaves the stomach more rapidly than smaller volumes and that a cold drink (5° C) leaves more rapidly than a warm one (35° C). Most important, they found that heavily sugared drinks in combination with high intensity exercise (over 70% $\dot{V}O_2$ max) may combine to block gastric emptying. It should be realized, however, that relative needs for fluids and carbohydrates vary greatly with the athletic event and climatic conditions. In mild exercise in severe heat, fluid replacement is the major concern, and therefore the sugar should be minimized to assure good gastric emptying, but in very heavy exercise over long durations in a cool environment (for example, distance running), maintaining the energy stores may become more important than gastric emptying rate if dehydration is no longer a problem.

These findings seem to cast considerable doubt on the value of the commercially available athletic drink. A direct comparison of three of the more popular drink brands against plain water showed that none was as effective in gastric emptying as water and one that was heavily sugared significantly slowed gastric emptying (9).

Costill suggests the following guidelines for hot weather competition and training (8):

1. *Hot-weather competition*—Use cold, palatable drinks in volumes of 3 to 10 ounces that are low in sugar concentration (less than 2.5 gm/100 ml of water).
2. *Precompetition*—Drink 13.5 to 20 ounces (two to three glasses) of the above drink thirty minutes before the competition starts.
3. *During competition*—Drink 3 to 6 ounces at ten- to fifteen-minute intervals.
4. *Postcompetition*—Salt foods moderately and use fruit juices such as orange juice and tomato juice to replace electrolytes lost in sweat.

Use of the above guidelines, together with the monitoring of the athlete as suggested by the author in chapter 28, should protect the athlete who must train or compete in the heat.

Exercise at High Altitudes

Man's travels to find athletic competition often involve not only changes in temperature and humidity, but large changes in altitude as well. It has been known since the turn of the century and the advent of aviation that whenever humans ascend to higher altitudes they encounter lower atmospheric pressures. Because oxygen maintains a constant 20.93% of decreasing total pressure regardless of altitude, a gradually decreasing partial pressure drives oxygen into the blood. This decreasing availability of oxygen to the tissues would be expected to hamper physical performance, and indeed it does.

The O_2 saturation of arterial blood at sea level approaches 100%, even under conditions of exercise, but at 19,000 feet saturation is only 67% at rest, and exercise at this altitude may drop the value below 50% (31). We do not have to go to this extreme altitude, however, to find changes that may be of great importance in athletic competition. At 3,000 feet even acclimatized subjects have lost 5% of their aerobic capacity, and at 6,500 feet, 15% (3).

This decreased aerobic capacity (maximum O_2 consumption) is brought about by a combination of factors, probably the most important of which are reduction in O_2 saturation of arterial blood, decreased cardiac output, and the higher cost of increased lung ventilation. The impaired lung diffusion is the result of the lowered O_2 pressure gradient. The decreased cardiac output is undoubtedly due to the hypoxic myocardium. The lung ventilation is increased progressively with altitude. All this results from the necessity of breathing more air to attempt to get the same number of molecules of O_2. The increased effort of the respiratory muscles increases their O_2 consumption and lowers their efficiency.

Limitations in Performance at High Altitudes.   Not all athletic performances suffer because of the hypoxia of higher altitudes. Obviously, athletes participating in *one maximal effort* activities, such as the shot put, broad jump, and high jump, do not suffer because they do not depend on O_2 transport. Furthermore, events of less than one minute's duration, such as the 100- and 220-yard dashes, are also performed very largely anaerobically. Consequently performances are unimpaired, but recovery times are longer.

In any event that lasts one minute or more, aerobic capacity is more important, and this importance increases as duration increases (see fig. 18.8). Considerable losses in performance may be expected in such events unless athletes have had adequate time for acclimatization.

Acclimatization.   The need for artificially increasing the available oxygen at higher altitudes has been recognized by the U.S. Air Force. In aircraft that are not pressurized, regulations require breathing *aviator's oxygen* at altitudes above 10,000 feet. Experiments in low-pressure chambers (to simulate high altitude) have shown that without additional oxygen the average individual may remain conscious *at rest* (with varying degrees of impairment) for about thirty minutes at 18,000 feet, but for one minute or less at 30,000 feet. Exercise would obviously shorten these times greatly.

On the other hand, human ability to adjust to higher environments over a period of time is truly phenomenal. It has been reported (4) that a member of the 1924 British Mt. Everest expedition reached an altitude of 28,126 feet without oxygen equipment.

The acclimatizing process can be accomplished by various systems of physical conditioning, and at progressively higher altitudes if possible. If altitude cannot be increased systematically, a progressive conditioning program at the *game altitude* is undertaken in which cardiorespiratory endurance is gradually improved by progressively increasing demands.

Altitudes below 3,000 feet probably require no acclimatization, the problem is slight up to about 5,000 feet, and the problem of altitude is only academic above 10,000 feet since no serious competition occurs above that level. The area of concern for physical education and athletics in the U.S.A. is really for altitudes between 5,000 and 10,000 feet.

D.B. Dill (10) has provided evidence that suggests that the physiological adaptation to altitude occurs in four phases:

1. The *acute phase*—In the first thirty minutes of exposure to the altitudes of concern here the loss in maximal O_2 consumption and consequently in performances that depend on aerobic capacity is less than 10%.
2. The *second phase*—Decrements may be from 20% to 30% (in one to three days).
3. The *third phase*—Adaptation to the decrements of phase two requires several weeks.
4. The *fourth phase*—Adaptation depends on the increase of red blood cell volume, which reaches a maximum in about a year or more.

Over a period of years it is eventually possible to achieve sea level performance up to as high as 13,200 feet. Above 17,500 feet there is only deterioration, and no adaptation seems to occur. This represents a schema (fig. 24.4) that is widely variable from individual to individual with respect to both rate and capacity for adaptation to altitude.

Thus the coach whose athletes will compete at a site such as Mexico City at an altitude of 7,350 feet is faced with the choice of timing his trip to compete within minutes of arrival (impossible) or to arrive several weeks early to allow time for acclimatization (also usually impossible, except for the fortunate few in Olympic competition). Possibly the only real solutions are for flatlanders to limit their interscholastic competitions to other flatlanders or to accept the alternative of a predictable loss of performance in aerobic events gracefully.

The physiological mechanisms that bring about acclimatization have been well demonstrated, at least in part. Increases of 10%–50% in the number of erythrocytes and in the hemoglobin content of the blood have been reported.

Physiology of Training and Conditioning Athletes

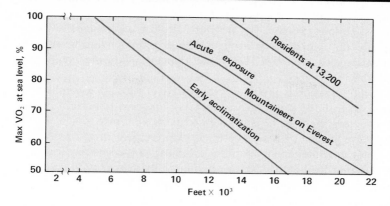

Figure 24.4 Decrement in capacity for supplying oxygen to tissues, $\dot{V}O_2$ max, at four stages of acclimatization. (From Dill, D.B. *J.A.M.A.* 205:753, 1968.)

Ability to increase the maximum ventilation rate has also been demonstrated, and there is a possibility that vascularization of lung and muscle tissue is also improved.

Administration of Oxygen to Improve Performance. There seems to be no evidence that breathing enriched mixtures of O_2 *before* an athletic event has a significant effect on the subsequent performance. Use of O_2 *during* work at high altitudes, however, is not only advantageous but absolutely necessary—at 18,000 to 20,000 feet—for most people without a long acclimatization period. (This, of course, is of no practical value for athletics.)

One use of O_2 for athletes that rests on sound theoretical and experimental bases is for shortening recovery times at altitude. In sports that involve rest periods between heavy endurance workouts, such as basketball and soccer, repayment of the O_2 debt can be hastened in unacclimatized athletes who have competed at an altitude substantially higher than the altitude they are used to.

SUMMARY

1. Constant human body temperature is maintained by striking a balance between heat gain and heat loss. Heat gain is due to metabolism and also to gains from radiation and convection when environmental temperatures are above skin temperature (92° F). Heat loss occurs by *conduction, radiation, convection,* and *evaporation* at temperatures below skin temperature, and by *evaporation* only when the environmental temperature is greater than skin temperature.

Environment and Exercise

2. In vigorous athletic events metabolic heat maintains an athlete's *core temperature* in all but the most severe cold.
3. The most serious problem for athletes in cold environments is having enough flexibility in dress to bring about heat retention during warm-up and rest periods, and yet allow heat dissipation during competitive periods.
4. Cold acclimatization is probably brought about through a combination of physiological and psychological factors. The most important physiological factors seem to be an increased metabolic rate and a greater temperature gradient between core and skin temperatures.
5. In hot, dry environments the ability to adjust to the severely increased cardiovascular load is most critically limited by *dehydration.*
6. In hot, wet environments the ability to adjust to the severely increased cardiovascular load is most critically limited by rising body core temperature due to inability to dissipate heat by evaporation.
7. For both hot and dry and hot and wet environments, acclimatization can be brought about by progressively increasing workouts in an artificial hot room over four or five days. This acclimatization will persist at least three weeks in cold weather.
8. Water replacement schedules should be set up to achieve both early replacement and overhydration to maintain performance at its highest level.
9. Exercise or competitive sport performance at altitudes higher by 3,000 feet or more than the home environment will be noticeably impaired by hypoxia if the activity depends largely upon aerobic energy (of one minute or greater duration).
10. Physiological adaptation to altitude appears to follow a time course of four phases: (a) during the acute phase of up to thirty minutes, performance is not greatly affected (up to 10% at 10,000 feet); (b) in one to three days performance falls off more severely; (c) over several weeks, acclimatization brings performance back to that of phase one; (d) red blood cell volume increases over a period of months, reaching its maximum after a year or more, with commensurate improvement in performance and eventual return to sea level performance at altitudes up to 13,200 feet.
11. Administration of oxygen to athletes at high altitudes should result in faster recovery times, but its use before an event cannot be expected to bring about large changes in performance.

Physiology of Training and Conditioning Athletes

REFERENCES

1. Adolph, E.F. *Physiology of Man in the Desert.* New York: Interscience Publishers, 1947.
2. American College Sports Medicine. Position statement on prevention of heat injuries during distance running. *Med. Sci. Sports* 7:vii, 1975.
3. Astrand, P-O. Physiological aspects on cross-country skiing at the high altitudes. *J. Sports Med. Phys. Fitness* 3:51–52, 1963.
4. Balke, B. Work capacity at altitude. In *Science and Medicine of Exercise and Sports,* ed. W.R. Johnson, chap. 18. New York: Harper & Row, 1960.
5. Botsford, J.H. A wet globe thermometer for environmental heat measurement. *Am. Ind. Hyg. Assoc. J.* 32:1–10, 1971.
6. Costill, D.L., and Saltin, B. Factors limiting gastric emptying during rest and exercise. *J. Appl. Physiol.* 37:678–83, 1974.
7. Costill, D.L.; Cote, R.; Miller, E.; Miller, T.; and Wynder, S. Water and electrolyte replacement during repeated days of work in the heat. *Aviat. Space Environ. Med.* 46:795–800, 1975.
8. Costill, D.L. Fluids for athletic performance: Why and what should you drink during prolonged exercise? In *Toward Understanding Human Performance,* ed. E.J. Burke. Ithaca: Mouvement Publications, 1977.
9. Coyle, E.F.; Costill, D.L.; Fink, W.J.; and Hoopes, D.G. Gastric emptying rates for selected athletic drinks. *Res. Q.* 49:119–24, 1978.
10. Dill, D.B. Physiological adjustments to altitude changes. *J.A.M.A.* 205:123–30, 1968.
11. Drinkwater, B.L.; Kupprat, I.C.; Denton, J.E.; Crist, J.L.; and Horvath, S.M. Response of prepubertal girls and college women to work in the heat. *J. Appl. Physiol.* 43:1046–53, 1977.
12. Drinkwater, B.L., and Horvath, S.M. Heat tolerance and aging. *Med. Sci. Sports* 11:49–55, 1979.
13. Fox, E.L.; Mathews, D.K.; Kaufman, W.S.; and Bowers, R.W. Effects of football equipment on thermal balance and energy cost during exercise. *Res. Q.* 37:332–39, 1966.
14. Gisolfi, C.V., and Cohen, J.S. Relationships among training, heat acclimatization and heat tolerance in men and women: the controversy revisited. *Med. Sci. Sports* 11:56–59, 1979.
15. Haymes, E.M.; Buskirk, E.R.; Hodgson, J.L.; Lundegren, H.M.; and Nicholas, W.C. Heat tolerance of exercising lean and heavy prepubertal girls. *J. Appl. Physiol.* 36:566–71, 1974.
16. Haymes, E.M.; McCormick, R.J.; and Buskirk, E.R. Heat tolerance of exercising lean and obese prepubertal boys. *J. Appl. Physiol.* 39:457–61, 1975.
17. Kang, B.S.; Song, S.H.; Suh, C.S.; and Hong, S.K. Changes in body temperature and basal metabolic rate of the ama. *J. Appl. Physiol.* 18:483–88, 1963.
18. Keatinge, W.R.; McIlroy, M.B.; and Goldfien, A. Cardiovascular responses to ice-cold showers. *J. Appl. Physiol.* 19:1145–50, 1964.
19. Kissen, A.T.; Reifler, C.B.; and Thaler, V.H. Modification of thermoregulatory responses to cold by hypnosis. *J. Appl. Physiol.* 19:1043–50, 1964.

20. Kotze, H.F.; van der Walt, W.H.; Rogers, G.G.; and Strydom, N.B. Effects of plasma ascorbic acid levels on heat acclimatization in man. *J. Appl. Physiol.* 42:711–16, 1977.

21. LeBlanc, J. Local adaptation to cold of Gaspé fishermen. *J. Appl. Physiol.* 17:950–52, 1962.

22. Milan, F.A.; Hannon, J.P.; and Evonuk, E. Temperature regulation of Eskimos, Indians, and Caucasians in a bath calorimeter. *J. Appl. Physiol.* 18:378–82, 1962.

23. Minard, D. Prevention of heat casualties in Marine Corps recruits. *Milit. Med.* 126:261–72, 1961.

24. Nadel, E.R. Control of sweating rate while exercising in the heat. *Med. Sci. Sports* 11:31–35, 1979.

25. Pandolf, K.B. Effects of physical training and cardiorespiratory physical fitness on exercise-heat tolerance: recent observations. *Med. Sci. Sports* 11:60–65, 1979.

26. Pugh, L.G.C.E. Tolerance to extreme cold at altitude in a Nepalese pilgrim. *J. Appl. Physiol.* 18:1234–38, 1963.

27. Roberts, M.F., and Wenger, C.B. Control of skin circulation during exercise and heat stress. *Med. Sci. Sports* 11:36–41, 1979.

28. Saltin, B. Circulatory response to submaximal and maximal exercise after thermal dehydration. *J. Appl. Physiol.* 19:1125–32, 1964.

29. Senay, L.C. Effects of exercise in the heat on body fluid distribution. *Med. Sci. Sports* 11:42–48, 1979.

30. Shoenfeld, Y.; Udassin, R.; Shapiro, Y.; Ohri, A.; and Sohar, E. Age and sex differences in response to short exposure to extreme dry heat. *J. Appl. Physiol.* 44:1–4, 1978.

31. West, J.B.; Lahiri, S.; Gill, M.B.; Milledge, J.S.; Pugh, L.G.C.E.; and Ward, M.P. Arterial oxygen saturation during exercise at high altitude. *J. Appl. Physiol.* 17:617–21, 1962.

32. Williams, C.G.; Bredell, G.A.G.; Wyndham, C.H.; Strydom, N.B.; Morrison, J.F.; Peter, J.; Fleming, P.W.; and Ward, J.S. Circulatory and metabolic reactions to work in the heat. *J. Appl. Physiol.* 17:625–38, 1962.

33. Yaglou, C.P. Temperature, humidity and air movement in industries: the effective temperature index. *J. Ind. Hyg.* 9:297–309, 1927.

25

Nutrition for Athletes

Long-Term Dietary Considerations and Requirements
Suggested Training Rules for Good Nutrition
Principles Involved in Pregame Nutrition
Pregame Procedure
Glycogen Supercompensation
(Carbohydrate Loading) for Endurance Athletes

In recent years, with ever-improving levels of competition, athletes and coaches have developed considerable interest in nutrition, but this, unfortunately, is an area in which the scientific efforts of trained nutritionists (and biochemists, who are the experts in this field) have often been obscured by clouds of misinformation generated by faddists and self-proclaimed experts.

Furthermore, athletes seem to be too easily influenced by the success of other athletes whose training regimens may have included such dietary fads as royal honey, kelp, blackstrap molasses, or other substances thought to have miraculous properties for improving athletic performance. More often than not, when these potions are tested by scientific methods in controlled experiments, it turns out that an athlete's success was achieved in spite of—not because of—his unusual dietary modifications.

Let it be clearly stated from the outset: *There is no scientific evidence at the present time to indicate that athletic performance can be improved by modifying a basically sound diet.* Furthermore, there are many different ways in which a nutritious diet can be obtained, and the best diet for one athlete will seldom be the best diet for all athletes. Individual differences exist in our senses of taste as well as in our enzyme systems, which are so necessary for digestion and absorption. In other words, "one man's meat may be another man's poison."

It should also be recognized that for explosive or short-duration events skill is the all-important factor. Even in endurance events where the total energy supply and the rate of energy supply are very important, the role of conditioning is infinitely more important than diet (if a diet is nutritionally sound). Still, in all athletic events, psychological as well as physiological factors affect performance, and there is no way to evaluate the psychological importance of eating steak when less expensive protein foods are just as nutritious. When ego and prestige factors enter the picture, science may fade into the background.

Although we may not be able to modify a sound diet to improve performance, athletes can go downhill very rapidly if their diet is less than optimum. Thus diet is still a very important consideration.

Long-Term Dietary Considerations and Requirements

Caloric Intake. Because this factor was discussed in chapter 15, it is enough to say that athletes must consume enough food daily to meet the energy demands of their training program. If they eat less than this, they will burn body tissues to make up the deficit and will approach "staleness" more rapidly. If they consume more food than they need, the result will be an increase in body weight with its accompanying mechanical disadvantages.

Physiology of Training and Conditioning Athletes

Proportion of Foodstuffs Ingested (Long-Term Diet). There are many opinions of what constitutes the proper proportion of carbohydrate, fat, and protein in human diet. None of the opinions, however, is supported by acceptable experimental evidence. In lieu of such evidence, we must apply theory and common sense.

As for theory, it is well known that in steady-state exercise (below the anaerobic threshold) the respiratory quotient goes up gradually from an average resting value of 0.85 to something like 0.90 or 0.95. This is interpreted as meaning that the organism prefers to burn carbohydrate for energy purposes during muscular activity, although it is also capable of utilizing fat. Furthermore, although fat produces more than twice as much energy per gram as carbohydrate, it requires more oxygen for each calorie (213 ml/cal of fat compared with 198 ml/cal of carbohydrate). In any athletic event where the work of the respiratory muscles is an important factor, there would seem to be an advantage of some 7.5% in favor of carbohydrate. As long ago as 1920, increased overall muscular efficiency of up to 10% had been shown experimentally for high carbohydrate diets (11). It had also been experimentally shown that fatigue occurred earlier on high fat diets (7).

The end products of protein metabolism are excreted in the urine, and by measuring the urinary nitrogen an estimate of protein metabolism can be made. Because urinary nitrogen increases very little with exercise under normal conditions, it is not considered an important source of energy. Only under conditions of starvation, when the carbohydrate and fat stores have been completely utilized, is the protein of body tissue consumed for energy. Protein is needed mainly for building new body tissue. Thus for growing children, and for strength athletes whose training is severe enough to result in increased muscle mass, the demands for protein are increased. This is not true for mature endurance athletes.

On the basis of these theoretical considerations, the proportions shown in table 25.1 seem to be sensible when the energy expenditure is not excessive—up to 3,000 kcal daily.

If energy requirements rise to very high values, where muscle mass is also likely to increase, it would seem reasonable to increase the percentage of protein to as much as twenty, with a proportional decrease in the fat consumed.

Table 25.1
Suggested Proportion of Basic Foodstuffs

	Grams	Kcal	Percentage of Total Kcal
Carbohydrate	435	1,740	58
Fat	100	900	30
Protein	90	360	12

Table 25.2

Proportion of Basic Foodstuffs in Diet of 4,700 Olympic Athletes

	Grams	Kcal	Percentage of Total Kcal
Carbohydrate	800	3,280	46
Fat	270	2,510	35
Protein	320	1,320	19

In this regard it is interesting to note that Abrahams (1) reports that Schenk, who studied the diet of 4,700 competitors at the 1936 Olympic games, found an average daily consumption of over 7,000 kcal (see table 25.2).

For endurance athletes such as distance runners, cyclists, and cross-country skiers, performance can be optimized by enhancing glycogen storage through careful use of carbohydrate loading (glycogen supercompensation). This procedure will be described later in this chapter.

Quality of Protein. We have so far concerned ourselves only with the total quantity and with the proportions of the basic foodstuffs within that total. In regard to protein, the quality is also very important. All proteins break down to amino acids during the digestive processes, so that these may be considered the units or building blocks for the synthesis of the proteins found in the human body. Of the twenty-three amino acids normally present in animal protein, only thirteen can be synthesized in the cells. The other ten must be supplied in the diet, and they are therefore called *essential amino acids.*

Supplying the essential amino acids is no problem for those who eat meat and animal products. The use of complete proteins (those that include all the essential amino acids) from milk and eggs and a generous and varied use of meat solve the problem quite easily. For those who for religious or other reasons do not eat animal products, the problem is more complicated, but vegetarians can be well nourished if they include all the essential amino acids in their diet. This can be done by including a diversity of vegetable products such as leaves, seeds, roots, and fruits (15).

Vitamins. The need for vitamins in the human diet is well established and needs no comment here, but a question of recurring interest is whether athletes need vitamin supplements to their normal diet. It was at one time thought that the requirements for vitamins increased much more rapidly than the increase in metabolism due to exercise, but more recent work (6) indicates that vitamin needs increase only in approximate proportion to metabolic activity. Thus ingestion of larger amounts of food as daily workout levels increase automatically provides the needed increase in vitamins (if the diet is sound to begin with).

Physiology of Training and Conditioning Athletes

Some investigators have claimed that vitamin supplementation has improved athletic performance in their subjects. However, when the proper controls are instituted, assuring that subjects are on an adequate diet *before starting the experiment,* these improvements in performance can no longer be demonstrated. Thus in all likelihood the reported improvements in performance because of vitamin supplementation were the result of having improved previously inadequate diets.

One point should be made before we leave this subject. Trace quantities of mineral elements seem to be intimately connected with the body's proper use of certain vitamins (15). Furthermore, it seems likely that several unknown factors affect nutrition. For these reasons, common sense dictates that, wherever possible, vitamins should be obtained from their natural sources rather than from purified, synthetic sources.

Minerals. Table 25.3 shows the suggested minimum daily requirements for vitamin and mineral constituents (15). As with vitamins, there is no evidence that the need for minerals is increased in exercise (in comfortable climates) beyond the increase brought about by the increased daily food consumption needed for metabolic demands.

Since athletes (and the general population as well) do not always have optimal nutritional habits, we offer in table 25.3 the best available knowledge with respect to vitamin and mineral supplementation, when that seems desirable (15).

Table 25.3

A Vitamin Supplement Recommended by Dr. Roger J. Williams, An Outstanding Authority in Biochemistry and Nutrition

Vitamins		Minerals	
Vitamin A	10,000 units	Calcium	300.0 mg
Vitamin D	500 units	Phosphate	250.0 mg
Ascorbic acid	100 mg	Magnesium	100.0 mg
Thiamine	2 mg	Cobalt	0.1 mg
Riboflavin	2 mg	Copper	1.0 mg
Pyridoxine	3 mg	Iodine	0.1 mg
Niacinamide	20 mg	Iron	10.0 mg
Vitamin B_{12}	5 mcg	Manganese	1.0 mg
Pantothenic acid	20 mg	Molybdenum	0.2 mg
Tocopherols (vitamin E)	5 mg	Zinc	5.0 mg
Inositol	100 mg		
Choline	100 mg		

From R.J. Williams, *Physician's Handbook of Nutritional Science.* Springfield, Ill.: Charles C Thomas, 1975. Available from Bronson Pharmaceuticals, 4526 Rinetti Lane, La Canada, Calif. 91011.

Table 25.4
Evaluation of Daily Food Selection

	Food group	Amount	Scores		Daily Score S M T W T F S
1.	Milk	4 cups or more (1 cup = 8 oz.)	4 cups = 3 cups = 2 cups = 1 cup =	10 8 6 4	
2.	Meat* (also fish, poultry, eggs, legumes)	2 servings or more, including at least one of meat, poultry, or fish	2 servings, including 1 of meat, poultry, or fish = 1 serv. as above = 1 serv. eggs, legumes =	10 8 6	
3.	Citrus fruit (also tomatoes, raw cabbage, salad greens)	1 serving or more (1 piece of fruit or ½ cup juice)	1 serving =	10	
4.	Leafy green or yellow vegetable	1 serving or more (½ cup)	1 serving =	10	
5.	Potatoes and other fruits and vegetables	2 servings or more (1 serving = ½ cup)	2 servings = 1 serving =	10 6	
6.	Whole-grain or enriched cereals and breads†	2 servings or more	2 servings = 1 serving =	10 6 ·	
7.	Butter or fortified margarine	2 pats or 2 tbsp.	2 pats = 1 pat =	10 6	
8.	For *not* eating any candy, cake, pastry, or other sweets			10	
9.	For *not* eating any food fried in deep fat			10	
10.	For *not* drinking any cola, carbonated soft drinks, coffee, tea, or imitation fruit drinks			10	
	Total			100	

From H.A. deVries, *Health Science: A Positive Approach.* Courtesy of Goodyear Publishing Co., 1979.

Suggested Training Rules for Good Nutrition

It is obviously impossible, as well as undesirable, to belabor athletes with the specifics of diet; it is also unnecessary because a relatively simple set of rules will result in good nutrition without making a dietitian (and possibly a hypochondriac) of each athlete. The author has found the following rules to be workable and effective, not only for good athletic conditioning, but also for forming sound, life-long dietary habits.

1. Distribute the daily consumption of food over three regularly spaced meals. If weight gain (or prevention of weight loss) is desirable, an evening snack can be added.
2. Eliminate from the diet as much as possible the foods that furnish only calories without contributing their share of vitamins and minerals (candy, cake, carbonated beverages, etc.). Use fruit and fruit juices for desserts and snacks.
3. Eliminate tea, coffee, and alcohol. Not only do these drinks usurp the place of more nutritious food, they may cause undesirable pharmacological effects such as decreased muscular efficiency.
4. Avoid fatty foods; they slow peristalsis and therefore gastric emptying.
5. Eat two servings daily of fresh fruit (one to be citrus fruit or tomatoes).
6. Eat four servings daily of vegetables, including leafy green vegetables (salads) and roots and tubers (turnips, beets, potatoes).
7. Eat at least three slices of whole-grain bread daily.
8. Eat enough butter or fortified margarine to supplement the bread in item 7.
9. Drink at least three glasses of milk daily.

A study of twenty-eight athletes (6) from varsity teams of three Big Ten universities showed that only ten of the athletes followed sound diets. The foods most commonly omitted were the green and yellow vegetables, citrus fruits, eggs, and milk. This indicated that their diets were probably low in vitamins A and C and in calcium. If these dietary habits can be accepted as typical of American athletes, then coaches would be well advised to provide vitamin supplements for their athletes in the form of a multiple vitamin and mineral pill, in spite of the earlier discussion to the effect that athletes on a well-balanced diet do not require vitamin or mineral supplements because of heavy workouts. Table 25.4 provides a simple and quick means for evaluating athletes' diets, which the author used for many years with swimmers, divers, and water polo teams (8).

Table 25.4

*1 serving = 2–4 oz. lean cooked meat, poultry, or fish (not counting bone)
1 serving = 2 eggs
1 serving = 1 cup of cooked dry beans, peas, or lentils
1 serving = ½ cup peanuts or other nuts
†1 serving = 1 slice bread
1 serving = 1 oz. ready-to-eat cereal
1 serving = ½ cup cooked cereal, etc.

In situations that involve mature athletes, coaches may have to compromise their principles if firmly held dietary beliefs are in evidence. But, individual differences being what they are, it is conceivable that some individuals will thrive on diets that would be totally unsatisfactory for most athletes. Furthermore, it is all-important to maintain harmonious relationships and undisturbed psychological equilibrium for successful athletic efforts.

Principles Involved in Pregame Nutrition

The digestive functions of the stomach can be divided into two components: secretory and motor. The secretory function consists of the elaboration and discharge into the stomach of hydrochloric acid, digestive enzymes, and alkaline mucus. The motor function consists of the maintenance of a degree of tonus plus the peristaltic contractions found during digestive processes. In some individuals there are, in addition, muscular contractions related to hunger pangs. Any factor that interferes with either secretory or motor functions may cause nausea.

The most definitive work on the effects of exercise on the functions of the stomach in the human was done some time ago by Hellebrandt and Hooper (9). For the secretory cycle, it was found that severe exercise results in inhibition of the secretory response, and that the resulting hypoacidity lasted as long as one hour. In mild activity, the acidity (secretory activity) was either unchanged or only slightly increased.

In respect to stomach motility, they found that mild exercise during the digestion of a meal seemed advantageous in hastening the final emptying time. Violent or exhaustive exercise, however, was found to inhibit gastric peristalsis, although this inhibition was followed (after exercise was ended) by augmented activity that resulted in little alteration of the final emptying time of the stomach.

Some evidence for a psychic effect was provided in their series of experiments in that repetition generally decreased a subject's response to the same exercise stressor.

Experimental evidence for the effects of exercise on the other portions of the digestive tract is lacking, inconclusive, or has been performed only on animals under conditions that do not justify extrapolation of the conclusions to humans.

The objectives to be attained in the twenty-four- to forty-eight-hour period preceding competition are as follows:

1. Attaining the largest possible storage of carbohydrate in the liver and musculature.
2. Entering competition with the smallest possible stomach volume, so that the diaphragm may descend as far as possible in inhalation.
3. Preventing gastric disturbances from occurring during a competition.

Physiology of Training and Conditioning Athletes

4. Maintaining an optimum psychological attitude in the athlete while accomplishing the first three items.

Pregame Procedure

Liver and muscle glycogen can be increased by the methods to be discussed.

Breakfast on the day of competition may be relatively larger if the event is scheduled for the afternoon, and breakfast and lunch may be larger if the event is in the evening.

In any event the preevent meal should be light, and the two meals that immediately precede competition should be high carbohydrate meals: cereals such as oatmeal, toast with jam, honey, etc.

The final preevent meal has usually preceded competition by three or four hours, but there is evidence that if it consists of cereal and milk of no more than 500 kcal, no adverse effects are suffered if it is taken up to thirty minutes before competition (2, 3, 13, 14).

The Pregame Meal. Theory and common sense dictate that the following precautions be observed for the pregame meal.

1. Avoid foods that are even mildly distasteful to an individual athlete— no matter how well they may serve nutritional objectives. An athlete may get sick even though the food is excellent.
2. Avoid irritating foods, such as highly spiced foods and roughage.
3. Avoid gas-forming foods: onions, cabbage, apples, baked beans.
4. Avoid fatty foods; they slow peristalsis and therefore gastric emptying.
5. Hold protein foods to a minimum because their metabolism results in fixed acids; in large quantities, this could result in an undesirable acidosis.
6. Fluid can best be supplied by bouillon (which supplies sodium, which is excreted in perspiration during an event). Many athletes will prefer milk or juices, and if experience shows no ill effects it is probably wise to accede to this preference.

Glycogen Supercompensation (Carbohydrate Loading) for Endurance Events

The theoretical basis for the importance of glycogen storage in endurance-type exercise was laid in chapter 3. When work loads greater than about 70% of aerobic capacity must be borne for thirty to sixty minutes or more, the *rate* of work is limited by aerobic capacity, but the *duration* over which the load can be maintained depends very largely on the level of glycogen storage in the involved muscles. Apparently the muscle cell cannot use other energy substrates to any great extent at high levels of work. When glycogen depletion occurs, work can be continued on other energy substrates, but only at work loads considerably below 70%.

Karlsson and Saltin (10) demonstrated this effect quite clearly when they had ten subjects run the same 30-km race twice, three weeks apart, once after a carbohydrate-enriched diet and once after a mixed diet. They found the muscle glycogen level in the quadriceps for the high carbohydrate diet to be double that for the mixed diet, and every subject turned in the best performance after the high carbohydrate diet. Interestingly, identical pace was maintained after both diets in the early part of the race when glycogen content was high, but pace fell off earlier after the mixed diet as glycogen depots were emptied.

The work of the original investigators in this area, Bergstrom and colleagues (4), also provides clear-cut data for modifying the diet to best prepare for prolonged endurance-type events. To achieve the highest possible level of muscle glycogen for such events, the athlete must work the same muscles to exhaustion about one week prior to the event. For the next three days, the diet should be made up almost exclusively of fat and protein, since it was shown that low carbohydrate diet followed by high carbohydrate diet results in the greatest possible glycogen storage. About three days should now be left for a carbohydrate-rich diet with only very light workouts to produce the maximum possible glycogen storage in the muscles. The low carbohydrate diet consisted of 1,500 kcal protein and 1,300 kcal fat for a total daily energy expenditure estimated at 2,800 kcal. The high carbohydrate diet made up the same total with 2,300 kcal of carbohydrate and 500 kcal of protein (4).

Long-term data on such diet modifications are not available, and it is questionable if such procedures would be wise to follow where weekly endurance competitions are held since high protein and high fat diets may have adverse long-term effects due to their lowering pH and slowing gastrointestinal tract emptying. The best course seems to be following a basically well-balanced diet throughout most of the competitive season and reserving this modification for the one or two most important competitions of the season. In any event, the exhaustion of the involved musculature early in the week followed by relatively high carbohydrate diet and relatively lighter workouts is sound procedure.

An excellent review of carbohydrate loading from the point of view of distance running, which includes principles for application as well as sensible precautions, has been provided by Londeree (12).

REFERENCES

1. Abrahams, A. The nutrition of athletes. *Br. J. Nutr.* 2:266–69, 1948.
2. Asprey, G. M.; Alley, L. E.; and Tuttle, W. W. Effect of eating at various times on subsequent performances in the 440-yard dash and half-mile run. *Res. Q.* 34:267–70, 1963.
3. ———. Effect of eating at various times upon subsequent performance in the one-mile run. *Res. Q.* 35:227–30, 1964.

4. Bergstrom, J.; Hermansen, L.; Hultman, E.; and Saltin, B. Diet, muscle glycogen and physical performance. *Acta Physiol. Scand.* 71:140–50, 1967.

5. Bicknell, F., and Prescott, F. *The Vitamins in Medicine.* New York: Grune & Stratton, 1953.

6. Bobb, A.; Pringle, D.; and Ryan, A. J. A brief study of the diet of athletes. *J. Sports Med.* 9:255–62, 1969.

7. Christensen, E. H. Beitrage zur Physiologie Schwerer Korperlicher Arbeit I-IX. *Arbeitsphysiologie* 4:453–503, 1931.

8. deVries, H. A. *Health Science: A Positive Approach.* Santa Monica, Calif.: Goodyear Publishing Co., 1979.

9. Hellebrandt, F. A., and Hooper, S. L. Studies in the influence of exercise on the digestive work of the stomach. *Am. J. Physiol.* 107:348, 355, 364, 370, 1934.

10. Karlsson, J., and Saltin, B. Diet, muscle glycogen, and endurance performance. *J. Appl. Physiol.* 31:203–206, 1971.

11. Krogh, A., and Lindhard, J. The relative value of fats and carbohydrates as sources of muscular energy. *Biochemistry J.* 14:290, 1920.

12. Londeree, B. Pre-event diet routine. *Runners World* 9:26–29, 1974.

13. Singer, R. N., and Neeves, R. E. Effect of food consumption on 200-yard freestyle swim performance. *Res. Q.* 39:355–60, 1968.

14. White, J. R. Effects of eating a liquid meal at specific times upon subsequent performances in the one-mile run. *Res. Q.* 39:206–10, 1968.

15. Williams, R. J. *Physician's Handbook of Nutritional Science.* Springfield, Ill.: Charles C Thomas, 1975.

26

Special Aids to Performance

Very small improvements in athletic performance can make the difference between mediocre and championship achievement. Differences of 1% or 2% may have large meaning. An improvement of only 2% in a four-minute mile brings the time down to 3:55.2.

Because of the importance of small improvements, which are very difficult to obtain by normal training methods when performance approaches record times or championship levels, coaches and athletes have cast about for special aids to performance, sometimes called *ergogenic aids*. Manipulation of diet, use of various drugs, use of "miracle" foods, all have been areas of interest at various times. This search for methods to improve athletic achievement can be considered wholesome as long as (1) special aids are used to supplement, not to supplant, excellence in training and conditioning, and (2) the special aids constitute no hazard to the athletes.

Ergogenic aids can function in one of two ways: (1) by improving the capacity of the muscles and/or the O_2 transport system to do work or (2) by removing or reducing inhibitory mechanisms to allow use of previously untapped reserves. The first function must be considered the sounder approach because the second function must inevitably reduce the safety factor with which the organism has been provided.

In general, the use of drugs falls into the second category; furthermore, the use of any drugs to improve athletic performance is cause for disqualification by the International Amateur Athletic Federation, the Amateur Athletic Union, and the U.S. Olympic Association as contrary to the highest ideals of sportsmanship. Even more important, some drugs that have reportedly been used by athletes (such as the amphetamines) can be habituating and can have other harmful effects. Although the ergogenic effects of some drugs are discussed in this chapter, this should in no way be construed as support for their use; this discussion is included for academic purposes only.

Alkalinizers

The amount of oxygen debt attainable by an athlete is a very important factor in heavy endurance work. The size of the O_2 debt, in turn, is very likely limited by the blood pH, which depends on the alkaline reserve (the capacity for buffering the lactic acid formed during work).

Early workers in this area established the feasibility of displacing the pH of the blood upward (prior to exercise) by the ingestion of alkaline salts, so that a heavy workout resulted only in a return to the normal pH value instead of a displacement toward more acidic values, which usually occurs (38). Dennig and associates (16), working at the Harvard Fatigue Laboratory, demonstrated a decreased ability to accumulate O_2 debt in acidosis brought about by ingestion of acid salts, and it was inferred from this that alkalosis should improve the possibility for buffering an increased O_2 debt. Dill and coworkers (18) demonstrated this increased O_2 debt capability; a runner in an

Physiology of Training and Conditioning Athletes

alkaline state ran 6:04 minutes to exhaustion (on a treadmill), compared with 5:22 minutes from a normal state. The O_2 debt was about 20% greater in the first case, which agrees roughly with the increased time of running.

Dennig (17) continued this line of experimentation in Germany with a well-controlled study on ten subjects who worked to exhaustion on a treadmill and bicycle ergometer. In all cases, his subjects were able to increase their endurance by 30%–100% when they started in an alkaline state. His procedure, after many experiments, consisted of ingestion of a mixture of sodium citrate (5.0 gm), sodium bicarbonate (3.5 gm), and potassium citrate (1.5 gm) in two to four doses per day taken after mealtime. (This procedure should start two days before an event and should cease at least five hours before the event.) Dennig pointed out that the effect will be lost over longer periods because the organism adjusts to the artificial alkalinization (it would also be undesirable from a health standpoint). Some of his subjects experienced moderate side effects, such as stomach gas and loose bowels.

Dennig's experiments furnish strong evidence for the value of alkaline salts in *his* subjects, who were only moderately trained. Whether this effect can be demonstrated on highly trained athletes, who might have already improved their alkaline reserve through their training, is questionable. Only one study has been conducted subsequently on highly trained runners, and in this case no significant changes were found (28). The salt mixture was given four hours before the event, however, and this timing was shown by Dennig to be ineffective.

More recently a well-controlled study by Jones and colleagues (29) has shown clearly improved endurance performance on a bicycle ergometer at 95% $\dot{V}O_2$ max after ingestion of sodium bicarbonate. Acting as their own controls, subjects showed a 41% decrement in performance after taking ammonium chloride to acidify blood reaction (lowered pH) and a 62% improvement after alkalinizing with bicarbonate, which was administered in small portions over a three-hour period for a total dose of 0.3 gm/kg body weight. Such a dose would amount to four to five level measuring teaspoonfuls of bicarbonate for an average-sized male. Their subjects apparently tolerated this amount well, but in some people the amount might induce symptoms such as diarrhea and vomiting. In view of the large benefits reported, this procedure would seem worthy of further investigation, especially for application to athletes whose events require high fractions of $\dot{V}O_2$ max for periods of time up to four to five minutes.

Amphetamine (Benzedrine)

This drug, in all its various forms (primarily d-amphetamine sulfate and its European relative Pervitin) is a sympathomimetic amine, and it is used by the medical profession as a central nervous system stimulant. Pharmacology texts warn against its use as a remedy for sleepiness or fatigue, or to increase capacity for work, because (1) there is a danger of addiction, (2) it removes the warning of impending overstrain, (3) its vasopressor effects are undesirable, and (4) cases of collapse have been reported. Obviously, this discussion is academic, as the use of such drugs by athletes is to be strongly discouraged.

The literature provides evidence that amphetamine sulfate inhibits fatigue as measured by voluntary contractions on an ergograph (2), improves hand and arm coordination (33), improves the strength of forearm flexion (25) and handgrip (25,33), and improves the athletic performance of swimmers, runners, and weight throwers (43). Pervitin was found to increase work output on a bicycle ergometer (32).

On the other hand, some investigators have been unable to verify these results (9,23,30,47), and we must conclude that the ergogenic effects are—at best—debatable and that the dangers involved are considerable.

Anabolic Steroids

Evidence provided some years ago suggested that administering testosterone (a male sex hormone) to animals (36) and humans (42) results in an increase in muscle weight (hypertrophy) and strength. Testosterone is a steroid that has both androgenic (producing masculine characteristics) and anabolic (nitrogen retention-protein building) qualities. Recently steroids have been synthetically developed that are chemically related to testosterone, but in which structural changes in the molecule have increased the anabolic effects while decreasing the androgenic effects. Various commercial preparations of the *anabolic steroids* are in vogue among strength athletes. This, again, is a drug of considerable physiological potency, with many undesirable side effects, and thus it must be prescribed by a physician.

The health hazards incurred by the use of the anabolic steroids are such that the American College of Sports Medicine has issued the position statement given in table 26.1.

The position by the ACSM stated in table 26.1 was not taken lightly. It must be realized that while the benefits to muscle strength, body mass, and performance are questionable at best, the health hazards are very real and can be extremely serious. Alterations of normal liver function have been found in as many as 80% of sixty-nine patients treated with oral anabolic-androgenic steroids, and five reports document the occurrence of hepatitis in seventeen patients treated with these steroids of whom seven died of liver failure (3). In the male athlete, endogenous hormone production is suppressed, and the production of sperm cells may be reduced to the point of sterility. In the female, these steroids may cause masculinization, disruption of normal growth pattern, voice changes, acne, hirsutism, and enlargement of the clitoris (3). The trade-off of real health hazards for scientifically unproven performance benefit seems imprudent.

Table 26.1

Position Statement on the Use and Abuse of Anabolic-Androgenic Steroids in Sports

Based on a comprehensive survey of the world literature and a careful analysis of the claims made for and against the efficacy of anabolic-androgenic steroids in improving human physical performance, it is the position of the American College of Sports Medicine that:

1. The administration of anabolic-androgenic steroids to healthy humans below age 50 in medically approved therapeutic doses often does not of itself bring about any significant improvements in strength, aerobic endurance, lean body mass, or body weight.

2. There is no conclusive scientific evidence that extremely large doses of anabolic-androgenic steroids either aid or hinder athletic performance.

3. The prolonged use of oral anabolic-androgenic steroids (C_{17}-alkylated derivatives of testosterone) has resulted in liver disorders in some persons. Some of these disorders are apparently reversible with the cessation of drug usage, but others are not.

4. The administration of anabolic-androgenic steroids to male humans may result in a decrease in testicular size and function and a decrease in sperm production. Although these effects appear to be reversible when small doses of steroids are used for short periods of time, the reversibility of the effects of large doses over extended periods of time is unclear.

5. Serious and continuing effort should be made to educate male and female athletes, coaches, physical educators, physicians, trainers, and the general public regarding the inconsistent effects of anabolic-androgenic steroids on improvement of human physical performance and the potential dangers of taking certain forms of these substances, especially in large doses, for prolonged periods.

From American College of Sports Medicine, *Med. Sci. Sports* 9:xi, 1977.

Aspartates

Aspartic acid is a dicarboxylic amino acid that is known to form one of the links between protein and carbohydrate metabolism. Its conversion to oxalacetic acid places it in the citric acid cycle, which provides energy from carbohydrate breakdown (chap. 3). It has been shown that the respiration of a minced pigeon breast muscle can be increased by the addition of aspartic acid.

These well-known facts of biochemistry led to experimentation by the medical profession with aspartic acid salts for the relief of fatigue. In a group of 200 patients, all of whom complained of fatigue (postinfluenza, neurosis, gastrointestinal problems, menopause, old age, etc.), administration of potassium and magnesium aspartates resulted in subjective relief in a large percentage of cases (31). A more objective study on rats showed that the swim time to complete exhaustion was increased 15% in a group of thirty-six on aspartates, compared with a similar control group (39). It is of interest that in this experiment the improvement was seen most clearly in the low-endurance group of rats; the performance of "athletic" rats was altered very little.

In an investigation of fatigue in 163 subjects that comprised a blind study (subjects did not know whether they were on aspartates or placebo) and a double blind crossover trial (neither subjects nor investigator knew, and each group had a course of aspartates and placebo), subjective and objective evidence of relief of fatigue was presented (41).

On the other hand, Consolazio and his co-workers (7,35) at the U.S. Army Medical Research and Nutrition Laboratory were unable to verify these results on animals or on men. Fallis and associates (22) ran an experiment on twenty-six penitentiary-inmate weight lifters who regularly engaged in athletic activities, and they reported no significant differences in eight different measures that involved weight lifting and endurance. It is of interest, however, that in six of the seven events that could be considered as having a muscular endurance factor, the results favored the aspartate trials. The lack of statistical significance of the differences could conceivably be the result of a real difference, which was obscured by a large variability and the small number of subjects.

Since the aspartic acid salts can be considered foods rather than drugs, there would be no danger (with sensible doses) in further experimentation. This seems advisable in view of the lack of agreement.

Blood Doping

In two different studies, Ekblom and co-workers (19,20) showed that drawing 800–1200 ml of blood and then reinfusing subjects with their own red blood cells three to four weeks later (after hemopoiesis had already restored the red blood cell count to normal values) resulted in 8%–9% increases in $\dot{V}o_2$ max. Although some other investigators have had difficulty reproducing this phenomenon (12,44,46), the only other group that verified an increase of blood O_2-carrying capacity was able to find similar improvements in $\dot{V}O_2$ max (6), thus corroborating the work of the Ekblom group. In any event, this procedure, although of considerable physiological importance, has no practical value for coaches and athletes because this practice falls under the category of "doping" and is therefore prohibited by the bodies that govern athletic competition.

Caffeine

Caffeine is used in medicine as a central nervous system stimulant, particularly for psychical functions; it is also used as a diuretic. Medicinal dosage ranges from 100 to 500 mg. A cup of coffee usually contains 100 to 150 mg, and tea contains slightly less.

Graf (24) has reported on experiments in Germany during World War II to find stimulants suitable for improving physical and mental efficiency in combatting the stressful conditions of war. It was found that, although caffeine was a strong mental stimulant, it resulted in a very undesirable impairment of motor coordination (in target shooting, writing, and simulated auto driving). There was also a hangover effect, in which mental efficiency, after having been improved, fell off below normal values from one to three hours after the stimulant was taken.

More recent data have shown an interesting effect of caffeine during heavy endurance exercise in shifting energy consumption toward a greater utilization of fat (free fatty acids). This would of course result in a glycogen sparing effect (see chap. 3), which should improve endurance performance. Ivy (26,27) and Costill (11) and their co-workers at Ball State have shown this to be the case for trained cyclists riding for one to two hours. In their most recent experiment, they found that 250 mg caffeine (two to three cups of coffee) ingested one hour prior to the ride followed by ingestion of an additional 250 mg taken at fifteen-minute intervals over the first ninety minutes of exercise increased work output by 7.4% and $\dot{V}O_2$ by 7.3% compared with control (27). Since the perceived exertion remained unchanged, it is possible that at least part of the improvement is related to the psychological effects of caffeine.

Glucose

As was discussed in chapter 3, the duration of moderately intense work that can be sustained depends on the initial glycogen concentration of the working muscle and its rate of utilization. When the glycogen reserves in the active muscles are depleted, work can only be continued at a lower power output, which allows utilization of free fatty acids as energy substrate. On this basis one would not expect improved performance in athletic events of short to moderate duration from glucose feeding. However, if exogenous glucose (that fed orally) could substitute for some of the energy substrate requirement normally provided by muscle glycogen, then the point of depletion of endogenous muscle glycogen should be delayed with a consequent enhancement of performance in the later stages of endurance events such as cross-country running, marathons, and cross-country skiing.

Both animal studies and human investigations have shown that skeletal muscle can take up large amounts of glucose and that when it is supplied from an external source, it can exert a glycogen sparing effect on the liver and the working skeletal muscles (1,4).

It has also been shown that the ergogenic effect of glucose feeding is not seen until at least 20 to 30 minutes after ingestion (10,27,37), and probably reaches its maximum between 60 and 120 minutes later. If the event lasts four hours, about 100 gm of glucose may be utilized, and this may represent as much as 55% of the overall carbohydrate metabolism, with peak utilization occurring after ninety minutes (37). Therefore it would be advisable to repeat glucose intake every sixty to ninety minutes or, even better, in smaller amounts administered more frequently.

Glucose feeding has been shown to improve endurance time by 19% and efficiency by 6% in a simulated bicycle road race of 100 miles that elicited 67% of the $\dot{V}O_2$ max in eight trained racing cyclists (5).

However, in the application of these principles one must not overlook the effect of glucose upon gastric emptying if fluid replacement is a potential problem. For heavy exercise in the heat, fluid replacement must take precedence over energy substrate availability so that glucose solutions should be held below 2.5% (see chap. 24).

Improving Lactate Tolerance by Lactate Ingestion

It has recently been suggested that since removal of the lactate generated by exercise constitutes one limit of physical fitness therefore performance might be augmented by a diet that would induce the enzymes necessary for lactate clearance through gluconeogenesis (34). Preliminary studies have suggested that both fitness and the rate of clearance of blood lactate were augmented by feeding lactate in man, but the experiments provided no evidence as to whether these effects resulted from a change in membrane transport of lactate or from the induction of some limiting enzyme in gluconeogenesis (34). This hypothesis has interesting implications, but much more investigation is needed.

Oxygen and Vitamins

The use of oxygen to improve athletic performance was discussed in chapter 9. Vitamin supplementation as an ergogenic aid has also been discussed; the reader is referred to chapter 25.

Wheat-Germ Oil

Wheat-germ oil (WGO) contains several factors that seem to have biological activity: (1) vitamin E (alpha, beta, and gamma tocopherols), (2) fatty acids, such as linoleic acid, and (3) octacosanol, an alcohol that can be synthetically prepared. Cureton and his co-workers (13,14) have provided evidence of an improved training effect on middle-aged men when the physical training was supplemented by WGO; however, a later study by Cureton (15) on young men provided only statistically nonsignificant differences.

In a dietary study on guinea pigs that lasted twenty-eight days and ended with a swim test to exhaustion, it was found that all animals on a natural (control) diet drowned within ten minutes; 25%–33% of those on a corn-oil (vitamin E) supplemented diet were still swimming at sixty minutes; and 60%

of those fed WGO were still swimming at sixty minutes. Weanling rats who were fed on WGO showed no difference in swimming ability from those supplemented with corn oil (21).

In another study on swimming rats, no differences in performance were observed between those on WGO, vitamin E, or octacosanol as compared with controls (8).

Well-controlled studies on adolescent human swimmers have also shown no effect due to vitamin E supplementation (40).

It must be concluded that neither an ergogenic principle in WGO nor an ergogenic effect of the whole oil has been conclusively established. Cureton's early work (13,14) is persuasive, and the failure to achieve significant differences in the later work might be explained on the basis of age differences in the subject populations. Further experimentation seems justified.

SUMMARY

1. A survey of the literature on ergogenic aids leaves the distinct impression that even if "doping" with drugs were legal, ethical, and nonhazardous, the practice could not be justified on experimental evidence. On the other hand, use of some proposed ergogenic aids that are nonhazardous and that can be considered normal hygienic procedures to help an athlete gain an extra 1% or 2% improvement in performance may be justified if every effort has been exerted to bring training and conditioning to a peak.

2. A further disadvantage of ergogenic aids is that an athlete may become psychologically addicted, and if at a critical moment the aid is unavailable, a decrement from normal performance can occur.

3. In interpretation of research data, the finding of no positive results can never be conclusive because a single experiment is never capable of seeing all the possible changes that may occur. For example, an investigator using a simple magnifying glass might deny the existence of bacteria, which are clearly seen under a high-powered microscope; similarly, a research design is not all-encompassing.

4. In some experiments the differences that favor the working hypothesis were disregarded because they did not achieve statistical significance. This is as it should be, but the rigor of our method must not obscure the fact that even a real difference may remain statistically nonsignificant if (a) the difference is small, (b) the number of subjects is small, and (c) the variability within or between subjects on the parameter of interest is large.

This concept was clearly demonstrated in a study that showed the advantage of using expert swimmers instead of nonexperts for evaluation of the effects of ergogenic aids (45). Obtaining as much precision with nonexperts as was obtained with fifteen experts would have necessitated (because of greater intrasubject variability) an increase of the nonexpert sample size from fifteen to about eighty.

5. In view of the experiments cited, it seems that further experimentation is justified for such ergogenic aids as the alkalinizers, aspartates, caffeine, glucose, and wheat-germ oil. Used judiciously, none of these should be hazardous, and further research is needed.

REFERENCES

1. Ahlborg, G., and Felig, P. Influence of glucose ingestion on fuel-hormone response during prolonged exercise. *J. Appl. Physiol.* 41:683–88, 1976.

2. Alles, G.A., and Feigen, G.A. The influence of benzedrine on work decrement and patellar reflex. *Am. J. Physiol.* 136:392–400, 1942.

3. American College Sports Medicine. Position statement on the use and abuse of anabolic-androgenic steroids in sports. *Med. Sci. Sports* 9:xi–xii, 1977.

4. Bagby, G.J.; Green H.J.; Katsuta, S.; and Gollnick, P.D. Glycogen depletion in exercising rats infused with glucose, lactate or pyruvate. *J. Appl. Physiol.* 45:425–29, 1978.

5. Brooke, J.D.; Davies, G.J.; and Green, L.F. The effects of normal and glucose syrup work diets on the performance of racing cyclists. *J. Sports Med. Phys. Fitness* 15:257–65, 1975.

6. Buick, F.J.; Gledhill, N.; Froese, A.B.; Spriet, L.; and Meyers, E.C. Double blind study of blood boosting in highly trained runners. Abstract in *Med. Sci. Sports* 10:49, 1978.

7. Consolazio, C.F.; Nelson, R.A.; Matoush, L.O.; and Isaac, G.J. Effects of aspartic acid salts (Mg + K) on physical performance of men. *J. Appl. Physiol.* 19:257–61, 1964.

8. Consolazio, C.F.; Matoush, L.O.; Nelson, R.A.; Isaac, G.J.; and Hursh, L.M. Effects of octacosanol, wheat germ oil, and vitamin E on performance of swimming rats. *J. Appl. Physiol.* 19:265–67, 1964.

9. Cooter, G.R., and Stull, G.A. The effect of amphetamine on endurance in rats. *J. Sports Med. Phys. Fitness* 14:120–26, 1974.

10. Costill, D.L.; Bennett, A.; Branam, G.; and Eddy, D. Glucose ingestion at rest and during prolonged exercise. *J. Appl. Physiol.* 34:764–69, 1973.

11. Costill, D.L.; Dalsky, G.P.; and Fink, W.J. Effects of caffeine ingestion on metabolism and exercise performance. *Med. Sci. Sports* 10:155–58, 1978.

12. Cunningham, K.G. The effect of transfusional polycythemia on aerobic work capacity. *J. Sports Med. Phys. Fitness* 18:353–58, 1978.

13. Cureton, T.K. Effects of wheat germ oil and vitamin E on normal human subjects in physical training programs. *Am. J. Physiol.* 179:628, 1954.

14. Cureton, T.K., and Pohndorf, R. Influence of wheat germ oil as a dietary supplement in a program of conditioning exercises with middle-aged subjects. *Res. Q.* 26:391–407, 1955.

15. Cureton, T.K. Improvements in physical fitness associated with a course of U.S. Navy underwater trainees with and without dietary supplements. *Res. Q.* 34:440–53, 1963.

16. Dennig, H.; Talbot, J.H.; Edwards, H.T.; and Dill, D.B. Effect of acidosis and alkalosis upon capacity for work. *J. Clin. Invest.* 9:601–13, 1931.

17. ———. Ueber steigerung der korperlichen leistungsfahigkeit durch eingriffe in den saurebasenhaushalt. *Dtsch. Med. Wochenschr.* 63:733–36, 1937.

18. Dill, D.B.; Edwards, H.T.; and Talbott, J.H. Alkalosis and the capacity for work. *J. Biol. Chem.* 97:58–59, 1932.

19. Ekblom, B.; Goldbarg, A.N.; and Gullbring, B. Response to exercise after blood loss and reinfusion. *J. Appl. Physiol.* 33:175–80, 1972.

20. Ekblom, B.; Wilson, G.; and Astrand, P-O. Central circulation during exercise after venesection and reinfusion of red blood cells. *J. Appl. Physiol.* 40:379–83, 1976.

21. Erschoff, B.H., and Levin, E. Beneficial effect of an unidentified factor in wheat germ oil on the swimming performance of guinea pigs. *Fed. Proc.* 14:431–32, 1955.

22. Fallis, N.; Wilson, W.R.; Tetreault, L.L.; and La Sagna, L. Effect of potassium and magnesium aspartates on athletic performance. *J.A.M.A.* 185:129, 1963.

23. Golding, L.A., and Barnard, J.R. The effects of d-amphetamine sulfate on physical performance. *J. Sports Med.* 3:221–24, 1963.

24. Graf, O. Increase of efficiency by means of pharmaceutics (stimulants). In *German Aviation Medicine, W.W. II,* vol. 2, p. 1080. Washington, D.C.: U.S. Government Printing Office, 1950.

25. Hurst, P.M.; Radlow, R.; and Bagley, S.K. The effects of d-amphetamine and chlordiazepoxide upon strength and estimated strength. *Ergonomics* 11:47–52, 1968.

26. Ivy, J.L.; Costill, D.L.; Fink, W.J.; and Lower, R.W. Role of caffeine and glucose ingestion on metabolism during exercise. Abstract in *Med. Sci. Sports* 10:66, 1978.

27. ———. Influence of caffeine and carbohydrate feedings on endurance performance. *Med. Sci. Sports* 11:6–11, 1979.

28. Johnson, W.R., and Black, D.H. Comparison of effects of certain blood alkalinizers and glucose upon competitive endurance performance. *J. Appl. Physiol.* 5:577–78, 1953.

29. Jones, N.L.; Sutton, J.R.; Taylor, R.; and Toews, C.J. Effect of pH on cardiorespiratory and metabolic responses to exercise. *J. Appl. Physiol.* 43:959–64, 1977.

30. Karpovich, P.V. Effect of amphetamine sulfate on athletic performance. *J.A.M.A.* 170:558–61, 1959.

31. Kruse, C.A. Treatment of fatigue with aspartic acid salts. *Northwest Med.* 60:597–603, 1961.

32. Lehman, G.; Straub, H.; and Szakall, A. Pervitin als leistungssteigerndes mittel. *Arbeitsphysiologie* 10:680–91, 1939.

33. Lovingood, B.W.; Blyth, C.S.; Peacock, W.H.; and Lindsay, R.B. Effects of d-amphetamine sulfate, caffeine and high temperature on human performance. *Res. Q.* 38:64–71, 1967.

34. Mann, G.V., and Garrett, H.L. Lactate tolerance, diet and physical fitness. In *Nutrition, Physical Fitness and Health,* eds. J. Pariskova and V.A. Rogozkin. Baltimore, Md.: University Park Press, 1978.

35. Matoush, L.O.; Consolazio, C.F.; Nelson, R.A.; Isaac, G.I.; and Torres, J.B. Effects of aspartic acid salts (Mg + K) on swimming performance of rats and dogs. *J. Appl. Physiol.* 19:262–64, 1964.

36. Papanicolaou, G.N., and Falk, E.A. General muscular hypertrophy induced by androgenic hormone. *Science* 82:238–39, 1938.

37. Pirnay, F.; LaCroix, M.; Mosora, F.; Luyckx, A.; and LeFebvre, P. Glucose oxidation during prolonged exercise evaluated with naturally labeled (^{13}C) glucose. *J. Appl. Physiol.* 43:258–61, 1977.

38. Ronzoni, E. The effect of exercise on breathing in experimental alkalosis by ingested sodium bicarbonate. *J. Biol. Chem.* 67:25–27, 1926.

39. Rosen, H.; Blumentahl, A.; and Agersborg, H.P.K. Effects of the potassium and magnesium salts of aspartic acid on metabolic exhaustion. *J. Pharm. Sci.* 51:592–93, 1962.

40. Sharman, I.M.; Down, M.G.; and Sen, R.N. The effects of training and vitamin E supplementation on the performance of adolescent swimmers. *Br. J. Sports Med.* 7:27–30, 1973.

41. Shaw, D.L., Jr.; Chesney, M.A.; Tullis, I.F.; and Agersborg, H.P.K. Management of fatigue: a physiologic approach. *Am. J. Med. Sci.* 243:758–69, 1962.

42. Simonson, E.; Kearns, W.M.; and Enzer, N. Effect of methyl testosterone treatment on muscular performance and central nervous system of older men. *J. Clin. Endocrinol. Metab.* 10:528–34, 1944.

43. Smith, G.M., and Beecher, H.K. Amphetamine sulfate and athletic performance. *J.A.M.A.* 170:542–57, 1959.

44. Videman, T., and Rytomaa, T. Effect of blood removal and autotransfusion on heart rate response to a submaximal workload. *J. Sports Med. Phys. Fitness* 17:387–90, 1977.

45. Weitzner, M., and Beecher, H.K. Increased sensitivity of measurements of drug effects in expert swimmers. *J. Pharm.* 139:114–19, 1963.

46. Williams, M.A.; Goodwin, A.R.; Perkins, R.; and Bocrie, J. Effect of blood reinjection upon endurance capacity and heart rate. *Med. Sci. Sports* 5:181–86, 1973.

47. Williams, M.H., and Thompson, J. Effect of variant dosages of amphetamine upon endurance. *Res. Q.* 44:417–22, 1973.

27

The Female
in Athletics

Athletic competition at the higher levels for women is a fairly recent development; it awaited the emancipation of the fair sex from its enslavement to puritan concepts and from clothes that were unsuited for comfortable movement, let alone athletic performance. It is indeed amusing to attempt to visualize present-day performances in swimming or running in the athletic costumes of the nineteenth century. Women's athletics—worthy of the name—did not exist prior to World War I, and women began Olympic competition only in 1928.

As a consequence we are only beginning to learn the specialized physiology involved in the reaction of the female at different ages to the various stressors in athletic competition. Furthermore, women's athletics have developed around modifications of existing men's sports, and whether these activities are best suited to the unique interests and the physiological, psychological, and sociological needs of girls and women has not really been investigated. Nevertheless, participation by girls and women in competitive athletics is increasing, and every physical educator and coach should be aware of the available knowledge about the special problems of the female in competitive sports. Obviously, women's athletics should be directed and supervised by professional physical educators because the problems require even greater concern for the principles of anatomy, physiology, and kinesiology than do men's sports.

Structural Sex Differences

One most obvious and important difference between the sexes in regard to sports performance is the ratio of strength to weight, which after puberty is normally much greater in the male. This factor is most important in activities in which the weight is supported by the relatively smaller muscles of the arms and shoulder girdle, as in gymnastics. It is also a consideration wherever the mass of the body must be accelerated rapidly, as in jumping.

The reason for the poorer strength-weight ratio is, of course, the smaller proportion of muscle in relation to the considerably larger amount of adipose tissue (chap. 15) in the female. The larger stores of fatty tissue are not an unmitigated disadvantage, however; in swimming, for example, this results in better buoyancy and less heat loss to cold water.

Not only are the proportions of various tissues different in the female, but also the chemical constituents within each tissue are different. The female tissues, for example, contain much greater amounts of sulfur (23% more in skeletal muscle), and the creatinine coefficients are also different. More research is required to establish the significance of these facts.

A structural difference that has very significant physiological implications for athletic performance is the difference in the ratio of heart weight to body weight in the sexes. From the age of ten to the age of sixty, the average value for women is only 85%-90% of the value for men (9). After age sixty, however, the ratio is similar for men and women.

Physiological Sex Differences

Although there are many physiological differences that have general significance, only those that apply directly to athletic performance will be considered
here.

Blood Constituents. On the average in the age group twenty to thirty,
men have approximately 15% more hemoglobin per 100 ml of blood and about
6% more erythrocytes per cubic millimeter (9). The combination of these two
factors should mean greater oxygen-carrying capacity for men.

Microcirculation. When the reddening of skin in reaction to ultraviolet
radiation was used as a measure of capillary function, men were found to be
less vulnerable through the entire age range (9). The resistance of the capillary
wall to breakdown from mechanical manipulation was also found to be greater
in the male. This very likely is the reason for the greater susceptibility to
bruises in the female.

Metabolic Rate and Efficiency. From just before puberty, and through
the rest of the life span, basal metabolic rate (BMR), as customarily measured
and normalized for body surface area, is higher for the male than for the
female. When BMR is evaluated in relation to muscle mass instead of surface
area, however, the sex difference disappears (9).

Recent evidence has shown that the O_2 cost of running was lower in male
than in female runners of college age. This difference was significant in comparisons of trained as well as untrained groups and existed at all speeds at
which comparisons could be made (6). However, factors other than sex, such
as level of training or mechanics of running, could have influenced the findings.

Oxygen Pulse. This is a widely used measure of the efficiency of the heart
as a respiratory organ, and it is calculated as the O_2 consumption in milliliters
per heartbeat. For equal work loads, boys and girls are about equal on this
measure for ages twelve to fifteen. However, there is a rapid improvement in
O_2 consumption in the male to a value about twice as high at ages twenty-one
to twenty-five, while the female's oxygen pulse remains constant at the twelve
to fifteen age value (34). This has implications that will be discussed below.

Maximum O_2 Consumption. The classic work of Astrand (2) has shown
that girls reach a high point in their maximum O_2 per unit weight between
eight and nine years of age; this figure declines slowly until about age fifteen,
after which it remains constant through young adulthood. Boys reach their
peak later, at about fifteen or sixteen years of age, and maintain this peak
through young adulthood.

Thus in the younger age groups (seven to thirteen) sex differences grow
larger with each increasing year. At ages seven to nine the differences are
small and probably not significant. By age twelve or thirteen, however, differences favoring boys of 13% to 16% in maximal O_2 consumption normalized
for body weight have appeared (44). McNab, Conger, and Taylor (29) have

shown in a direct comparison of twenty-four male and twenty-four female college physical education majors that the difference at this age has grown to 32% when measured as maximal O_2 per kilogram weight, as in the data above. Even when the increasing adiposity of the female is taken into consideration by expressing the data as O_2 per unit fat-free weight the difference still favors the male by 18%. These differences were, of course, highly significant. More recent data from Dill and colleagues (12) on high school age boys and girls provide similar findings with $\dot{V}O_2$ max being 15% lower in the girls even when expressed as $\dot{V}O_2$ per unit fat-free weight.

Cardiac Output. In view of the smaller O_2-carrying capacity of the blood in the female due to lower levels of hemoglobin and red blood cells, one would expect that the female would have to provide more cardiac output at any given level of O_2 consumption at submaximal work loads. This does turn out to be the case for young women (4,5,18). However, for women beyond child-bearing age this difference is no longer significant (5). The physiological basis for this age change remains to be elucidated.

With respect to maximal cardiac output, the best available data show a value of 18.5 liters/min for young women compared with 24.1 liters/min for men (4). Thus there appears to be a 30% difference between the sexes, but it is unlikely that levels of training were equal, and therefore the true sex difference may not be quite that large.

Phenomenal Success of Young Girl Swimmers. In the light of the foregoing facts about O_2 pulse and maximum O_2 consumption, the success of young American girl swimmers in national and international competition becomes understandable. It would seem that all physiological functions essential to competitive swimming have achieved peak values by age twelve to fourteen in the female, whereas these are delayed in the male to late high school and college age. When we add to this the factors of (1) very early commencement of training and (2) absence of social pressures, the accomplishment of our young girl swimmers is entirely comprehensible.

Neuromuscular Functions. Thus far we have discussed the physiology of endurance-type sports, and it is time to consider some of the factors that collectively make up the *skill* of motor performance. It has been reported from Germany that women generally have greater manual skill and dexterity than men (24). In the U.S.A., Pierson and Lockhart (38) have shown there is no significant sex difference in reaction time to a visual stimulus, although men have faster movement times.

A review of the literature in this area seems to indicate that there are probably no real sex differences in either motor learning rate or capacity, unless strength is a factor.

Physiological Adjustments to Heavy Training

The physiological adjustments to heavy training in girls and women are only now getting the attention they deserve. Astrand and his co-workers (3) in Stockholm studied thirty girl swimmers, twelve to sixteen years of age, for one year. The girls trained from 6,000 to 71,500 yards (six to twenty-eight hours) per week and were examined extensively—medically and physiologically—during this period. It was shown that large differences existed in such important measures as maximal O_2 consumption when these girls were compared with average, untrained girls. Furthermore, the differences were highly correlated to the volume of training for each girl. More recently, Brown and associates (7) studied the effects of training for competitive cross-country running upon preadolescent girls. They found maximal O_2 consumption increased by 18% at six weeks and 26% at twelve weeks. Heart rate at submaximal loads declined, and no detrimental effects were seen. Kilbom (22) studied the effects of conditioning on mature females with the bicycle or walking routines using work loads that represented 52%–77% of maximal O_2. In the young group (nineteen to thirty-one years), aerobic capacity improved 12% and cardiac output 11%; in the middle-aged, the improvements were 11% and 10%, and in the older women (fifty-one to sixty-four), 8% and 10%. Systolic blood pressure dropped by 15 mm in the older group, and serum cholesterol declined by 10%. They saw no orthopedic training complications since they used the bicycle and walking-type exercise. However, they did find that serum iron levels declined by 25% in all groups. This finding was thought to be due to greater iron usage in the enhanced erythropoiesis that accompanies vigorous exercise.

More recent work does not support the need for iron supplementation for women in moderately heavy training such as basketball (10) or cycling workouts at 70% $\dot{V}O_2$ max for twenty-five to thirty minutes three times per week (47). Further investigation of this question is needed for women who train at very heavy work loads for national and international competition.

Nature of the Cardiovascular Adaptations. A recent summary of available data (39) suggests that in the young, unfit male, training produces improved $\dot{V}O_2$ max by a combination of improvements in stroke volume and arterio-venous O_2 difference. However, in the young, unconditioned female, it appears that the initial changes over a period of some nine weeks are almost entirely due to central changes (stroke volume and cardiac output), with peripheral adaptations occurring only after the early central changes have taken place (11,26). The increased left ventricular dimension at the end of diastolic filling, which is a typical training response in the young male, is also seen in the young female (48).

Adaptations in $\dot{V}O_2$ max. As pointed out earlier, the physiological functions necessary to success in competitive swimming are well established in young girls twelve to fifteen. The question may be raised, however, Are girls also capable of better training adaptation at this age? A well-controlled study of eight girls (twelve to thirteen) compared with eight young women (eighteen to twenty-one) showed that not only did the two age groups increase in $\dot{V}O_2$ max by a similar magnitude, but they also demonstrated similar rates of change in $\dot{V}O_2$ max during the training period (15).

Recent data suggest the possibility that the magnitude of the training adaptation with respect to $\dot{V}O_2$ max is related to the level of serum testosterone (27). More data on endocrine function in relation to training parameters are badly needed.

Adaptations to Strength Training. Occasionally girls are concerned about the possibility that heavy training may result in increased growth rates, or that on cessation of activity unseemly weight increases may detract from their appearance. No real evidence has been found to substantiate these fears.

Another factor of great concern to young girls contemplating athletic participation is that of becoming less feminine in appearance because of larger or more bunchy muscles. Klaus and Noack (24), two of the foremost experts on the effects of athletics and exercise on the female, feel the available evidence suggests that *properly designed* exercise programs improve rather than hinder femininity. The observation that some girl athletes are very muscular is undoubtedly due to the fact that muscular girls are more apt to be successful in such sports as track and field athletics, and therefore they are more likely to elect to participate in such competition.

Recently three different investigations have shown that even heavy resistance weight training does not result in any substantial hypertrophy of the exercised muscles in women, although strength gains are similar on a relative basis to those of men (8,32,45). Wilmore (45) showed that when men and women used the same weight training techniques and worked to the same fraction of maximum capacity, the degree of muscle hypertrophy was substantially greater in the men than in the women. Muscle hypertrophy in the women was less than one-quarter inch for the upper body girths, which must be considered an insignificant amount for all practical purposes. Wilmore also showed that, whereas the correlation between strength and girth for the men was strong ($r = .63-.77$), there was virtually no real relationship between strength and girth for the women ($r = .09-.42$). Brown and Wilmore (8) showed that while women responded to maximum resistance training with large and significant strength gains, their subjects showed only 0.4% and 2.9% increases in thigh and arm girth, respectively, after six months of training. Indeed these gains were no greater than the controls who did no weight training. Thus it would seem that even the heaviest of weight training programs does not bulk up the female athlete or have any so-called masculinizing effect! The explanation for this sex difference in adaptation to heavy resistance training undoubtedly lies in the fact that both serum testosterone levels and production rates are from twenty to thirty times less in the female, and that therefore the female is unable to achieve the same degree of hypertrophy as the male.

Gynecological Problems

The effects of strenuous exercise programs on the sexual and reproductive functions of the female have been a matter of some concern in the past, although not on the basis of scientific evidence. Observations of 729 Hungarian female athletes showed there was no disturbance of the onset of menarche (16). Nor was there evidence of dysmenorrhea of any consequence as the result of athletic participation, much less moderate physical exercise (3,16).

More recent data on American girls showed menarche occurring significantly later in athletes (13.58 years) compared with nonathletes (12.23 years) (30). The latest study by Malina and co-workers (31) corroborates this difference in onset of menarche and also suggests significant differences related to the type of sport and intensity of training. They also found that the athletes reported a greater incidence of dysmenorrhea and menstrual irregularity, although the difference was not statistically significant.

On the other hand, Astrand and associates (3) reported a slightly earlier onset of menarche in highly trained, young Swedish swimmers compared with a Swedish reference group. Since swimmers do not typically reduce body fat to the extent that athletes involved in heavy running activity do (46), it is interesting to hypothesize a relationship between early onset of menarche and body fat, but much further investigation is needed.

In the case of girl swimmers, bacteriological examination disclosed the presence of pathogenic organisms in the vagina of a third of the subjects (3). In spite of this, there were no signs of any infection of reproductive organs, except in one case of colpitis (inflammation of the vagina). In view of these observations, it would seem undesirable from the medical standpoint for girls to train for swimming during the menstrual period, although this is now common practice in this country.

Female Limitations in Athletics

Table 27.1 shows a comparison of world records in various activities between male and female competitors. Among other things, the table shows the undesirability of competition between the sexes. Most important, however, is the comparison of how close girls come to men in the various types of activity. For lack of better information at this time, it may be inferred from table 27.1 that those activities in which women approach men's records most closely are those in which women suffer the smallest physiological disadvantage, and thus these activities may be considered more suitable, at least until better evidence is available.

Table 27.1
Comparison of World Records for Men and Women as of August 24, 1978

Event	Women	Men	Ratio of Performance 1978, %	1963, %
Swimming (meters)				
100 freestyle	55.41	49.44	89	90
200 freestyle	1:58.53	1:50.29	93	90
400 freestyle	4:06.28	3:51.56	94	89
800 freestyle	8:30.53	8:01.54	94	90
1500 freestyle	16:14.93	15:02.40	93	91
100 breaststroke	1:10.31	1:02.86	89	86
200 breaststroke	2:31.42	2:15.11	89	89
100 butterfly	59.46	54.18	91	86
200 butterfly	2:09.87	1:59.23	92	86
100 backstroke	1:01.51	55.49	90	88
200 backstroke	2:11.93	1:59.19	90	88
Running (meters)				
100	10.88	9.95	91	89
400	49.02	43.86	89	85
800	1:54.9	1:43.4	90	86
1500	3:56.0	3:32.2	90	—
3000	8:27.1	7:32.1	89	—
Marathon (unofficial)	2 hr 34:47.5	2 hr 08:33.6	83	—
Field events				
High jump	2.00 M	2.44 M	82	84
Long jump	7.07 M	8.90 M	79	80

In general, it can be seen that the events that depend upon explosive power (such as the high jump and long jump) show the greatest sex difference. The freestyle and backstroke swimming events and the short runs seem to be the best suited to the feminine physiology if we can assume that all events attract equal numbers of participants (a rather doubtful assumption).

The percentage ratio of performance for men and women is also given in table 27.1 for 1963 to allow comparison with 1978. Over this period of time great changes have occurred in the involvement of women in high-level competitive sport. Thus, where the standings of women relative to men have not improved, the difference probably represents a real sex difference. This appears to be the case for the high jump and long jump. However, for the other fourteen events for which comparisons can be made, twelve show considerable improvement in female performance relative to male, and we may hypothesize on this basis that these differences are not entirely real sex differences, but in some part are due to the poorer selection process for female athletes, which is rapidly improving.

Physiology of Training and Conditioning Athletes

Trainability of the Female. In relation to strength, Hettinger (21) has shown that the female is much less responsive to training than the male. At the age of greatest trainability, twenty to thirty, women respond to training with only 50% the rate of improvement of men. However, Wilmore (45) found similar relative gains in strength for males and females in a well-controlled study using weight training. Since Hettinger used isometric training, it is possible the divergent findings are related to method.

For endurance, the experiments of Klaus and Noack (24) on physical education students showed that at the end of an eighteen-week training program, the men's capacity was about one-third better than that of the women. This is to be expected in the light of the physiological measurements discussed above. Klaus and Noack suggest, on the basis of their experiments, that competitive distances for women in running events should not exceed 1,000 meters, although training procedures may well go beyond this distance.

On the other hand, the data of table 27.1 seem to indicate that endurance per se is not a factor, since in freestyle swimming events, where power (strength) is not a large factor, girls do relatively as well at 1,500 meters as at 100 meters. In running events, a distinct drop in relative performances is seen between the sprints and the middle distances. Whether this is due to a sex difference or to the lesser number of female participants in the middle distances remains to be demonstrated. In any event, the pure power events (high jump and long jump) show the greatest sex difference, which undoubtedly is a reflection of lesser strength in the female.

The Menstrual Cycle and Athletics

Participation in Sports during the Period. It is well known that female athletes in the upper levels of competition seldom allow the menstrual cycle to interfere with their training, but this does not, in itself, justify the practice. There is no unanimity of opinion by medical authorities on this matter. The only scientific evidence available is that of Astrand and co-workers (3) whose work indicates that swimming training during the period is undesirable because of the presence of pathogenic bacteria in the vagina and because of reports of lower abdominal pain in about a third of the girls.

It is probably as unwise to *prohibit* participation in physical education or athletics as it would be to *require* it. In either of these extremes, undue emphasis would be placed on a physiological function that girls should learn to accept as normal. Undoubtedly, the preferred course of action until more scientific evidence is available is to *allow* participation on a voluntary basis, with no undue comments about possible undesirable consequences.

Rhythmicity as a Criterion of Training Progress. Regular, asymptomatic menstruation is usually considered to be a measure of general good health in the female after a regular rhythm has been established. Conversely, medical authorities feel that any deviation from the normal rhythmic pattern may be one of the first indications of overtraining (16,24).

Menstrual irregularities seem to be reported with greater frequency as women become more involved in high intensity, year-round training programs. Wilmore (46) quotes Foreman (unpublished survey) as having found among a group of forty-seven nationally ranked long-distance runners that 19% had irregular menstrual periods and 23% had severe oligomenorrhea or amenorrhea. Such problems could result from the stress effect of training on the hypothalamic-pituitary-gonadal axis, or from reduction of body fat. With presently available evidence, the latter hypothesis seems more attractive because (1) women who lose weight from a variety of causes are prone to amenorrhea (19), and (2) highly trained swimmers who typically maintain normal fat values (46), but probably endure equal training stress, seem to have few problems (3). Further investigation is needed in this area.

It would be good practice for coaches and physical educators of women to advise the members of their teams and classes to maintain accurate records of their menstrual cycles, and they should encourage consultations whenever deviations occur. This practice should result in better health and performance for the athletes.

Effect of the Menstrual Cycle on Performance. In spite of the widespread impression of impaired performance during certain periods of the menstrual cycle, there is no agreement among the investigators who have attacked this problem. Some have found no effect of the menstrual cycle on motor performance (1,13,28,35,36,37,40,41); others report that performance is best in the postmenstrual phase or intermenstrual phase, and at its worst in the two or three days preceding menstruation (16,24,33) or during menstruation (43).

Pregnancy, Childbirth, and Athletics

It was formerly suggested that participation in athletic competition, training, or vigorous sports should be forbidden during pregnancy. The reasons for this are discussed by Klaus and Noack (24), who point out that the work of the right heart is increased threefold and the work of the left heart is increased twofold, even in the nonpregnant female, by a moderate work load. During pregnancy, such an increase—with the demands of fetal circulation—could be considered hazardous for the right heart and for lung circulation. The danger would be especially great if unrecognized heart defects are present. In addition, it should be recognized that the kidneys and liver function with very little reserve capacity during pregnancy.

However, more recent findings do not seem to support this reasoning entirely. Knuttgen and Emerson (25) studied thirteen normal pregnant women during rest and moderate exercise. They found no evidence of ventilatory impairment or dyspnea and concluded that exercise does not constitute a more severe physiological stress during pregnancy, if lifting work and possible encumbrance of the fetal tissues are minimized.

Erkkola (17) investigated sixty-two healthy young women in their first pregnancy beginning in the tenth or fourteenth week of pregnancy and lasting until term. One-half of the group was encouraged to exercise with strenuous physical exertion. The increase in physical work capacity during pregnancy was 17.6% greater in the training group, who showed no negative effects from the training program.

Dressendorfer (14) measured $\dot{V}O_2$ max in one healthy young woman during the course of two of her pregnancies and subsequent lactation periods over a period of four years. Training mileage averaged only five to ten miles per week in the first trimester because of nausea, but thereafter the subject was able to run an average of fifteen miles a week up to delivery. Training was gradually increased to twenty miles per week at four months postpartum. It was concluded that during *normal* pregnancy and lactation, $\dot{V}O_2$ max and endurance performance cannot only be maintained, but even improved by physical training without harmful effects on the mother or child.

In the absence of careful medical supervision and approval, such procedures could not be recommended, but the evidence does seem to suggest greater functional capacities in the healthy pregnant young woman than had been previously suspected.

Effects of Heavy Exercise Programs on Labor and Delivery. Although it was once believed that athletic women developed tense (unyielding) abdominal walls that hindered normal delivery, the results of many investigations in more recent years indicate that athletic women have quick and easy deliveries (24). Erdelyi (16), who has studied many Hungarian women athletes, found a smaller incidence of complications (especially toxemia) during pregnancy and 50% fewer cesarean sections performed in women athletes when compared with controls. It was also found that the duration of labor was shorter than the average in 87.2% of the women athletes.

It would therefore appear that there is no need for concern about the effects of strenuous exercise upon subsequent pregnancies or childbirth. Indeed, physical conditioning seems to be a valuable prophylactic procedure.

Effect of Pregnancy and Childbirth on Subsequent Athletic Performance. Noack (33) took histories of fifteen German champion women athletes who bore children during their athletic careers. Of the fifteen, five gave up sports because of their new responsibilities; of the remaining ten, two maintained equal performance and eight made definite objective improvements after childbirth. All of the women agreed that after childbirth they were "tougher" and had more strength and endurance.

The Female in Athletics

It has been pointed out that pregnancy, far from being an illness, should be considered an intensive, day and night, nine-month period of physical conditioning because of the increased demands on metabolism and the entire cardiovascular system (24).

Athletic Injuries in Women

Even a cursory study of anatomy reveals a considerable difference in the locomotor structures of the female compared with the male. On the average, bones, muscles, tendons, and ligaments are more delicately constructed, although body weight is not decreased proportionately because of the greater percentage of fatty tissue in the female. On this basis, a sex difference in incidence of athletic injuries is to be expected, and indeed is found.

It has been shown in studies involving comparable groups of men and women that the overall incidence of athletic injuries in women was almost double that in men (23). Furthermore, the incidence of injuries involving *overstrain*—such as contractures, inflammations of tendons, tendon sheaths, bursae, foot deficiencies, and periosteal injuries—was almost four times more common in women than in similarly trained men.

The distribution of injuries according to the sport activity is of interest. In women, by far the greatest percentage of all injuries is found in sports that require explosive efforts: short runs (53%) and the long jump (31%).

Emotional Factors in Women's Athletics

The comment is frequently heard that the female is less well suited to competitive sport than her brothers because of a more emotional nature and that highly competitive situations might elicit unfavorable responses, but there is no acceptable scientific evidence to support this assertion. Ulrich (42), who used eosinophil count and cardiorespiratory response to measure the stressfulness of various competitive situations, found that measurable stress reactions occurred not only in response to participation in class, intramural, and interscholastic basketball games, but in written test situations as well! She concluded that stress was much more closely related to the psychological than the physiological components of a situation. It is of interest that her study showed lesser levels of stress occurred as the result of experience, which suggests that girls successfully adjust to the stress of competition.

Astrand and associates (3), in a year-long study of girl swimmers twelve to sixteen (a supposedly emotional, labile period), could not find a single case of demonstrable nervous symptoms that could be attributed to training or to participation in competitive events.

SUMMARY

1. Probably one of the best criteria for the value of competitive athletics for women can be gained from the opinions of former female athletes in regard to their daughters' participation in sports. In regard to swimming competitions, at least, eighty-four former top-level Swedish swimmers had a positive attitude on the subject (2).
2. The anatomic and physiological differences between the sexes bear directly on physical education and athletics, and these should be carefully considered in planning programs for girls.
3. There seems to be athletic-injury evidence that the interests of girls and women would be best served if sports and competitions were selected and designed specifically for the female. The adoption and modification of men's sports for women has undoubtedly resulted in the latter's participation in some activities that are not well suited to the female's body structure.

REFERENCES

1. Allsen, P.E.; Parson, P.; and Bryce, G.R. Effect of the menstrual cycle on maximum oxygen uptake. *Physician Sports Med.* 5:53–55, 1977.
2. Astrand, P-O. *Experimental studies of physical working capacity in relation to sex and age.* Copenhagen: E. Munksgaard, 1952.
3. Astrand, P-O.; Engstrom, L.; Eriksson, B.; Karlberg, P.; Nylander, I.; Saltin, B.; and Thoren, C. Girl swimmers—with special reference to respiratory and circulatory adaptation and gynecological and psychiatric aspects. *Acta Paediatr.* (Stockholm), suppl. 147, 1963.
4. Astrand, P-O.; Cuddy, T.E.; Saltin, B.; and Stenberg, J. Cardiac output during submaximal and maximal work. *J. Appl. Physiol.* 19:268–74, 1964.
5. Becklake, M.R.; Frank, H.; Dagenais, G.R.; Ostiguy, G.L.; and Guzman, C.A. Influence of age and sex on exercise cardiac output. *J. Appl. Physiol.* 20:938–47, 1965.
6. Bransford, D.R., and Howley, E.T. Oxygen cost of running in trained and untrained men and women. *Med. Sci. Sports* 9:41–44, 1977.
7. Brown, C.H.; Harrower, J.R.; and Deeter, M.R. The effects of cross-country running on preadolescent girls. *Med. Sci. Sports* 4:1–5, 1972.
8. Brown, C.H., and Wilmore, J.H. The effects of maximal resistance training on the strength and body composition of women athletes. *Med. Sci. Sports* 6:174–77, 1974.
9. Burger, M. Zur pathophysiologie der geschlechter. *Munch. Med. Wochenschr.* 97:981–88, 1955.
10. Cooter, G.R., and Mowbray, K. Effect of iron supplementation and activity on serum iron depletion and hemoglobin levels in female athletes. *Res. Q.* 49:114–18, 1978.
11. Cunningham, D.A., and Hill, J.S. Effect of training on cardiovascular response to exercise in women. *J. Appl. Physiol.* 39:891–95, 1975.

The Female in Athletics

12. Dill, D.B.; Myhre, L.G.; Greer, S.M.; Richardson, J.C.; and Singleton, K.J. Body composition and aerobic capacity of youth of both sexes. *Med. Sci. Sports* 4:198–204, 1972.

13. Doolittle, T.L., and Engebretsen, J. Performance variations during the menstrual cycle. *J. Sports Med.* 12:54–58, 1972.

14. Dressendorfer, R.H. Physical training during pregnancy and lactation. *Physician Sports Med.,* vol. 6, February 1978.

15. Eisenman, P.A., and Golding, L.A. Comparison of effects of training on $\dot{V}O_2$ max in girls and young women. *Med. Sci. Sports* 7:136–38, 1975.

16. Erdelyi, G.J. Gynecological survey of female athletes. *J. Sports Med.* 2:174–79, 1962.

17. Erkkola, R. The influence of physical training during pregnancy on physical work capacity and circulatory parameters. *Scand. J. Clin. Lab. Invest.* 36:747–54, 1976.

18. Freedson, P.; Katch, V.L.; Sady, S.; and Weltman, A. Cardiac output differences in males and females during mild cycle ergometer exercise. *Med. Sci. Sports* 11:16–19, 1979.

19. Frisch, R., and McArthur, J. Menstrual cycles: fatness as a determinant of minimum weight for height necessary for their maintenance or onset. *Science* 185:949–51, 1974.

20. Garlick, M.A., and Bernauer, E.M. Exercise during the menstrual cycle: variations in physiological baseline. *Res. Q.* 39:533–42, 1968.

21. Hettinger, T. *Physiology of Strength.* Springfield, Ill.: Charles C Thomas, 1961.

22. Kilbom, A. Physical training in women. *Scand. J. Clin. Lab. Invest.* vol. 28, suppl. 119, 1971.

23. Klaus, E.J. The athletic status of women. In *International Research in Sport and Physical Education,* eds. E. Jokl and E. Simon. Springfield, Ill.: Charles C Thomas, 1964.

24. Klaus, E.J., and Noack, H. *Frau and Sport.* Stuttgart: Georg Thieme Verlag, 1961.

25. Knuttgen, H.G., and Emerson, K. Physiological response to pregnancy at rest and during exercise. *J. Appl. Physiol.* 36:549–53, 1974.

26. Kollias, J.; Barlett, H.L.; Mendez, J.; and Franklin, B. Hemodynamic responses of well-trained women athletes to graded treadmill exercise. *J. Sports Med. Phys. Fitness* 18:365–72, 1978.

27. Krahenbuhl, G.S.; Archer, P.A.; and Pettit, L.L. Serum testosterone and adult female trainability. *J. Sports Med. Phys. Fitness* 18:359–64, 1978.

28. Loucks, J., and Thompson, H. Effect of menstruation on reaction time. *Res. Q.* 39:407–8, 1968.

29. MacNab, R.B.J.; Conger, P.R.; and Taylor, P.S. Differences in maximal and submaximal work capacity in men and women. *J. Appl. Physiol.* 27:644–48, 1969.

30. Malina, R.M.; Harper, A.B.; Avent, H.H.; and Campbell, D.E. Age at menarche in athletes and nonathletes. *Med. Sci. Sports* 5:11–13, 1973.

31. Malina, R.M.; Spirduso, W.W.; Tate, C.; and Baylor, A.M. Age at menarche and selected menstrual characteristics in athletes at different competitive levels and in different sports. *Med. Sci. Sports* 10:218–22, 1978.

32. Mayhew, J.L., and Gross, P.M. Body composition changes in young women with high resistance weight training. *Res. Q.* 45:433–40, 1974.

33. Noack, H. Die sportliche leistungsfahigkeit der frau im menstrualzyklus. *Dtsch. Med. Wochenschr.* 79:1523–25, 1954.

34. Nocker, J., and Bohlau, V. Abhangigkeit der leistungsfahigkeit vom alter und geschlecht. *Munch. Med. Wochenschr.* 97:1517–22, 1955.

35. Petrofsky, J.S.; LeDonne, D.M.; Rinehart, J.S.; and Lind, A.R. Isometric strength and endurance during the menstrual cycle. *Eur. J. Appl. Physiol.* 35:1–10, 1976.

36. Phillips, M. Effect of the menstrual cycle on pulse rate and blood pressure before and after exercise. *Res. Q.* 39:327–33, 1968.

37. Pierson, W.R., and Lockhart, A. Effect of menstruation on simple reaction and movement time. *Br. Med. J.* 1:796–97, 1963.

38. ———. Fatigue, work decrement, and endurance of women in a simple repetitive task. *Aerospace Med.* 35:724–25, 1964.

39. Rowell, L.B. Human cardiovascular adjustments to exercise and thermal stress. *Physiol. Rev.* 54:75–159, 1974.

40. Sloan, A.W. Effect of training on physical fitness of women students. *J. Appl. Physiol.* 16:167–69, 1961.

41. ———. Physical fitness of college students in South Africa, U.S.A., and England. *Res. Q.* 34:244–48, 1963.

42. Ulrich, C. Measurement of stress evidenced by college women in situations involving competition. Doctoral dissertation, Physical Education, USC, 1956.

43. Wearing, M.P.; Yuhasz, M.D.; Campbell, R.; and Love, E.I. The effect of the menstrual cycle on tests of physical fitness. *J. Sports Med.* 12:38–41, 1972.

44. Wilmore, J.H., and Sigerseth, P.O. Physical work capacity of young girls, 7–13 years of age. *J. Appl. Physiol.* 22:923–28, 1967.

45. Wilmore, J.H. Alterations in strength, body composition, and anthropometric measurements consequent to a 10-week weight training program. *Med. Sci. Sports* 6:133–38, 1974.

46. Wilmore, J.H.; Brown, C.H.; and Davis, J.A. Body physique and composition of the female distance runner. *Ann. N.Y. Acad. Sci.* 301:764–76, 1977.

47. Wirth, J.C.; Lohman, T.G.; Avallone, J.P.; Shire, T.; and Boileau, R.A. The effect of physical training on the serum iron levels of college age women. *Med. Sci. Sports* 10:223–26, 1978.

48. Zeldis, S.M.; Morganroth, J.; and Rubler, S. Cardiac hypertrophy in response to dynamic conditioning in female athletes. *J. Appl. Physiol.* 44:849–52, 1978.

28

The Unified Athlete:
Monitoring Training Progress

The scientific method often requires analytic procedures that break down the whole into its component parts for greater ease of study. This is typical of the *systems approach* in physiology, and we have quite artificially fragmented athletes into their many systems and even by elements of performance such as strength, endurance, and speed. This analytic approach is of course absolutely essential because the immense complexity of the human organism defies human comprehension on any other basis. However, the analytic process must be followed by a synthetic process to encourage a unification of principles that will allow application of theory to practice in athletics.

For example, we need to know the individual contributions of the heart, blood vessels, lungs, muscles, endocrine glands, and nervous system to meeting the increased metabolic demand of an exercise bout. But in reality each response of each system is intricately interwoven into the responses of each other system, affecting the others and being in turn affected by them.

The same may be said for the artificially contrived divisions of the elements of performance. The elements of strength and endurance of muscle tissue, for instance, while they do have separate identities, are so closely related that one can scarcely train for one without affecting the other, as experiments have shown.

Thus, having benefited from analysis to the extent of understanding detail better, let us now synthesize and unify our thinking; let's try to get it all together for the benefit of the athlete and for athletics.

Basically, the ultimate in human performance depends on maximizing two factors: (1) state of overall health in the athlete and (2) capacity for physiological response to the challenge of the game situation. These are not one and the same. For example, it is not uncommon for the young inexperienced coach to pursue the second goal at the expense of the first and ultimately lose out with respect to both. Too great a demand in the training regimen may load up the total stress on the athlete to the point where health suffers and performance goes downhill.

We are often faced with such practical questions as How long should the practice be? How hard can I work my athletes? How much is enough? It is the purpose of this chapter to provide, not ready-made answers, but a *sound approach* that is based on knowledge gained in the physiology of exercise laboratory and that is practical and feasible on the field.

Training, Conditioning, and Stress

At the outset, we must recall the earlier discussion on the *stress syndrome* and realize that the whole training-conditioning process is nothing more or less than devising appropriate levels of stress to bring about the best possible combination of responses (training effect) on the part of the athlete. The problem becomes complex for several reasons:

1. We usually deal with fairly large numbers of athletes in a team situation, and physiological responses differ from athlete to athlete.
2. Each sport has a different combination of demands.
3. Even within one sport, there are differences in demand from one point in the season to the next.

Furthermore, we as coaches must realize that stress is cumulative and athletes are meeting other stressors in their life beyond that of the athletic conditioning program. Any given athlete may be faced with differing combinations of total stress resulting from a combination of athletics with any one or more of the following:

1. Academic problems
2. Home problems
3. Other extracurricular activities
4. Outside work to help support self or family
5. Problems of social interaction
6. Sex life

In all likelihood, athletes who succumb to every respiratory bug, whose performance falls off in mid or late season, maybe even those who suffer injury are being overstressed. That is, the sum total of stresses exceeds their adaptational capacity.

In the light of this discussion, it is patently impossible to arrive at any given combination of intensity-duration or any arbitrary length of workout that will be optimal for all athletes on any given team. Obviously we must base physiological demands on physiological capacities, and this is where the art and science of coaching must merge. Science, in this case, physiology of exercise methods, can supply the data on the athletes' level of response to stress (their training progress), but only the coaches' art (and good common sense) can help them modify the demands on individual athletes to optimize the rate of improvement in performance.

Monitoring Training Progress

In essence, to get the best possible performance from our athletes the training stress must be optimized on an individual basis. To do this we must keep records of training progress so that we can "see the forest for the trees." The items of paramount interest fall into four categories:

1. Performance data in sports, when this can be measured
2. Physiological capacities, particularly physical working capacity (PWC)
3. Nutritional status
4. Sleep and rest habits

The frequency at which observations are made and recorded should vary with the variability of the measurement. The following schedule is suggested as a basis:

A. Evaluation once per season
　1. Diet evaluation
　2. Total effort expended by the athlete
　　a. Athletics
　　b. Home life
　　c. Academic load
　　d. Extracurricular load
　　e. Outside work (jobs)
　　f. Social problems
　　g. Problems related to sex life
　3. Patterns of sleep and rest
B. Once per week
　1. Resting heart rate and blood pressure
　2. Estimate of PWC by Astrand test (chap. 11)
　3. Blood pressure under load of PWC test
C. Daily record
　1. Weight (nude when suiting up for workout)
　2. Performance data (for example, time for selected distance run, height or distance jumped)

Getting it All Together — Administration

Having been involved in ten years of age-group coaching, seven years of high school coaching, and seven years of college coaching, the author can well anticipate the question forming in the reader's mind: Where in the devil do I find time for all this record keeping? The answer is, You don't. You become an efficient administrator, finding and motivating skilled help and delegating responsibilities.

To begin with, the actual data collection should be handled by a student manager. This may entail giving an additional sports letter to entice a suitable prospect. In college, a physical education major with an active interest in physiology of exercise makes the ideal choice. In high school, one looks for an individual with a good IQ and a history of responsibility.

Next comes the matter of training the manager to collect reliable data. The training should be accomplished by the coach, working in conjunction with the most interested health professional available. Ideally this should be the team physician or school physician. If neither can make the time available, then the school nurse or the athletic trainer may assume this responsibility with the coach. In any event, all of the health professionals mentioned (whether directly involved or not) should be made aware of this aspect of the athletic program. Those who are interested should also be provided with brief summaries of the testing results. On such efforts is rapport with the medical community and the community at large established!

In most situations, the direct supervision of the data collection should be handled by the coach or athletic trainer since no other personnel would be sufficiently available at the necessary time.

Methods and Record Keeping

Diet Evaluation. Each athlete is issued seven copies of the form shown in table 28.1 at a preseason team meeting (1). The group is taught how to fill out the form, one copy for each day of the one-week sampling period. These are collected, and the data are transferred to the food selection scorecard shown in table 25.4. The scorecard is self-explanatory and should be completed by the most interested of the health professionals available (school dietitian if available) or by the coach. At a later team meeting the results of this survey should make an excellent springboard for discussion of good nutrition and its importance in athletics. Every athlete should shoot for a 100% diet score.

Table 28.1
Record for Meals
Keep a complete record of all food and beverages consumed at and between meals for a seven-day period. Use the following form for your data. One for each day. Under kind of food, specify whether raw or cooked, if fruit or vegetable, and how prepared. For example, a vegetable salad may be made of cooked carrots and beans and raw celery; potatoes may be creamed. Amounts should be expressed as definitely and accurately as possible, servings of meats in measurements, servings of vegetables in cups, slices of bread in number and size, etc.

Time and place of meal	Food	Kind	Amount	Calories

From M.S. Chaney and M.L. Ross. *Nutrition,* 7th ed., 1966. Courtesy of Houghton-Mifflin Co., Boston.

Table 28.2

Total Effort and Sleep-Rest Patterns

List times spent on your weekly activities below, including sleeping or resting and eating, as accurately as you can. This form will be used by the coach to get the best possible results from your training program in terms of your health and athletic achievement.

Time A.M.	Mon	Tues	Wed	Thurs	Fri	Sat	Sun
12:00- 1:00							
1:00- 2:00							
2:00- 3:00							
3:00- 4:00							
4:00- 5:00							
5:00- 6:00							
6:00- 7:00							
7:00- 8:00							
8:00- 9:00							
9:00-10:00							
10:00-11:00							
11:00-12:00							
P.M.							
12:00- 1:00							
1:00- 2:00							
2:00- 3:00							
3:00- 4:00							
4:00- 5:00							
5:00- 6:00							
6:00- 7:00							
7:00- 8:00							
8:00- 9:00							
9:00-10:00							
10:00-11:00							
11:00-12:00							

Total Effort and Sleep-Rest Evaluation. The form shown in table 28.2 is used for evaluating athletes' total effort and sleep-rest patterns, and is largely self-explanatory. The coach cannot evaluate this chart quantitatively without great difficulty; therefore it is suggested that this form be applied only in a rather informal manner and that the coach look for the individual athlete who may be grossly overloaded with respect to the total week's activity. Such athletes would be counseled on an individual basis, and those for whom a gross overload is unavoidable should be counseled out of athletics in the interest of the individual's health as well as the welfare of the team. Such an individual would probably be retrogressing in performance when needed most by the team—in late season.

Once-a-Week Physiological Testing. This most important testing can be accomplished in little more than ten minutes per athlete when procedures have become routine. The only equipment required is an inexpensive bicycle ergometer which every corrective or adaptive physical education department should have anyway. Purchase may be more easily justified in this fashion. Also needed are a stethoscope, metronome, and two stopwatches. The procedure is briefly described in this text in chapter 12 and described in detail in the lab manual that accompanies this text (2). In any event, the load should be selected to achieve a heart rate of 150 beats per minute, plus or minus 10 beats, and this same load should be used on the individual athlete throughout the season. Thus changes in heart rate and blood pressure response to a standard load can be graphed and trends will become obvious over a period of weeks. Athletes should report for testing at least fifteen minutes prior to the actual test so that reasonably representative values of resting heart rate and blood pressure can be obtained after fifteen minutes rest in the lab and just before the test begins. Resting heart rate and blood pressure and exercise heart rate and blood pressure are then recorded on the form as shown in figure 28.1. Athletes should be encouraged to calculate their own estimated maximal O_2 consumption from a table provided for their use (see chap. 12, tables 12.3 and 12.4). A copy of the norms for maximal O_2 (table 12.5) should also be posted so that they can evaluate their own status. All of the precautions mentioned in chapter 12 such as temperature control must be closely observed so that the data will be reliable. Only systolic blood pressure is recorded since diastolic pressure is both difficult to take and notoriously unreliable under exercise conditions. Consequently resting heart rate and blood pressure plus the exercise values can all be entered in one graph (fig. 28.1), allowing easier visual identification of undesirable trends.

Daily Measurements. Coaches will probably wish to devise their own methods of charting performance in accord with their sport and coaching techniques. The suggested form for graphing weight changes is shown in figure 28.2.

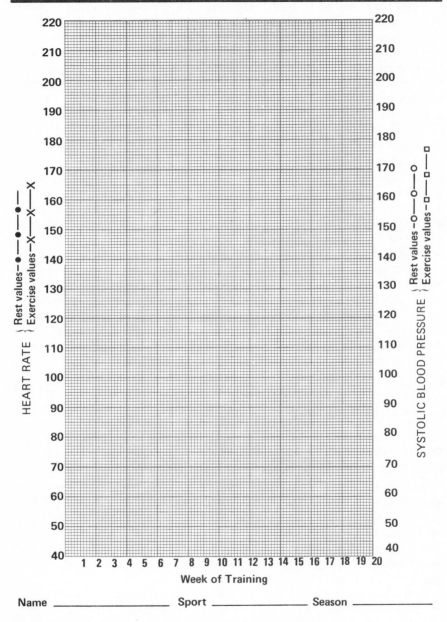

Figure 28.1 Form for recording weekly heart rate and blood pressure.

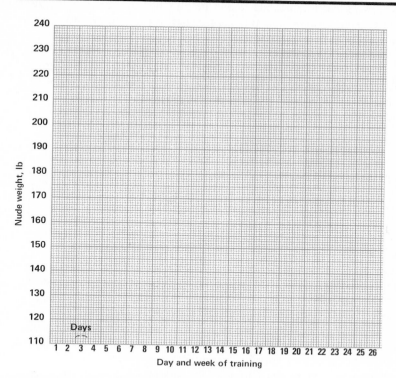

Figure 28.2 Suggested form for recording athletes' body weight changes on a five-day week basis throughout season.

Interpretation of Physiological Data

It must be realized from the outset that all the measurements taken have considerable variability due to unavoidable errors of measurement and due to normal biological variability. This, of course, is the major reason for graphing the data. Meaningful trends that would otherwise be lost due to meaningless variability will emerge when the data points are connected.

In general, a diagnosis of overtraining should not be based on any single measurement along one parameter alone. The order of importance of the various measurements is as follows:

1. Exercise heart rate
2. Exercise blood pressure
3. Body weight
4. Resting blood pressure
5. Resting heart rate

The Unified Athlete: Monitoring Training Progress

Furthermore, no great importance should be attached to a change observed on only one occasion. Only when a trend is seen over three or more observations are the data likely to be meaningful.

The typical picture seen in overtraining would consist of a rise in exercise heart rate, exercise blood pressure, resting heart rate, and resting blood pressure, combined with a loss in weight. Such a pattern should be considered a clear-cut indication for easing up on the training load, and the first discernible sign of such a trend should be cause for caution and some lessening of the load.

Obviously, individual counseling is of the utmost importance. For example, continued weight loss over two to three weeks in the absence of other trends would suggest that the coach question the athlete about possible changes in diet or other behaviors. *Science* can only supply better data to improve the *art* of personal interactions between coach and athlete!

Benefits to be Gained From the Scientific Effort

No claim is made here that losers can become winners by following the procedures under discussion. However, there is every reason to believe that applying what we know of exercise physiology can make a considerable contribution to better athletic performances. This can be assumed on at least two bases: (1) the *physiological* basis, which has been fairly well defined through the course of this text, and (2) the so-called *Hawthorne effect,* which is the name given to performance benefits arising in experiments where the only possible explanation lies in the fact that the subjects have perceived some change in experimental conditions that is interpreted by them as "concern for their welfare."

There is probably little question that attention to the details of optimizing the general health of athletes will result in better performances, if only due to the lessening of workout time lost to respiratory tract infections and other "bugs."

Of equal importance is the resulting increased rapport of the coach with members of the athlete's family, the academic institution, the medical community, and the community at large.

REFERENCES

1. Chaney, M.S., and Ross, M.L. *Nutrition,* 7th ed. Boston: Houghton Mifflin Co., 1966.
2. deVries, H.A. *Laboratory Experiments in Physiology of Exercise.* Dubuque, Iowa: Wm. C. Brown Co. Publishers, 1971.

Index

Index